Practical Optics

William P. Ewald
W. Arthur Young
Richard H. Roberts

Practical Optics

A comprehensive video presentation about optical engineering principles

IMP® IMAGE MAKERS OF PITTSFORD

Practical Optics

Produced and Published by Image Makers of Pittsford
6 Wood Gate, Pittsford, New York 14534 U.S.A.

Videotape Presentations

Program and Lectures by William P. Ewald

Directed by Richard H. Roberts

Textbook

Text by W. Arthur Young

Editor: Richard H. Roberts

Editorial Staff: David R. Young
Catherine M. Mooney
Stanford Perry
Geraldine W. Roberts

Technical Illustrations by Rich Sensenbach
based on drawings by W. Arthur Young

Cover Design and Photographs: Richard H. Roberts

Library of Congress Cataloging in Publication Data

Ewald, William P.,
Practical optics.

Includes index.
1. Optics—Study and teaching—Audio-visual aids. 2. Optical instruments—Design and construction—Study and teaching—Audio-visual aids.
I. Young, W. Arthur.
II. Roberts, Richard H.
III. Title.
QC363.E85 1982 535 82-15792
Printed in the United States of America

ISBN 0-911705-00-7

Preface

This text was written to be used in conjunction with a series of videotape lectures on PRACTICAL OPTICS presented by Mr. William P. Ewald. As Mr. Ewald's lectures assume a certain background knowledge of optics on the part of students, the first five chapters are designed to fulfill this need for those who are new to the field of optics or those who want a review of fundamentals. These chapters include a preliminary vocabulary, discussion of the types of optical elements and the materials from which they are made and introduction to the design of optical and photographic instruments.

The later chapters follow the subject matter of the lecture videotapes. Mr. Ewald's drawings are reproduced in this text to eliminate the need for detailed note taking during lectures. There are some differences in the lecture and text material: the lectures present the more introductory portions of the topic while the text is more complete. Together, the lectures and the text combine to give a thorough and comprehensive knowledge of the field of optics.

After watching the video presentation, students are given an opportunity to answer questions and work out problems located at the end of each chapter. This effort reinforces their understanding of the subject material and assures student involvement in the program. Upon completion of the course, this text will serve as an invaluable reference in each student's library.

The early lessons are relatively simple; the later lessons are more complex as they build on the information previously given. Therefore it is important to take all subjects in sequence and to master each lesson before going on to the next. If a tape lecture is missed, the text material can serve as a temporary substitute until the lecture is seen and heard.

R.H.R.

Contents

Practical Optics

Introduction

Optics is one of the fastest growing and one of the most dynamic branches of physics today. Since the late 1800's optics had been considered by many to be a completed field of study and little was being done to expand on its principles. Renewed interest in geometric optics and new discoveries in physical optics started things rolling in the late 1950's, and by the 1970's optics was entering the "Age of Electronics." One of the biggest advancements in optics, however, started in the late 1970's and early 1980's with the wide application of the laser. Laser optics has an infinite number of possibilities and its applications are growing daily.

It is this demand for up-to-date information and the continually changing requirements in the field of optics that inspired the creation of PRACTICAL OPTICS, an original training program designed for color television presentation combined with a modern approach to written material. While focusing on the basics necessary for beginners, this program contains information which can be especially useful to individuals with prior experience in the field.

The field of optics is an integral part of many rapidly advancing fields such as laser fusion, electronics, medicine, communications, solar energy, surveying and leveling. PRACTICAL OPTICS presents a solid foundation on which to build a variety of challenging professions dependent on the field of optics.

PRACTICAL OPTICS can be used for the training of those new to the field of optics as well as those experienced in it. In addition to covering a variety of topics dealing with light, lenses, mirrors, and prisms, the program includes a number of original exhibits, demonstrations, and previously unpublished solutions to a variety of optical problems. After each optical component is identified, a challenging application of this component is shown. The student is guided through a series of steps for analyzing a proposed application and confirming the application with ray tracing. A concluding demonstration verifies the entire procedure and leaves the student with confidence and the ability to apply the principle to on-the-job applications.

Program Format

This instructional program consists of 16 videotape lessons and an accompanying textbook which offers comprehensive exercises and serves as a reference book when the program has been completed.

The video lessons are on ¾-inch tape cassettes, which provide convenient, high quality color images accompanied by high quality sound. This is a standard professional video format, and equipment used for playing these tapes is readily available in schools, and industry.

PRACTICAL OPTICS is designed specifically for closed circuit television presentation and is recorded under carefully controlled conditions to permit the presentation of the material in a highly personalized manner. Each viewer becomes involved with the instructor on a one-to-one basis.

The textbook provides definitions of terms and applications of examples covered in a videotape program. Questions or problems are offered at the end of each chapter to provide the students with an opportunity to review and test their knowledge of the material covered.

About the Authors

William P. Ewald has been fascinated by the field of optics for over fifty years. In 1941 Mr. Ewald entered the Institute of Optics at the University of Rochester (New York). World War II interrupted his formal education but not his training in optics. He continued in the field as an optical specialist for the U.S. Navy. In the Submarine Service Mr. Ewald developed a submersible optical reflex gunsight for deck guns. After the war he returned to Rochester where he obtained his degree from the University of Rochester. Mr. Ewald joined Eastman Kodak Company in Rochester where he had the good fortune to work directly for Joseph Mihalyi and Otto Wittel, internationally recognized optical experts. For twelve years he taught a popular course, "The Mechanical Design of Optical Instruments", at the University of Rochester. Mr. Ewald is a respected authority in the field of optical science with over twenty patents to his credit. He was awarded the highly prestigious Richardson Medal by the Optical Society of America for his many distinguished contributions to applied optics. Mr. Ewald is a member of the Society of Photographic Scientists and Engineers, the Society of Motion Picture and Television Engineers, and the Optical Society of America.

W. Arthur Young graduated cum laude from Knox College's honors program of physics in 1939. He started work at Kodak's optical plant in 1941 and worked there for nearly 25 years. Mr. Young held positions as an engineer, a supervisor in optics manufacture, and as a quality control engineer. He is credited with a number of inventions in optical manufacture, optical testing, and photographic testing. Since the early sixties, Mr. Young has worked at the Kodak corporate headquarters in technical training and educational development, and as a technical editor and contributing editor of the Encyclopedia of Practical Photography. Mr. Young is a successful exhibiting photographer with an Associateship in the Photographic Society of America. His photographs and articles on light, optics, photographic lenses, perspective, and vision have been published in *Studio Light, Popular Photography,* and *The Professional Photographer.* He has won a number of photographic awards, including the coveted George Eastman Medal. Mr. Young and Mr. Ewald have been friends and business associates for more than thirty years.

Richard H. Roberts graduated from Rochester Institute of Technology in 1951. He began his career by producing motion pictures and videotapes for the U.S. Army Signal Corps Pictorial Center, Long Island City, where the majority of the Army's orientation and training films were created. After completing his military service, Mr. Roberts pursued graduate work at Rochester Institute of Technology and also served on its faculty. He worked for many years in the training organization of Eastman Kodak Company where he had responsibility for the design, testing, production and teaching of photographic and optical programs for engineers and technicians. He has also conducted extensive studies and experiments related to adult learning which focused on design, motivation, and retention. Mr. Roberts' standards for accurate detail and high quality are evident in the educational soundness and professional excellence of this videotape program and textbook.

Chapter 1
Light

Summary

Light is a form of kinetic energy that comes from electromagnetic radiant sources and radiates in all directions. Light is the only form of radiant energy which we can see. When tiny, moving charged particles, such as electrons, are forced to change their speed or direction of travel, electromagnetic radiant energy is emitted. The energy of light travels in straight lines in a given transparent medium at a speed of about 186,000 miles per second. As it travels, light exhibits the characteristics of both particle and wave motion. The wavelengths of light are 400 to 700 nanometres. White light, an equal mix of all light wavelengths, can be refracted by a prism to form a visible spectrum that appears colored. When all the forms of electromagnetic radiant energy are arranged sequentially by wavelength, the electromagnetic spectrum is formed.

What Is Light?

Light is a form of kinetic, or moving energy. Specifically, it is the form of energy to which the human eye is sensitive.

Light energy, acting on the light-sensitive cells in the retina of the eye, causes vision. Light also acts on the light-sensitive emulsion of a photographic film, forming a latent (invisible) image that can be chemically processed to become a photograph. Light, acting on the leaves of a plant, converts carbon dioxide from the air, and water taken by the plant from the ground, into carbohydrates, a basic food of plants and animals. Infrared radiation evaporates water from the oceans and lakes, forming the clouds that supply water in the form of rain to the land. The energy of heat helps cause the winds which are an obvious form of energy. Light is visible radiant energy.

The Source of Light

All light comes from electromagnetic radiant sources. Matter is made of tiny moving particles such as atoms, electrons and protons; and some of these particles are electrically charged. Electrical current is the movement of negatively-charged particles and electrons through conductive material such as metal.

Under the right condition, moving electrically charged particles are forced to change their speed or direction and energy is given off in the form of light or other forms of radiant energy. For example, electricity moves easily through a low resistance wire such as one made of copper. If the copper wire is connected to a high resistance wire, such as the tungsten filament of a light bulb, the electrons are forced to slow down. Energy in the form of light

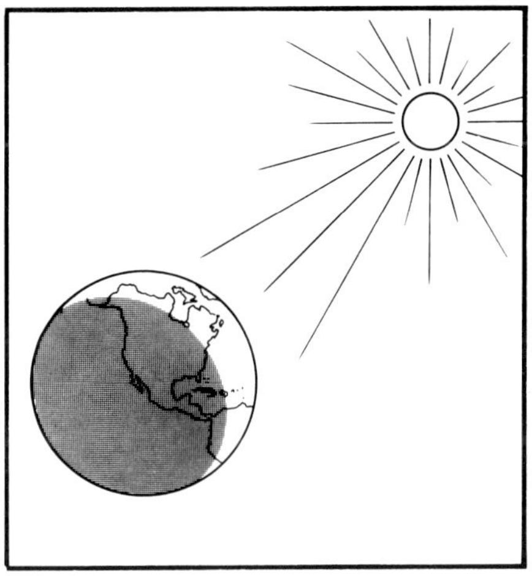

Figure 1-1. Earth's main source of electromagnetic energy.

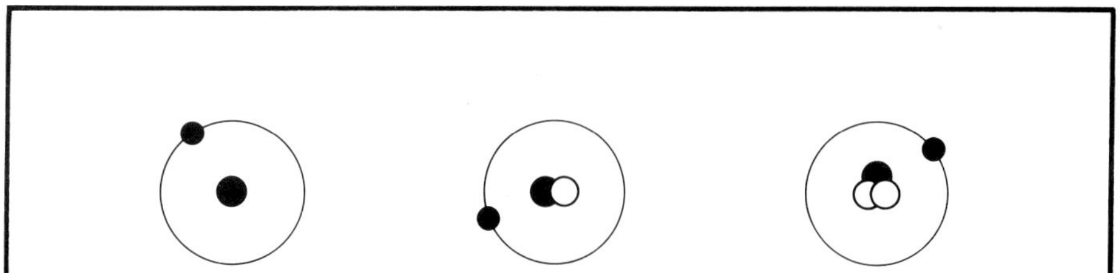

Figure 1-2. Three Isotopes of Hydrogen

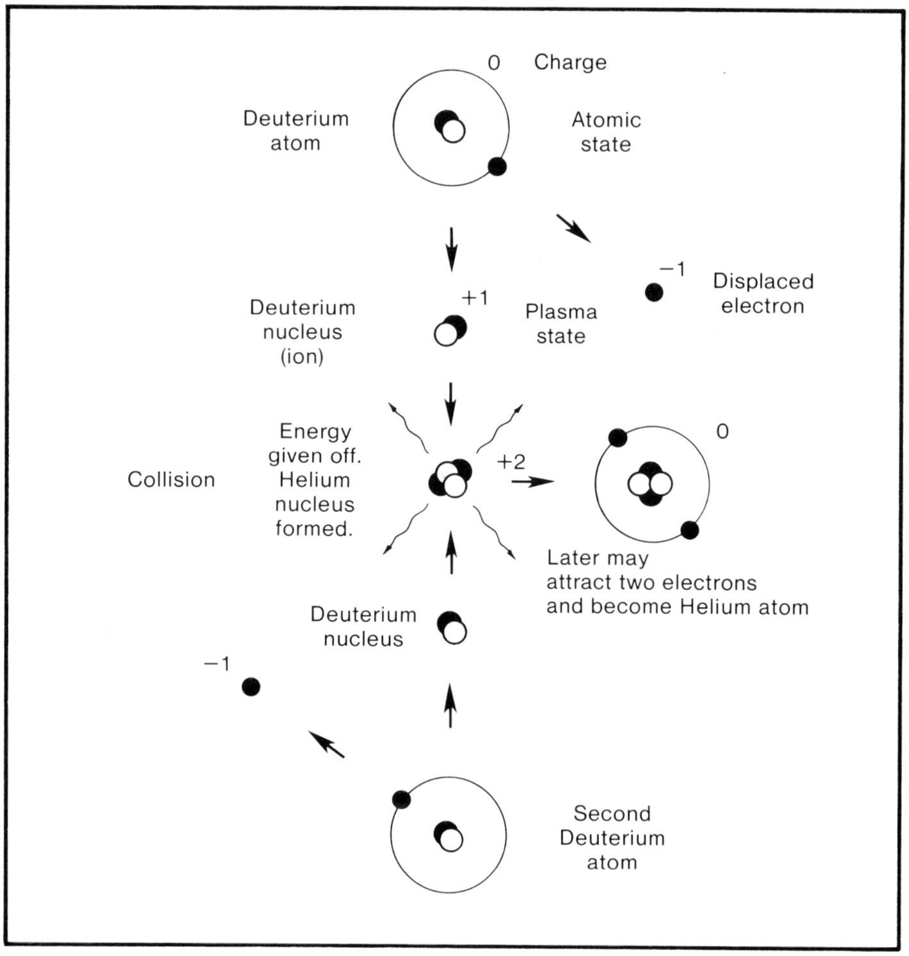

Figure 1-3. Nuclear Fusion.

and heat is given off.

The sun, our primary energy source, is a seething mass of gas undergoing nuclear fusion (Figure 1-1). Much of the gas is hydrogen. When four atoms of hydrogen are fused into one atom of helium, some of the mass of the hydrogen is changed into electromagnetic energy, some of which is light. So much energy is created in the sun that it loses about four million tons of mass each second, but it has enough mass to last for billions of years.

An atom of hydrogen consists of a nucleus made of a simple proton with a single electron in orbit around it as shown in Figure 1-2. Heavy hydrogen or deuterium has a neutron added to the nucleus.

Because of the extremely high temperature and pressure in the sun, many electrons get stripped off the hydrogen atoms so that the nucleus becomes an ion with a positive charge. The freed electron has a negative charge. This condition of matter, where there are many free ions, electrons and other parts of matter, is called "plasma".

Nuclear ion particles move with great speed in the sun's plasma. Normally, positively charged particles would repel other positively charged particles and be attracted to negatively charged particles. However, some hydrogen nuclei collide because of their extreme speed and force. This is illustrated in Figure 1-3. As they collide they overcome the atomic force and once joined become a helium nucleus consisting of two protons and two neutrons. They are held together tightly by atomic force. When they collide, a great deal of energy, much of it electromagnetic radiant energy, is given off. This nuclear fusion is the source of much of the sun's energy some of it being light. The helium nucleus may later attract two electrons and become a helium atom.

A gram of deuterium (heavy hydrogen) nuclei being fused in this manner gives off several million times as much energy as a gram of gasoline being burned in an automobile engine.

There are atoms of other elements in the sun—nearly as many as on Earth. Some of them are formed by nuclear fusion. However, it is believed that most of the sun's energy comes from the fusion of hydrogen atoms.

Incandescent Light

Solid materials are generally made of molecules consisting of the atoms of two or more elements. Metals such as tungsten (Figure 1-4) and copper are single element molecules. Tungsten has a cubic lattice structure. Its atoms are arranged in a very regular fashion and are held in place by atomic forces. This is what distinguishes a solid from a gas or a liquid. In gases and liquids, the molecules can move around within

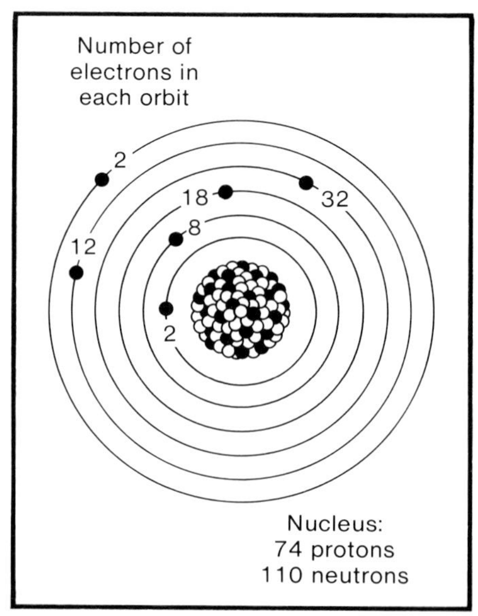

Figure 1-4. Tungsten Atom

the confines of the container, therefore, they are called fluid.

In all matter there is much more space than material. In tungsten the

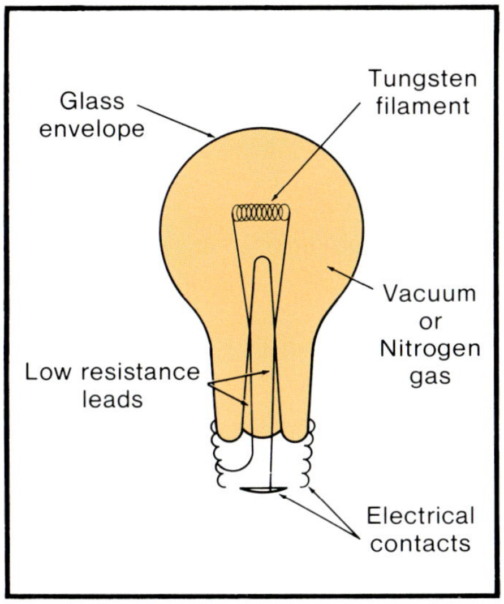

Figure 1-5. Light Bulb

molecules move back and forth in the space between the molecules, but keep their same relative positions. The distance and speed of their movement depend on their temperature. When the temperature is at -273°C (absolute zero) the motion stops. As the temperature increases, their travel distance and speed increase. When the temperature is increased to 3370°C, their motion becomes so great that the molecular force is greater than the atomic force (that holds them in their lattice structure) and the tungsten melts.

In a light bulb, a tungsten filament (thin wire) is attached to two low resistance copper support leads inside a glass bulb as shown in Figure 1-5. Electricity is made to flow through the tungsten filament which has a high resistance to electricity.

Electricity is electrons in motion. In the copper support wire, the electrons slip in and out of the orbits around the copper atomic nuclei and move from atom to atom easily as they are forced by the voltage along the wire. However, they do not slip easily into orbit in tungsten atoms. Tungsten has a high resistance which causes the electrons to slow down. Since this is one of the criteria for generating electromagnetic radiation, it is one of the two causes of light in a light bulb.

The other is the movement of the tungsten atoms. The electrons' motion is kinetic energy. This energy causes the tungsten atoms (molecules) to increase their speed and length of "swing" in the intermolecular space. As they reach the end of their "swing" they are forced to slow down by the molecular force. Because they have extra electrons in their outer orbit they are charged particles. This increase in motion raises the temperature, and when the temperature reaches about 2500°C, light is emitted. Light bulbs are designed so that their filaments can be heated to about 2500°C to be efficient light producers and still not melt.

Tungsten would burn (oxidize) if heated to these temperatures in air where there is oxygen. This is why the filament is placed in a sealed glass envelope in a vacuum or in an atmosphere of nitrogen.

The Nature of Light

As light and the other forms of electromagnetic energy radiate from a source, the energy generally travels in straight line paths at a high rate of speed, over 30,000 kilometres (186,300 miles) per second in space. This speed is so fast that, for all practical purposes in this course, we can consider its travel to be instantaneous.

There are exceptions to the straight-line paths in which light rays travel. When light travels near an enormous mass such as a star, the extremely small mass of light particles is attracted by the gravity of the mass, bending the path slightly. This is illustrated in Figure 1-6.

As will be discussed later in more detail, when a light ray goes very near the edge of an opaque object, it is *diffracted* or bent slightly, as shown in Figure 1-7. When light

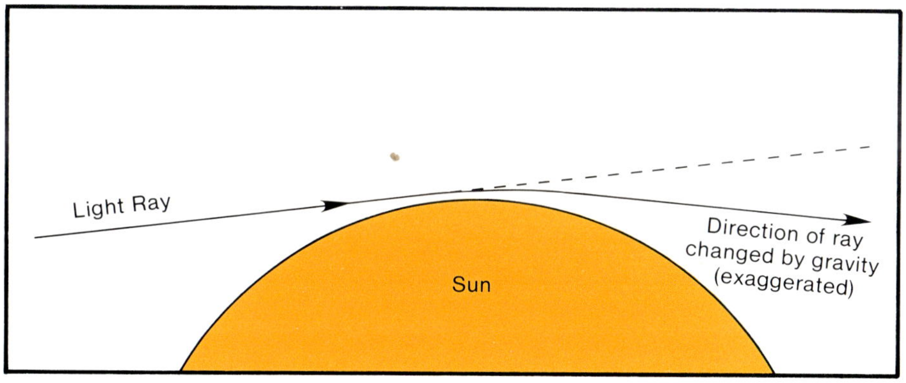

Figure 1-6. Light Ray Bent by Gravity

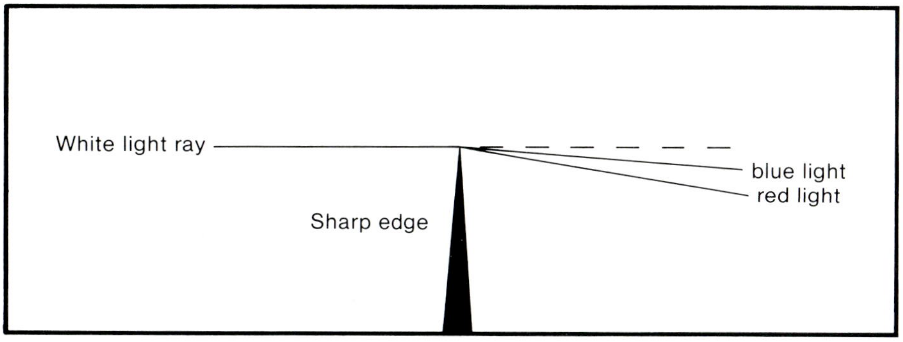

Figure 1-7. Light Ray Bent by Diffraction

rays are reflected or go from air into glass or water, their paths are changed.

Finally, the paths of light rays going through heated air are bent slightly causing mirages and the twinkling of stars.

When discussing the distance of stars, the speed of light becomes important. However, this is of importance to astronomers, not to optical engineers. The greatest path length in our largest optical instruments seldom exceeds 50 feet. Light travels 50 feet in about .000,000,005 seconds, which can be considered instantaneous. There is one situation in which the relative speed of light will be considered. When light goes from one transparent medium to another one of greater density (air into water or glass) its speed is changed slightly. It slows down a bit. This change is important because it explains how lenses form images. However, while the relative speed change is significant, the time taken by light to form images is still measured in hundredths of millionths of a second—still instantaneous for all practical purposes.

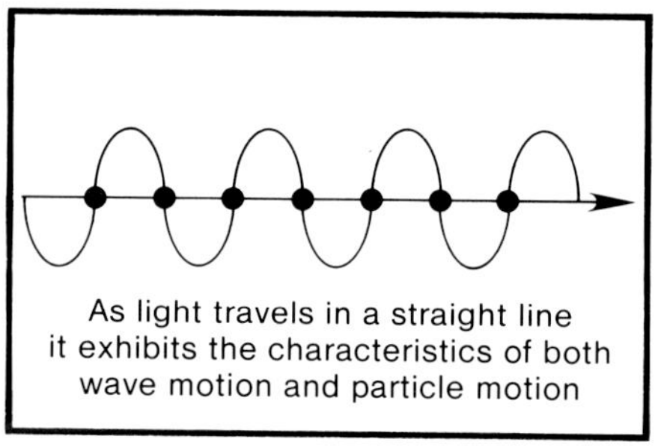

Figure 1-8. Light Travel

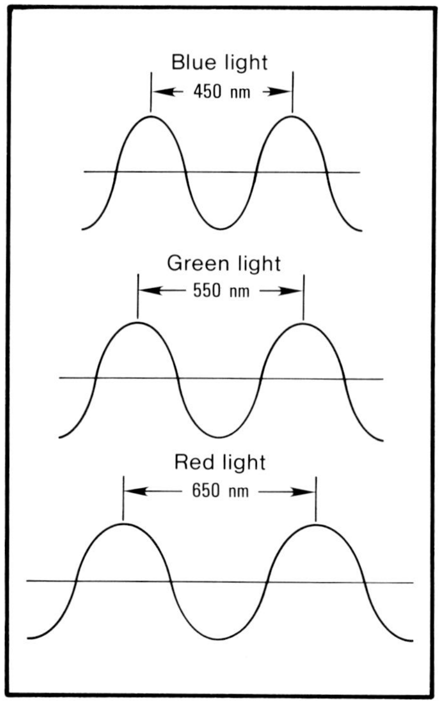

Figure 1-10. Wavelengths of Blue, Green, and Red Light.

As it travels in straight-line paths at a high rate of speed, light exhibits both the characteristics of wave motion and corpuscular (particle) motion (Figure 1-8). Wave motion can be visualized as the motion of water waves which, particularly in a storm, carry a great amount of energy. To visualize corpuscular or particle motion, think of the energy carried by a baseball. The tiny "corpuscles" that carry the energy of light are called photons. Photons are so small that they are normally considered to have no mass, but their energy is relatively great because of their very high velocity.

Wave Motion

In any form of wave motion, the distance from the crest of one wave to the crest of the next wave is called the wavelength (Figure 1-9). The wavelengths of waves on a lake are likely to be measured in feet. The wavelengths of light are so short they are measured in nanometres; (one nanometre is one billionth of a metre). A wavelength of light is about 1/50,000 of an inch.

If the wavelength and the speed are known, the frequency, which is the number of waves that pass a point in a given time, can be found. The relationship between the three factors is:

$$\text{Speed} = \text{Frequency} \times \text{Wavelength}$$

or

$$\text{Frequency} = \frac{\text{Speed}}{\text{Wavelength}}$$

Since the speed of light is so great and the wavelengths are so short, the frequencies of light are enormous. About 300 trillion waves per second go past a point. Three hundred trillion is 300,000,000,000,000. Obviously, it is more convenient to discuss light in terms of wavelengths rather than frequencies.

Light's wavelengths range from about 400 to 700 nanometres (Figure 1-10). Since light is defined in terms of what people can see, and the visual apparatus of one person differs from that of another, the limits of the light range must be approximate. White light which comes from the sun or from a light

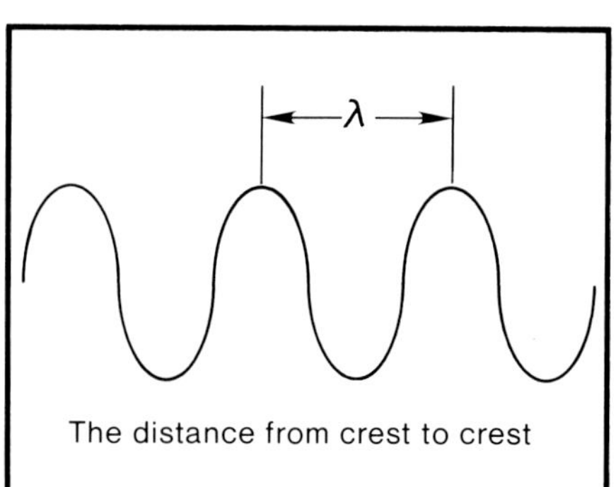

Figure 1-9. Wavelength

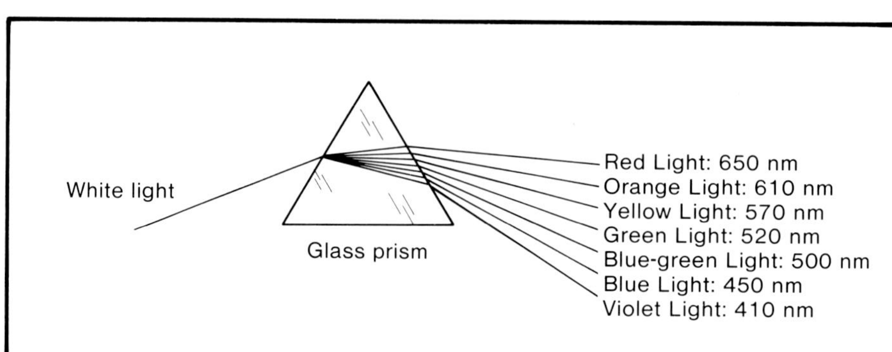

Figure 1-11. The Visible Spectrum

bulb is a nearly equal mixture of all wavelengths from 400 to 700 nanometres.

When a narrow band of light is allowed to pass through a prism, the light is refracted (bent) in inverse proportion to its wavelength so that the light leaving the prism forms a visible spectrum of many colors as shown in Figure 1-11. The shortest wavelengths are violet light and are bent most by the prism. The longest wavelengths are red light and are bent least by the prism. Light of intermediate wavelengths, hence of other colors, is bent less than the violet and more than the red and appears between these two colors in the spectrum.

The Electromagnetic Spectrum

This program is primarily concerned with optical devices used with light. However, light is but one type of electromagnetic radiant energy. The other types differ from light fundamentally because they have different wavelengths. However, each band of wavelengths has its own characteristics. All the forms can be shown in a diagram in which the various types of electromagnetic radiant energy are arranged by wavelength. Such a diagram is called an *electromagnetic spectrum* (Figure 1-12). Optical instruments using glass elements can be used with only a small portion of the electromagnetic spectrum. Parts of the ultraviolet (UV) and the infrared (IR) spectra can be refracted to form images by typical optical devices. These images can be recorded on film but UV and IR wavelengths are invisible to the human eye.

The illustration shows the electromagnetic spectrum from the extremely short waves associated with cosmic radiation to middle-long wavelengths of radio waves.

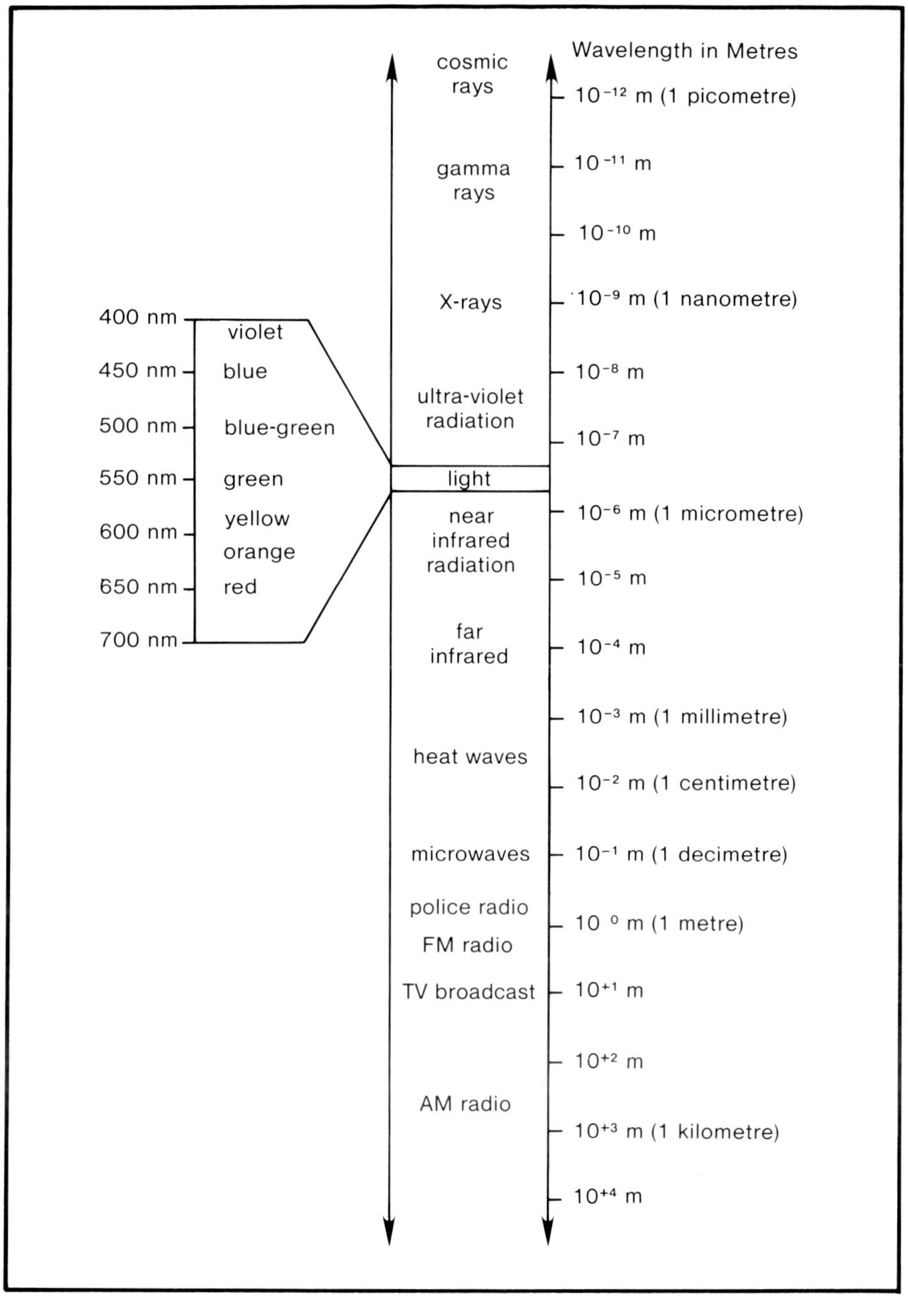

Figure 1-12. Electromagnetic Spectrum

International Systems of Units

pico	$= 10^{-12}$	0.000 000 000 001
nano	$= 10^{-9}$	0.000 000 001
micro	$= 10^{-6}$	0.000 001
milli	$= 10^{-3}$	0.001
centi	$= 10^{-2}$	0.01
deci	$= 10^{-1}$	0.1
deka	$= 10$	10
hecto	$= 10^{2}$	100
kilo	$= 10^{3}$	1000
mega	$= 10^{6}$	1000 000
giga	$= 10^{9}$	1000 000 000
tera	$= 10^{12}$	1000 000 000 000

Questions	True	False
1. Light is a form of potential energy.	_____	_____
2. All light comes from electromagnetic sources.	_____	_____
3. Small, moving particles must carry an electrical charge if they are capable of emitting light.	_____	_____
4. A high-resistance wire can be made to emit light because it can cause electrons to change their direction of travel.	_____	_____
5. Light travels in straight lines as it goes close to a sharp edge.	_____	_____
6. The speed of light is 186,000 miles per hour.	_____	_____
7. The travel of light exhibits the characteristics of both wave and corpuscular motion.	_____	_____
8. In wave motion, as the frequency increases the wavelength decreases when the speed is constant.	_____	_____
9. The wavelengths of light vary from 500 to 700 nanometres.	_____	_____
10. White light is a very nearly even mix of the light waves of all the light wavelengths.	_____	_____
11. For most practical optical purposes, light can be considered to move instantaneously.	_____	_____
12. Of all the other forms of electromagnetic radiant energy, those closest to light on the spectrum are ultraviolet and infrared radiant energy.	_____	_____
13. The chief differences between the various forms of electromagnetic energy are in their wavelengths and their speeds of travel.	_____	_____

Chapter 2

The Behavior of Light

Summary

All light incident on an object surface is reflected, absorbed, or transmitted (I = R + A + T). Light is transmitted in an orderly fashion through *transparent* materials. Light is *diffusely* transmitted through *translucent* materials. If all wavelengths are transmitted evenly, the transmission is *achromatic.* If some wavelengths are transmitted less than others, the transmission is chromatic. An object that reflects or *absorbs* all light, transmitting none, is *opaque.* If light rays are reflected in an orderly fashion from a smooth, polished surface, the reflection is *specular*. However, if the surface is microscopically rough, the light is reflected in a disorderly manner causing a diffuse reflection. There are *achromatic* and *chromatic* reflections. The law of reflection is: the angle of reflection equals the angle of incidence. Light rays that enter a transparent material at an angle to the surface are *refracted.* Snell's law: $n \sin I = n' \sin I'$. Different wavelengths of light are refracted at different angles; this is *dispersion.* Light going near an edge is diffracted; different wavelengths are bent slightly more than others. *Interference* occurs between closely spaced layers of thin transparent materials. Light passing through very thick layers of air is *scattered,* while light reflected specularly from nonmetallic surfaces is *polarized* at or near Brewster's angle, which is defined as the angle of maximum polarization. Inverse square law: the intensity of light varies inversely with the square of the distance from the source.

The RAT Formula

As noted earlier, light travels at high rates of speed in space and air. Space and air through which light travels are said to be *transparent.* There are many other transparent materials. Glass, some plastics, water, many liquids, and nearly all gasses are transparent. Even the wings of some insects and parts of mammals, such as the cornea of the eye, are transparent.

When light radiates from a source through a homogeneous transparent medium it travels in straight lines. As it travels, light frequently strikes an object. If the object is transparent, most of the light is transmitted through the object. However, some of the light striking the surface of the object is *reflected.* This means that it bounces off the object's surface rather than passing through it. If the material is glass and if the light strikes the glass normal to the surface (perpendicular to it), about 95% of the light will pass through it, while about 5% will be reflected. However, if light strikes the surface at very small angles to the surface, nearly all of the light will be reflected and very little or no light will enter the glass. When the light is passing through the glass, some of it is *absorbed.* That is, some of the photons of light get trapped in the molecules of the glass. The trapped light energy is transformed into heat energy which raises the temperature of the glass slightly.

If light strikes the surface of a highly polished metal such as aluminum or silver, nearly 90% of the light is reflected. Since the metal is not transparent, the other 10% of the light is absorbed and no light is transmitted. Objects that do not transmit any light are called *opaque.*

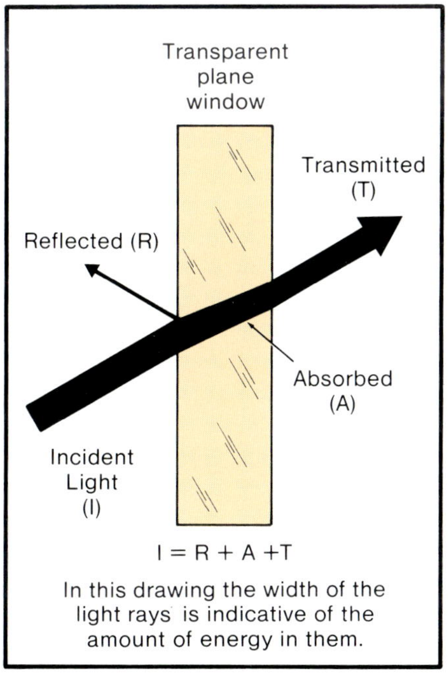

Figure 2-1. The RAT Formula.

Light falling on the surface of an object is called the *incident* light. All the incident light is either reflected, absorbed or transmitted. This is shown in an equation called the RAT formula.

$$I = R + A + T$$

Incident energy must equal the sum of the energies that are reflected, absorbed and transmitted (Figure 2-1).

Light Rays

The straight-line paths that light takes as it travels are called *light rays*. In the diagrams that are used to illustrate various examples of what happens to light, straight lines are drawn to represent the light rays (Figure 2-2). The paths that light follows as it is transmitted, absorbed, or reflected can be shown by the use of diagrams with straight lines representing the light rays. These straight lines are referred to as *schematic rays*.

Figure 2-2. Light rays radiate from their source. The rays are straight-line paths that light follows.

Transmission

Transparent materials that are colorless, such as clear glass, transmit light *achromatically* (colorlessly). They transmit all wavelengths equally. The color of the light being transmitted is not changed by the material. Some transparent materials transmit light *chromatically* (colored) that is, they absorb some wavelengths more than others as the light goes through the material. The light going through material is changed in color as shown in Figures 2-3 and 2-4. When certain metal salts are mixed with molten glass, the glass becomes colored and transmits light chromatically (glass in stained-glass windows is made this way).

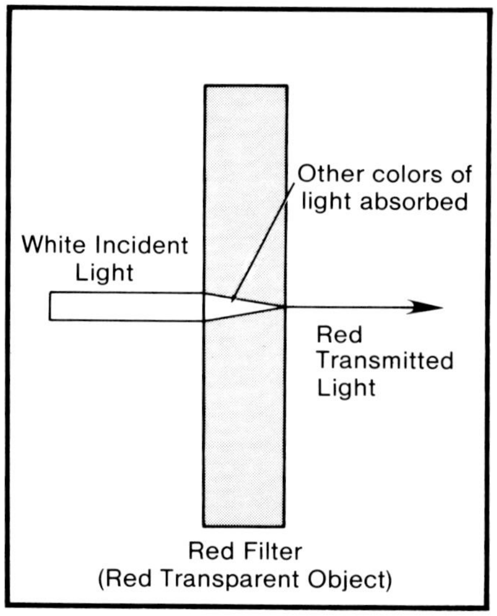

Figure 2-3. Chromatic Transmission

A piece of glass with plane, parallel polished surfaces is called a *plane window* or *optical flat*. Light is transmitted freely through windows. Because the surfaces of a window are smooth and polished (microscopically smooth), the light rays are transmitted in an orderly manner. If the incident light is composed of parallel rays, the transmitted rays will still be parallel. Images can be seen through windows because they are transparent. If the window transmits light chromatically, it changes the color of the light passing through it. Such windows are called *filters*. Filters are used to change the color of light without changing the image-carrying capabilities of the light being transmitted. They are often used in front of camera lenses to control the color rendition of photographs.

If one of the surfaces of a window is microscopically rough, as in ground glass, the light is not transmitted in an orderly pattern, but is randomly transmitted in all directions. Such glass is called *translucent* which means that it transmits light but images cannot be seen through it (Figure 2-5). If an image is focused on the ground surface by a lens, it can be seen as when the ground glass viewing screen of a camera is used for focusing. A glass of milk is also translucent because it transmits light but not images. The transmission of translucent windows can be achromatic or chromatic. Most pieces of glass in stained-glass windows are chromatic translucent transmitters of light.

Reflection

Light is reflected *specularly* or *diffusely*. If an object's surface is smooth and polished and the incident rays are orderly, light rays are

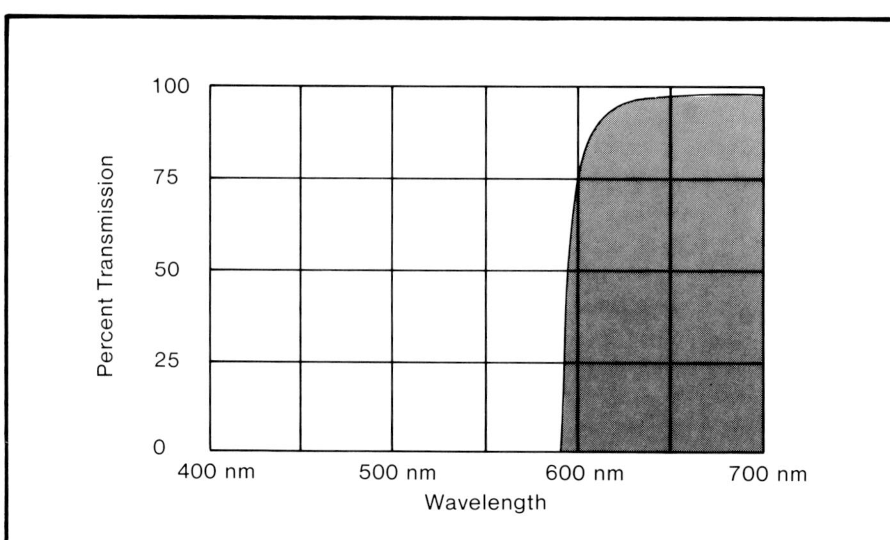

Figure 2-4. Spectral Transmission Curve for a Red filter

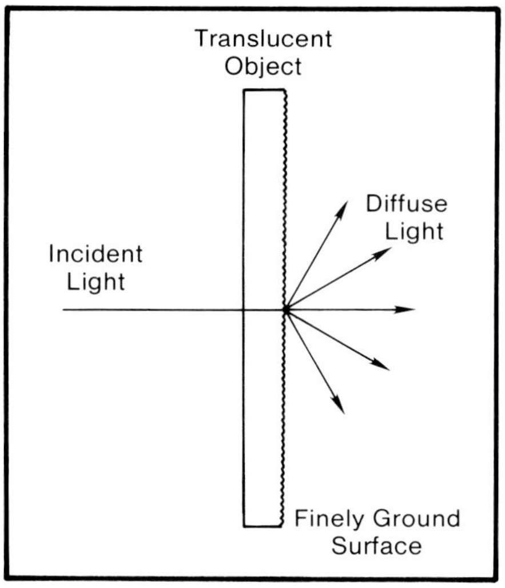

Figure 2-5. Diffusion

reflected in an orderly fashion. Such a reflection is called a specular reflection. If the surface is a flat plane, light rays which are parallel when they are incident upon the glass, will be parallel when they are reflected as illustrated in Figure 2-6. If the surface is highly reflective, as it is if it has a thin layer of metal such as silver or aluminum on it, it is called a *mirror*.

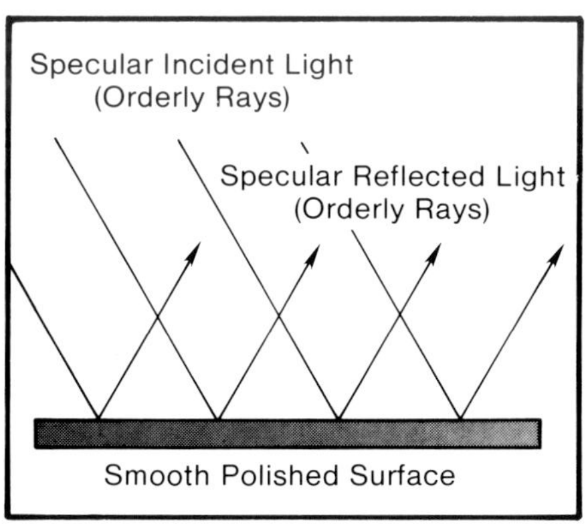

Figure 2-6. Specular Reflection

Not all mirrors are plane or flat. A concave spherical mirror is one whose surface is a section of a sphere, and whose center of curvature is in the direction from which the light rays come. A concave spherical mirror causes parallel light rays to be focused to a point (nearly). The parallel incident light rays *converge* after being reflected (Figure 2-7). Converging light rays are orderly but they are not parallel. Conversely, rays from a tiny light source, such as a flashlight, which converge at the focal point of the reflector will form parallel reflected rays.

A convex spherical mirror also reflects light rays in an orderly fashion, but it *diverges* the light rays rather than converging them. No image is formed (Figure 2-8).

If the reflecting surface is microscopically rough, it reflects light *diffusely*. That is, it reflects orderly incident rays of light in many directions in a disorderly fashion. Flat wall paint is a diffuse reflector because it reflects light rays in many directions; there is no shine to it (Figure 2-9).

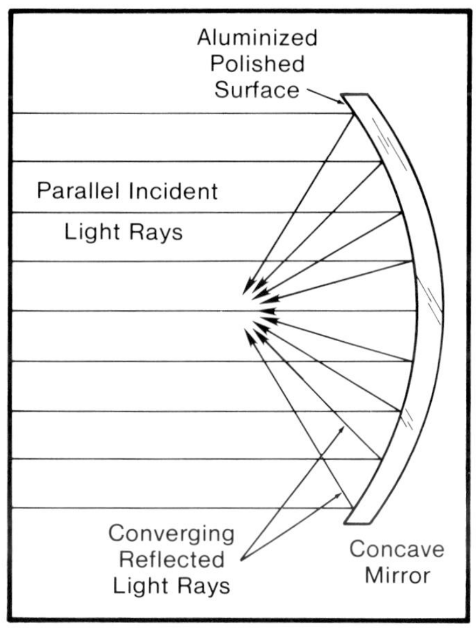

Figure 2-7. Reflection from a concave surface

Most object surfaces are both specular and diffuse reflectors. The outer surface is somewhat smooth and transparent, and reflects specularly. Just under the outer surface is a diffuse surface that reflects diffusely. A shiny nose exhibits both types of reflections. The shine comes from an oily layer on the skin, while the skin color comes from a diffuse reflection from the skin surface just under the oil layer as illustrated in Figure 2-10.

Many objects reflect light rays *achromatically*, that is they reflect all wavelengths of light equally. Such objects appear neutral in color, that is, white, gray or black. Where there is a high degree of achromatic, diffuse reflection in an object surface, we see the object as white. When the amount of reflection is low, barely visible and achromatic, the object is seen as black. Where the level of reflection is in between, the surface is called gray. This can be seen in Figures 2-11 and 2-12.

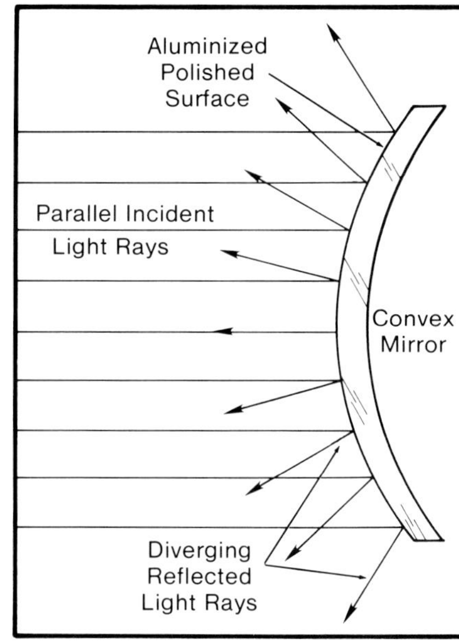

Figure 2-8. Reflection from a convex surface

On the other hand, when a surface diffusely reflects some wavelengths more than others, the reflection is *chromatic*, and the object is seen as colored. If the wavelengths most strongly reflected are the short, blue

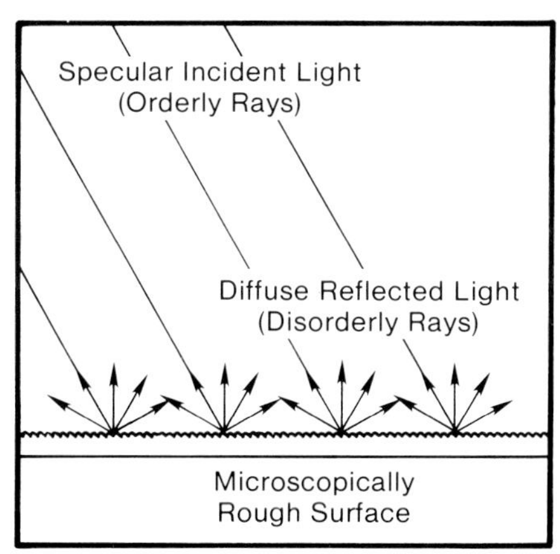

Figure 2-9. Diffuse reflection
(Reflection from an uneven surface)

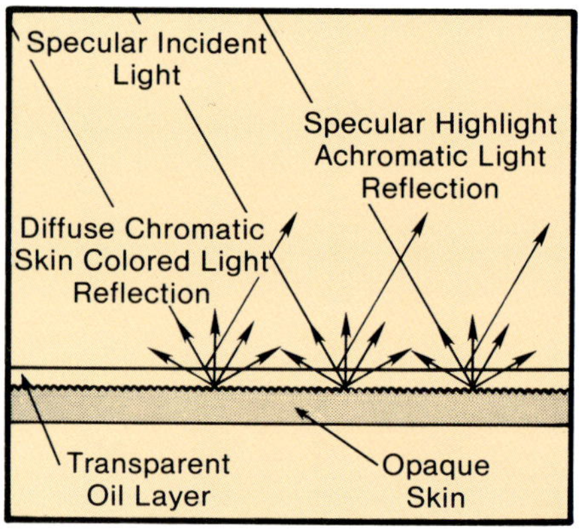

Figure 2-10. Reflection from a shiny nose

wavelengths (between 400 and 500 nanometres) the color of the object is seen as blue as demonstrated in Figures 2-13 and 2-14.

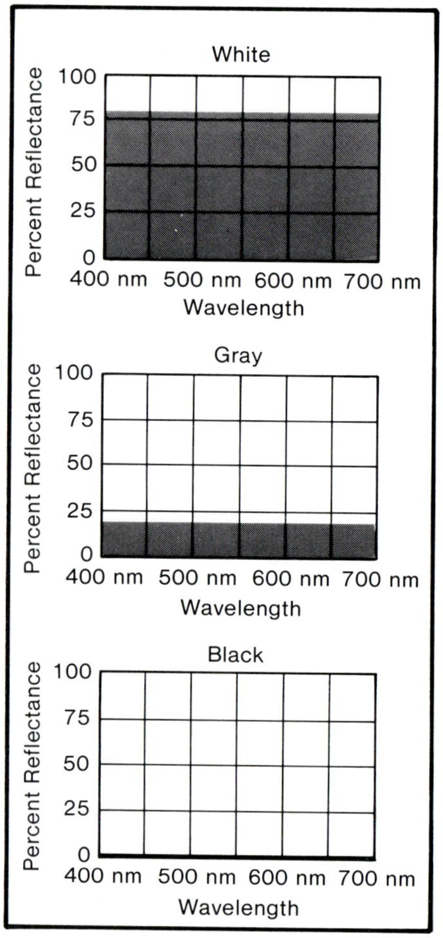

Figure 2-12. Spectral Reflectance Curves

Except for the specular reflections from colored metals such as copper and gold, nearly all specular reflections are achromatic. Diffuse reflections can be either chromatic or achromatic. A shiny red apple is seen as shiny because the highlight on it is a specular reflection of the light source. The highlight is seen as white, which is achromatic. The red of the apple is the diffuse chromatic reflection from the subsurface of the apple skin. This is illustrated in Figure 2-15.

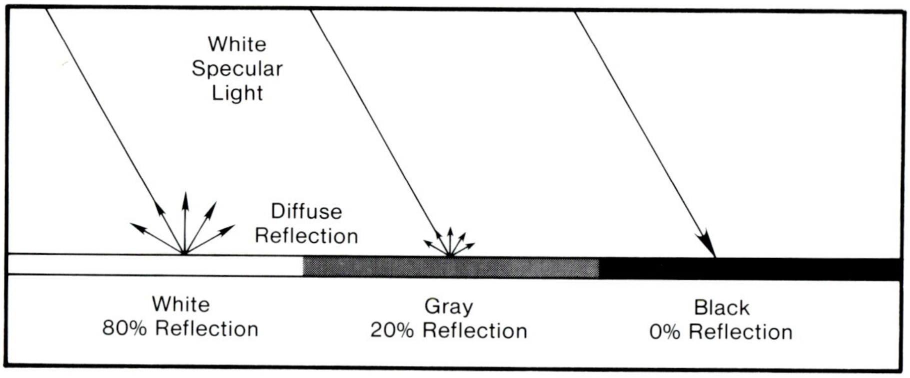

Figure 2-11. Achromatic Diffuse Reflections

If the light incident on a shiny surface is not white, but is colored (chromatic), the light specularly reflected will be chromatic, even though the reflection is achromatic. Achromatic reflections do not change the mix of wavelengths in the incident light when they reflect it. If an apple is seen in the late afternoon sunlight, the highlight will appear yellow-orange in color, the color of the sun. The apple will still appear red.

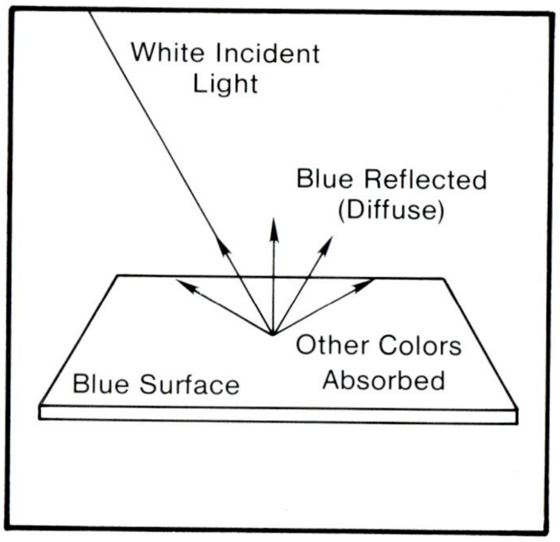

Figure 2-13. Chromatic Reflection

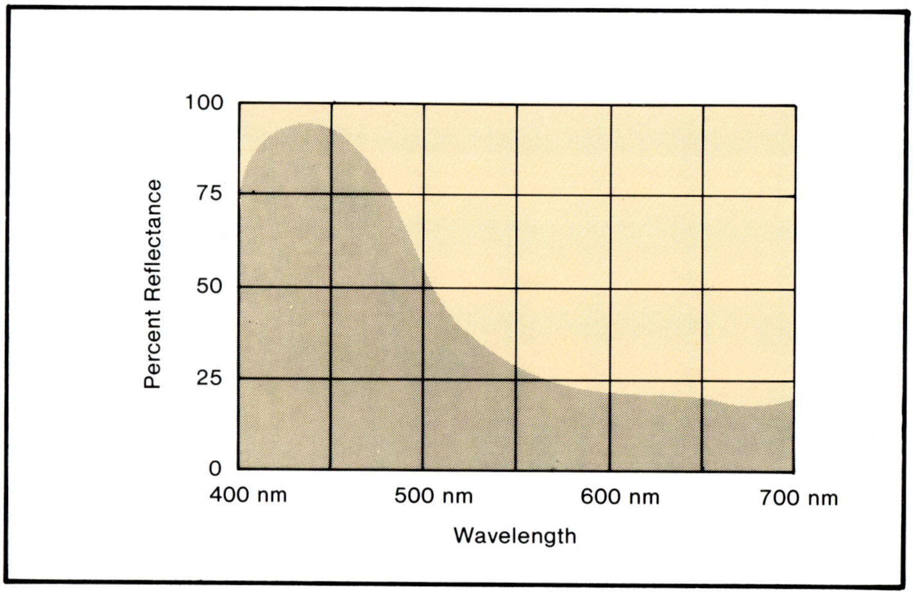

Figure 2-14. Spectral Reflection Curve of Blue Diffuse Surface

Figure 2-15 Red Apple.

Law of Reflection

The angle at which each light ray reflects from a surface is equal to the angle of incidence upon the surface as shown in Figure 2-16. The angles of reflection and incidence are measured to the normal (perpendicular) rather than to the surface itself. When the surface is polished smooth, the rays leave the surface in an orderly fashion as previously demonstrated in Figure 2-6. When the surface is microscopically rough, each light ray follows the reflection law, but because the roughness presents tiny surfaces at different angles to the incident rays, the rays are reflected in many directions and the reflection is *diffuse* (Figure 2-17, 2-18).

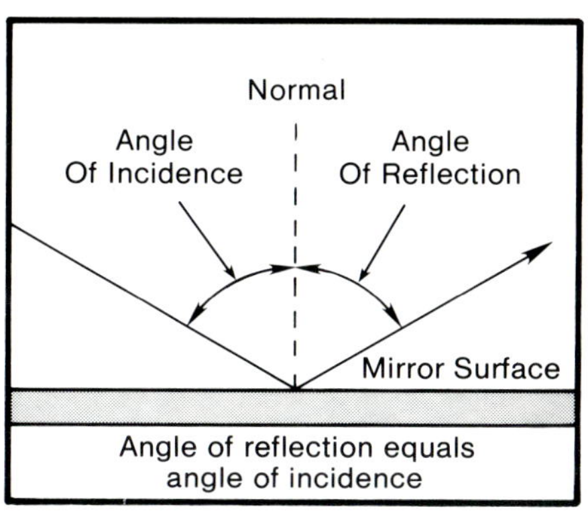

Figure 2-16. Law of Reflection

Refraction

The speed of light in air is about 186,000 miles per second. It is less than that in denser transparent materials such as glass and water. As light enters a denser transparent material from air at an angle, this change in speed causes the light to bend at the air-glass interface. It is this bending, or *refraction*, that makes optical elements such as lenses and prisms useful. The speed of light in air* divided by the speed of light in the optical medium is called the *index of refraction.* The speed of light in water is about 0.75 as fast as it is in air, or about 140,000 miles per second. Index of refraction of water is:

$$\frac{186{,}000}{140{,}000} = 1.33$$

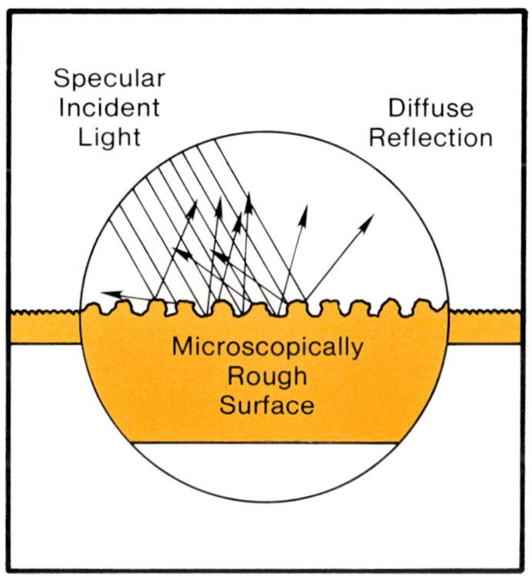

Figure 2-17. Diffuse Reflection

This is shown in Figure 2-19.

The speed of light in ordinary glass is about 65% that of its speed in air, or about 124,000 miles per second. Index of refraction of common glass can be shown as:

$$\frac{186{,}000}{124{,}000} = 1.50$$

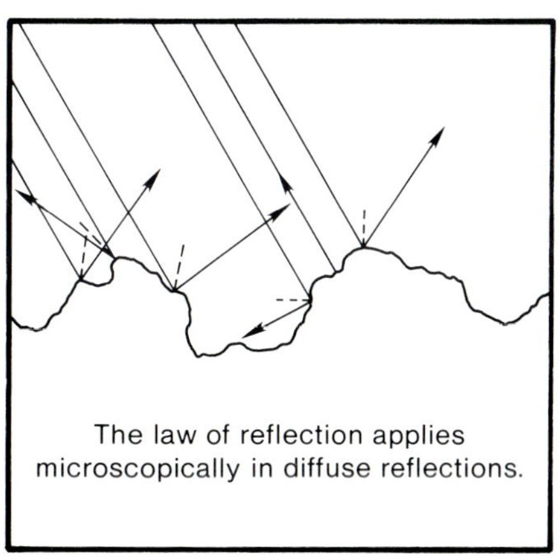

Figure 2-18. Diffuse Reflection

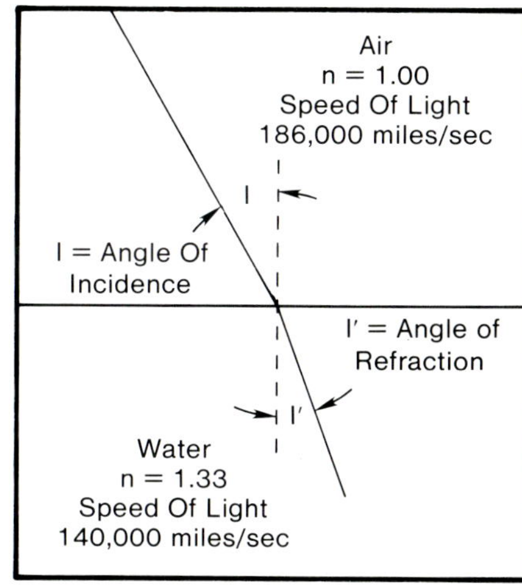

Figure 2-19. Refraction

The law that describes how much bending occurs as a ray passes from one medium into another is known as Snell's law.

Light goes from air with an index of refraction of 1.00, into glass with a higher index of refraction as stated in the following equation:

$$n' \text{ (index of glass)} = \frac{\sin I}{\sin I'}$$

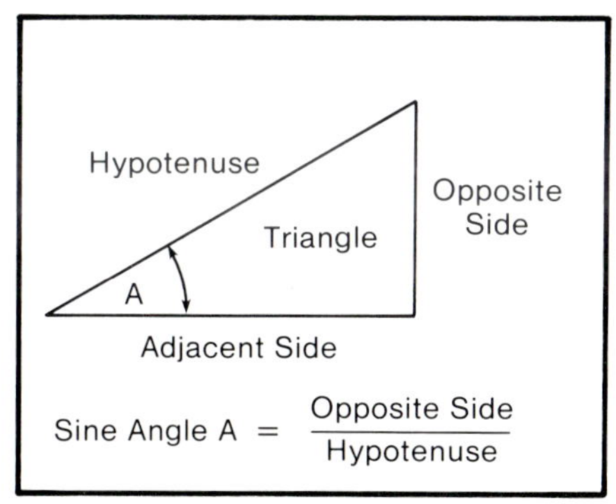

Figure 2-20 Definition of Trigonometric Sine.

where I is the angle of incidence (measured from the normal to the glass surface) and I′ is the angle of the new path in the glass (also measured from the normal to the glass surface). For those not familiar with trigonometry, the sine is a ratio of two sides of a right triangle. This is shown in Figure 2-20.

*Actually in a vacuum or space, but since the optical instruments covered in this program are used in air, it is convenient to use this convention. The difference is slight. The index of air compared to space is 1.0016.

$$\text{sine of the indicated angle} = \frac{\text{length of side opposite the angle}}{\text{hypotenuse}}$$

The sine values of angles can be found in the tables of trigonometry functions in many mathematics books or chemistry and physics handbooks. The sine values of angles are built into many scientific pocket calculators as well. A more general statement of Snell's law is:

$$n \sin I = n' \sin I'$$

Where n is the index of refraction of the first optical material, and when this is air, n is considered to be 1.00. Usually the angle of incidence is known, as well as the two indices. The law then becomes:

$$\sin I' = \frac{n \sin I}{n'}$$

Example: A light ray enters a prism whose index is 1.600. The angle of incidence is 10°. What is the angle of refraction?

$$I' = \frac{1 \times \sin 10°}{1.600} = \frac{.17365}{1.600} = .10853$$

$$I' = 6.23°$$

If the glass surface is curved, as in a lens, the angles are measured to the normal drawn perpendicular to a tangent to the curved surface at the point of entry. This is illustrated in Figure 2-21.

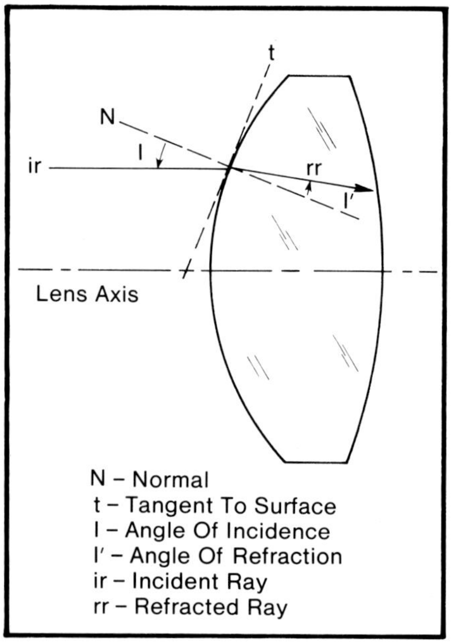

Figure 2-21 Diagram of Snell's Law at a lens surface.

As light goes from the glass (dense medium) back into the air, it is bent away from the normal. The index of the glass then becomes n, while the index of air becomes n'. The angle in the glass is I, while the refracted angle in the air becomes I'. This is demonstrated in Figure 2-22.

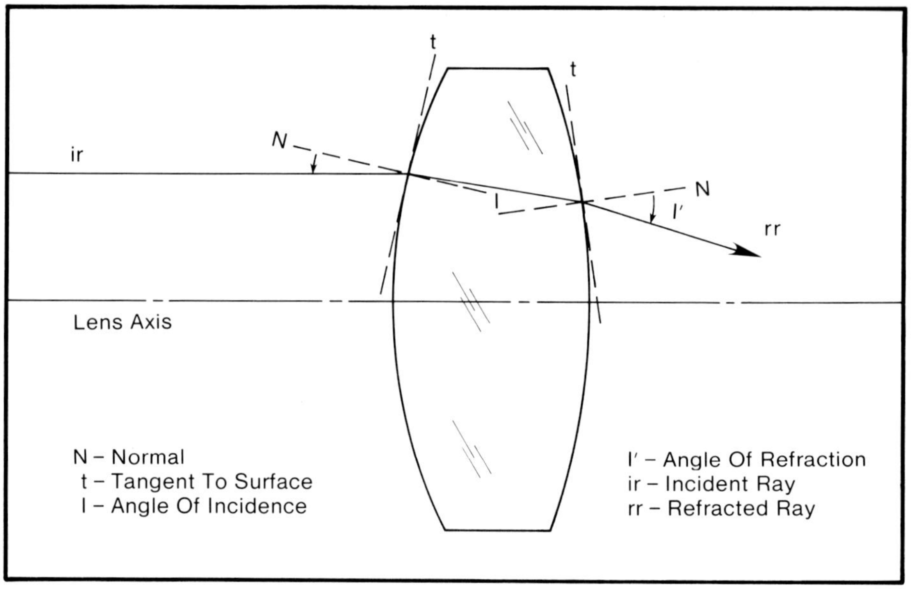

Figure 2-22 Diagram of Snell's Laws at the lens's second surface.

Dispersion

Optical materials have different indices of refraction for light of different colors because the different wavelengths have different speeds in the same medium. Only in space, which has an index of refraction of 1.000,000, do all wavelengths travel at the same speed. The different speeds in denser mediums cause the light rays to bend different amounts as they enter and leave the medium at an angle. This behavior of light is called dispersion. When white incident light passes through a glass prism the light is separated by wavelengths to form a spectrum and we have *dispersion* (Figure 2-23).

The following table shows the variation of the index of refraction by wavelength of a dense flint optical glass.

Color	*Wavelength*	*Index of Refraction*
Violet	405 nm	1.685
Blue	486 nm	1.664
Green	546 nm	1.655
Yellow	589 nm	1.650
Red	656 nm	1.644
Deep Red	671 nm	1.643

The index of refraction most commonly used to trace rays of thin lenses is the index for yellow light, 589 nm, and this index is called n_D because that is the wavelength of the sodium vapor spectral D line. Sodium vapor lamps give off radiant light found in the middle of the spectrum. Scientists use these lamps because of their availability and constant standard of light which makes measuring easier.

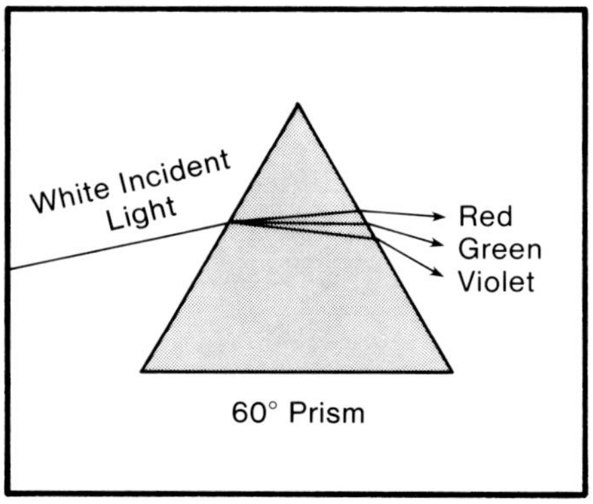

Figure 2-23. Dispersion

Diffraction

One exception to the straight line travel of light occurs when light rays pass very near a solid edge. When this happens, the light is bent very slightly. We are not visually aware of this phenomenon because the amount of the bend is so minute, however, it is of some importance in certain optical situations. This bending of light as rays pass an opaque edge is called *diffraction*, as illustrated in Figure 2-24. Different wavelengths of light are bent different amounts so that a spectrum can be formed by diffraction. In dispersion, light of short wavelengths (blue and violet) is bent the most; in diffraction it is just the opposite, the longest rays, red wavelengths, are bent the most.

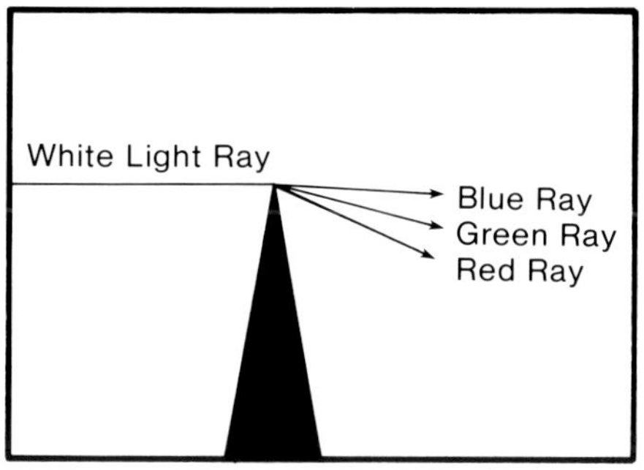

Figure 2-24. Diffraction

In highly corrected photographic lenses, diffraction is one of the limits of a lens's ability to image sharp detail. This is especially apparent in small lens apertures as demonstrated in Figure 2-25.

A flat piece of optical glass can be pressure ruled with many opaque fine parallel lines so that light is diffracted at both edges of each groove. Such a *diffraction grating* creates a very good spectrum that is of use in scientific laboratories. Replica diffraction gratings, which are plastic molded in contact with a ruled master grating, are used as specialty filters over camera and television camera lenses to create spectra that emanate from light sources in the picture area: thus "stars" with four or eight ray-like spectra are formed. Reflection gratings are molded for use as bumper stickers and sign elements.

Interference

When light goes through an extremely thin (from a fraction of light wavelength to several wavelengths in thickness) layer of optical material, the light waves that reflect from the back surface of the thin film of material *interfere* with the incoming waves and cause banding of the reflected light and the transmitted light. If the light is white, the banding is spectral in color. Thin soap bubbles and oil on water are examples of interference.

If the light is *monochromatic*, that is, if the incident light is all of the same wavelength, alternate light and dark bands can be seen. The banding is a measure of the variable thickness of the thin spacing layer. If the layer is exactly even in thickness, the bands become so broad that only one is seen.

The thin layer can be a higher index material in air such as the soap bubble, but it can also be a thin layer of air between the two surfaces of a medium of higher index of refraction. Two pieces of clean glass placed in close contact will show interference patterns called Newton's rings. The colors seen in a peacock's feathers and the blue in a bluebird's feathers are caused by interference.

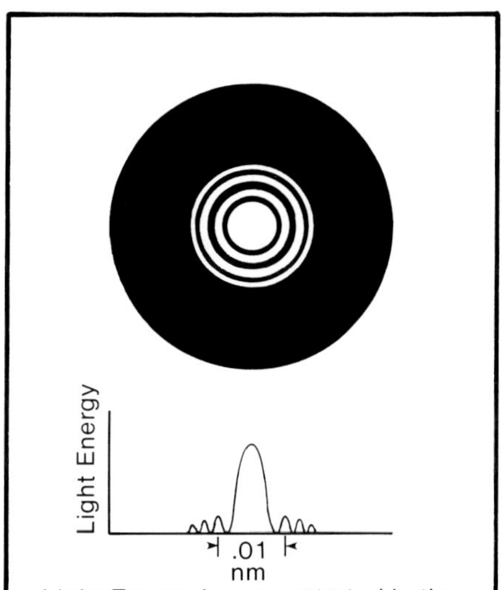

Light Energy is concentrated in the central area. Light rings are caused by diffraction, and reduce the sharpness of the lens image.

Figure 2-25. Greatly Magnified Star Image Diffraction Rings

Camera lenses are coated with a very thin layer of a mineral substance such as magnesium fluoride to create an even thickness interference layer to reduce the reflectance. The coating also allows more image-forming light to enter the lens and reduces ghost images and flare spots at the glass-air surface. The index of refraction of the coating is lower than that of the index of the glass, but higher than that of the air. Such coatings reduce unwanted reflection from the glass surface from about 5% to 0.5%, which improves the lens transmission. Therefore, the lens surfaces of modern camera lenses show a variety of colors formed by interference. Coating alternate layers of high and low index material creates dichroic filters, which reflect some colors of light and transmit others.

Scattering

When light from the sun goes through thick layers of air, photons are absorbed by the air molecules and other very small particles which increase their energy level. While the light coming from the sun is specular, the air molecules release the extra energy in the form of blue light in all directions. This release of light is called *scattering*. The diagram shows how this absorbtion and release of light energy causes the sky to appear blue (Figure 2-26).

At sunrise and sunset the sun is near the horizon. The light we see at these times passes through an extra thickness of air. There is so much of the shorter wavelengths such as the blue, and even some green light that is scattered (subtracted) from the white light that we see the remaining long wavelengths of red, orange and yellow light. Therefore, the orange-red color of sunrises and sunsets is caused by scattering.

When the blue and green wavelengths are caused to go in all directions by this molecular action, they

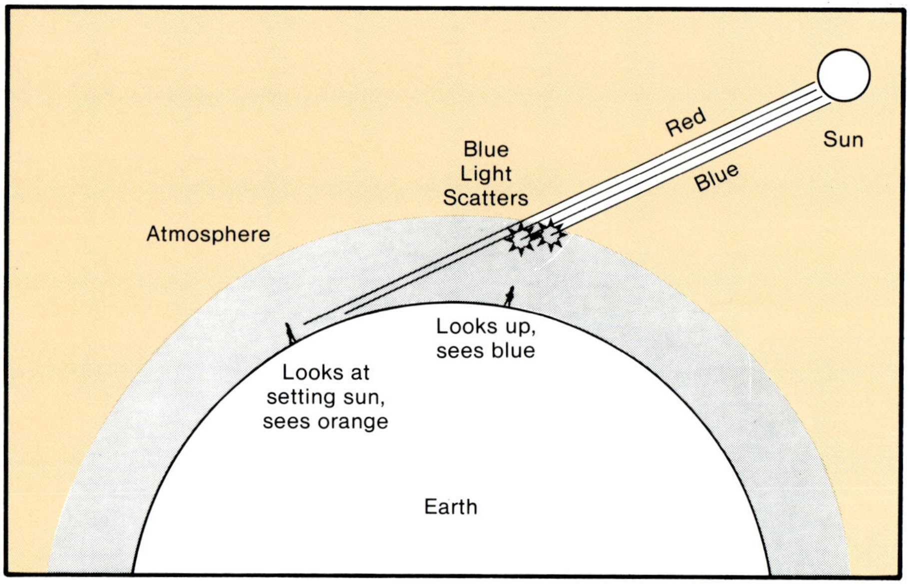

Figure 2-26. Scattering

are scattered. When it is caused to go in all directions by passing through a translucent material, or by reflection from a non-shiny surface, it is *diffused.* Scattering causes a color change whereas diffusion does not.

Polarization

When skylight is scattered, some of it is *polarized.* Ordinary light rays (un-polarized) act as if their waves vibrate in all directions around the light ray path. When light is polarized, the waves act as though they are vibrating in only one direction.

Light reflecting at certain angles from non-metallic, shiny surfaces is partially polarized, because both scattering and reflection can polarize light as shown in Figure 2-27.

Polarizing screens (filters) also have the ability to polarize light (Figure 2-28)—that is, to change the light rays from vibrating in all directions around the paths to a one directional vibration. However, if the light reaching a polarizing filter is already polarized, the polarizing screen will not transmit the polarized light if its angle of polarization is at 90° to the direction of polarization of the incident polarized light. If the filter is rotated so that its direction of polarization is the same as that of the incident light, the light is transmitted. The light remains still polarized as it exits from the filter.

Polarizing filters are used in photography because they can be positioned to absorb the polarized part of skylight. This results in the darkening of the blue sky in color photographs. The color of many objects is lowered in purity because of the colorless specular reflections mixed with the colored diffuse reflections of colored surfaces. If a polarizing filter is used to absorb the polarized specular reflection, the colored diffuse reflections can be photographed with increased color purity. With cameras that focus on a ground glass, such as view cameras and single-lens reflex cameras, the darkening of the sky and the increase in color purity can be seen in the image. The polarizing filter is rotated until the effect is optimized.

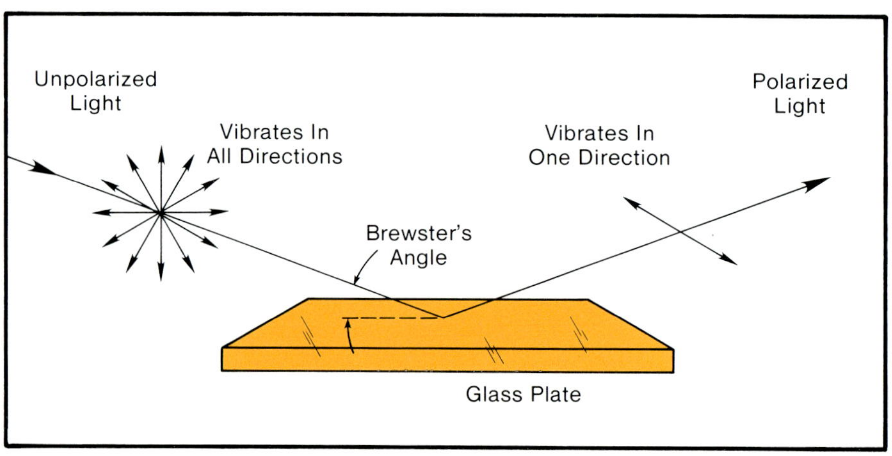

Figure 2-27. Polarization by Reflection

Inverse Square Law

As light is emitted from a source, it radiates from the source in all directions. As the distance from the source increases, the intensity of the light decreases. This is because the surface area on which the light falls increases as it gets farther from the source. The general rule is called the *inverse square law,* which can be stated: the light intensity varies inversely with the square of the distance from the source. This is illustrated in Figure 2-29. This equation shows the relationship:

$$\text{Illuminance} = \frac{1}{\text{distance}^2}$$

This is true only when the correct units are used. The illuminance in lux (metre candles) is found by this equation if the source is a standard candle and the distance is in metres.

In practice, it is common to compare the illuminance of light falling on two surfaces at different distances from the source.

$$\frac{\text{Illuminance a}}{\text{Illuminance b}} = \frac{\text{distance b}^2}{\text{distance a}^2}$$

If the illuminance on a surface 10 ft. away from a source is 10 f.c. what is the illuminance at 20 ft.?

$$\frac{\text{10 f.c.}}{\text{Illuminance b}} = \frac{\text{20 ft.}^2}{\text{10 ft.}^2}$$

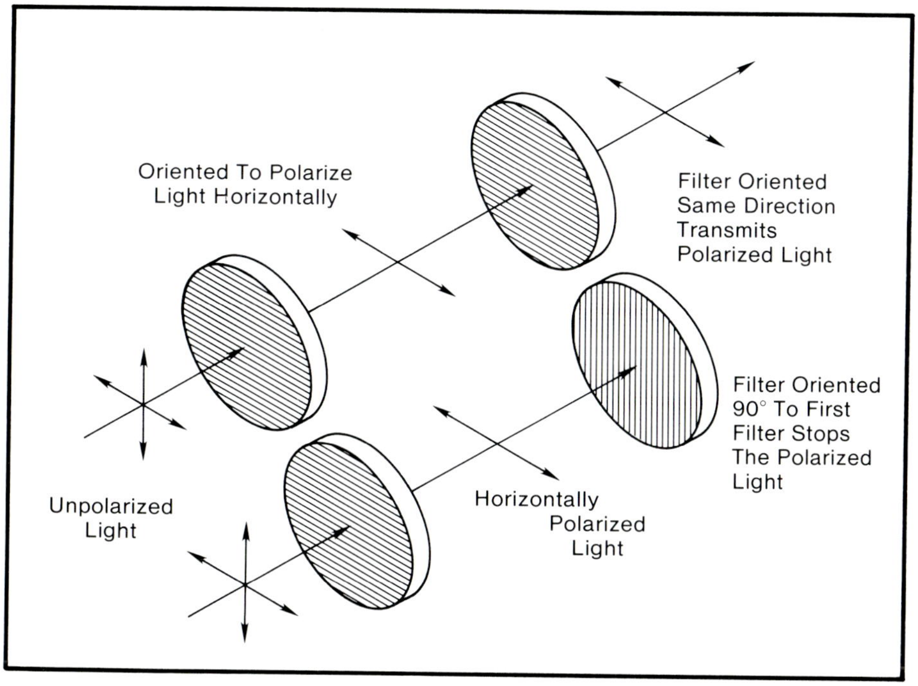

Figure 2-28. Polarizing Filters

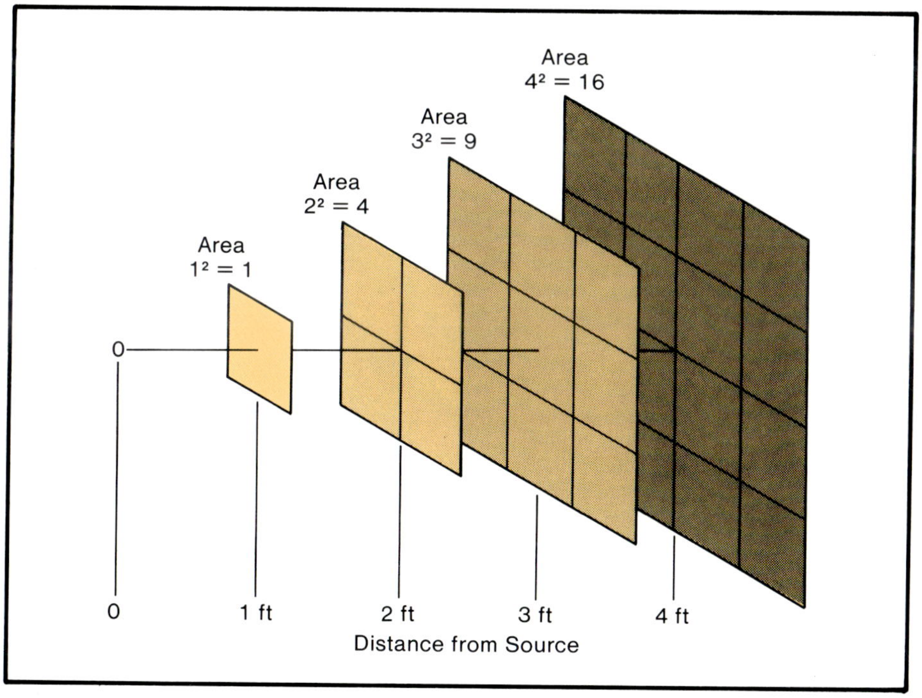

Figure 2-29. Inverse Square Law

Illuminance b x 20 ft.^{2} =
10 f.c. x 10 ft.^{2}
Illuminance b x 400 = 1,000
Illuminance b = 1,000 ÷ 400
or 2.5 f.c.

Illuminance equals 2.5 f.c.

The inverse square law is strictly true only with point sources of light. In practice it is used when the distances are large compared to the size of source.

The sun is a large source of light. But, because it is such a great distance from the earth, it can be considered a practical point source. Why, then, doesn't the surface of the earth closest to the sun have a higher illuminance than the surface farther away? The reason is that earth distances are so small compared to the distance from the sun to the earth. Let's compare the illuminance between two locations 1,000 miles apart.

$$\frac{7{,}200 \text{ f.c.}}{93{,}001{,}000 \text{ miles}^2} = \frac{\text{luminance b}}{93{,}000{,}000 \text{ miles}^2}$$

$$\text{luminance b} = \frac{8{,}649{,}000{,}000{,}000{,}000 \times 7{,}200}{8{,}649{,}186{,}001{,}000{,}000} = 7{,}199.85 \text{ f.c.}$$

Therefore, the difference between the illuminance on two points 1,000 miles apart on the earth's surface due to the inverse square law is about 0.15 foot candle, a change of only about .002%.

Questions

1. Place the letters that identify the most appropriate answers in the blanks.

1. RAT formula	_____	a. Material that transmits light rays but changes their condition from orderly to disorderly.
2. Transmission	_____	b. The bending of light as it enters a transparent medium.
3. Specular light	_____	c. Light rays traveling in a disorderly fashion.
4. Diffuse light	_____	d. Material that does not allow light to pass through.
5. Transparent	_____	e. Colored.
6. Translucent	_____	f. Bending of light passing an edge.
7. Opaque	_____	g. Thin layer optical phenomenon.
8. Achromatic	_____	h. Accounts for all light striking a surface.
9. Chromatic	_____	i. Effect of thick air layer on light rays.
10. Refraction	_____	j. Without color–neutral.
11. Reflection	_____	k. Bouncing of light rays off a surface.
12. Dispersion	_____	l. The passing of light through a material.
13. Diffraction	_____	m. Spreading of light into a spectrum by a prism.
14. Interference	_____	n. All waves vibrating in one direction.
15. Scattering	_____	o. Light rays traveling in an orderly fashion.
16. Polarization	_____	p. Lessening of illuminance with distance.
17. Inverse square	_____	q. Material that allows light rays to pass through in an orderly fashion.

2. Underline all of the following words. Use one line under the examples of orderly light ray travel and two lines under the examples of disorderly light ray travel: parallel, diffuse, converging, specular, diverging, scattered

3. Fill in the blanks: All light that strikes a material surface is either__________, __________, or __________.

4. Incident light is that light which __________________________.

5. Clear glass transmits light __________________________.
Colored glass transmits light __________________________.
(Use the words achromatically and chromatically.)

6. Light rays strike two surfaces at the same angle. One surface produces a diffuse reflection, the other produces a specular reflection. In which case can the angle of reflection easily be determined?__________________________.

7. In the space beside each type of surface write whether the reflection of white light will be achromatic, chromatic, or both.

White blotter__________	Black felt__________
Copper penny__________	New black car__________
Whole peach__________	Green blotter__________
Skinned peach__________	Silver coin__________
Still water__________	Unglazed brick__________

8. Which is the correct light path for the light ray in the glass block, a or b? __________

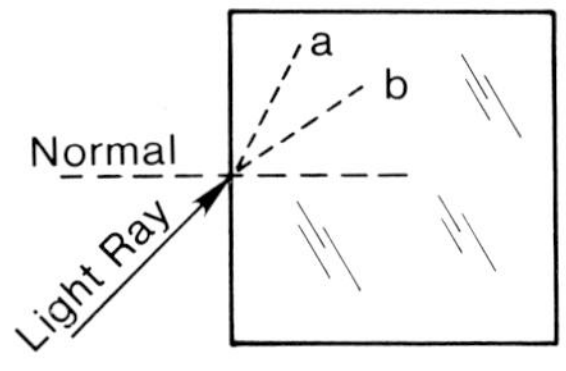

9. Water is allowed to stand on the polished surface of a barium crown glass lens. The water reacts with the glass surface so that its index of refraction is lowered a few wavelengths into the glass. When cleaned, there is a multicolored stain on the glass surface. Which behavior of light causes the colored appearance of the stain?

____

10. One wall, which is 5 feet from a bare light bulb, has an illuminance of 54 foot candles. Another wall is 15 feet from the same bulb. What is its illuminance?

____

Chapter 3

Optical Instruments

Summary

Optical instruments are made using various types of optical elements such as windows, reflectors, prisms and lenses. Most optical elements are made of glass. Optical glass is made in the form of pot glass, slab glass, rolled glass, drawn glass, fire-polished rods and cast glass blanks. Depending on the constituents, optical glasses are classified as common crown, borosilicate crown, barium crown, flint and rare earth glasses. Windows, usually made with plane-parallel surfaces, are primarily used as protective devices. Reflectors are mirrors usually with first-surface reflection and are used in optical instruments. Reflectors have plane or curved surfaces; plane-surfaced mirrors are used to change the direction or orientation of light beams. Concave mirrors converge light rays to form real images and convex mirrors diverge light rays to form virtual images. Prisms are multi-faceted blocks of glass also used to change the direction and/or the orientation of light beams. Lenses are used to form, relay or orient images. Positive lens elements are plano-convex, bi-convex, or positive meniscus. Negative lens elements are plano-concave, bi-concave, or negative meniscus. Compound lenses are made up of two or more lens elements assembled together with mounts to form corrected images.

Introduction

In the early days of man's history, light was used only for direct vision. Still water could be used as a mirror in which one's likeness could be seen. As technological knowledge began to develop, it was found that brass could be polished to make more permanent mirrors. Later, man realized that fire converted sand into rough glass spheres that could magnify, and lenses were discovered. During the past several centuries, these discoveries were developed into more complex optical instruments that allowed man to view the stars in all their enormity, and also to view the almost invisible world of microbes. Today telescopes are sent into space to image celestial bodies that cannot be imaged on earth. Messages are sent by light through tiny optical fibers, and optical techniques are used to produce electronic circuits on tiny chips with conductors so small they cannot be seen with the naked eye. The study of optics is primarily the study of optical instruments to acquire an understanding of their construction and their operation.

Optical Elements

Optical instruments consist of various optical elements suitably mounted. The following section introduces the common types of elements and describes them.

Optical Materials

Most optical elements are made of optical glass which is made in several ways. The basic ingredients of most glasses are sand, which is silicon dioxide combined with other minerals, including sodium oxide, calcium oxide and soda ash (sodium carbonate or washing soda). The ingredients are mixed and heated to a very high temperature in order to melt and combine them to form glass. Often special ingredients are added to change the optical characteristics of the glass, such as its indices of refraction and dispersion. The name of the optical glass is derived from its method of manufacture. The two main types are crown glass and flint glass. Crown glass is harder and has a lower index of refraction and dispersion whereas flint glass is softer and has a higher dispersion and refraction index. Both types are used in the production of compound lenses.

Pot glasses are made by melting the ingredients in a large clay pot or crucible where they are heated and stirred carefully to avoid the formation of bubbles. After mixing, the glass is allowed to cool slowly until it hardens. The pot is broken away, and the large glass chunk is broken into pieces. The outer pieces are discarded because they are contaminated by the clay pot. The inner pieces are cut into slabs with diamond saws. The slabs are later cut into cubes from which lens blanks are molded. Pot glass is very costly because of the waste and this process has been mostly replaced by other methods.

Rolled glass is made in a crucible as is pot glass, but this glass is only allowed to cool to a honey-like consistency before it is poured onto a flat surface and rolled to thickness with steel rollers. Platinum lined crucibles are sometimes used to eliminate contamination of the glass. Since rolled glass is made in a variety of thicknesses, it can be cut into cubes without saw-slabbing.

Polished plate glass is rolled glass that is made in large widths, perhaps fifteen feet, and ground and polished in extremely long, continuous-process machines. While most plate glass goes into making store, building and car windows, it is also used as window and mirror material in optical instruments.

Drawn glass is ordinary window glass drawn up from large tanks of molten glass into large sheets. The surface cools to a fairly flat surface that is adequate for windows in a house, but not flat enough for use with the image-forming light beams formed in optical instruments. However, some drawn glass is used as raw material from which optical flats are made by grinding and polishing the two surfaces to a much higher degree of flatness.

Float glass is similar to drawn glass but it is flatter. This glass is melted in a tank and poured onto a hot mercury bed. It hardens with two fire polished surfaces that are flatter than drawn glass, but not as flat as polished plate glass.

Rare earth glass is entirely different from other optical glasses and is made in a somewhat different way. It is made of boric acid mixed with the salts of such rare chemical elements as thorium and lanthanum. Because of the high cost of the ingredients, rare earth glasses are made in smaller batches in platinum-lined crucibles to avoid contamination. The ingredients are mixed, heated and stirred with platinum stirrers. The glass is poured, cooled and cut much like rolled glass.

Glass rod is made of regular crown glass, but is extracted directly from large tanks of molten glass. The surface of the rod is fired to produce a polished surface. Such rods are used to make viewfinders and condenser lenses with one molded surface and one ground and polished surface although occasionally both surfaces are molded. In negative front finder lenses it is often desirable to have an aspheric surface to reduce distortion of the image. Aspheric surfaces can easily be molded. Condenser lenses with one aspheric surface can often provide a better distribution of light in projection systems. The rod glass is heated in a furnace until one end softens to a taffy-like consistency. It is then pressed in a mold that has a polished surface with the curve that is needed. The lens takes a polished surface without further work. The other surface is conventionally ground and polished to a plano (flat) or spherical curve.

Cast blanks are sometimes made directly from molten glass. The molten glass, either silica or rare earth type, is poured into molds to become blanks for prisms or large lenses. The blanks are cast oversize to allow for the material that is to be removed in the grinding and polishing operations.

Clear plastic can be molded directly into lenses using molds with highly polished surfaces. The limited selection of plastics with usable optical characteristics limits the number of indices of refraction available, and the molded surfaces cannot be made as precisely as the ground and polished surfaces on conventional lenses. However, the technology is being improved, and plastic elements are now being used as compound lenses in fairly sophisticated optical devices.

Types of Optical Glass

Optical crown glass is common silica glass in which the indices of refraction and dispersion are carefully controlled by the type of sand and the proportions of the ingredient mix. Additional ingredients may be added to reduce the typically greenish color (seen in bottle glass) which is caused by traces of iron salts. Crown glass usually has an index (n_D) of about 1.500 and has a low dispersion. This is illustrated in Figure 3-1.

Borosilicate crown glass is made by adding borosilicates to the mix of crown glass. Borosilicate (BSC) glass has about the same index as optical crown glass, but has slightly lower dispersion. BSC glass is physically tough glass and does not stain easily, so it is often used to make prisms and optical flats where these physical characteristics are desirable. It is also used for lenses.

Barium crown glass is made by adding barium salts to a crown glass mix. Barium crown glasses have much higher indices of refraction than crown glass (up to $n_D = 1.625$) with about the same dispersion. This glass is very useful for optical instrument lenses because it gives more power (light ray bending) with weaker curves. Lenses with weaker curves are less costly to manufacture and they permit the manufacture of wider aperture camera lenses with improved sharpness.

However, barium glass stains easily; and is quite soft and prone to scratches. It is easily etched with hydrofluoric acid, and is often used as the base material for reticles, the crosshair optics used to aim telescopes.

Flint glass is made by adding lead salts to crown glass mix. Flint glasses have high indices of refraction (up to $n_D = 1.720$) but also have high dispersions. Flint glass is used primarily to make negative lens elements. These lenses are used in combination with crown positive

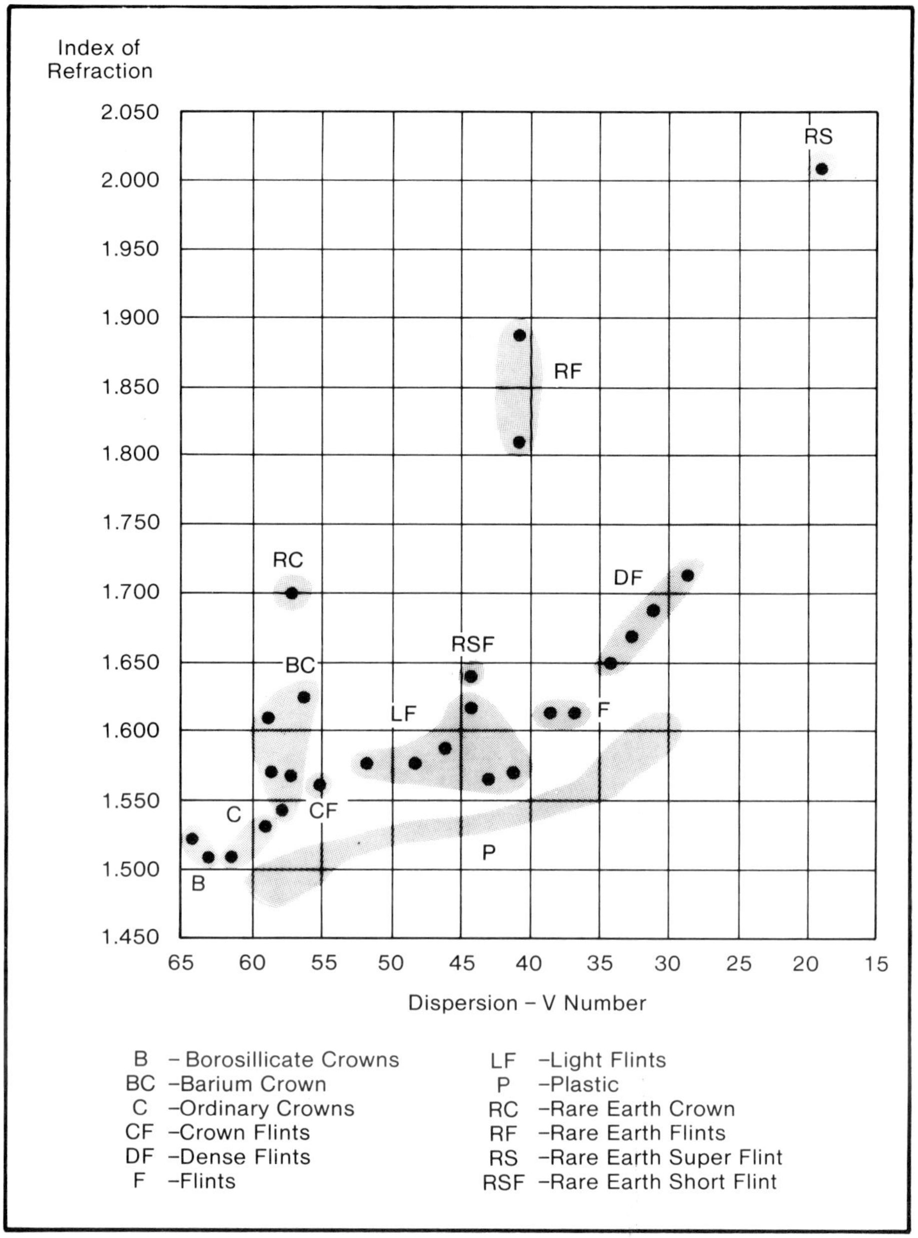

Figure 3-1. Optical Characteristics of a few Optical Glasses and Plastics.

lens elements to correct for lens aberrations.

Rare earth glasses are usually made with crown glass characteristics but can also be made as flint glasses. A big advantage of such glasses is that they can be made with indices of refraction up to n_D = 1.700 with low dispersion, and up to n_D = 2.00 with higher, flint-glass type of dispersion. This compares with n_D = 2.417 for diamonds which have an extremely high dispersion.

Special Optical Materials

Glass transmits all wavelengths of light, and some wavelengths of ultraviolet and infrared radiation. Certain optical devices are made to work in regions of the electromagnetic spectrum other than light and may require optical elements that transmit wavelengths to which glass is opaque.

Rock salt transmits wavelengths far into the infrared region, and is sometimes used to make elements for instruments designed to work in this region. Since rock salt is very soluble in water and is very soft, such elements must be carefully protected.

Pure quartz transmits further into the ultraviolet region than glass does and is used for elements in instruments that require ultraviolet transmission. Quartz is a relatively hard and stable material and transmits wavelengths down to about 185 nm. When transmission of even shorter wavelengths is needed, fluorite (calcium fluoride) which transmits down to about 120 nm is used. Lithium fluoride transmits even further into the ultraviolet region to wavelengths of 108 nm. Such materials are not usually encountered in optical instruments.

Pellicles are thin optical mirrors used for splitting beams of light to create two images from one. They are usually made of a thin layer of collodion stretched on an optically flat frame. The pellicle is coated with a partial mirror surface which both reflects and transmits some of the light (Figure 3-2). If a glass plate is used as a beam splitter, an offset weak image is reflected by the second surface of the glass. With a pellicle, the second image is offset by perhaps less than a thousandth of an inch, and cannot be distinguished from the main reflected image. Pellicles made of collodion are ex-

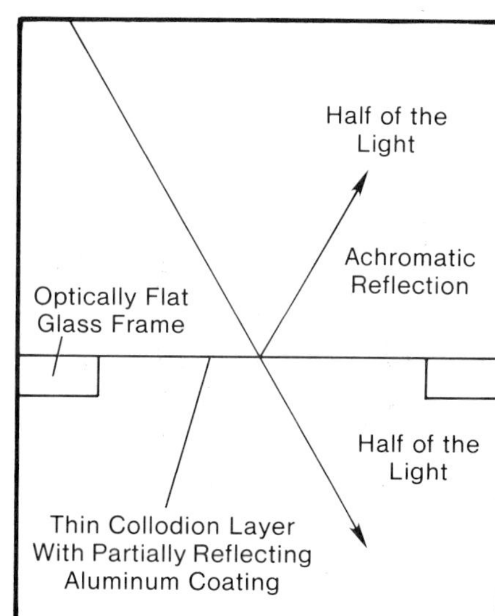

Figure 3-2. 50-50 Pellicle Beamsplitter

tremely fragile and are always mounted inside the instrument where they are fully protected.

Reflectors

Reflectors are optical elements with highly polished, highly reflective surfaces, consisting of a thin layer of metal, usually aluminum or silver, on a glass base. Where extreme toughness is required, the reflector base may be metal.

Most reflectors used in optical instruments are first-surface mirrors, that is, the light beam does not go through the glass, as shown in Figure 3-3. The glass acts as a support for the reflecting surface. It is ground and polished and the metal reflective surface is applied by evaporation in a vacuum. Because the glass surface is polished the evaporated metal forms a polished surface as it is applied and does not require further polishing. Most evaporated metals are extremely soft and easily scratched. To prevent this, reflecting surfaces are usually overcoated with a very thin (0.5 wavelength) layer of a mineral such as silicon monoxide, which protects the surface and enhances its reflectivity.

Second-surface mirrors are made with the reflecting surface on the back of the glass so that the light goes through the glass to be reflected (Figure 3-4). The front surface of the glass also reflects an image, which, while weaker than the primary image, is still visible. This usually limits the use of second-surface mirrors to light-controlling uses rather than imaging uses. The silver reflecting surface of second-surface mirrors is usually applied in a chemical bath and can be copper plated and enameled to provide considerable protection. Cleaning the exposed glass surface is simple since the glass does not scratch easily.

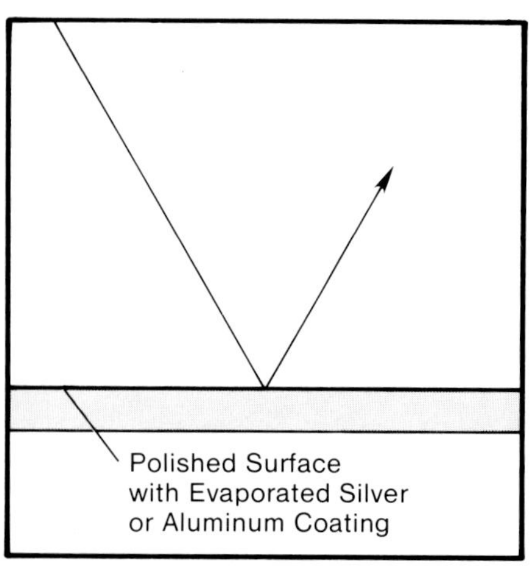

Figure 3-3. First-Surface Mirror

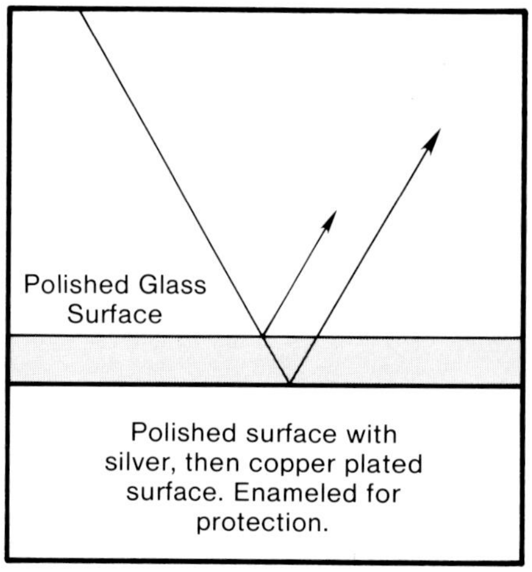

Figure 3-4. Second-Surface Mirror

Most mirrors are designed to reflect as much light as possible. Sometimes, however, it is desirable to reflect some wavelengths and to transmit others. Other times, as was discussed earlier when describing pellicles (Figure 3-2), it is desirable to reflect part of the light and transmit part of it achromatically. This is accomplished when a thin layer of aluminum or silver is evaporated on glass. As mentioned, such a reflector is called a beam splitter. When multiple layers of minerals and metals are evaporated onto glass, surfaces that reflect some wavelengths and transmit others can be made. Such surfaces are called dichroic reflectors. These utilize interference of light waves to chromatize the light (Figure 3-5). For example, dichroic reflectors are made to reflect visible wavelengths and transmit infrared wavelengths. Such reflectors are used to concentrate the light of the bulb filament for projection systems (Figure 3-6a). By reflecting light and transmitting the infrared, the film being projected does not get as hot. A reflector that does just the opposite, that is, transmits the light and reflects the infrared is used between the lamp and the film gate and keeps the heat of the infrared rays from reaching the film. The first type is known as a cool reflector, while the second is called a hot or heat reflector as shown in Figure 3-6b.

Similar dichroic coatings which divide the visible spectrum are used in superimposed-image rangefinders to provide color contrast for the movable image.

Plane Mirrors

Plane mirrors are used primarily to change the direction and/or the orientation of image forming beams of light. While the reflection does change the direction of the light beam, it does not change its charac-

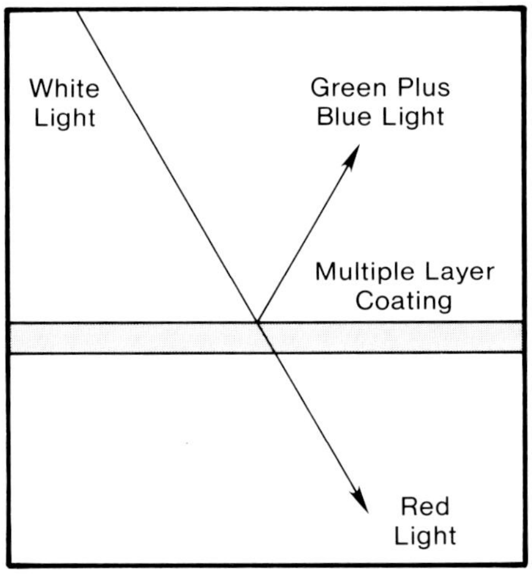

Figure 3-5. Dichroic Filter

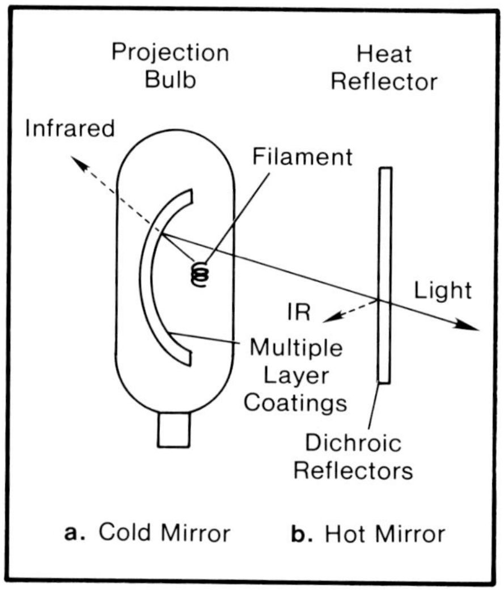

Figure 3-6. Hot and Cold Mirrors.

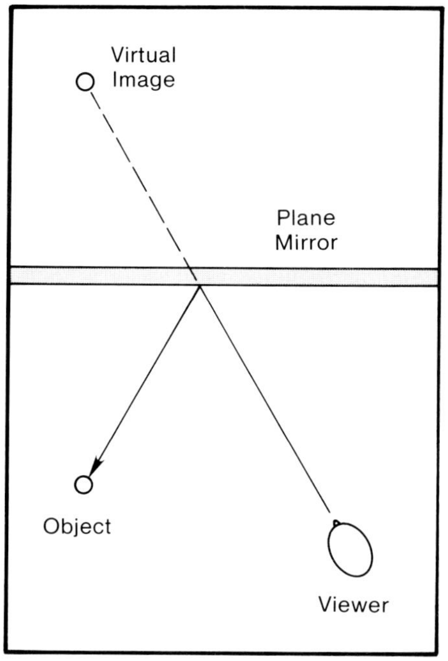

Figure 3-7. Virtual Image Seen in Plane Mirror.

teristics. For example, if the rays that compose the beam are parallel incident to the mirror surface, they are still parallel after the reflection. If they are orderly (either diverging or converging) at a certain angle to each other, the reflected beam retains the same angle of divergence or convergence.

A plane mirror does not form a *real* image. The images seen in household mirrors are *virtual* images, that is, they do not focus on an image plane without the use of an additional optical device such as a lens (Figure 3-7).

Curved Mirrors

Concave mirrors are used to form images and to concentrate light. The curves on such mirrors are usually spherical, parabolic, or elliptical. Spherically-curved mirrors may be used as light concentrators in microscopes or projection systems. Sometimes they are used as image formers in telescopic lenses such as Schmidt cameras or catadioptric lenses for 35 mm cameras. Convex spherical mirrors are seldom used except as secondary mirrors in such instruments as the Cassegrain reflecting telescope. Concave elliptical mirrors are used to concentrate light in stage spotlights. Concave parabolic mirrors are used to form images in reflecting telescopes and telescopic cameras, and to send beams of parallel light long distances in outdoor spotlights (Figure 3-8).

Windows

Optical flats have plane, parallel surfaces and are usually used as protective windows as illustrated in Figure 3-9. Some filters are made by cementing a layer of dyed gelatin between two optical flats. Other filters are made by coloring the glass with metallic salts when it is being made, and then manufacturing optical flats from the colored glass.

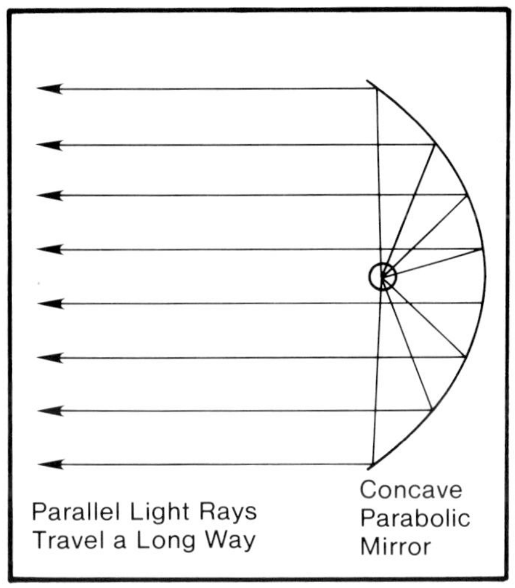

Figure 3-8. Military-Type Searchlight

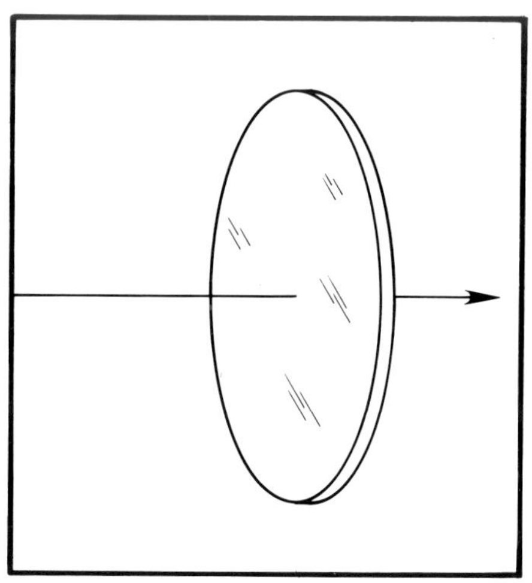

Figure 3-9. Plane Window – Optical Flat

While most windows are flat, some are curved. Cover glasses for some types of dials and meter needles are spherically curved. They may be ground and polished as plane mirrors. The mirrors are heated to the proper degree of softness and then drawn into spherical molds by a vacuum. If the surface requirements are rigid, they may be ground and polished into shape just like a lens. Spherical windows have concentric spherical surfaces. Strongly curved windows that approach the shape of hemispheres are called domes (Figure 3-10).

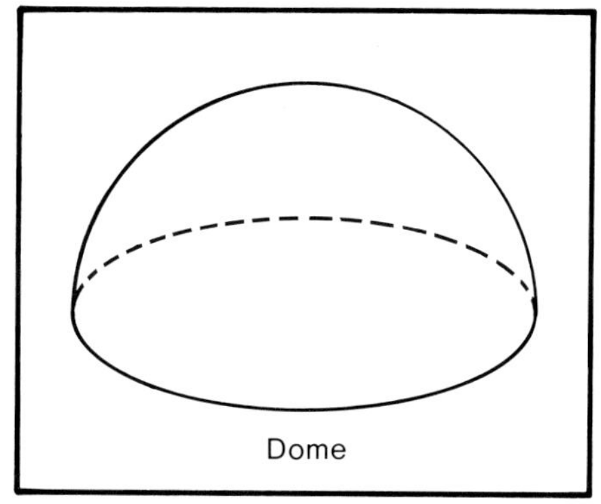

Figure 3-10. Hemisphere Window

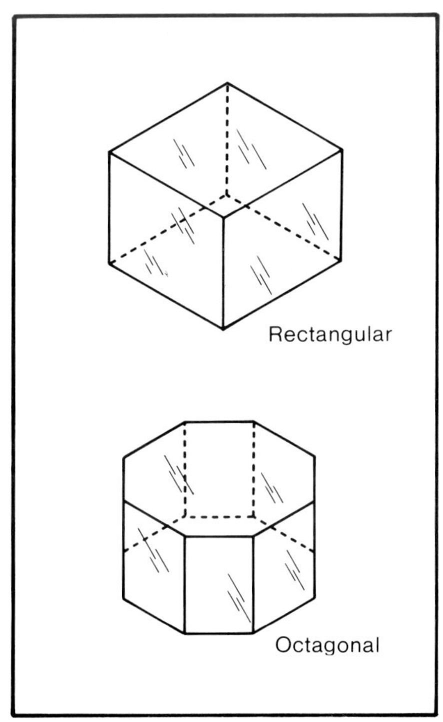

Figure 3-11. Optical Block Prisms

Prisms

Prisms are blocks of glass with surfaces at various angles. They are usually used to change the direction and/or orientation of light beams in optical instruments.

An *optical block* is a very thick window whose thickness approaches its length and width dimensions. A rectangular block is one in which all the angles are 90°. A rotating optical block is used to match the image movement with the film movement in shutterless, high-speed motion picture cameras (Figure 3-11).

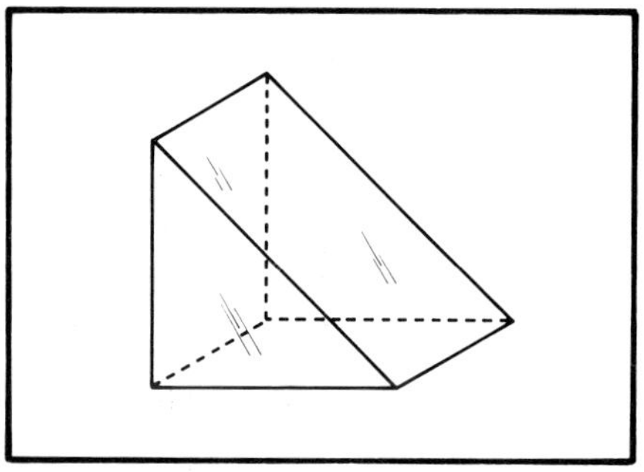

Figure 3-12. 90° Prism

A *90° prism* has one 90° angle and two 45° angles. It is used to change the direction of light beams in optical instruments by 90° (Figure 3-12).

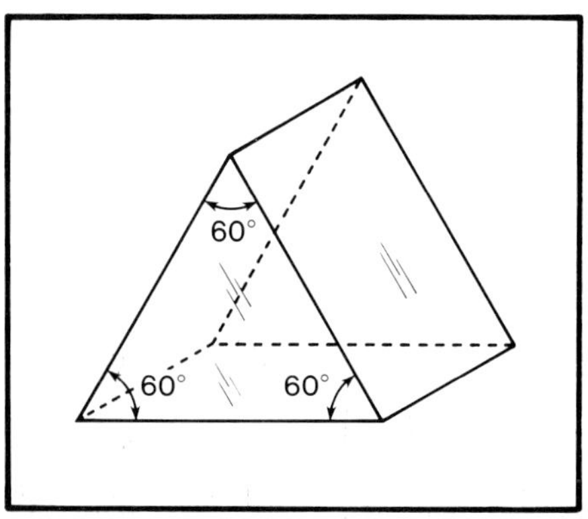

Figure 3-13. 60° Prism

A *60° prism* has three 60° angles. It is also called an equiangular prism. It may have two or three polished surfaces. Generally it is used to measure glass index and dispersion, and to create spectra (Figure 3-13).

A *Porro prism* is a special 90° prism with rounded ends that is used in binoculars and telescopes (Figure 3-14).

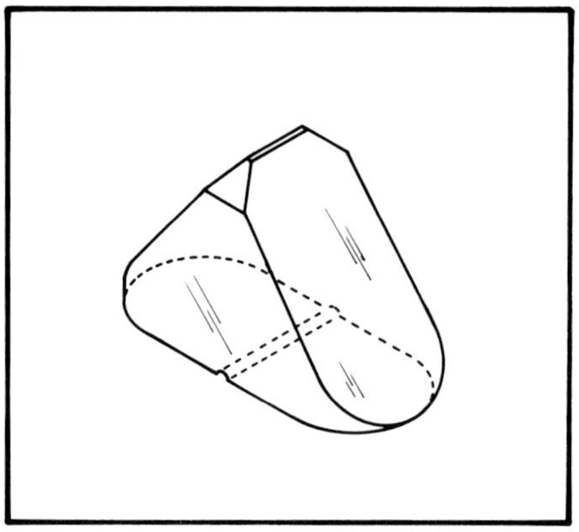

Figure 3-14. Porro Prism

A *roof prism* is a 90° triangular prism in which the hypotenuse (long side) is replaced by two faces at 90° that intersect to form a sharp-edged roof on the hypotenuse side. A roof prism inverts the image beam at the same time that it changes its direction 90° (Figure 3-15).

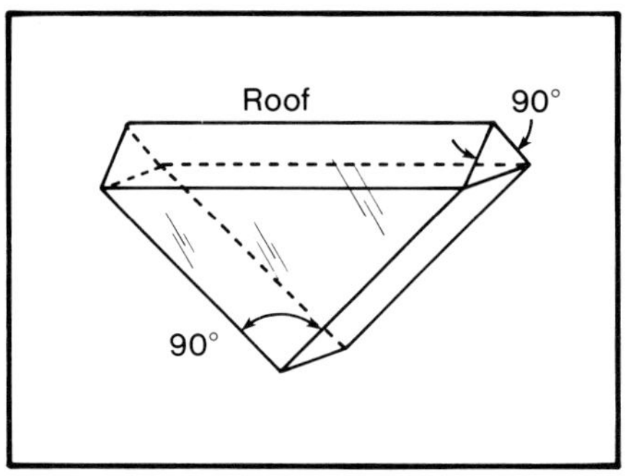

Figure 3-15. 90° Roof Prism

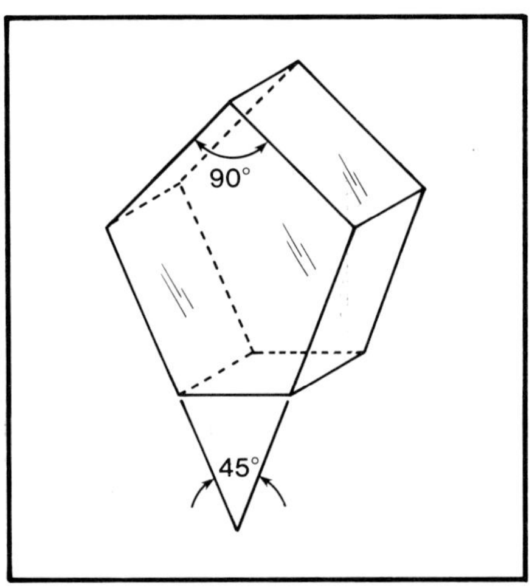

Figure 3-16. Penta Prism

A *penta prism* is a five-sided prism that changes the direction of an image-forming beam of light at the same time that it inverts it as illustrated in Figure 3-16.

A *penta roof prism* is a prism that has a two-sided roof replacing one of the surfaces. It is used in single-lens reflex cameras to make the ground-glass image appear right-side up and oriented correctly left-to-right (Figure 3-17).

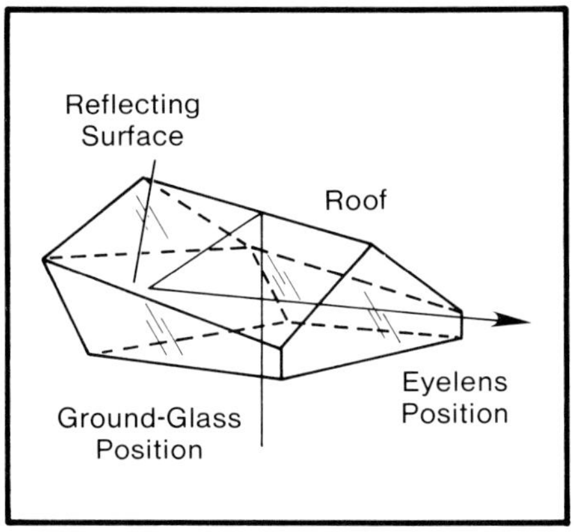

Figure 3-17. Penta-Roof Prism as used in single-lens-reflex cameras.

A *dove prism* is a section of a right-angle prism with a polished hypotenuse. The image-forming beam passing through the prism rotates twice while the prism is rotated once. It is used to retain correct orientation of an image while the position of the instrument is being changed (Figure 3-18).

Lenses can be described as pieces of transparent optical material that have two curved polished surfaces. The curves are usually spherical surfaces. In this definition, plano surfaces are considered spheres with an infinite radius of curvature. In some cases, one of the surfaces will be aspheric rather than spherical.

Positive lenses are thicker in the center than at the edge. Positive lenses form real images and they converge parallel light rays. There are three basic forms of positive lenses: biconvex (two convex surfaces); planoconvex (one convex surface and one plano surface); positive meniscus (one convex surface and one concave surface). The convex surface has a shorter radius of curvature and is stronger than the

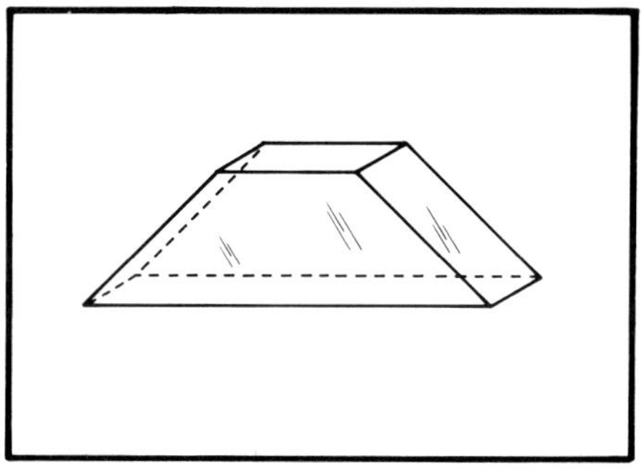

Figure 3-18. Dove Prism

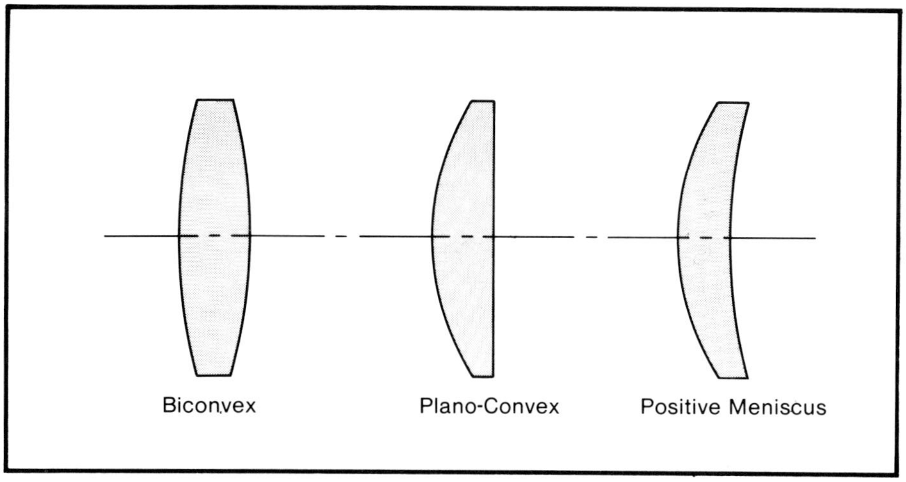

Figure 3-19. Positive Lenses (Cross Sections).

concave surface (Figure 3-19).

Negative lenses are lenses that are thinner in the middle than at the edge. Negative lenses form virtual images. They are usually used in conjunction with stronger positive lenses to form real images. Compound lenses (two or more lens elements) are designed to reduce the effects of lens aberrations*, so that the images formed by compound lenses are sharper than those formed by single lenses. There are three basic forms of negative lenses: biconcave (two concave surfaces); plano-concave (one concave surface and one plano surface); negative meniscus (one convex surface and one concave surface). The concave surface has a shorter radius of curvature (Figure 3-20).

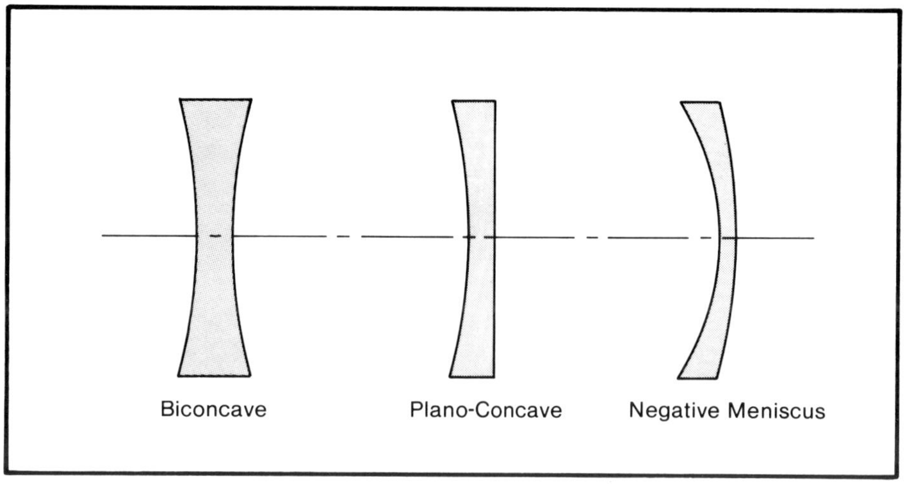

Figure 3-20. Negative Lenses (Cross Sections)

Cemented lens assemblies are two lens elements that are cemented together to form a single lens called a doublet. It is important that the two curves that are cemented together match, that is, both curves must have the same radius of curvature even though one curve must be concave and the other convex. Thus, when cemented, the layer of cement between them is very thin and of even thickness throughout. The one exception, of course, is that two plano curves may be cemented together, and they are identical. In a very few cases, three lenses are cemented together.

The outer surfaces (those not cemented) may be convex, concave or plano, and the doublet (or triplet) may be positive or negative as shown in Figure 3-21.

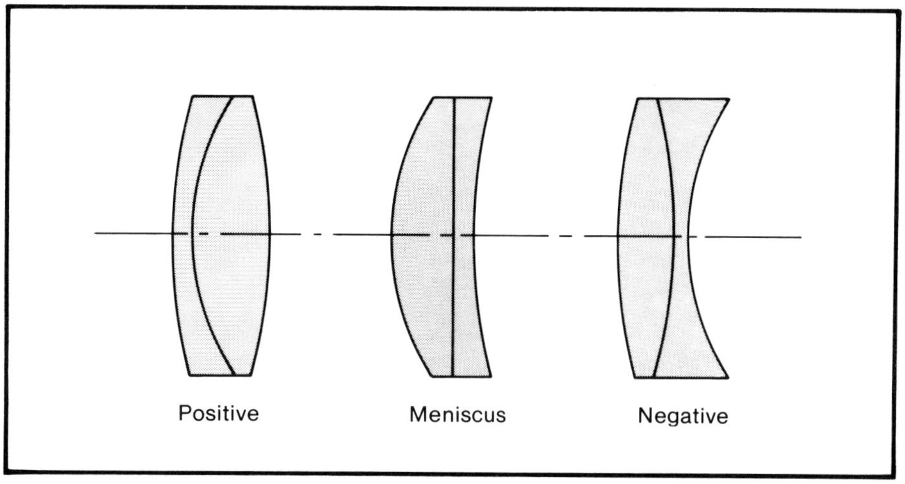

Figure 3-21. Cemented Lens Assemblies

*Aberrations will be covered in detail in a later section. Aberrations are image imperfections caused by the way light goes through lenses. A simple lens element has serious aberrations. By putting several carefully designed lens elements together to form a compound lens, the effects of the aberrations are reduced considerably.

Questions

1. The four major types of optical elements are ____________, ____________, ____________, and ____________.

2. The material used to make most optical elements is ____________, which is usually made of two ingredients, ____________ and ____________.

3. Match the following by placing the most appropriate letters in the blanks.

1. Pot glass	_____	a. Finished from rolled glass
2. Slab glass	_____	b. Poured into molds
3. Rolled glass	_____	c. Sawed from pot glass
4. Drawn glass	_____	d. Extruded from tank of glass
5. Polished plate	_____	e. Made in clay crucible and broken
6. Glass rod	_____	f. Has imperfect plane surfaces
7. Cast blanks	_____	g. Cheaper and less loss than slab glass

4. The basic ingredients of rare-earth glasses are ____________ and ____________.

5. Window glass is called ____________ glass because of the manner in which it is made.

6. ____________ glass is used to mold lenses with one fire-polished surface.

7. A material used to mold inexpensive lenses that do not require further polishing is ____________ ____________.

8. Match the following by placing the most appropriate letters in the blanks.

1. Optical crown	_____	a. Extremely high index of refraction
2. Flint	_____	b. Basic glass
3. Barium crown	_____	c. Stains easily
4. Borosilicate crown	_____	d. Transmits infrared radiation
5. Rare earth	_____	e. Contains lead salts
6. Pure quartz	_____	f. Extremely thin beam-splitters
7. Rock salt	_____	g. Physically strong and stable
8. Collodion	_____	h. Transmits near ultraviolet
9. Fluoride	_____	i. Transmits far ultraviolet

9. The reflectors used in optical instruments are usually called ____________ surface ____________.

10. Such reflectors are made by coating two thin layers on polished glass. Identify the layers.
    Material on the glass ____________ or ____________.
    Overcoat____________

11. The overcoat serves two functions:
    a. ____________
    b. ____________

12. Reflectors with several layers coated on them to produce chromatic reflections are called ____________ reflectors.

13. Plane mirrors are primarily used in optical instruments to change the ____________ or the ____________ of image forming beams of light.

14. Mirrors used to form real images are: a. plane, b. convex, or c. concave? ____________

15. Windows are usually optical flats used for ____________ purposes.

16. Identify the following by placing the proper letter in each blank.

1. Optical block ______
2. 90° prism ______
3. 60° prism ______
4. Roof prism ______
5. Porro prism ______
6. Penta prism ______
7. Penta-roof prism ______
8. Dove prism ______

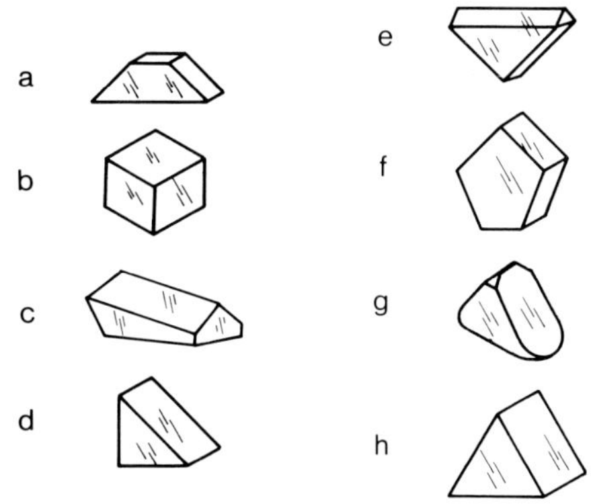

17. Identify the following by placing the correct letter in each blank.

1. Bi-convex lens ______
2. Plano-convex lens ______
3. Positive meniscus lens ______
4. Bi-concave lens ______
5. Plano-concave lens ______
6. Negative meniscus lens ______
7. Doublet (cemented lens) ______

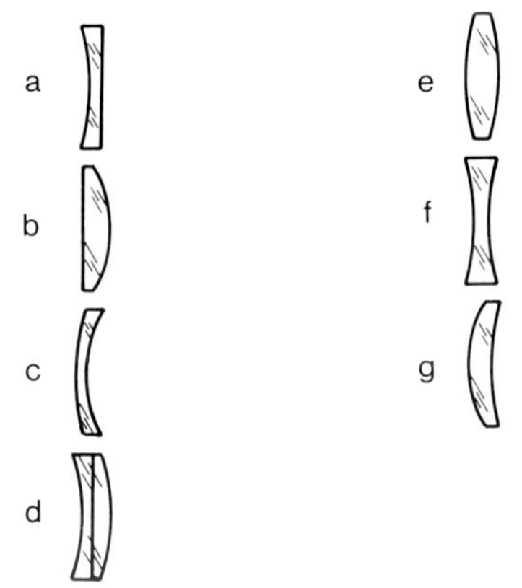

Chapter 4

Optical Element Manufacture

Summary

Optical elements usually start as optical glass which is cast or molded into blanks. They are annealed (heated and slowly cooled) to eliminate strain. The blanks may be milled, after which they are mounted on blocks for first-side finishing. The blocks of blanks are ground with rough emery or diamonds, then with finer grits to leave the surfaces very smooth. Optical surfaces are then polished with polishing blocks surfaced with felt, pitch or plastic, using rouge and water as the polishing agent. The remaining polished surfaces are finished in a similar manner. The accuracy of each polished surface is tested using master glasses with monochromatic light; the Newton's Rings give a measure of the accuracy of the curve. Lenses may be centered, and then coated to reduce reflection. Reflectors are usually coated with aluminum or silver, which is applied by evaporation in a vacuum. They are usually overcoated for protection and to increase the reflectivity. Some surfaces of prisms may be coated, and others silvered, while still other surfaces may be left untreated. Some lenses may be cemented together to form doublets, while prisms may be cemented together into clusters. Lens edges are usually inked black while the silvered surfaces of prisms may be plated and enameled for protection. Unpolished prism surfaces may be left in their ground condition, or may be blackened.

The Need for Tight Tolerances

Optical elements must be manufactured to very precise tolerances if they are to work satisfactorily. Polished surfaces are tested against master glasses under monochromatic light and the number and shape of the Newton's Rings formed between the two surfaces indicate the accuracy of the polished surfaces. The difference between the radius of curvature of the polished surface and that of the test glass is indicated by the number of rings. Tolerances run from about ten rings down to less than one-tenth of a ring for critical applications. Since Newton's Rings measure in fractions of light wavelengths, such tolerances test surface precision to less than a millionth of an inch. Lens thicknesses and diameters are usually held to tolerances of several hundredths of a millimetre, while the lens circumference often must be concentric to the optical center of the lens and have the same degree of accuracy. Prism angles and dimensions must also be manufactured to extremely tight tolerances.

Optical viewfinder lenses used in cameras, condenser lenses used in enlargers and projectors, and lenses used in a few other applications do not require such a high degree of precision in their manufacture.

The Manufacture of Lens Blanks

Optical parts start as optical glass in slab, rolled, or cast form. Most elements of moderate size are made from rolled glass which is cut into rectangular cubes (Figure 4-1). The

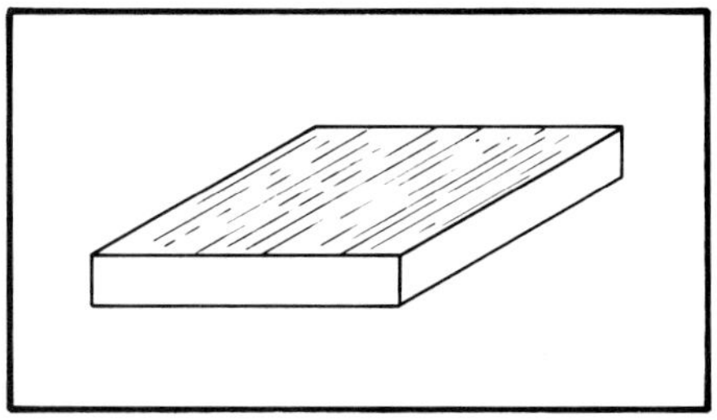

Figure 4-1. Cut cube made from rolled glass.

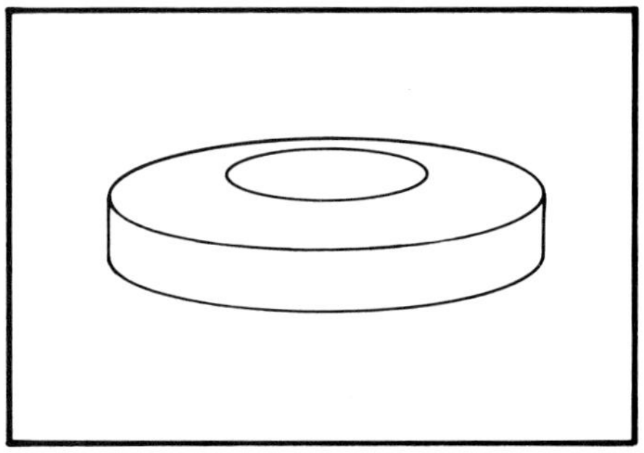

Figure 4-2. Molded lens blank.

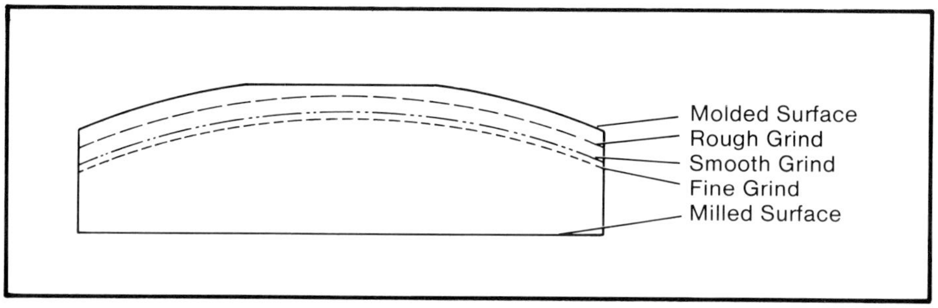

Figure 4-3. Lens blank cross-section showing how the three grinds remove successive amounts of glass.

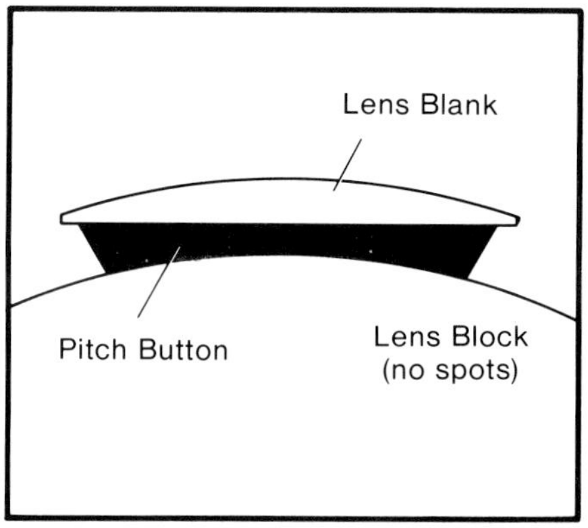

Figure 4-5. Pitch-Button blocking.

weight of the cubes is controlled by adjusting the length and width to insure the correct volume of glass in each cube for succeeding operations. Very large blanks are cast, while very small blanks may be cut into squares of the correct thickness, blocked into "sticks", and rounded by grinding between centers using a carborundum wheel. They are then separated for molding.

The cubes are placed in a furnace and heated to a taffy-like consistency. The hot blank is then placed in a mold and pressed to form the approximate shape of the lens (Figure 4-2). Extra thickness is allowed for the glass removal in succeeding manufacturing steps, and, if the lens is to be centered, the molded diameter is made somewhat larger than the finished diameter (Figure 4-3).

Because glass develops a permanent internal strain if it is allowed to cool rapidly, molded blanks are placed in an annealing oven where they are brought up to a near molding temperature, and then cooled to room temperature very slowly. Cooling often takes several weeks. Annealing produces strain-free glass which is unlikely to break during manufacture, and which retains the accurately polished surfaces without warping. Strain-free glass is required in image-forming elements because strain can make light rays deviate from their intended paths.

Usually one side of the molded blanks is left plano, and this side may contain dust contamination from the molding furnace. The blanks are held in a fixture while the plano side is milled (with a diamond faced mill) to remove the contaminated glass. The milled side becomes the weak-curve side of the lens.

Grinding and Polishing

The blanks are placed on a support called a lens block for the grinding and polishing operations. Most blocks are spotted and have recesses which hold the lenses in place (Figure 4-4). The number of lenses on each block is determined by the radius of curvature and the diameters of the lenses. There may be only one lens on a block if radius of curvature is quite short compared to the lens diameter. There may be many lenses if the radius of curvature is long, and the lens diameter small.

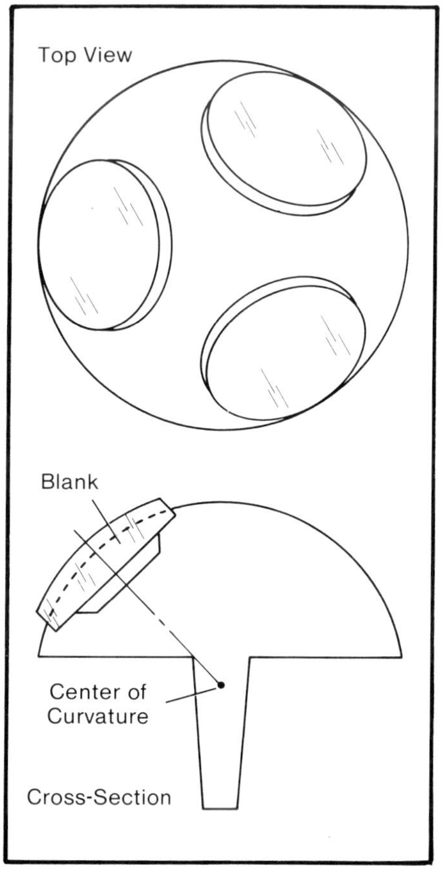

Figure 4-4 3-Spot Lens Block

Lens blocks are spherically curved and the lenses are placed in such a position on the blocks so that their final polished surface will be slightly higher than the lens block surface. The lenses are held on the block with pitch or wax.

If the lenses are to be made relatively thin, they may be put on pitch buttons and mounted on a block without seats. This will enable each lens to be supported across its entire surface while being ground and polished. This avoids "springing" which would spoil the precision of the curve (Figure 4-5).

Prisms may be ground to approximate shape by hand using emery on a flat grinding shell. They are then cast into a block of plaster-of-paris for the finishing grind and polishing steps. Prism blocks with rows of seats that hold large quantities of prisms may be used when production requirements are great. Careful placement of the prism blanks in precisely manufactured seats allows grinding and polishing of the prisms to close angular and dimensional tolerances (Figure 4-6).

The coventional method of grinding lens surfaces involves three grinding stages using rough, smooth, and fine emery. The rough grind may be done in a grinding sink with the blocks of lenses handled individually. The smooth and fine

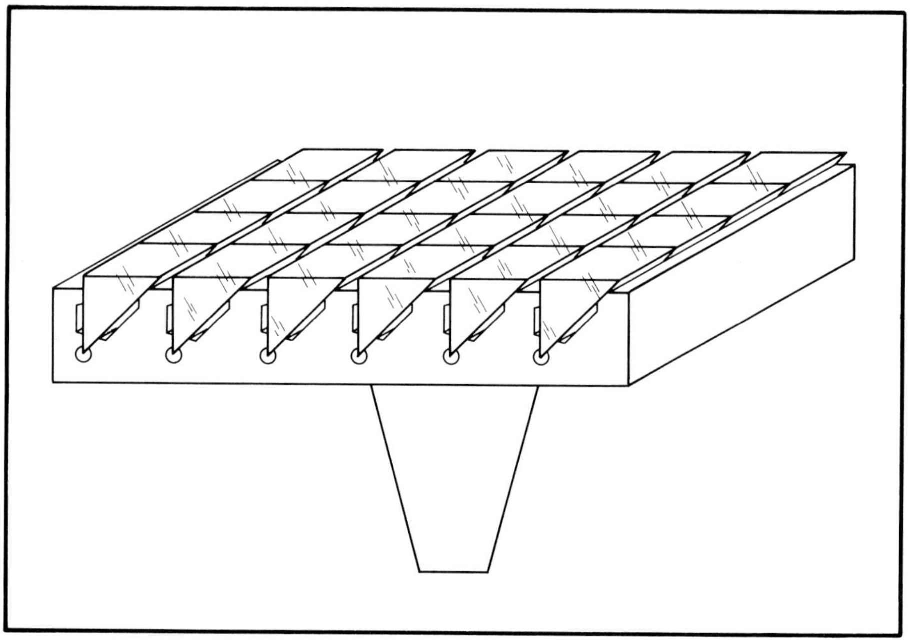

Figure 4-6. Slotted Block for 90° Prisms.

grinding operations are usually done on grinding machines with a long row of grinding spindles (Figure 4-7). One operator can oversee the grinding of many blocks. Water is mixed with the emery to provide lubrication and to cool the grinding surfaces. Adhesion holds the water-emery slurry between the glass and the iron grinder surfaces. The emery is completely washed off after each stage so that the lens blocks are left with extremely fine ground surfaces. Three grinding shells are used to grind each surface, one for each grit of emery. Each shell has a slightly different radius of curvature to compensate for the glass that is removed at each grinding stage. The curvatures of the surfaces are checked with gauges that have been cut to the correct curves, and also with a spherometer which is a special type of dial gage.

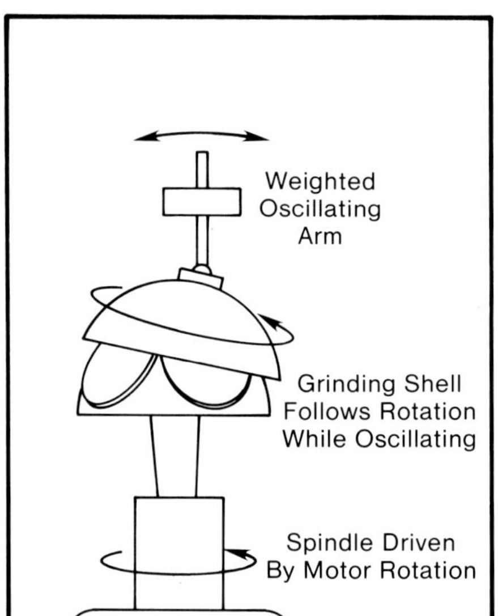

Figure 4-7. Grinding or Polishing Spindle, conventional process.

Polishing machines are similar to grinding machines. Many spindles are watched by one operator who applies water and rouge as needed.

Surfaces that do not require the finest degree of polishing are often polished with felt polishers. The felt is fastened to a metal polishing shell and rouge and water are used as the polishing medium. White rouge, which is powdered cerric oxide, is used because it polishes quite rapidly and produces a highly polished surface. Felt is the fastest polishing surface but it does not leave as good a surface as pitch-surfaced polishers. Pitch polishers are generally used for high quality, conventional-process polishing. Polishing with pitch is a slow process, often taking several days.

The time needed to produce lenses can be considerably reduced by using diamond abrasive rather than emery. When grinding with emery, and when polishing with the conventional method, the block of lenses is rotated by power, and the grinding shell or polishing shell rotates in the same direction by friction as it oscillates back and forth generating the curve.

With diamond grinding, the diamonds are fixed in the grinding surface of the shell, rather than being applied in a slurry as with emery. The block of lenses is power driven in one direction while the grinding or polishing shell is power driven in the opposite direction. Great surface speed between the glass and the grinding surface combines with the hardness and sharpness of the diamond-dust abrasive and reduces the grinding time from hours to minutes or seconds (Figure 4-8). The polishing shell is lined with a synthetic pitch strong enough to stand the power-polishing action.

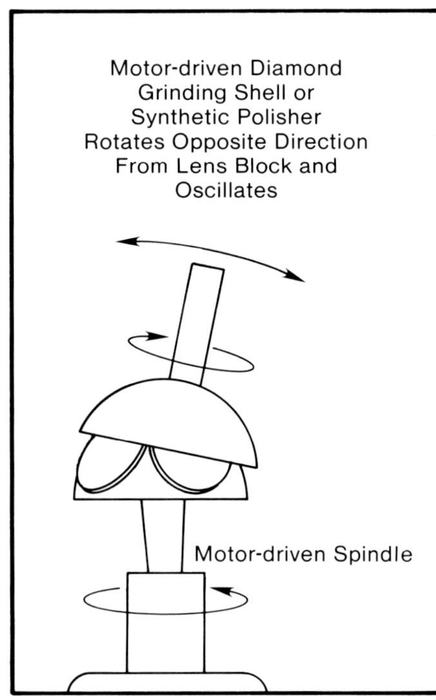

Figure 4-8. Grinding or Polishing Spindle, high-speed process.

When the operator has finished polishing a block of lenses, he or she cleans them, inspects them for scratches and the completeness of the polishing, and tests them using a test glass under monochromatic

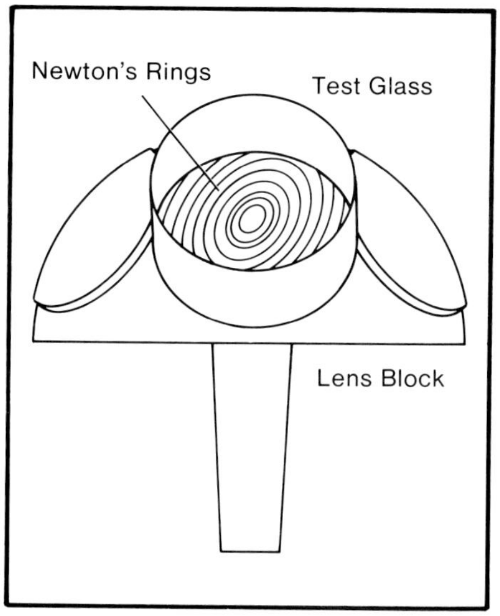

Figure 4-9. Testing surface fit with Test Glass in monochromatic light.

light (Figure 4-9). If they are satisfactory, the block is heated to soften the blocking pitch and the lenses are removed from the block. The blocking pitch is removed from the lenses, usually in solvent degreasers and with ultrasonics. The second surfaces of lenses are finished in a similar manner, as are the other surfaces of prisms. Unused surfaces of prisms are ground but not polished.

Other Optical Manufacturing Operations

When the lenses have had both surfaces ground and polished, the optical center of the lens may not align exactly with its circumference. A lens is "centered" by grinding its circumference concentric with its optical center.

The lens is mounted on a centering chuck with wax or pitch and aligned by repeated repositioning of the lens until the reflection of a lamp ceases to "dance" as the lens is aligned with the axis of the centering chuck. The wax is cooled with water so that it hardens and holds the lens firmly, and the edge and bevels are ground with a grinding wheel (a copper wheel with imbedded diamond dust) while the chuck-mounted lens slowly rotates. Water, sprayed on the cutting edge, cools the wheel and lens during the grinding operation.

An alternative method is called cup centering (Figure 4-10). The cups are the ends of polished stainless steel spindles. One spindle moves along its axis, allowing a space to open up between the cups in which the lens is placed. The movable spindle is slowly closed upon the lens, which finds its center between the cups. Pressure is applied to the spindle to hold the lens firmly, while the spindles rotate slowly with the lens. A grinding wheel is brought up against the lens edge, grinding its diameter to the proper size while other wheels grind the bevels. Water is again used as coolant.

Many lenses are coated to reduce the reflection of light from their polished surfaces and to reduce the flare level in finished compound lenses. Prior to coating, the lenses are cleaned thoroughly and then placed in seats cut into an aluminum dome shaped holder. The domes are placed in a fixture on the coating machine, and a bell jar is lowered over the dome, lenses and other fixtures required for the coating operation (Figure 4-11). A very high vacuum is drawn in the bell jar. A glow discharge is activated, in which a high voltage filament shoots streams of electrons at the lens surfaces, driving off microscopic particles of dirt and grease. The lenses are then heated electrically to aid adhesion of the coating.

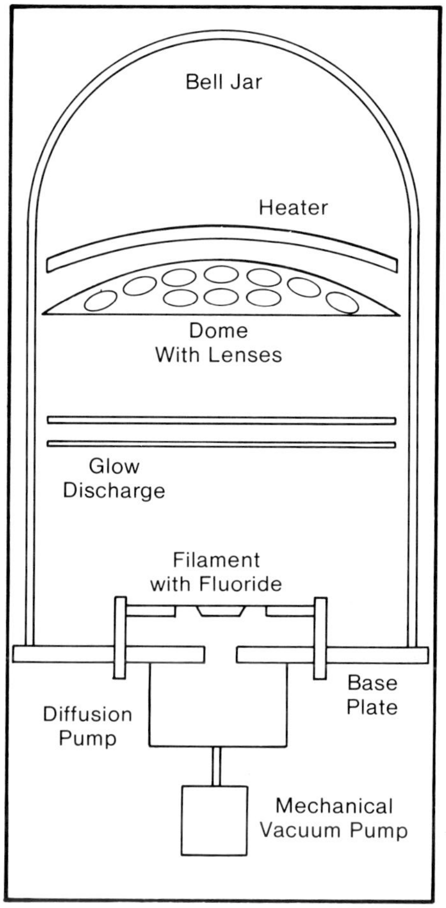

Figure 4-11. Vacuum Coating

When the lenses have reached coating temperature and their surfaces are absolutely clean, a current is passed through a boat-shaped filament containing the coating material which is usually magnesium fluoride. The fluoride melts and

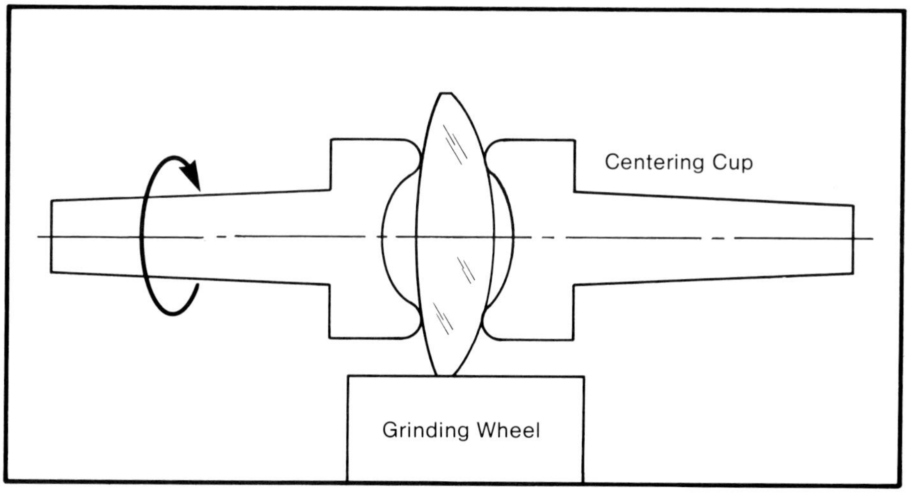

Figure 4-10. Cup Centering.

evaporates, sending streams of molecules rapidly through the vacuum. The molecules adhere to the glass, building a thin layer of fluoride on the lens surfaces. When the coating has reached the correct thickness, usually one-fourth of an average wavelength of light, the coating is finished. Most lenses are coated on both sides, but lens surfaces to be cemented are left uncoated. If the surfaces are to be multi-coated, as with dichroic coatings, the various materials are applied successively from several different filaments located near each other.

For many years Canadian balsam, the purified resin from a fir tree, was used to cement lenses together. Today synthetic cements such as methacrylates are used. Some cements are thermoplastic which means they are solid at room temperature, but soften when heated. When cementing with a thermoplastic cement, the lenses are heated, and the warm cement is applied to one of the first lens surfaces. The second lens is placed on the cement and moved about to make a thin layer of cement between the two lenses, and to drive out any air bubbles. A fixture is used to hold the lenses in alignment while the lenses cool, setting the cement.

Thermosetting cements have a syrupy consistency at room temperature, and harden when heated. Lenses are cemented in much the same way as with thermoplastic cement, except that the lenses and cement are at room temperature when the cement is applied. The lenses, aligned in a fixture, are placed in an oven and baked in order to set the cement.

Most lens elements and cemented doublets have their edges blackened with a lens ink so that they will not reflect light.

A thin layer of silver is sometimes applied to mirror surfaces and to the reflecting surfaces of prisms by chemical deposition, but this method has largely been replaced by vapor coating. When the reflection is inside the glass, as in second-surface mirrors or as in all prisms, the silver is usually plated with copper to prevent the silver from tarnishing, and enamelled to provide physical protection. Aluminum is usually used for first-surface reflectors and it is applied by evaporative coating in a vacuum. The aluminum does not tarnish, as does silver, but it is very soft, and scratches easily during cleaning. A half-wavelength thick layer of silicon monoxide is usually coated over the aluminum, to provide protection from scratching, as well as to increase the reflectivity of the mirror's surface.

Questions	True	False
1. Cast lens blanks are made by heating the glass cubes and pressing them in a mold.	_____	_____
2. The weighing of cut cubes is the key to controlling the thickness of molded lens blanks.	_____	_____
3. The curved side of the cubes is milled to remove contamination.	_____	_____
4. The "spots" in lens blocks are seats for the lens blanks.	_____	_____
5. In conventional lens-grinding, the lens blocks are usually rough-ground one block at a time.	_____	_____
6. Water is used during grinding primarily to keep the lenses clean.	_____	_____
7. Many lenses can be blocked on each block if the lens radius is short compared to the lens diameter.	_____	_____
8. Pitch is used for blocking and for polishing lenses.	_____	_____
9. Smooth and fine grinding are generally done on multiple spindle machines.	_____	_____
10. Synthetic plastic, pitch, and felt are used as the polishing surfaces in the conventional process.	_____	_____
11. In conventional polishing, both the lens blocks and the polishing shells rotate in the same direction.	_____	_____
12. A mixture of rouge and water is the conventional polishing medium.	_____	_____
13. Diamonds are used in lens grinding primarily to increase the speed of grinding.	_____	_____
14. In high-speed polishing, both the lens blocks and the polishing shells rotate in the same direction.	_____	_____
15. The precision of the curves of the polished lens surfaces are tested with brass gauges or a spherometer.	_____	_____
16. Lenses are inspected on the block for scratches and to determine that the polishing is complete.	_____	_____
17. Lenses are edge ground during the centering operation because the molded lens edges are not perfectly round and are not finished enough to be used in mounts.	_____	_____
18. The preparation of lenses for evaporative coating consists of cleaning.	_____	_____
19. Lens coating requires high vacuum and heat.	_____	_____
20. Magnesium fluoride is used to increase the reflectivity of mirror surfaces.	_____	_____
21. The thickness of lens coatings is measured in fractions of light wavelengths.	_____	_____
22. Dichroic coatings are made of several very thin layers.	_____	_____
23. There are two methods for applying aluminum to mirror and prism surfaces: chemical deposition and evaporation.	_____	_____
24. Thermoplastic cement is hard at room temperature and must be heated for use.	_____	_____
25. Thermosetting cement must be heated to harden it.	_____	_____
26. Lenses are inked to increase flare which improves their performance in optical instruments.	_____	_____
27. The silvered surfaces of prisms are enamelled to protect the thin layer of metal.	_____	_____

Chapter 5

Basic Design of Optical Instruments

Summary

There are numerous elements used in the production of optical equipment. The most common of these is the *magnifying lens* which is a positive lens which produces an enlarged virtual image of an object placed at or near its front focal point. A *simple camera lens* is a positive lens having one surface convex and the other concave and is used in most cameras. A more complex usage of the convex and concave lenses is found in the *Galilean telescope* which acts as a refracting telescope and forms an erect image. A simpler and more familiar form of the Galilean telescope is found in *field glasses*. A *refracting telescope*, sometimes called a simple transmission telescope, inverts the image and is nearly always used for astronomical purposes where the inversion of the image is not psychologically disturbing to the viewer. *Terrestrial telescopes* are telescopes used on land and sea to make distant objects plainly visible. These telescopes combine an image erecting lens system with a refracting telescope to invert the upside down image formed by the objective lens. *Mirror reflecting telescopes* use a parabolic mirror to serve as the objective lens. These telescopes are nearly always used for astronomical purposes. *Binoculars* consist of two parallel terrestrial telescopes, in which Porro prisms are used to erect the image. Many instruments used in optics require *optical viewfinders* for viewing and framing objects. Eye-level optical viewfinders are reversed Galilean telescopes with a negative finder lens and a positive eye lens. Whereas optical viewfinders are used to find the line of sight, a *periscope* is an optical device used to displace the line of sight. A simple periscope consists of two mirrors in a simple mount used to see over things. *Rangefinders* use triangulation principles to find the distance from the observer to a distant object. Sometimes it is necessary to form a bright, enlarged image of an object on a screen. *Projection systems* use two lens systems; one for distributing the light from the filament of a bulb over the area of the film image and the second, a projection lens, to form the image of the film on the screen.

Basic Requirements

All optical instruments that provide visual images require that light rays

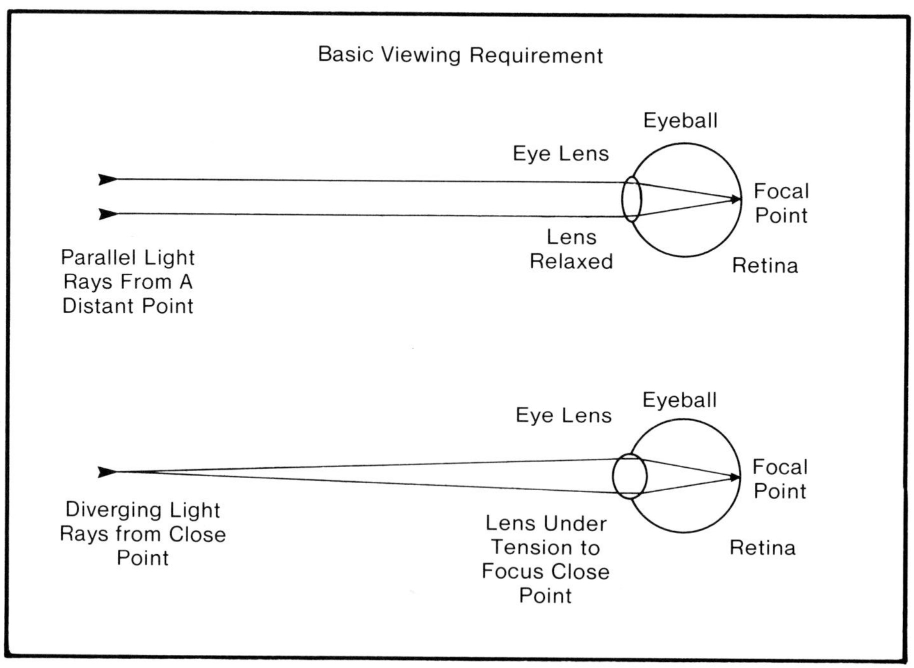

Figure 5-1. Parallel or slightly-divergent light rays create the normal conditions for viewing.

coming to the eye must be parallel to each other (as if they were coming from an object at infinity) or slightly divergent (as if they were coming from an object no closer than 10 inches from the eye). The reason for this is that the eye's lens must be able to focus the light on the retina for a sharp image to be seen (Figure 5-1). All optical instruments can be focused, either by changing the object-to-lens distance as in a microscope, or by changing the distance between lenses within the instrument as in a telescope. There are two types of camera lenses: those that have a fixed focus at an intermediate distance, and those which provide an adjustable focus in which the lens to film distance is adjusted.

In describing the optical instruments in the following section some terms are used that have not yet been defined. These terms can be looked up in the glossary in the back of this book.

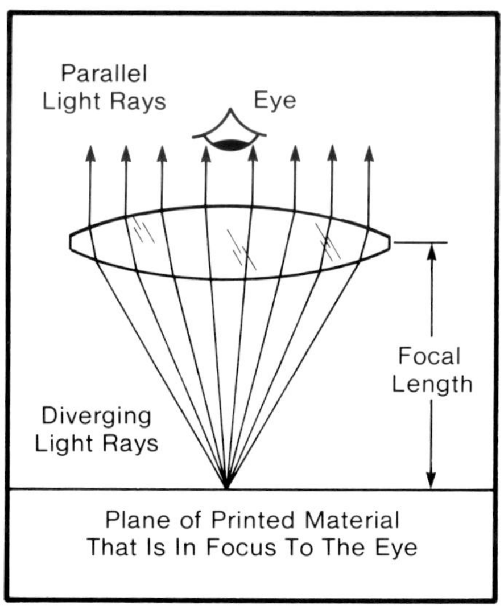

Figure 5-2. Magnifying Lens.

Magnifying Lens

A positive lens (one that is thicker in the middle than at the edges) placed at a distance equal to its focal length from an illuminated object magnifies the image of the object (Figure 5-2). The magnification, or magnifying power, can be found by using either of the following equations:

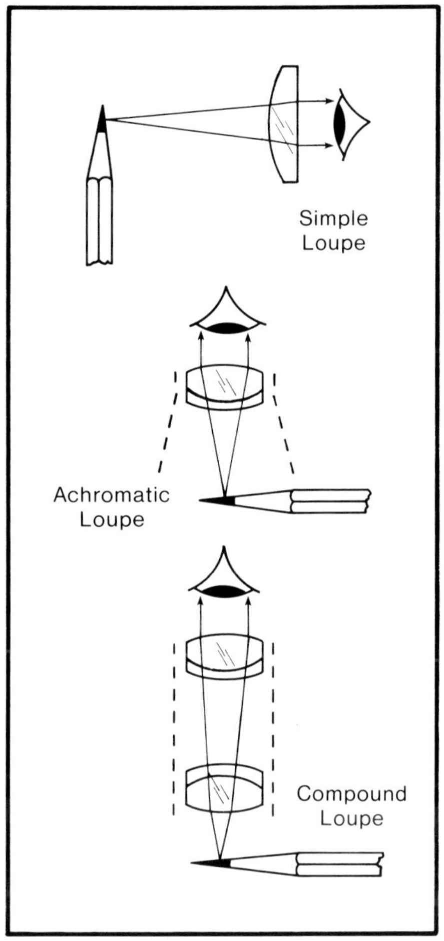

Figure 5-3. Loupes.

$$m = \frac{250 \text{ mm}}{\text{focal length in mm}}$$

or

$$m = \frac{10 \text{ inches}}{\text{focal length in inches}}$$

The focal length of a positive lens is the distance between the lens and the image it forms of a distant object.

Loupe

A magnifying lens corrected to give sharper images over a greater field size than a magnifying glass is called a loupe. It is usually a cemented doublet (two lens elements cemented together) or two cemented doublets (Figure 5-3).

Simple Camera Lens

A positive meniscus lens can be used as a simple camera lens. It is commonly used with a fixed aperture of from *f*/11 to *f*/16 to give adequately sharp images. The aperture, or stop, is the round lens opening that allows the light to enter the camera. The *f* number is the focal length divided by the diameter of the lens opening. Used as a rear meniscus lens it provides a relatively flat field, but requires a relatively long camera body (Figure 5-4a). Used as a front meniscus, it provides a shorter camera length, but images a curved field (Figure 5-4b). Corrected camera lenses composed of several elements can form brighter images.

Microscope

Microscopes are used for magnifications greater than those provided by magnifying lenses and loupes. Relatively fast, short-focal-length objective lenses form real images that are observed with magnification

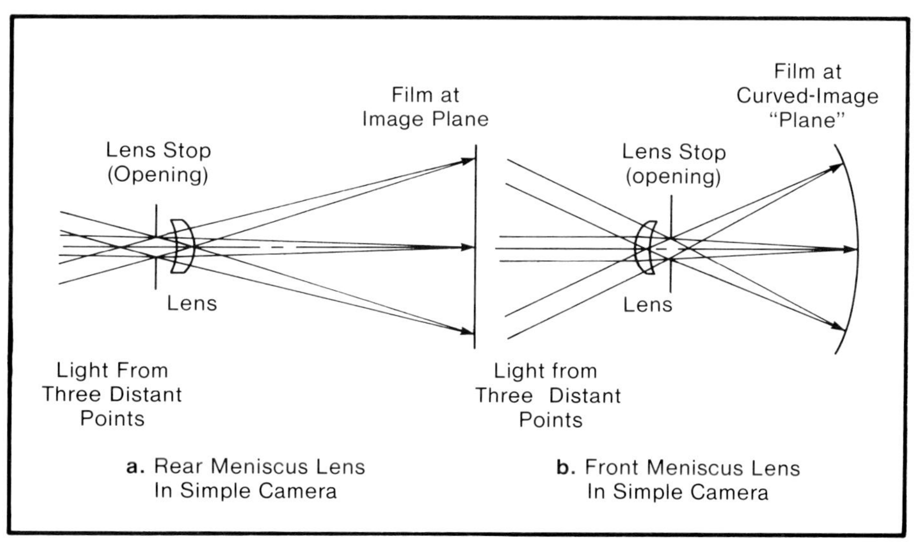

Figure 5-4. Meniscus Camera Lens Positions.

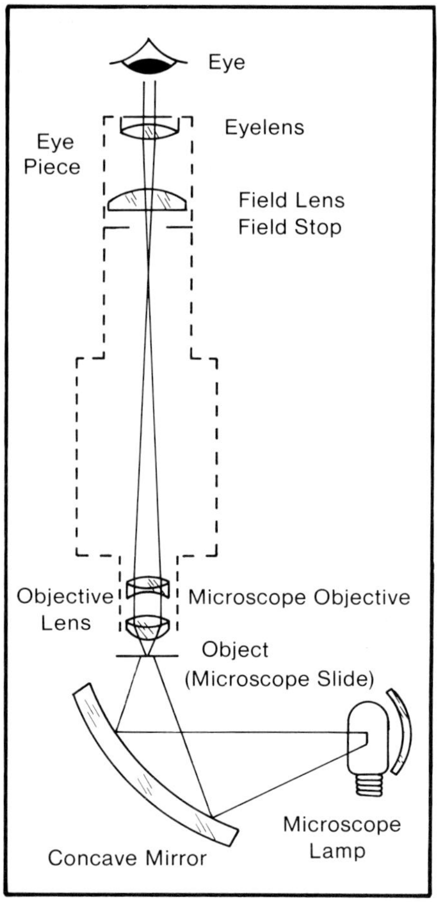

Figure 5-5. Microscope Optics

through eyelenses which are essentially magnifying lenses. The microscope tube length, which determines the image distance of the objective lens, is normally fixed at 160 mm (6.3 inches) which, along with the objective focal length, determines the magnification of the image before it is magnified by the eyelens (ocular) as seen in Figure 5-5. The magnification of a microscope is found by multiplying the magnification of the objective lens by the magnification of the eyelens. The magnification of the objective lens is found by the following equation:

$$m_{obj.} = \frac{\text{160 mm (tube length)}}{\text{focal length of objective lens in mm}}$$

The magnification of the eyelens is found by using the method for determining the magnification of magnifying lenses given above. The magnification of each microscope lens is marked on its mount so that the user does not have to measure or calculate it. Microscopes can yield satisfactory images up to magnifications of about 2000X.

Galilean Telescopes

A Galilean telescope is a simple instrument that is often used in pairs to form field glasses or opera glasses. This type of telescope has an objective lens with a focal length of several inches and a negative eyelens that has a negative focal length approximately equal to the focal length of the human eye (which cancels the power of the lens in the eye) as shown in Figure 5-6a. The image formed by the object lens is therefore formed on the retina of the eye.

The magnification of a Galilean telescope is found by dividing the focal length of the objective lens by the focal length of the lens in the eye.

$$m = \frac{\text{focal length of the objective lens in mm}}{\text{22 mm}}$$

Field Glasses and Opera Glasses

Field glasses are two Galilean telescopes with parallel axis and focusing eyepieces spaced at a distance equal to the interpupillary distance. The spacing is usually adjustable (Figure 5-6b).

Simple Transmission (Refracting) Telescope

Because this type of telescope inverts the image, it is nearly always used for astronomical purposes, where inversion of the image is not psychologically disturbing to the viewer. It consists of an objective lens and an eyelens (Figure 5-7). The objective lens forms an inverted image which is magnified by the eyelens. The magnification of a refracting telescope is found by dividing the focal length of the objective lens by the focal length of the eyelens:

$$m = \frac{F}{f}$$

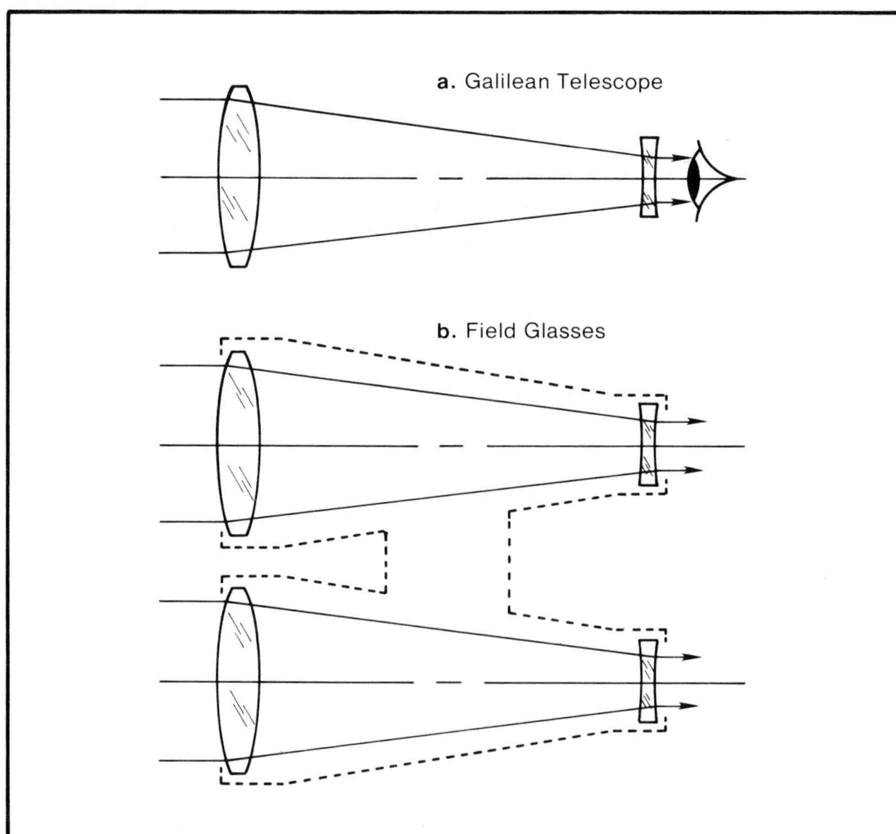

Figure 5-6. Galilean Telescope and Field Glasses optics.

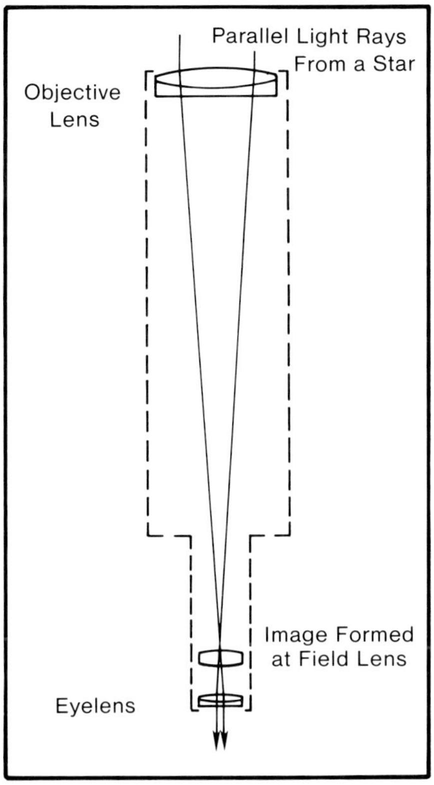

Figure 5-7. Refracting Astronomical Telescope.

Because of the added magnification of the eyelens, the refracting telescope gives more magnification for a given tube length than does a Galilean telescope.The focal lengths of objective lenses are often several metres long. Such scopes require substantial mounts to keep the viewed images steady. Many mounts are equipped with movements, one of which is driven with a clock so that the telescope moves with the motion of the earth, thus following a given star. The magnification is usually limited to 50X per inch of objective diameter to maintain image brightness. Therefore, a 6 inch objective lens would be limited to a 300X power. The largest existing refracting telescope (1979) has an objective lens 40 inches in diameter.

Terrestrial Telescope

The terrestrial telescope is a refracting telescope with an image-erecting system added to invert the upside down image formed by the objective lens. Thus, objects are seen right-side-up. The erector lens is a symmetrical lens system (compound) that is corrected and placed to form a same-size image of the objective lens image (Figure 5-8). The eyelens magnifies the image formed by the erector lens. Such telescopes are used on land and sea to make distant objects plainly visible, and are used as scopes on guns.

Mirror (Reflecting) Telescope

In a mirror telescope, a parabolic mirror serves as the objective lens (Figure 5-9). Such telescopes are nearly always used for astronomical purposes because of the inverted image formed. The image is formed in front of the mirror so a prism or small plane mirror set at 45° is used to reflect the image forming beam to the eyepiece. Mirror telescopes are classified by the diameter of the mirror, and range from 2-inch telescopes for home use to the 200-inch reflecting telescope at the Palomar Observatory in California. A 90-inch telescope mirror is now being made to send into space.

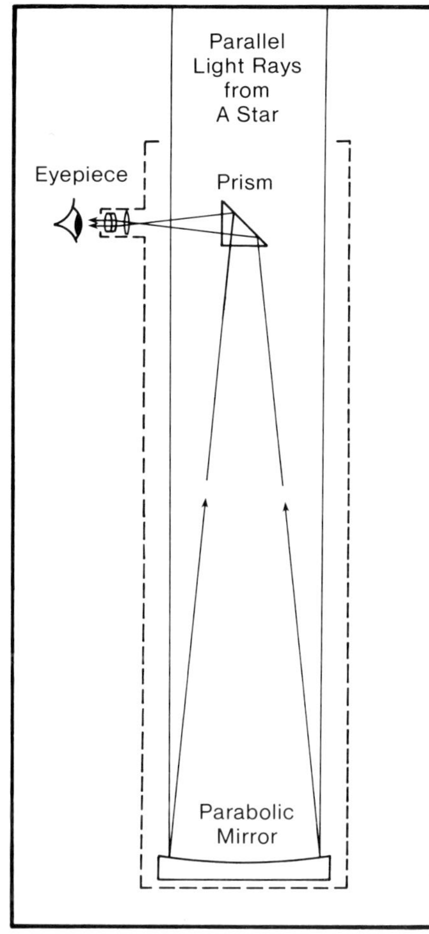

Figure 5-9. Reflecting Telescope.

The Cassegrain reflecting telescope is another mirror telescope with a somewhat different design (Figure 5-10). The objective mirror has a hole in it. The light from the stars reflects from this parabolic concave mirror to a small, convex mirror with a hyperbolic surface that reflects the light to an eyepiece located just behind the hole in the objective mirror. Parabolic and hyperbolic curves are aspheric curves that must be hand polished in the mirror glass. Cassegrain telescopes are used both as astronomical and terrestrial telescopes and, in slightly different form, as telescopic camera lenses.

Binoculars

Binoculars consist of two parallel terrestrial telescopes in which Porro

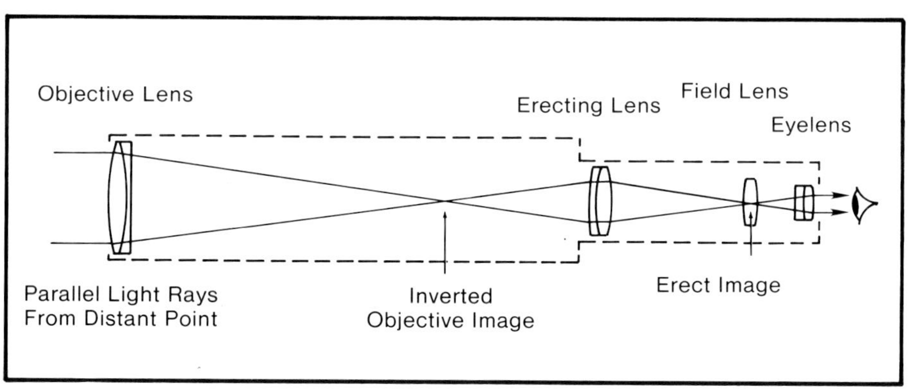

Figure 5-8. Terrestrial Telescope (Spy Glass).

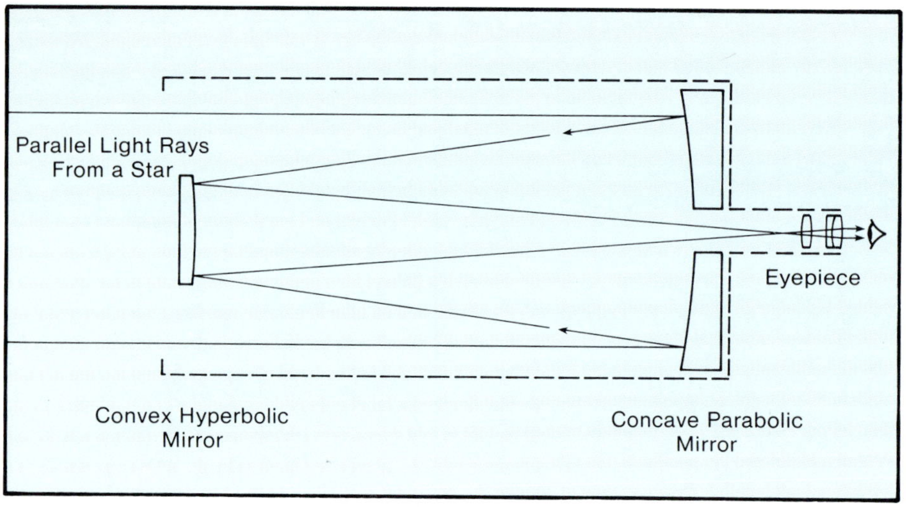

Figure 5-10. Cassegrain Telescope.

prisms are used to erect the image (instead of erector lenses) as shown in Figure 5-11. There are two Porro prisms cemented together (or clamped together in a mount) at right angles to each other to erect and to give correct left-to-right orientation to the images. These prism clusters allow the two objective lenses to be spaced farther apart than the interpupillary spacing of the eye lenses, which enhances the stereo effect (stereopsis) of the images seen through the binoculars. They also shorten the length of the binoculars. Binoculars are classified by two numbers such as 6X30. The first number identifies the power (the magnification of the binoculars) while the second number gives the diameter of the objective lens opening in millimetres, which is a measure of the brightness of the images.

Optical Viewfinders

Many cameras have optical finders for viewing and framing. Eye-level optical viewfinders are reversed Galilean telescopes with a negative finder lens and a positive eyelens. They have a fractional magnification, which means they make objects look smaller than when seen with the unaided eye (Figure 5-12a).

In an Albada finder, the concave curve on the negative finder lens is made partially reflective with a metalized coating (Figure 5-12b). This reflects a bright line rectangle that is located on a black mask that surrounds the positive eyelens. The viewer sees this line as projected into space superimposed over the view; the line indicates the field of view. Use of this method eliminates the parallax caused by change of position of the eye behind the eyelens.

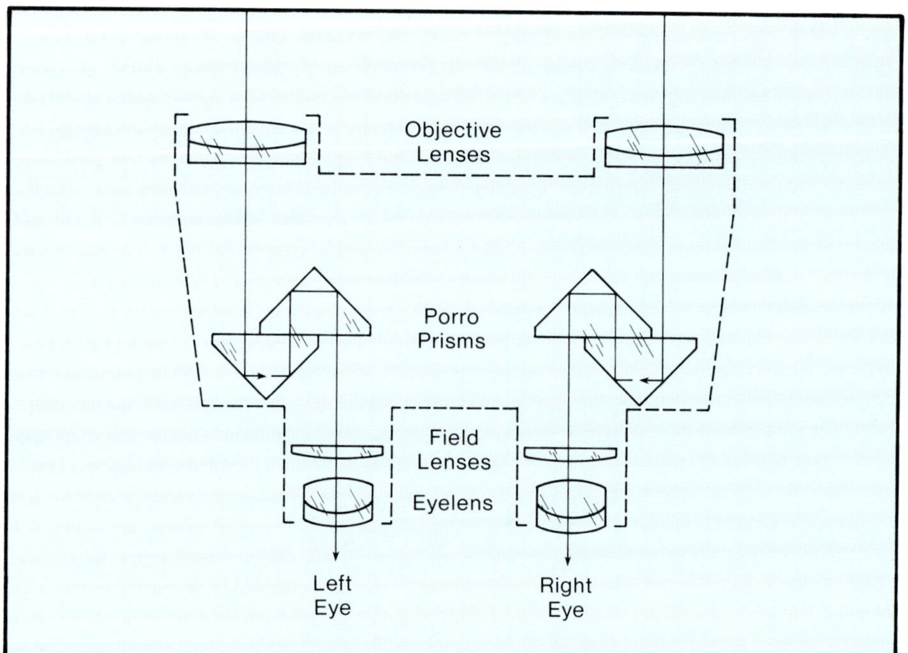

Figure 5-11. Prism Binoculars.

Reflex Finders

In a ground glass reflex finder, a positive lens images views on a ground glass placed at right angles to the lens axis via a 45° mirror (Figure 5-13a). The image is seen right-side-up, but is reversed left-to-right. The finder is arranged to have about the same angle of view as the camera. When used in twin-lens reflex cameras, the ground glass is made the same size as the film format, and the well corrected finder lens is made the same focal length as the camera lens (Figure 5-14). The camera and viewfinder lenses are mounted so that they move in unison. When the viewing image is focused on the ground glass, the camera lens is simultaneously focused on the film. Usually a magnifying lens is provided, attached to a hood over the ground glass to aid in critical focusing. The images on ground glass reflex finders are somewhat dim due to light being diffused by the ground glass, so a version of the reflex finder called a brilliant finder, is sometimes used (Figure 5-13b). The construction is similar, but a rectangular positive lens replaces the ground glass. Optically, this creates a compound positive lens, and the image seen is a virtual image.

The reflex finder is modified in several ways when used in single lens reflex cameras (Figure 5-15). The camera lens serves two purposes; finding (aiming and framing) and taking. The mirror is hinged

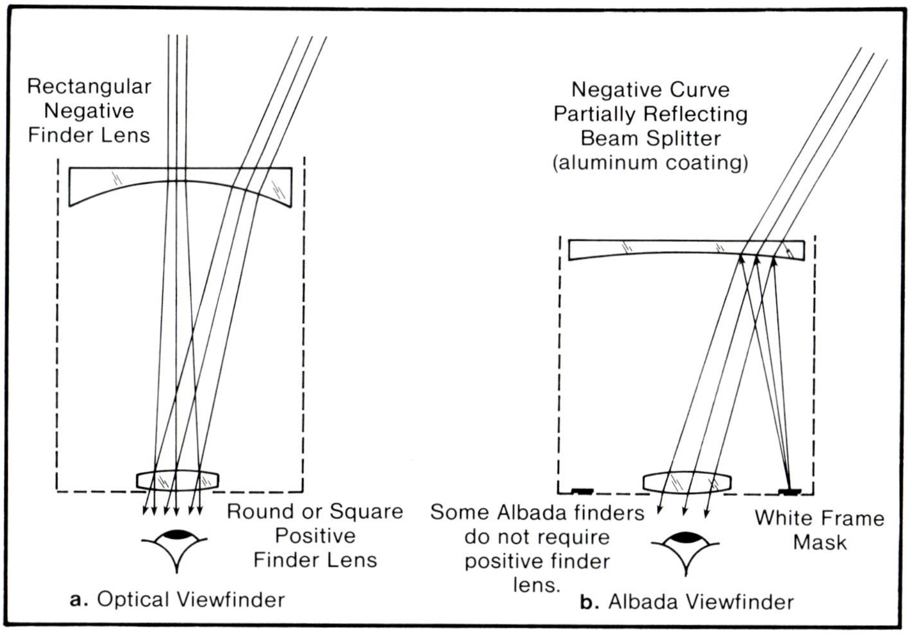

Figure 5-12. Optical and Albada Viewfinders.

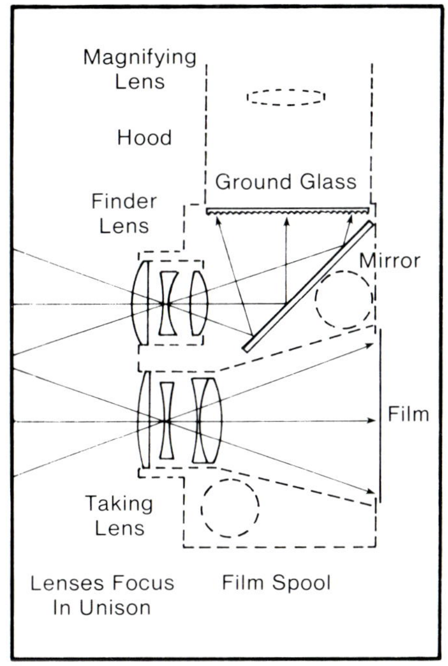

Figure 5-14. Reflex Finder on Twin-Lens Reflex Camera.

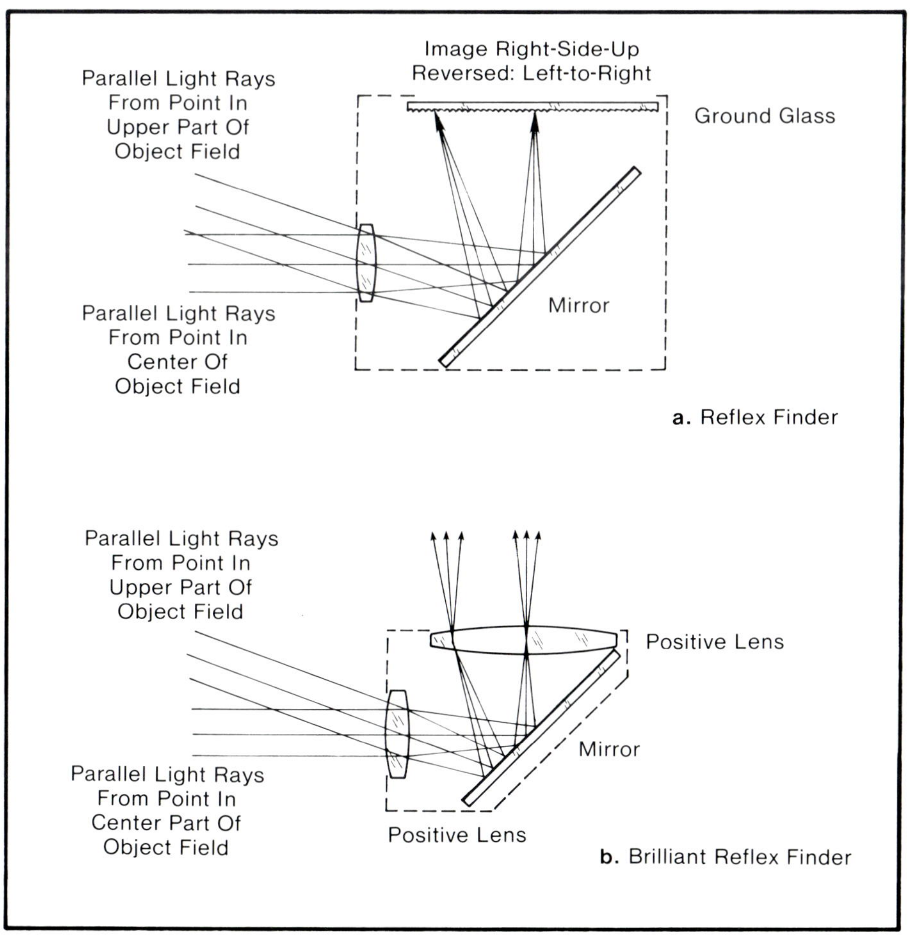

Figure 5-13. Reflex and Brilliant Reflex Finders.

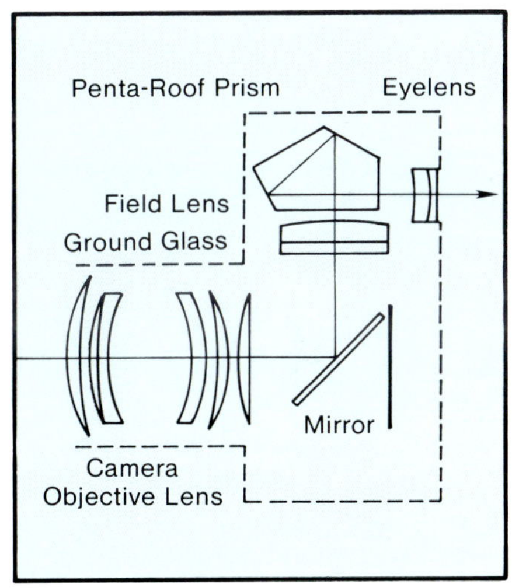

Figure 5-15. Single-Lens-Reflex Finder

along the top edge, so that when finding is complete, the mirror can swing up out of the way for taking the photograph. A roof penta prism is placed just above the ground glass to provide eye-level viewing of the finder image, and to reverse the image so that it is oriented correctly, left to right.

An eyelens focused on the ground glass magnifies the image, but since its focal length must be approximately equal to the light path in the prism, its magnification is limited.

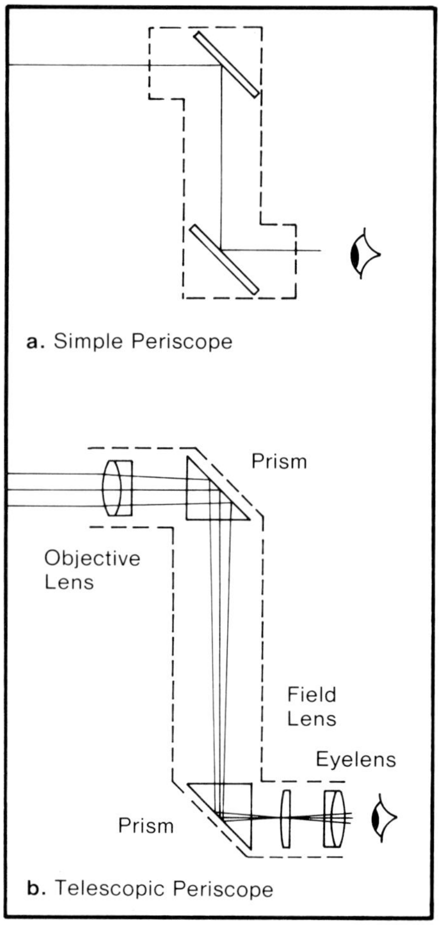

Figure 5-16. Simple Periscope and Telescopic Periscope Optical Systems.

Modern focusing screens are molded of plastic with opposed wedges acting as rangefinders, fresnel lenses to brighten the corners of the image, a doughnut-shaped ring of very fine "ground" surface for cirtical focusing, and a very weakly ground surface in the outer area for brilliance. In some cameras, the focus screens are interchangeable, so that the user can select different viewing characteristics.

Periscopes

A periscope is an optical device used to displace the line of sight. A simple periscope consisting of two mirrors in a mount is used to see over the heads of a crowd at parades and sporting events (Figure 5-16a).

A telescopic periscope is used in submarines and military aircraft to permit viewing outside the body of the vehicle (Figure 5-16b). In a submarine, the periscope permits both observation and the aiming of torpedoes. In some military aircraft the periscope permits the aiming of guns from the midsection of the fuselage. In such periscopes 45° prisms are substituted for the mirrors, and telescope optics are added to the basic periscope to permit wide angle or enlarged viewing. To permit the wider angle of view, a relatively short focal length objective lens is used, while relay lenses (similar to erector lenses in a terrestrial telescope) are used to carry the image down the long tube of the periscope. Such periscopes are equipped with calibrated reticles so that they can be used to aim the weapons.

Rangefinders

Rangefinders are optical devices that use triangulation principles to find the distance from the observer to a distant object. In warfare, they are used for finding the distance to artillery targets, for adjusting the missile trajectory, and for timing the explosion of the projectile. In photography, they are often coupled to the camera lenses so that when the distance is found with the rangefinder the lens is automatically focused on that distance.

In split-field rangefinders, (Figure 5-17a), mirrors (or prisms) are arranged to provide triangulation. Fixed mirrors split the view into two beams at the eyelens. Mirrors or prisms at the ends of the assembly direct the beams toward an object at infinity. Closer objects are seen as displaced or split in halves. Moving one outer mirror or rotating a wedge shifts one half of the image so that the closer object can be made to appear whole, that is, the two halves are brought together. When used in cameras, the rotation of the movable mirror is coupled to the camera lens so that the lens is in focus for each distance at which the triangulation makes a whole image. In military rangefinders, the movable rangefinder may have a distance scale attached to the movable mirror or wedge, or it may be connected elec-

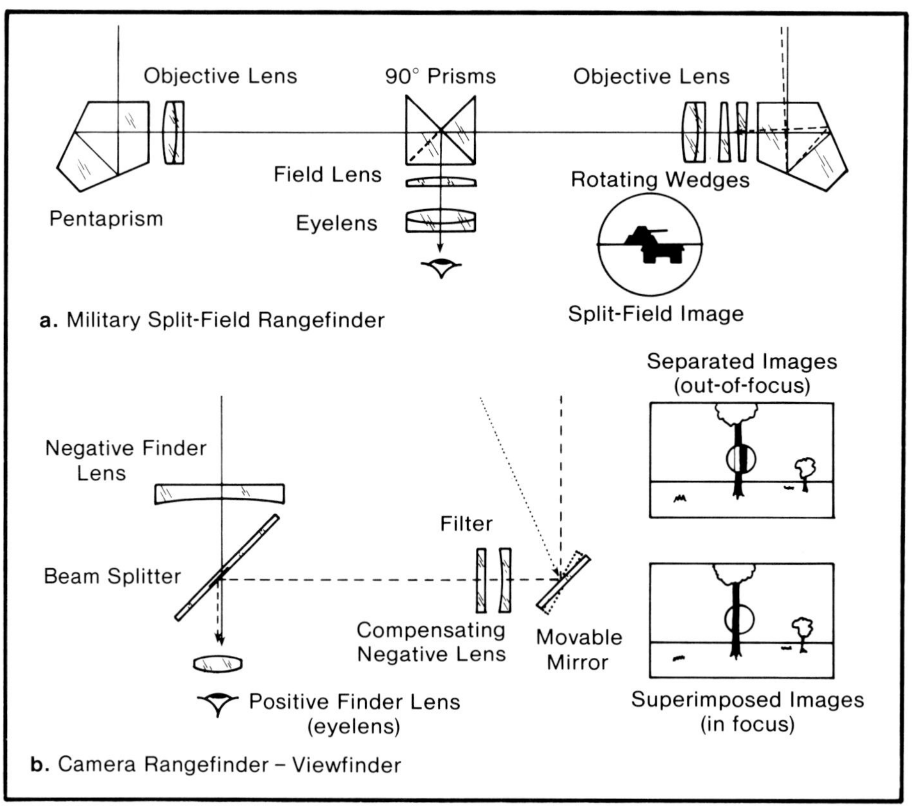

Figure 5-17. Camera Range/View Finder and Military Split-Field Rangefinder.

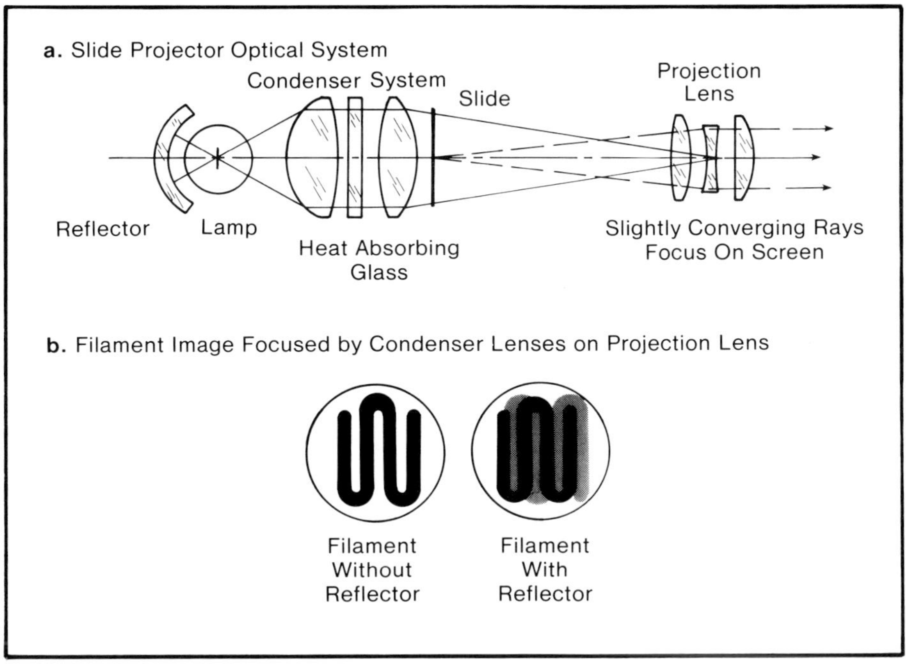

Figure 5-18. Slide Projector Optical System.

trically to the weapon's aiming system.

Superimposed image rangefinders on cameras have the advantage that they can be combined with the viewfinder, so that focusing can be done at the same time and through the same eyepiece as framing (Figure 5-17b). In this type of rangefinder the fixed mirror has an elliptical area in the center that is partially aluminized or coated with a dichroic coating that transmits one color and reflects another. The movable mirror is a fully reflecting, first-surface mirror as in the split image rangefinder. In front of the movable mirror is a negative lens that has the same power as the negative finder lens so that the two images will be the same size. Looking through the viewfinder, the field is seen with a circular colored area in the center. Because the central spot is partially reflecting, two images are seen in the spot, one by transmission, which is the viewing image, and one by reflection, which is ranging or the focusing image. When an object is at a distance that is not triangulated, two superimposed images are seen out of register. As the movable mirror is rotated, the colored image is seen to move in relation to the fixed image, and when the two images are exactly superimposed, that is, seen as one, the camera lens is in focus for that object.

Projection Systems

Projectors are used to form bright, enlarged images of film transparencies on a screen for viewing. They usually employ two lens systems. The first is a condenser system for distributing the light from the filament of a lamp over the area of the film image. The second is a projection lens to form the image of the film on the screen (Figure 18a).

The lamps used are clear glass, usually with several coil filaments separated by a distance equal to the coil diameter, and arranged to form a square source of light (Figure 5-18b).

Condenser lenses are relatively large, strong, positive lenses with a diameter larger than the diagonal of the film format. Usually two condenser lenses are used. In small-systems one is adequate, while in some systems, three or more are used. In some modern Super-8 motion picture projectors, a reflector is built as part of the lamp and takes the place of the condenser lenses.

A reflector is used with nearly all projection systems. The reflector is positioned so that it images the filaments back upon themselves, but displaced slightly so that the image of a filament coil is placed between two real coils, filling the space with light. In this way light from the back of the coils is sent forward to add to the illumination coming from the front of the coils. The projection lens "sees" a solid square of light.

Two conditions must be met to provide the greatest amount of light on the screen for the bulb and projection lens used. The first is that the condenser lens system must focus the image of the filament on the rear of the projection lens. To keep the size of projectors within reason, the focal length of the condenser lens system must be very short (to keep the overall length of the projection system reasonably compact). This allows the lamp to be placed close to the projector lens and reduces the overall length of the optical system. This accounts for the strong curves on the condenser lenses. To provide even more light, and to make the illumination even across the image fields, one of the curves of one of the condenser lenses is often made aspheric. Since condenser lenses are often molded from glass rod rather than being ground and polished, they are economically priced.

The second condition for light efficiency is that the condenser lens system must have the same or a smaller f-number than the projection lens. For example, if the condenser lens system is $f/2.8$, and the projection lens is $f/2.0$, the screen image will be no brighter than if the projection lens is $f/2.8$. On the other hand, if the projection lens is $f/4$, the brightness will be less than with either of the other lenses.

Projection lenses must have the same high degree of image forming performance as camera taking lenses. This is to insure sharp images on the screen. Such lenses are relatively large aperture lenses to provide adequate screen brightness. Slide projectors usually have an aperture of $f/2.8$, while motion picture projectors have apertures of $f/2.0$ to $f/1.4$.

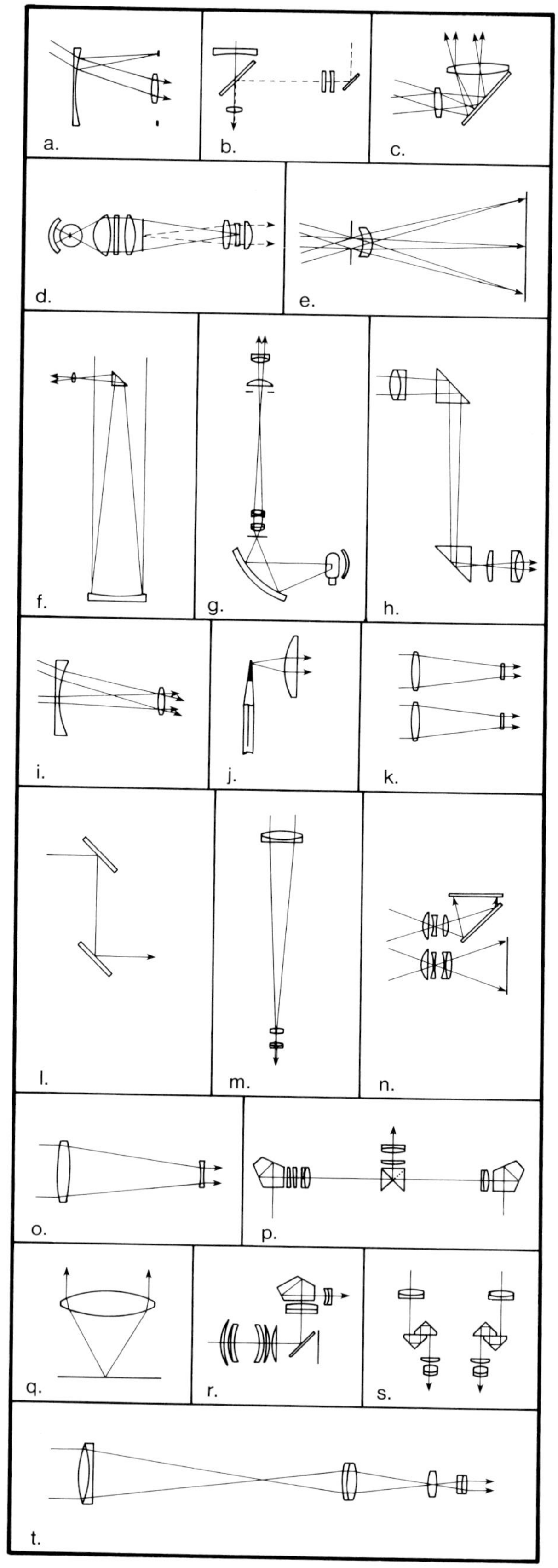

Questions

Match the letter of the most appropriate drawing with the name of the optical instrument.

1. Magnifying glass _____
2. Loupe _____
3. Simple camera lens _____
4. Microscope _____
5. Galilean telescope _____
6. Field glasses _____
7. Refracting telescope _____
8. Terrestrial telescope _____
9. Newtonian mirror telescope _____
10. Binoculars _____
11. Eye-level optical finder _____
12. Albada finder _____
13. Twin-lens reflex camera _____
14. Single-lens reflex camera _____
15. Brilliant finder _____
16. Simple periscope _____
17. Telescopic periscope _____
18. Split-field rangefinder _____
19. Superimposed rangefinder _____
20. Slide projector _____

Chapter 6

VIDEO LESSON 1

Introduction to Geometric Optics

Summary

The study of optics consists of three main parts: Physical Optics, Physiological Optics, and Geometric Optics. Schematic ray tracing is a basic tool of optical engineers and is an important part of Geometric Optics. Light rays coming from distant object points are essentially parallel. Rays of light can be imaged by a pinhole. Any object is a reflector of light, and each point in the object reflects light in many directions. It is called a secondary light source. Some of the light rays coming from secondary sources can be imaged. To show how they are imaged, a few selected rays are traced (drawn) on paper. Thin lenses are representative of real lenses, and for the purposes of design, can be assumed to have zero thickness. Such lenses are drawn as a line perpendicular to the optical axis and represent the equivalent refracting plane. Rays are traced from an infinitely distant object point on the lens axis. These rays are parallel to the axis and are refracted by the lens at the refracting plane. The refracted rays then cross the axis at the focal point. Light coming from other object points at great distances from the lens, forms image points at the same distance from the refracting plane but not on the axis. All these rays form the optical image which is assumed to lie in a plane perpendicular to the lens axis. The distance from the equivalent refracting plane to the focal plane is called the lens focal length. Lenses have sizeable lens openings that allow bundles of light rays from each object point to enter the lens and to be refracted to form image points, thus making the optical image relatively bright compared to a pinhole image. Light rays traced from object points at finite distances from the lens which are greater than the focal length are focused by the lens at image distances greater than the focal length. The magnification of the image is found by dividing the image's length by the object's length. Light rays are considered reversible. They are refracted and follow the same paths going through the lens in either direction. Objects closer to a lens than its focal plane form virtual images.

Three Branches of Optics

This course is particularly concerned with the branch of optics called *Geometric Optics.* There are two other branches of optics study (Figure 6-1).

Physical Optics was briefly discussed in the preceding chapters as a general introduction to this program. The study of light and other electromagnetic wave radiation makes up much of this branch. The specific behaviors of light are interference, polarization, refraction, dispersion, diffraction, transmission and absorption. The movement of light is treated as wavefront propagation, rather than ray movement. The study of lasers and holography is part of this branch of optics. A few topics from the field of Physical Optics will be presented in this course.

Physiological Optics is the study of vision including the study of the eye, nerve, and brain structures and functions that are involved in vision. Physiological Optics includes such aspects of vision as hue, saturation,

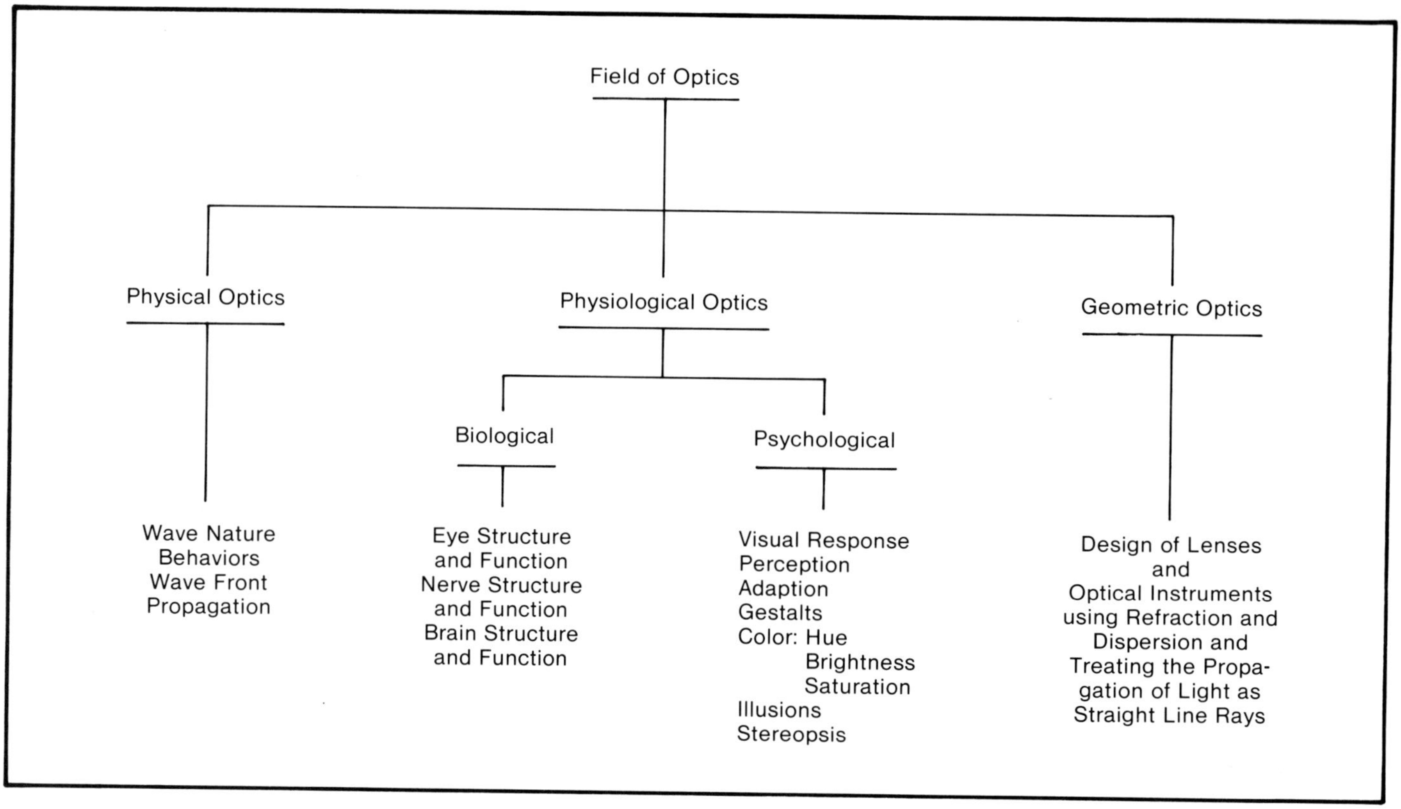

Figure 6-1. The Field of Optics

and brightness of color. It also includes special adaptations of vision such as after images, binocular vision, optical illusions, and perception. These are a part of Physiological optics called *Psychological Optics,* which will be touched upon when necessary for a better understanding of Geometric Optics.

Geometric Optics is the study of how light rays change their paths when they are refracted or reflected. Geometric Optics is a practical study since it applies the general principles of refraction and reflection to particular optical elements, and to groups of particular optical elements that are placed in a definite relationship to each other to accomplish particular optical results. In other words, Geometric Optics has to do with the design of optical instruments.

The Design of Lenses

As we have seen, lenses (single elements or compound lenses) are required for practical cameras and optical instruments such as microscopes, telescopes, rangefinders, and projectors. Such lenses are needed to produce high quality images of particular sizes, relatively free from aberrations. The design of lenses that will accomplish such purposes requires the knowledge of what happens to specific light rays as they encounter lenses made of specific optical glasses with particular curves on them. This course includes a simple introduction to some of the methods of lens design.

The Design of Optical Instruments

The primary design method used in this course is called *schematic ray tracing*. This consists of drawing the outlines of optical elements or their symbols in their proper relationships to each other, and drawing straight lines representing light rays. An understanding of each optical element and design principle can be obtained by tracing selected rays and by seeing where they go as they are refracted and reflected. The same basic method can be used to design optical systems that perform certain optical functions (Figure 6-2).

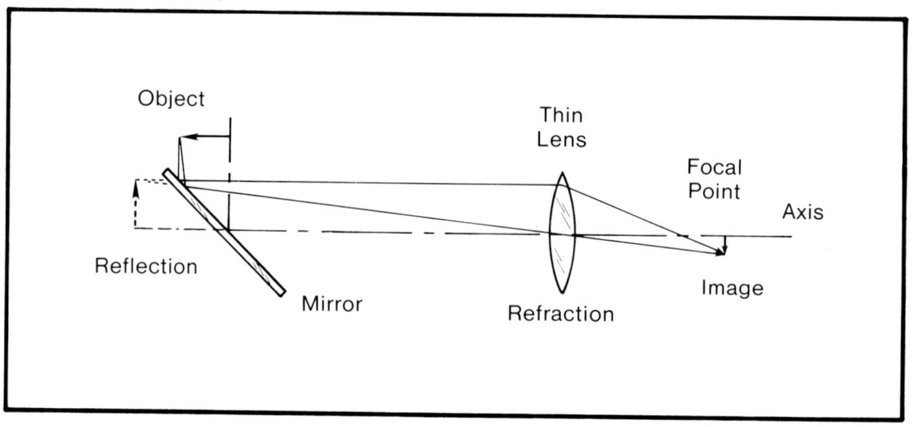

Figure 6-2. Schematic Ray Trace

Parallel Light Rays

Primary light sources are shown in ray trace diagrams only when sources such as light bulbs are part of an optical instrument. Most instruments do not have light sources in them. Usually they image objects illuminated by remote light sources and these objects are technically known as secondary light sources. They reflect rather than emit light.

Light radiates from a source such as the sun or a light bulb. This means that the light rays go evenly in all possible directions from each point in the light source. In ray tracing only those light rays that come to the object being imaged are considered (Figure 6-3).

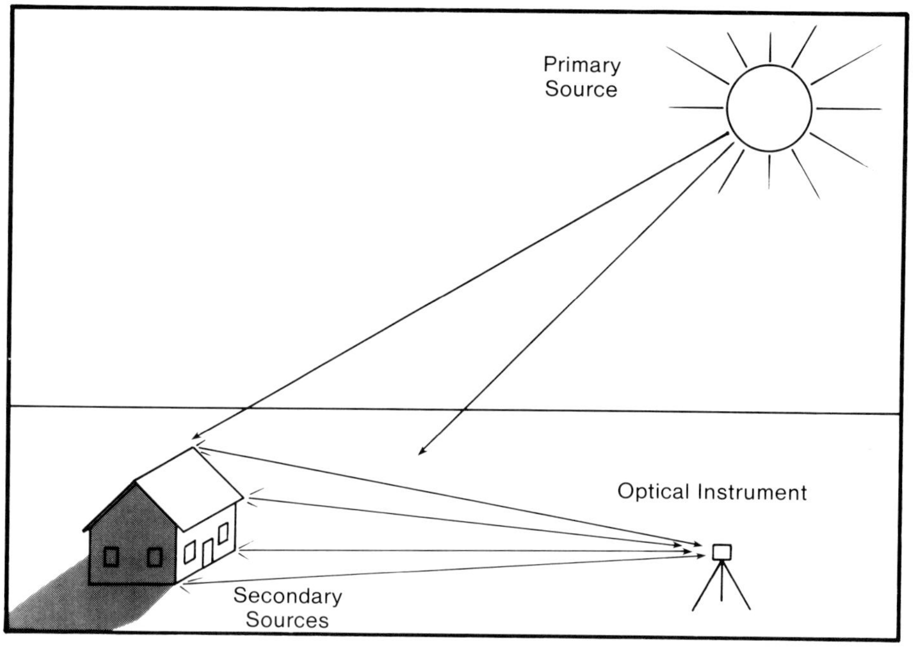

Figure 6-3. Light ray travel.

Light rays coming to an object from a source at a great distance are parallel (Figure 6-4). Stars are point sources of light; the light rays coming from each star are parallel.

Imaging with a Pinhole

Figure 6-5 shows a diagram of a box with a small pinhole at one end. The box is placed on the ground and aimed at the sky. It is a clear night, and there are two bright stars which are 10° apart in the sky. No matter where you place the box in a large open field, it is illuminated by light rays from the two stars. Parallel-light rays from each of the two stars falls everywhere on the side of the Earth facing the stars.

The parallel-light rays from one star are shown by light solid lines, while light rays from the second star are shown as dashed lines. The light rays from each star are parallel to the other light rays from the same star, but they are not parallel to the light rays from the other star. The light rays from the second star follow paths that form a 10° angle with the light paths of the rays of the first star.

Because of the small size of the pinhole, relatively few rays from each star can enter the box. Therefore each star has only a tiny beam of light reaching the image plane in the box. Each beam of light forms a tiny circle of light in the image plane (Figure 6-6). These circles are the optical images of the stars. If you measure the angle between the two beams of light within the box, it also will be 10°. If you put photographic film in the image plane and expose the film to the optical image of the stars and then process the film, you will have a picture (photographic image) of the stars.

Any object that is a great distance from an optical device is said to be at infinity; it is at an infinite distance from the optical device. The distance is infinitely large compared to the size of the optical device, and the light coming to the optical device that will form the image is essentially parallel. The closest distance that can be considered infinite varies with such factors as the *f*-number and the focal length of the objective lens of the device. For practical purposes it is assumed to vary from between 15 to 20 feet up to several miles.

If the image plane in the box is moved closer to the pinhole, the images of the two stars are formed closer together. When the image plane is moved farther away from the pinhole, the images are formed farther apart. The angle between the two beams of light stays at 10°.

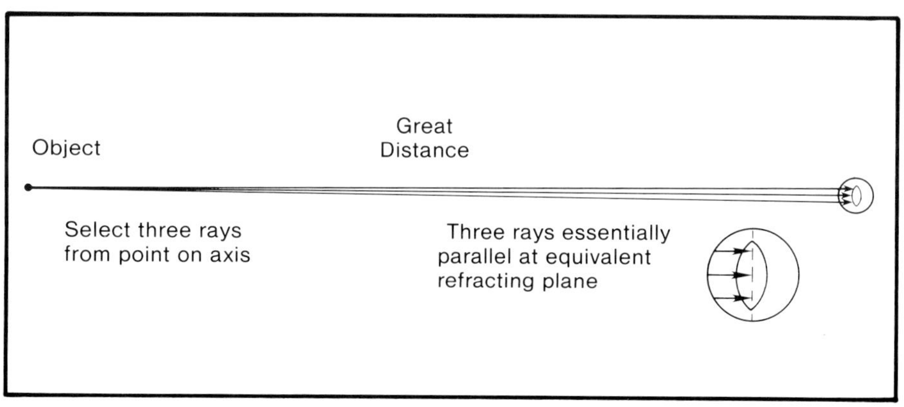

Figure 6-4. Parallel Light Rays

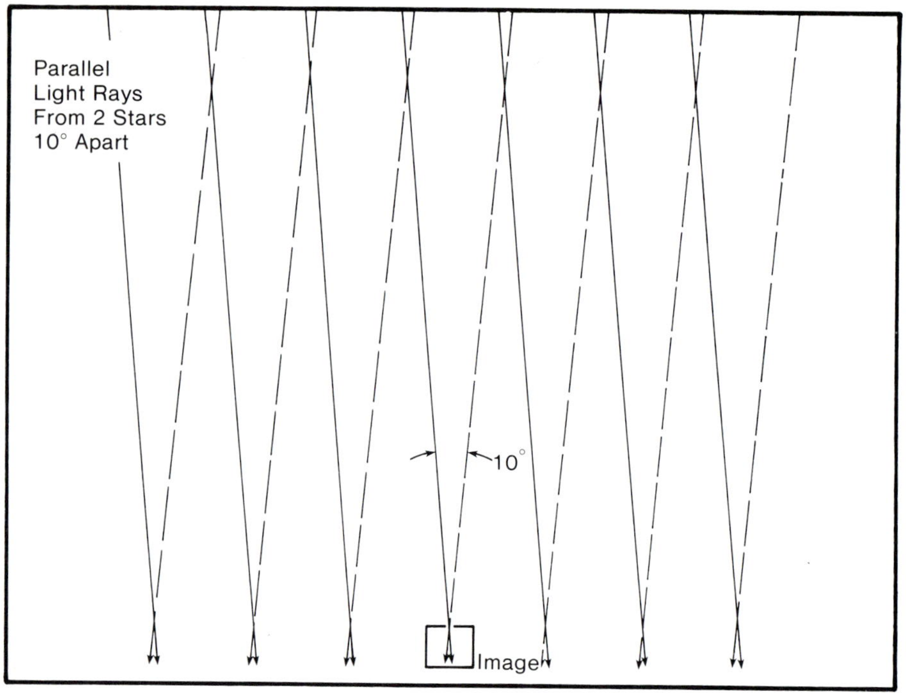

Figure 6-5. Pinhole Camera

Primary and Secondary Objects

In optics, an optical image is formed of an *object*. The stars in the example given above are the objects that are imaged on the film. Stars radiate light; objects which emit light are called *primary objects*. However, most objects are not primary objects; most objects reflect rather than emit light. Objects that reflect light are called secondary light sources or *secondary objects*. Each star in the above example is a *point source* of light: they are so distant that they have no discernable diameter from Earth. Each point on a reflecting object is considered a secondary point source of light and reflects light toward the optical device.

Most object surfaces reflect light diffusely or at least have a partial diffusion, a light ray coming to each point on the object is reflected in many directions. The only light rays of interest in ray tracing are those which are reflected to the optical device under consideration (Figure 6-7). If the source is a primary one, the only rays considered are those emitted directly to the optical device. In the star example, the entire field, indeed, the whole side of the Earth facing the stars received light from the stars. Only that light which came to the pinhole of the camera could form the optical image.

In order to simplify ray tracing even more, only certain selected rays coming to the optical device are selected for tracing. Those that contribute to the solving of the particular optical problem are chosen for tracing and sometimes as few as two rays lead to the solution.

In most schematic ray tracing, an arrow is used as an object. This serves two purposes. Two rays, one from each end of the arrow, are enough to determine the image size. The direction of the arrow head serves to indicate whether the image is erect (right-side-up) or inverted (up-side-down). Light rays from objects at an infinite distance are, of course, shown as traveling in parallel paths.

Thin Lenses

All real lenses have thickness. When light rays pass through real lenses, they refract at each of the two lens surfaces. However, there is a plane at which the light *appears* to bend. This is called the *equivalent refracting plane*. Figure 6-8 shows this plane, and how it is derived by extending the parallel rays from a point on a distant object located on the lens axis, and the bent rays that form the focal point. The equivalent refracting plane is the intersection of these two sets of light rays. In most real lenses this equivalent refracting plane is not a plane, but a curved spherical surface with a radius equal to the focal distance. For general purposes of thin lens ray tracing, it is practical to consider the principal plane to be flat. In the diagrams the equivalent refracting plane is shown as a dashed line (Figure 6-8, 6-9, etc.).

Since the bending of light rays is considered to occur at the equivalent refracting plane, this plane can represent a lens with no thickness—or what is commonly called a *thin lens*. The drawings also include two thin solid lines that represent a cross section of an actual lens so that the ray traces will be easier to interpret.

Image Formation with a Thin Lens

A thin biconvex lens, which is a positive lens, is shown. A dashed (long and short dashes) line representing the lens axis is drawn hori-

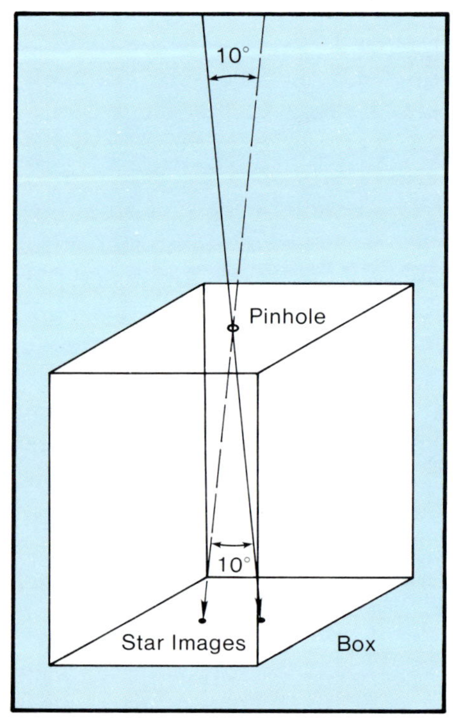

Figure 6-6. Pinhole Image Formation

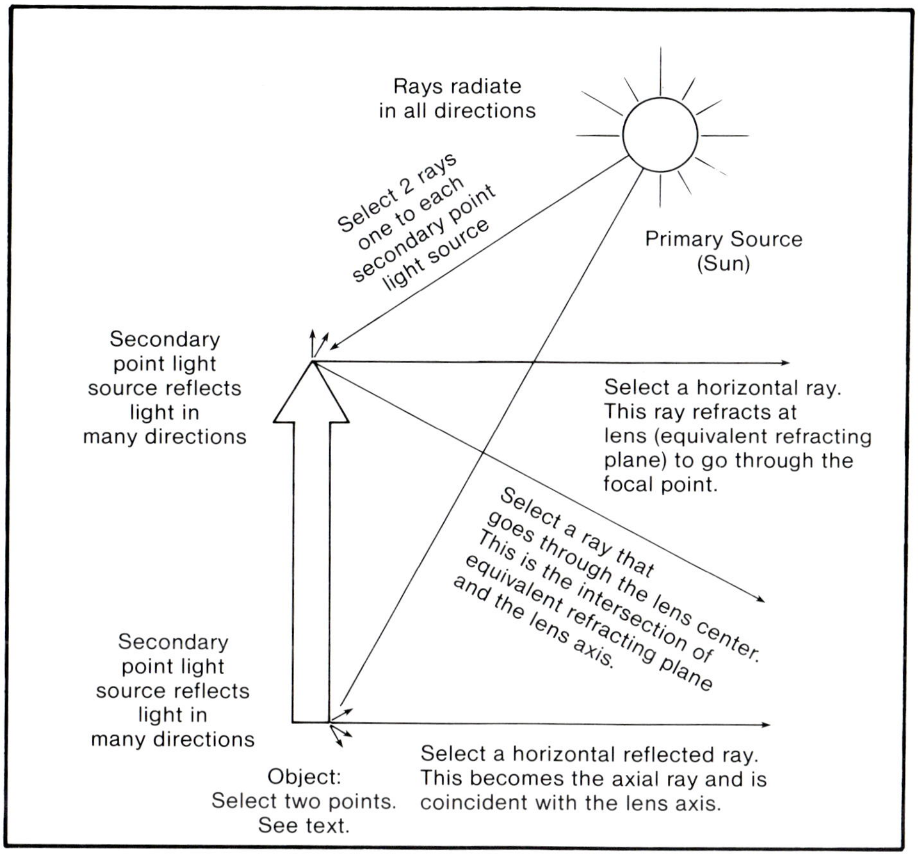

Figure 6-7. Primary and Secondary Light Sources as commonly used in Schematic Ray Tracing.

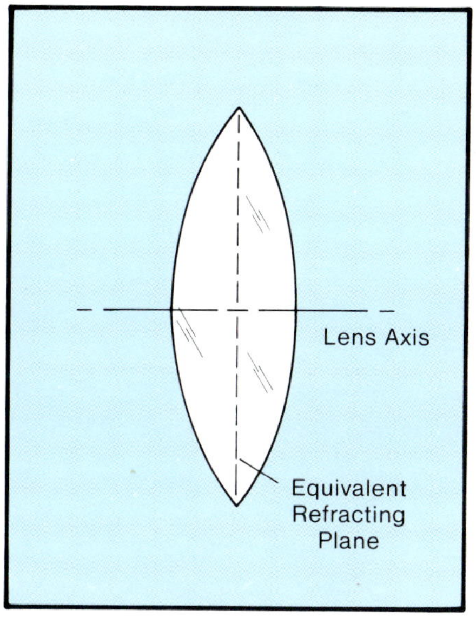

Figure 6-9. Thin-Lens Convention

zontally through the center of the lens. It represents the central ray of light that passes through the lens without bending because it passes through perpendicular to each lens surface. This *axial ray* goes through the thickest part of this lens (Figure 6-9).

Near the upper and lower edges of the lens, equidistant from the axial ray, two other rays are drawn parallel to it. These represent rays from a distant point on the axis. The light is coming from the left (a convention in ray tracing). Two more rays are drawn parallel to the others, and are halfway between the outer rays and the axial ray. These rays are not essential, but will help illustrate the point.

All the rays entering the lens, except the axial ray, are refracted (bent). Because the surface of the lens presents a greater angle to the outer rays than to the middle rays, the outer rays are bent more. As the rays leave the lens through the second surface, they are refracted again (all except the axial ray). Similarly, the outer rays are bent more than the middle rays. All the rays meet at a point to the right of the lens. This point is called the *focal point.* **The focal point is the image of an infinitely distant object point located on the axis.**

If all the parallel object rays coming from the left are extended with dotted lines, and if all the image rays coming to the focal point are extended to the left back through the lens, the dotted lines intersect within the lens. A vertical, straight, dashed line is drawn through the intersections and represents the equivalent refracting plane at which the rays seem to be bent at the lens. Therefore, equivalent refracting planes can be used in ray trace diagrams to represent thin lenses.

The drawings represent a cross section of the three-dimensional lens and the lens surfaces shown as

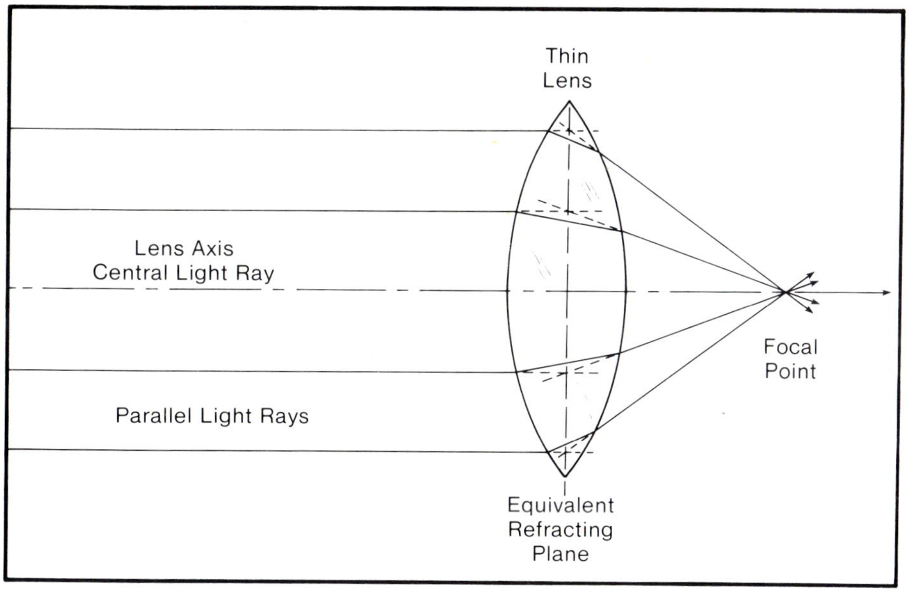

Figure 6-8. Thin Lens with Parallel Light Rays, Parallel to Axis, Forming Focal Point.

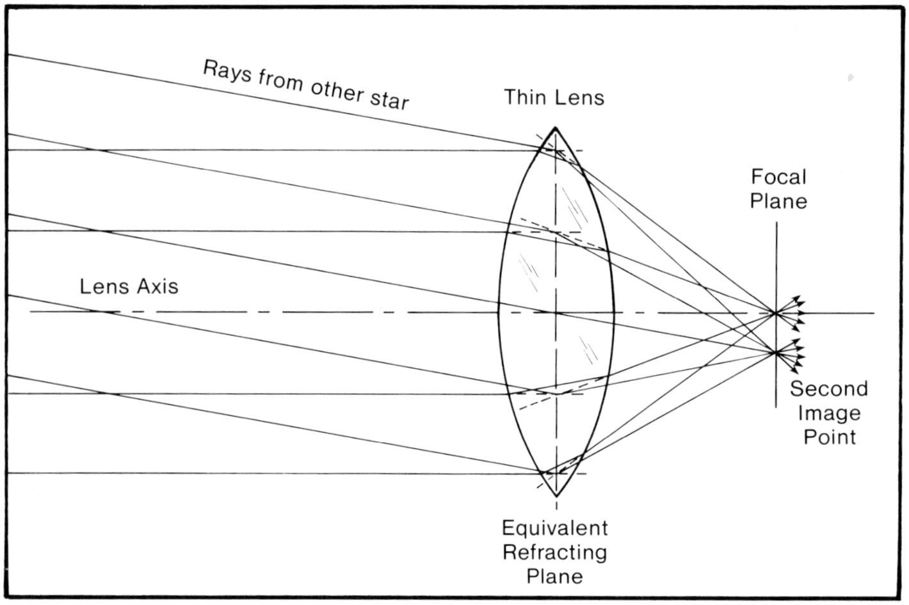

Figure 6-10. Thin Lens with Parallel Light Rays, 10° to Axis, Forming Second Image Point.

sections of a circle representing spherical segments that are the actual lens surfaces.

In Figure 6-10, five more rays have been added. They enter the lens and are bent at the equivalent refracting plane. These rays come from an object point at infinity that is 10° off axis. The new rays intersect at the right of the diagram and form the image point of the new object point.

A thin vertical line is drawn through this image point and through the focal point. This line represents the *focal plane*. **The focal plane is that plane in which all rays from object points at infinity will focus, to form image points.** In a camera, the focal plane represents the area where the film should be placed in order to get a sharp image of infinitely distant objects. The image formed by all the rays coming from all the object points is called the *optical image*. In a normal scenic photograph, the objects in the scene are composed of an infinite number of secondary source points that are focused in the optical image. The optical image is made up of an infinite number of image points, one for each object point.

Object points vary in color and brightness and therefore the light rays coming from the secondary sources vary in color and brightness (Figure 6-11). The optical image usually appears as a miniature of the object. It is light where the object is light and colored where the object is colored (Figure 6-12).

The materials and the surface curvature of each lens determine its refracting power. A strong lens bends the light rays more than a weak lens. When the rays are bent more, they come to focus closer to the lens and when they are bent less, they focus farther from the lens. The distance from the equivalent refracting plane to the focal plane when the object is at infinity is called the *focal length* of the lens (Figure 6-13). The focal length of a strong lens is relatively short, whereas the focal length of a weak lens is relatively long.

Bundles of Rays

It is usually necessary to select only a few rays in the schematic ray tracing procedure to solve most optical problems. To show all the rays would be impossible. However, it is important to keep in mind that the few rays drawn represent an infinite number of light rays. A *beam* of light is often called a *bundle* of light rays. The illustration suggests that coming from a single object point at infinity, parallel light rays cover a broad area in the general location of the lens (Figure 6-14). Working backwards (right to left) and using the periphery of the lens as limits, the bundle of light rays from the single object point is used in forming

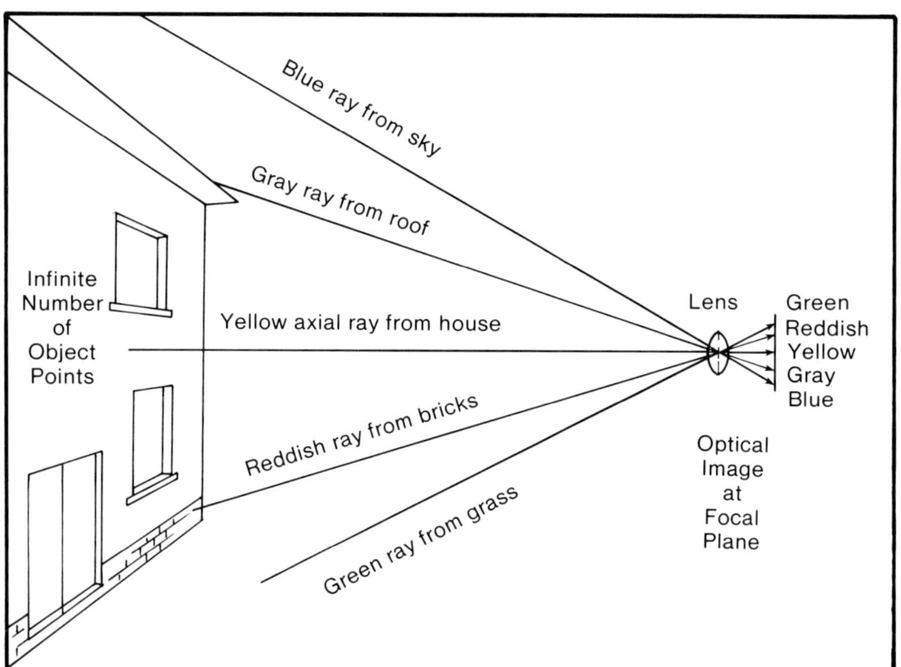

Figure 6-11. Image Formation using five selected rays.

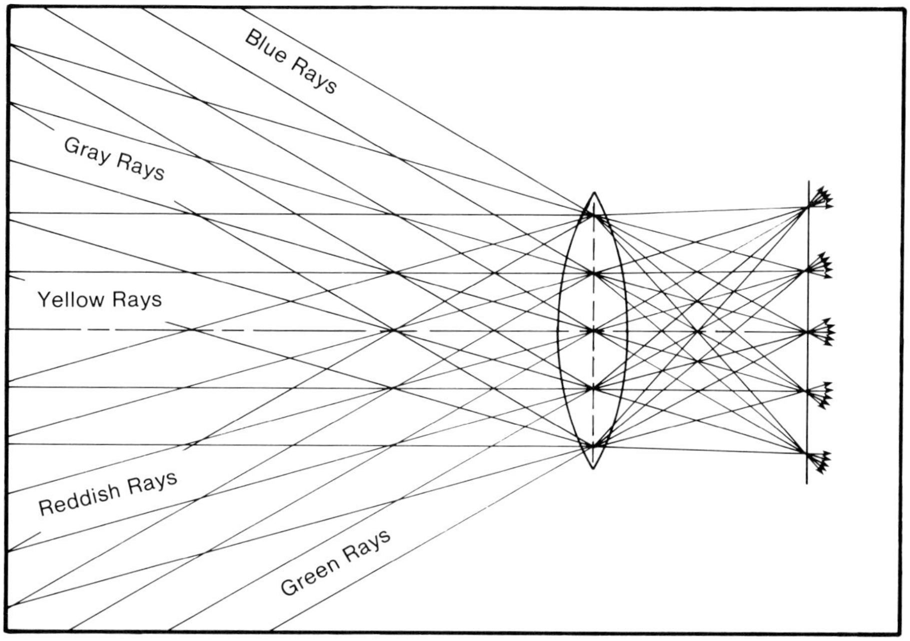

Figure 6-12. Light from five distant object points forming five image points. Real terrestrial objects have an infinite number of object points (secondary light sources).

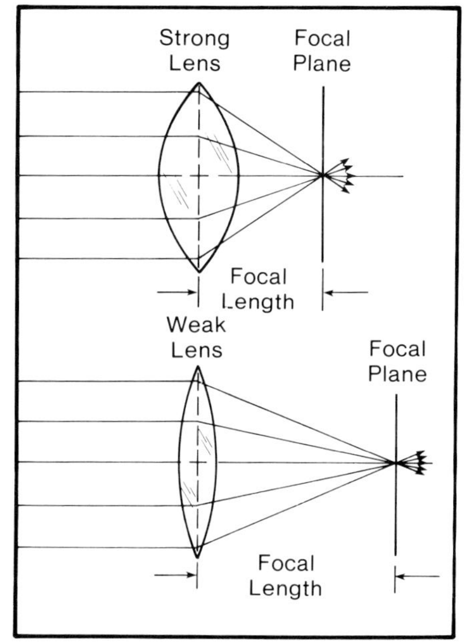

Figure 6-13. The focal length is the distance between the equivalent refracting plane and the focal plane when the object is at infinity.

the image of the point to be defined. It is out of this bundle that several rays were selected to show the imaging of the focal point. The axial ray is always one of those chosen, and the extreme, or peripheral rays are often used as the other rays of choice.

In the above examples, light rays were shown going to the focal plane and continuing through the plane. If there is film in this position the rays will not continue but will be absorbed.

When the eye is placed behind the lens where light from an infinitely far object point is being imaged on the axis, the focal point falls on the lens of the eye and all the light will continue into the eye striking the eye's retina. The eye will see the lens as fully illuminated by the light from the single object point (Figure 6-15). The entire bundle of light rays coming from the object point will have been converged by the lens into the eye.

Normally, where the entire object is composed of an infinite number of secondary light sources, each point in the optical image receives a converged bundle of light rays from the corresponding object point. The larger the diameter of the lens, the larger the bundle of rays converged to each image point and, given the same focal length, the brighter the optical image will be because it is composed of more light rays.

If the object is one of the stars discussed earlier and, if the eye is placed at the focal point, the lens will appear to be filled with white light. If the object point is one of many secondary light sources (object points) of a daylight scene, it may, for example, reflect red light. Then the eye will see the lens filled with red light. If the object point is dark, the lens will appear dark.

The Focus of Close Objects

As indicated before, an arrow is sometimes used to represent the object. More precisely, it is used to represent close objects. The arrow represents an object at a finite distance rather than an infinite distance from the lens.

Parallel-light rays were used to

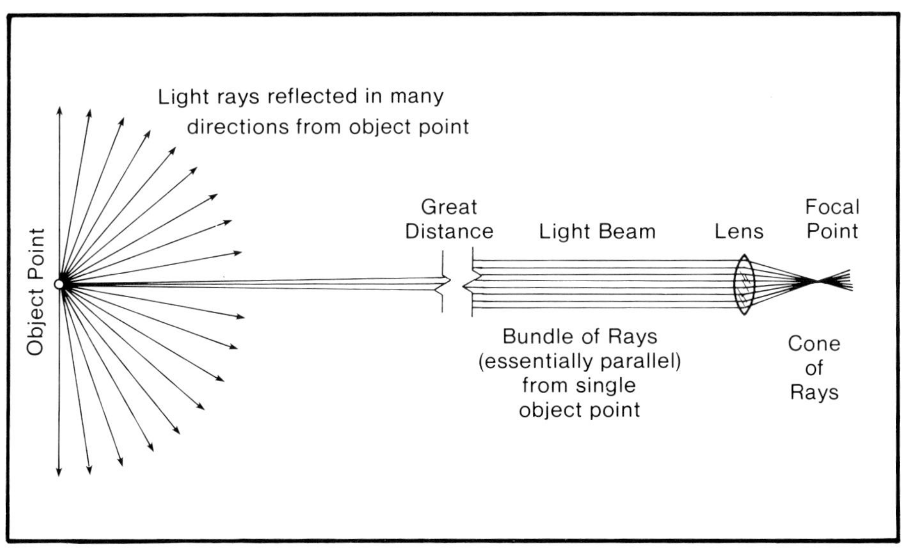

Figure 6-14. Light beam as a bundle of rays projected backward from periphery of lens.

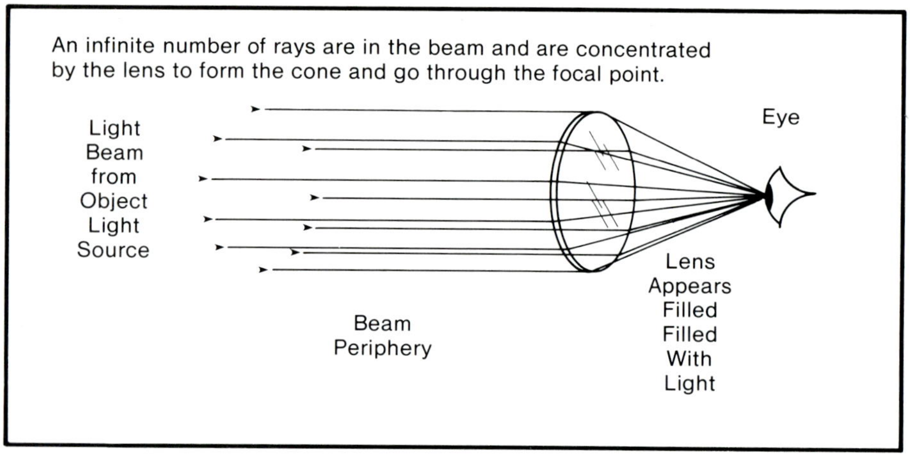

Figure 6-15. Lens diameter determines beam diameter and cone base diameter.

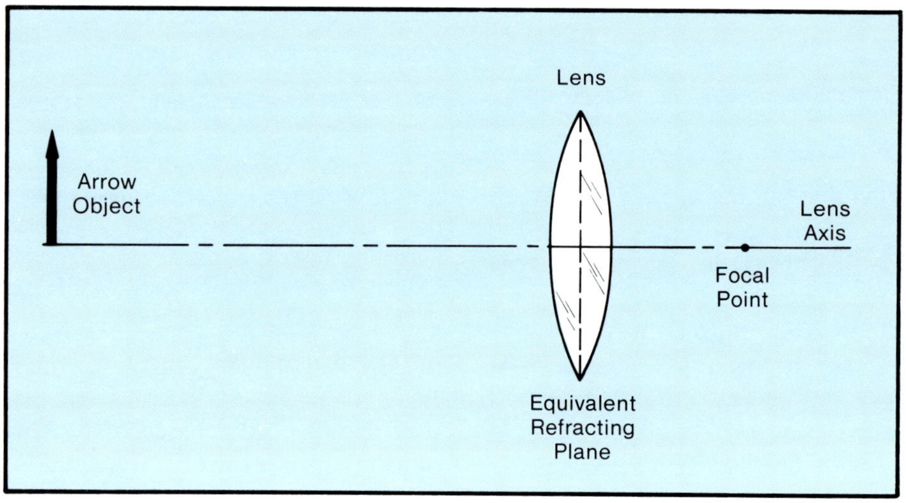

Figure 6-16. Typical drawing for schematic ray tracing image formation of finite objects.

find the focal plane of infinitely distant objects. The arrow can be used to find the focal planes of objects that are at finite distances from the lens.

First the lens and its equivalent refracting plane are drawn (Figure 6-16). Next the lens axis is drawn, noting that this also represents an axial light ray. The arrow object is drawn at the left of the lens with the bottom of the arrow placed on the axis and the arrow's head is drawn a reasonable distance above the axis (less than half the lens diameter). A light ray parallel to the axial ray is drawn from the arrow point to the lens (Figure 6-17). As was found when imaging objects at infinity, rays that are parallel to the lens axis are refracted so that they pass through the focal point. Therefore this ray is drawn from the equivalent refracting plane through the focal point and, because there is no film to stop the ray, it continues past the focal point.

Now a ray is drawn from the point of the arrow through the intersection of the lens axis and the equivalent refracting plane. Rays that go through the center of a thin lens are not bent, so the image ray is a continuation of the object ray. The intersection of the two rays from the arrow's point is the image point of the arrow's point. The image of the other end of the arrow is on the lens axis, which is also a ray that goes from the base of the arrow through the center of the lens. Careful selection of these three rays establishes both the size and the position of the image, as well as the fact that the image is inverted.

If the lens is a real lens, and the object is placed in the position shown by the arrow in the drawing, a ground glass can be placed where the image is established and a real image of the object can be seen. If a piece of film or photographic paper is placed in the position established as the image plane, the real image can be recorded. The image formed by the positive lens is a *real image*.

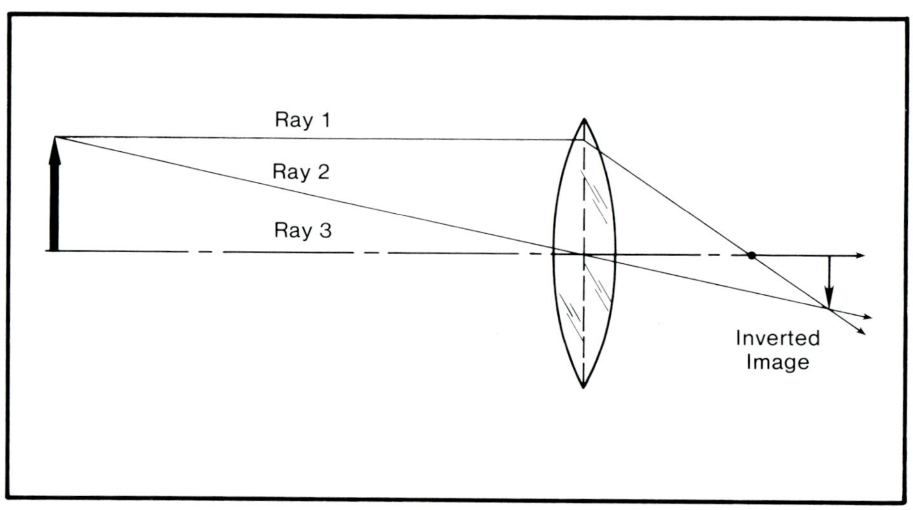

Figure 6-17. Locating the image of a finite object by selecting three rays.

A real image is formed by converging light and it can be seen by placing a ground glass in the image plane. It is upside down and reversed left to right, that is, it is rotated 180° and looks like a two-dimensional miniature of the object (Figure 6-18). If the eye is placed in different positions within the image plane (without the ground glass), the lens will seem to be filled with light of one color, corresponding to the color of the object point that is being imaged at the location of the eye's pupil.

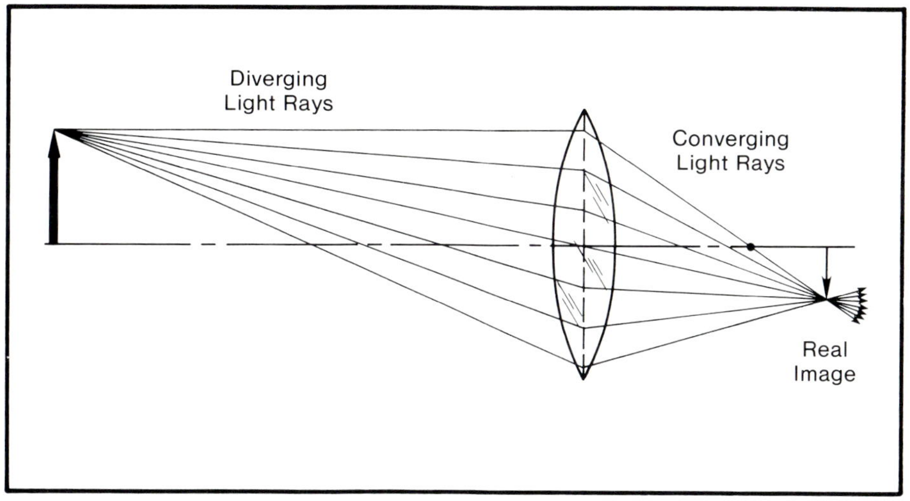

Figure 6-18. Many rays from each object point are refracted to form each image point. A real image is formed by converging light rays.

Magnification

If the size of the image arrow is measured and compared to the size of the object arrow, the magnification of the image can be calculated:

$$\text{Magnification} = \frac{\text{Image size}}{\text{Object size}}$$

This is illustrated in Figure 6-19.

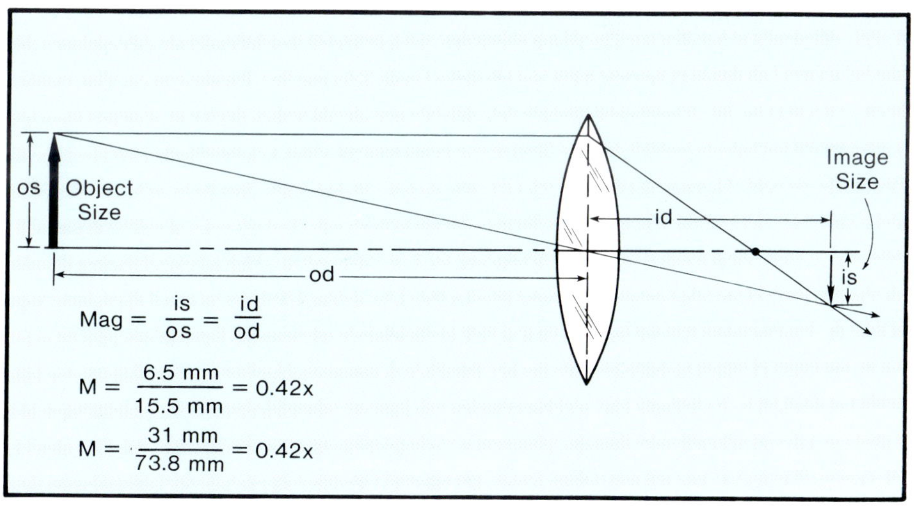

Figure 6-19. Magnification

There is also an important relationship between the image size and the object size, and the image distance and object distance:

$$\frac{\text{Image size}}{\text{Image distance}} = \frac{\text{Object size}}{\text{Object distance}}$$

Schematic ray tracing is often called geometric ray tracing and the above relationship can be shown by using geometry. The object arrow and the image plane are drawn perpendicular to the axis. The angles that the axial ray makes with the ray drawn from the arrow point through the axis are equal because the two rays are intersecting straight lines. When two angles of two triangles are equal, the triangles are similar. The above relationship is therefore true because equivalent sides of similar triangles are always proportional.

The Reversibility Principle

In the example shown in Figure 6-20, the lens has been shown to have a focal length to the right of the lens, and all light rays have been shown going from left to right. This is the convention, and is the normal situation encountered in ray tracing. On occasion it will be useful to reverse the normal procedure. Real lenses have the same focal length in either direction. This is shown in the diagram by placing a second focal point on the axis at the left of the lens. It is placed at a distance from the equivalent refracting plane equal to that of the right focal point. A lens has the same focal length left to right as it does right to left. The ray is traced from the image arrowhead point to the equivalent refracting plane parallel to the lens axis. Because all rays parallel to the axis must pass through the focal point, a ray is traced through the left focal point. Obviously this ray must also go to the object arrow point. This ray can now be thought of as a ray originating at the object point, passing through the left focal point, and being refracted to go parallel to the axis. Thus it is one of the rays that form the image point. On occasion, this ray, parallel to the axis on the image side of the lens, is a useful ray to trace.

Virtual Images

In the example just given, the object distance was chosen to be less than infinity and greater than the lens focal length. If the object is placed closer to the lens than the focal

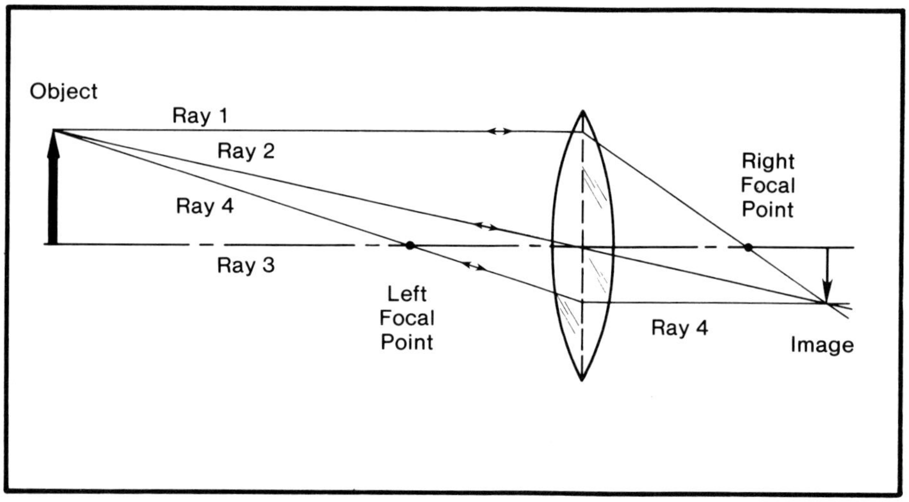

Figure 6-20. Reversibility of Direction

length, and several rays are traced, it will be seen that the light rays diverge instead of converge. The rays can form no real image; they form a *virtual image*, one that can be seen by placing the eye in the diverging beam of light as shown in Figure 6-21. The lens has become a magnifying glass and the position and size of the virtual image can be found by extending the imaging rays backward to the lens until they meet. To the eye, the arrowhead will appear to be located at this point and the bottom of the arrow will appear to be on the lens axis. The magnification can be found by dividing the length of the virtual image by the length of the object.

Using schematic ray tracing, the following relationships can be shown: when the object is close to the lens, the magnification is slight (Figure 6-22a). As the object is moved away from the lens, the magnification increases (Figure 6-22b). Just before the object reaches the focal point, the image becomes blurred. As it is moved farther away from the lens it again becomes sharp, but now it appears upside down. As it is moved still farther away, it becomes smaller. The position of the eye is important. Close to the lens, a sharp image can be seen when the object is between the lens and the focal point. The image reversal occurs only when the eye is several focal lengths away from the lens.

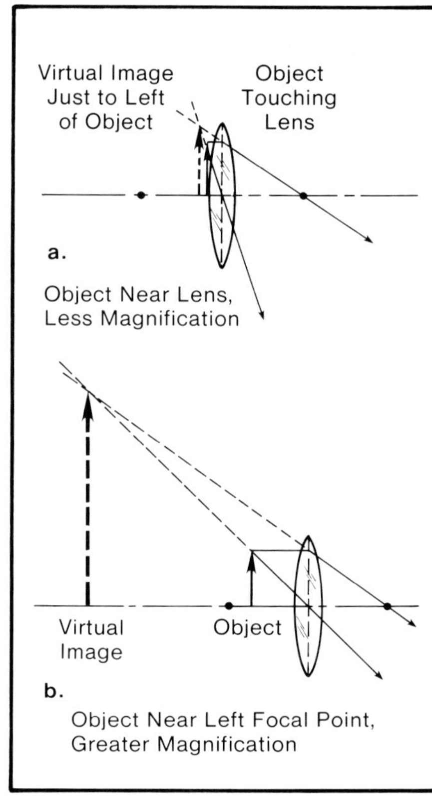

Figure 6-22. Virtual Image Magnification.

Unit Magnification

If the object is placed at an object distance of exactly twice the focal length, and the axial ray, the parallel ray, and the ray from the arrow point through the lens axis are traced, it can be seen that the image size is exactly the same as the object size (Figure 6-23). There is a magnification of one. It will also be seen that the image distance is exactly the same as the object distance, which is twice the focal length. The object-to-image distance is therefore four times the focal length when the magnification is one. This relationship can be used to find the focal length of a lens.

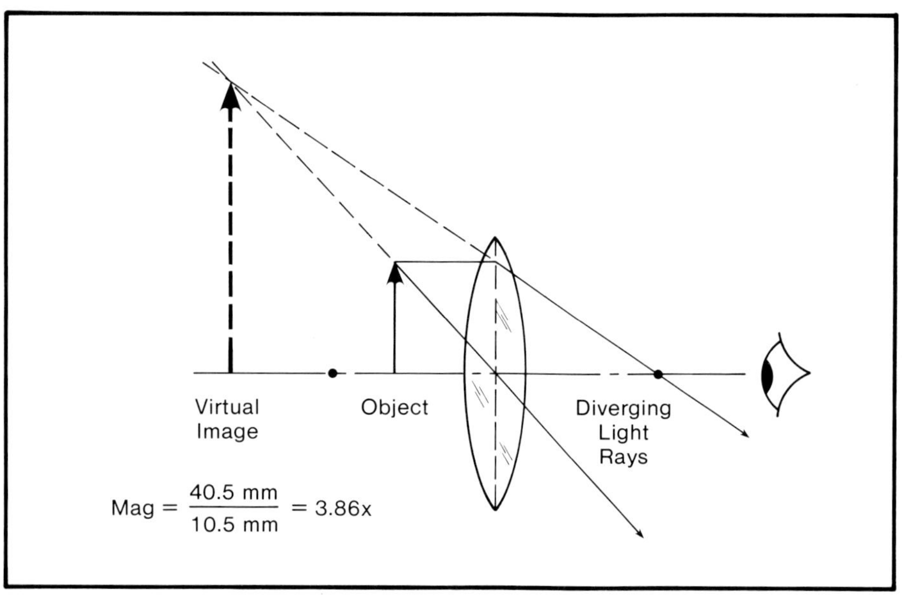

Figure 6-21. Virtual Image

Extending the Equivalent Refracting Plane

In the examples given to this point, the object size has been no larger than one half the lens diameter. The lens can, of course, image objects much larger than this. A useful convention used in schematic ray tracing is to assume that the lens diameter is as large as it needs to be to trace the rays. This is accomplished by extending the equivalent refracting plane as far as needed to trace the ray parallel to the axis (Figure 6-24).

In the diagram, the arrow is shown larger than one-half the lens diameter. The equivalent refracting plane has also been extended to be larger than the object. A ray is drawn parallel to the axis from the arrow point to the extended refracting plane, and refracted so that it goes through the focal point, as it must, because all rays parallel to the axis must go through the focal point. A second ray is drawn through the center of the lens. This ray is not refracted, so the object and image rays form a straight line. The bottom of the image arrow is on the axis. This extension of the equivalent refracting plane is useful when the object size is large.

Problems

In the following two drawings, identify the lens axis, the equivalent refracting plane, and focal point. Measure the focal length of the lens and the object distance and mark them on the drawings. Using one ray parallel to the axis and one ray through the center of the lens, draw the images. By measurement and arithmetic, show that the image-to-object size ratio is the same as the image to object distance ratio. What is special about the second drawing?

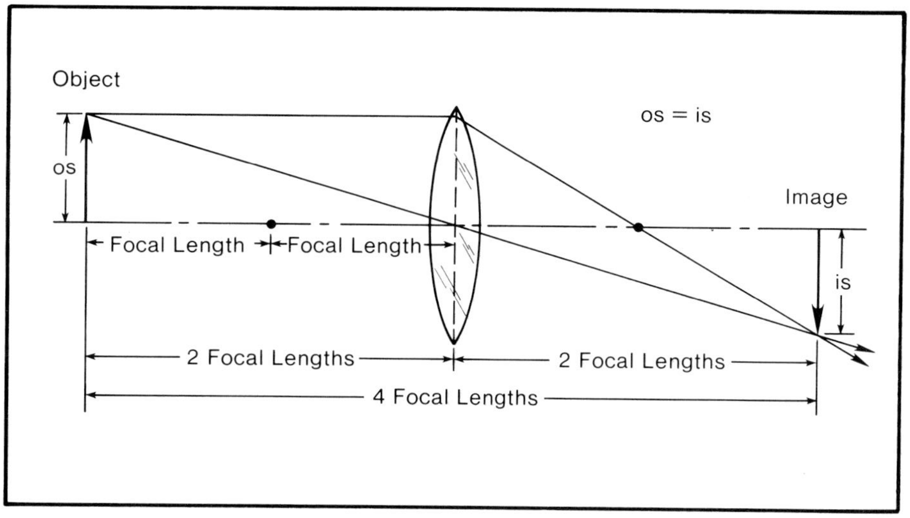

Figure 6-23. Unit Magnification.

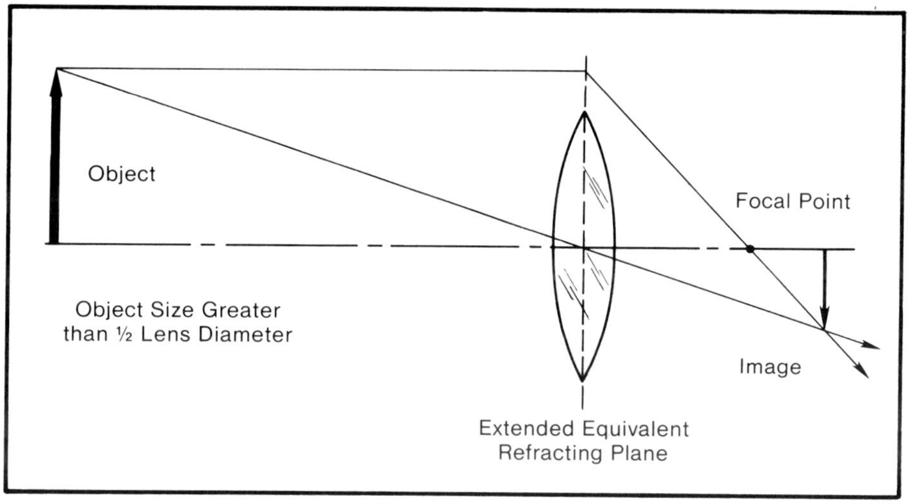

Figure 6-24. Treatment of Large Objects

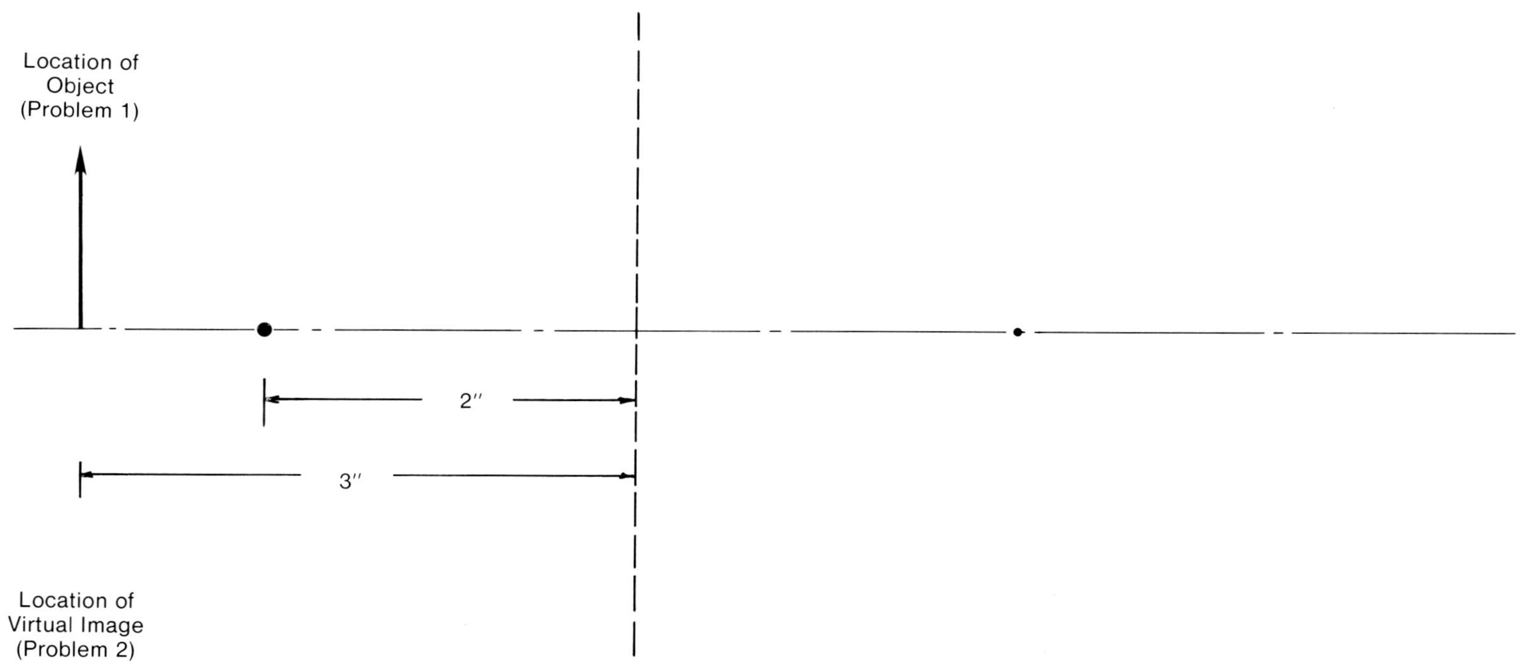

Problem
Chapter 6
Video Lesson 1

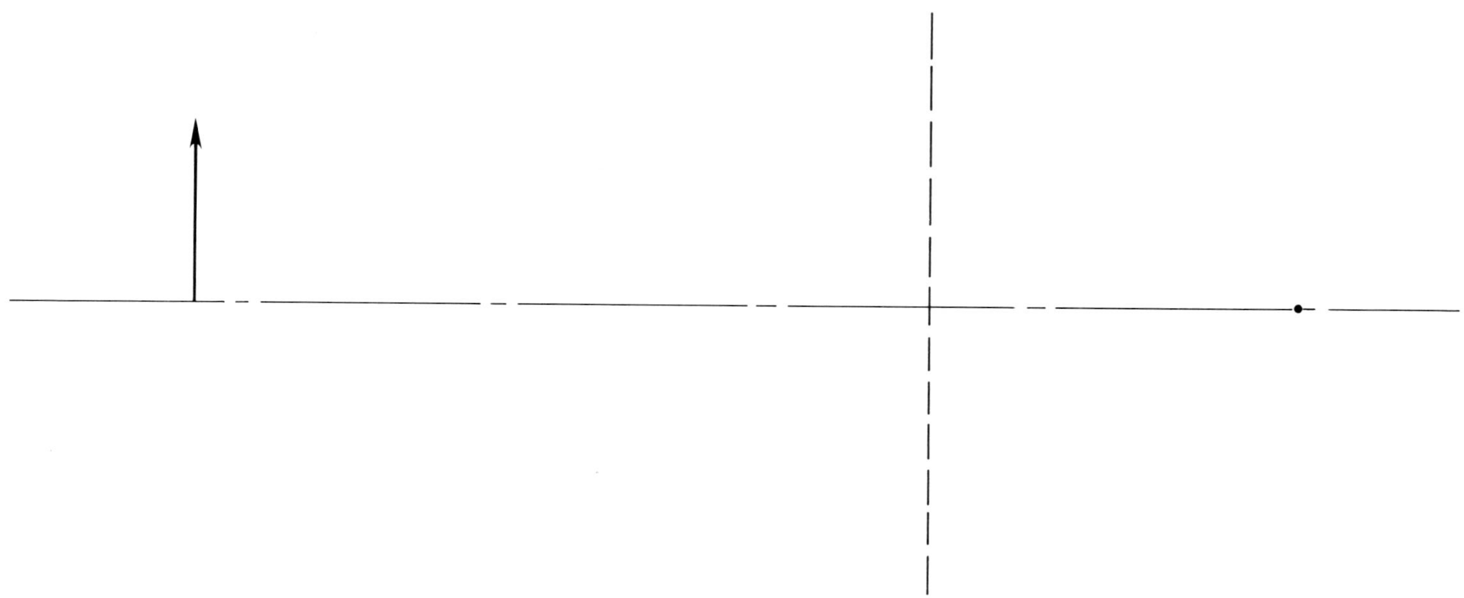

Supplemental Problem
Chapter 6
Video Lesson 1

Chapter 7

VIDEO LESSON 2

Magnifiers and Virtual Images

Summary

Lens forms are reviewed. Magnifying power is related to the natural close-viewing distance of the unaided eye. Young eyes are fairly adaptable, while older eyes lose their flexibility. It is sometimes convenient to use angular size when describing visual objects as images. The amount of detail the eye can see is related to its limits of resolution. Image magnification with a positive lens is shown by schematic ray tracing. The image formed is virtual. Virtual images of negative lenses can also be traced. When an object is at the focal point of a positive lens, the virtual image can be seen with the eye close to and far away from the lens. Close or nearby, the field coverage is large; farther out, the field coverage is small. The apparent size of the virtual image stays constant with distance. Light rays can be checked for parallelism with a camera or binoculars. Magnifying lenses generally produce images with pincushion distortion, but under certain circumstances, the image can be distortion-free. More than one lens in a system can be ray traced. A thin lens refracting telescope is illustrated to show that a telescope eyelens is actually a magnifier.

Lens Forms

Positive and negative lenses can be made in several forms as was discussed in chapter 3. In the schematic ray tracings done so far, the shape of the positive lens used has been considered relatively unimportant. In the study of magnifying lenses, the lens's shape is significant.

Positive lenses are thicker on axis than at their edges (Figure 7-1). They converge light rays and form real images. Biconvex and plano-convex lenses are obviously thicker in the middle than at their edges. A meniscus lens is positive only when the convex curve is stronger; that is, it has a shorter radius of curvature than the concave surface. This difference is not always obvious. Meniscus lenses are not commonly used as magnifying lenses.

In the earlier discussion of positive lenses used as magnifiers, such lenses were shown as biconvex with the same curve strength on both sides. Obviously the image is unchanged when a symmetrically shaped lens is inverted. The magnified image appears the same through either side of the lens.

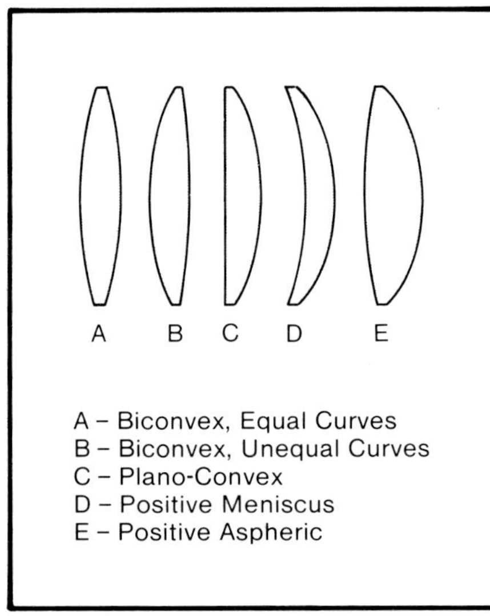

Figure 7-1. Positive Lens Shapes

Magnifying Power

One type of positive lens is used as a jewelers loupe. This loupe is a fairly strong lens mounted in a special frame that the jeweler can grip with the muscles around his eye (in the way an Englishman might wear a monocle). This is shown in Figure 7-2. The lens is positioned at the right distance from the eye for convenient viewing of tiny watch mech-

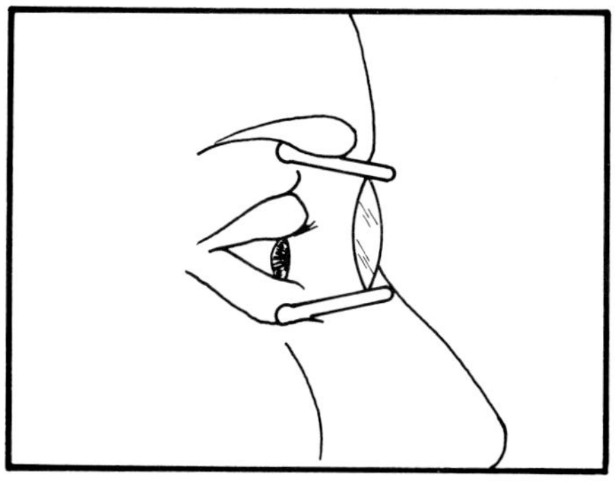

Figure 7-2. Jewelers' Loupe

anisms or jewelry, leaving both hands free to perform the delicate work. Such eye loupes have a magnifying power of about 5X. Another type of jewelers loupe having the same power is attached to a pair of eyeglasses.

A reading glass is larger in diameter and has a handle for holding (Figure 7-3). It may be a rectangular section of lens large enough so that both eyes can be used when reading through it. It might be called a two-eyed magnifier. Such "reading glasses" have less magnification than the eye loupe, but they serve a useful purpose especially for people with poor vision. In what may seem to be a paradox, such reading glasses seem to make things look larger, but actually they have a magnification power of about 1X. How the power is calculated is reviewed in the following section.

Figure 7-3. Binocular Reading Glass

Eye Adaption

Children have very flexible eye lenses. The muscles that change the focus of their eye lenses permit focusing on objects as close as two or three inches. As eyes grow older, some of this flexibility is lost and the closest distance at which they can focus increases. Some older people can see clearly at great distances, but need to read their evening paper at arm's length.

Somewhat arbitrarily, ten inches has been accepted as the normal close-viewing distance. Perhaps this distance was chosen because it is an average close limit distance for many people during their young adult years. The equation for finding the magnifying power (magnification) of reading glasses and loupes, given in an earlier chapter, is based

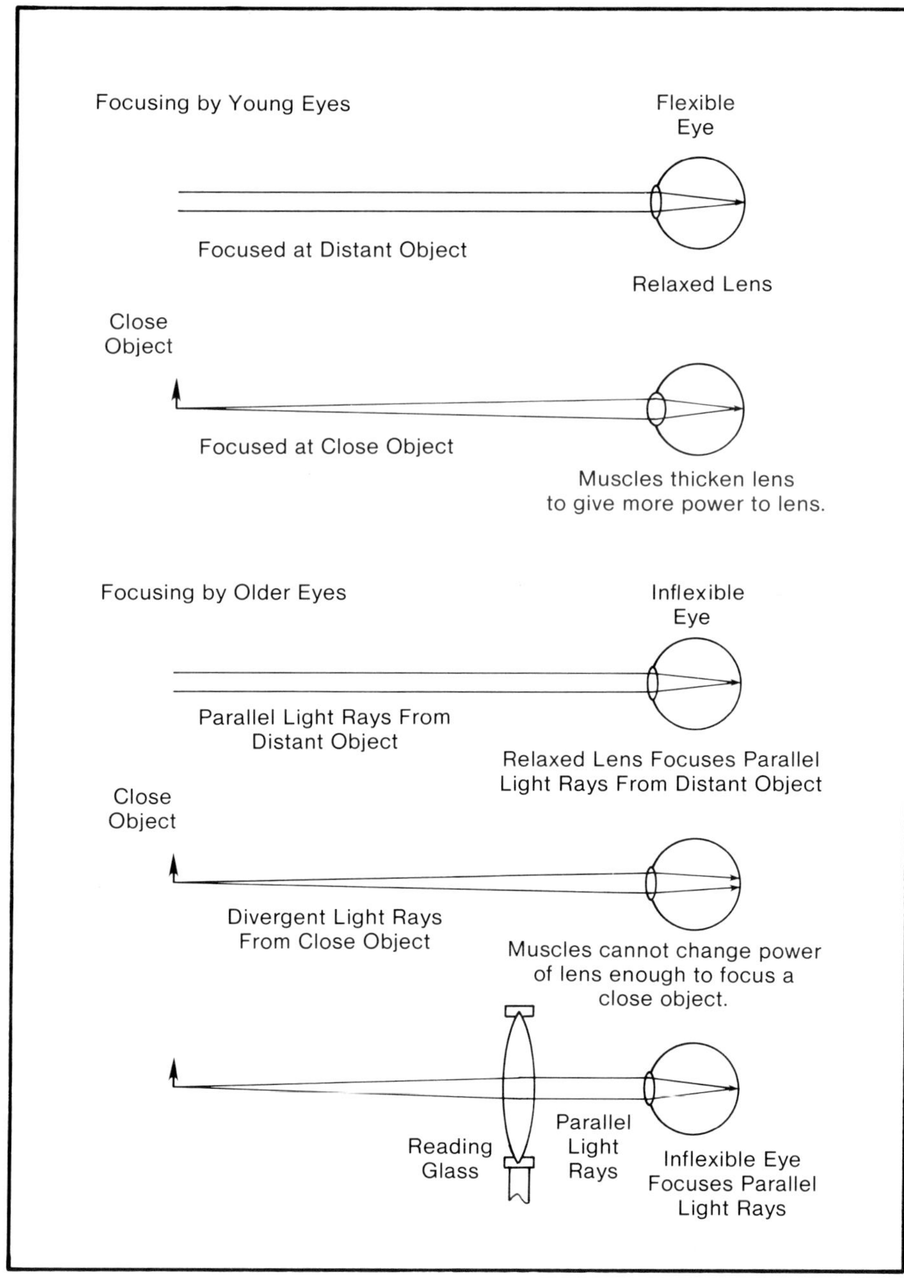

Figure 7-4. Eye Adaptation.

on this accepted ten inch value. Magnification equals 10 inches divided by the focal length in inches. Many reading glasses have focal lengths of about ten inches which means they have a magnification of 1X. Instead of magnifying, these glasses permit easy viewing at the 10 inch close viewing distance (Figure 7-4).

If a schematic ray trace is made of a ten inch reading glass held ten inches above an object, such as a printed page, it can be seen that the light rays leaving the glass are parallel (Figure 7-5). The older eye focuses these rays easily since they are the same as light rays coming from distant objects (objects at optical infinity). This is not only a possible viewing condition, it is a relaxed position for most older eyes. Reading with this glass is comfortable.

Some near sighted eyes require correction with glasses for distant vision. Such eyes, when corrected can use magnifying glasses just as well as eyes that do not require glasses.

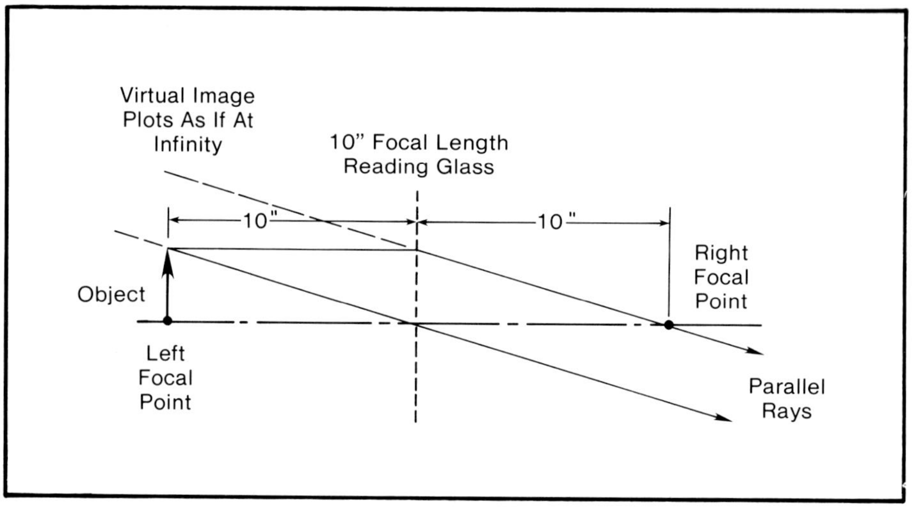

Figure 7-5. Reading Glass trace when object is at the left focal point.

Angular Size and Magnification

The size of objects is normally measured in units such as inches, feet, yards, or centimetres, metres, and so on. However, when dealing with *visual* size, it is often convenient to use angular terms.

A quarter (USA 25-cent coin) is about one inch in diameter. At a convenient close viewing distance of ten inches, a quarter subtends a viewing angle of about 5.7 degrees (Figure 7-6). If the coin is placed five inches from the eye, it subtends an angle of about 11.4 degrees, or about twice the ten-inch angular size. At two inches, where it can be viewed sharply with the jeweler's loupe, it subtends an angle of about 30 degrees, or about five times the angle that it does at ten inches. Thus, it appears 5X as large as it does at 10 inches and is the reason magnifying lenses seem to make objects appear larger.

The eye has a limit of resolving power of about one and a half minutes of arc which, roughly stated, means that the eye can see 40 units of detail in each degree of viewing angle. At ten inches the eye can see about 225 linear-unit details across the quarter. At five inches, 450 linear-unit details, and at two inches about 1200 linear-unit details.

At ten feet, the quarter subtends an angle of about a half degree so that only about 20 linear-units of detail can be seen. The total amount of detail per square inch can be found by squaring the values given above.

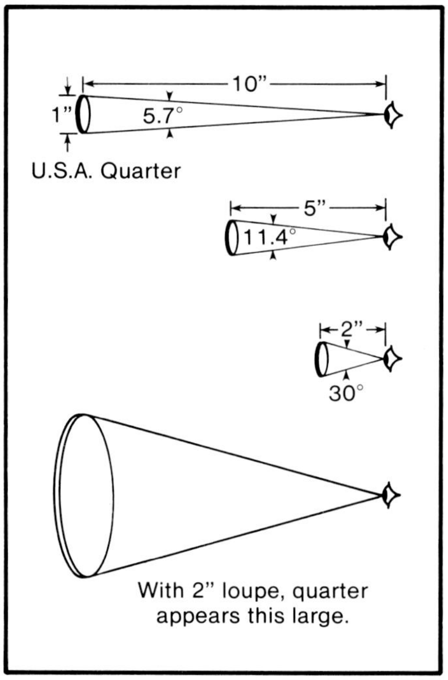

Figure 7-6. Image magnification with a loupe.

Distance	Linear Units	Square Units
2 inches	1,200	1,440,000
5 inches	450	202,500
10 inches	225	50,625
10 feet	20	400

It is easy to see how the jeweler gains by using his loupe. Over a million more details can be seen with the aid of a loupe than with a good unaided eye. The purpose of the magnifying glass is to permit comfortable viewing of close objects (closer than the unaided eye can focus) so that a greater viewing angle is subtended and finer detail can be seen.

Schematic Ray Tracing of Magnification

A trick is required to use ray tracing to show magnification. In Figure 7-7 the object is located at a distance to the left of the lens that is *less* than the focal length. The focal point is placed on the axis to the right of the lens. The usual two rays are drawn, one from the arrow point parallel to the axis and refracted at the equivalent refracting plane through the focal point. The second ray is drawn from the arrow point through the center of the lens. If the object was farther from the lens than the focal length, the two rays would intersect because they would be converging. In this case, however, they are diverging.

To locate the image, the diverging rays are extended to the left where the image is seen. These are shown as dashed construction lines. They are converging and cross at the head of the image of the arrow, shown in dashed lines. The length of the image divided by the length of the object gives the magnification.

The images that are formed by *converging* light rays are *real* images. Real images can be recorded by placing a film where the image is located, or can be seen by placing a ground glass at the location of the image. In most of the schematic ray tracing in this book, the real images are formed to the right of a positive lens. This is a commonly used convention.

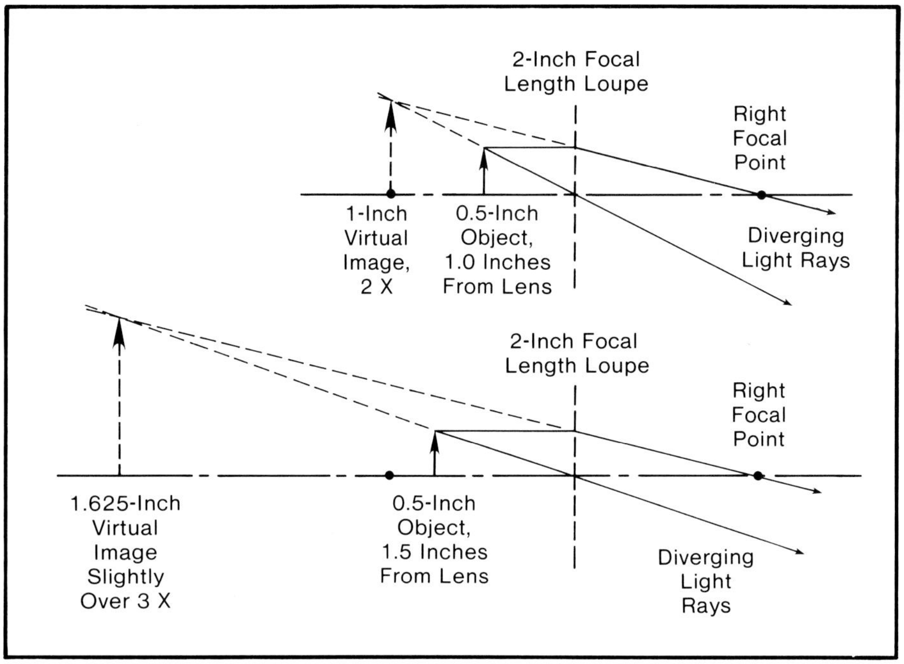

Figure 7-7. Virtual Image Size

When an image is formed by diverging light rays, and located by the use of construction lines, it is called a *virtual* image. It is an image that can be seen, but will not show up on a ground glass placed where it is located. The dashed arrow in the illustration is a virtual image. It is seen by looking through the lens. Another virtual image is the one you see when you look in a mirror.

In the special case where the object is located exactly at the left focal point the two rays are parallel and the virtual image is said to be at infinity. It might seem that the image would be infinitely large but as illustrated, (Figure 7-7, Figure 7-8) the virtual image size approaches a maximum as the object is moved toward the left focal point. When the focal length is 2 inches and the 0.5 inch object is at 1 inch from the lens, the virtual image is about one inch, a magnification of 2X. When the object is at 1.5 inches from the lens, the image is 1.625 inches long, a magnification over 3X. At 1.75 inches from the lens, the image is slightly over two inches long, a magnification of slightly over 4X. The magnification reaches 5X when the object is at the focal point, even though the rays are parallel.

Field Coverage by Magnifying Lenses

When the eye is near a single magnifying lens which is held at a distance of its focal length away from an object, the angular field that can be seen by the eye is relatively large. As the eye is moved away from the lens, the image can still be seen because the rays of image light coming from the lens are parallel (Figure 7-9). However, the farther the eye is moved away from the lens, the smaller the angular field becomes, and less of the magnified image can be seen. This is primarily because the lens aperture presents a smaller angular field to the eye at greater viewing distances. Because the rays

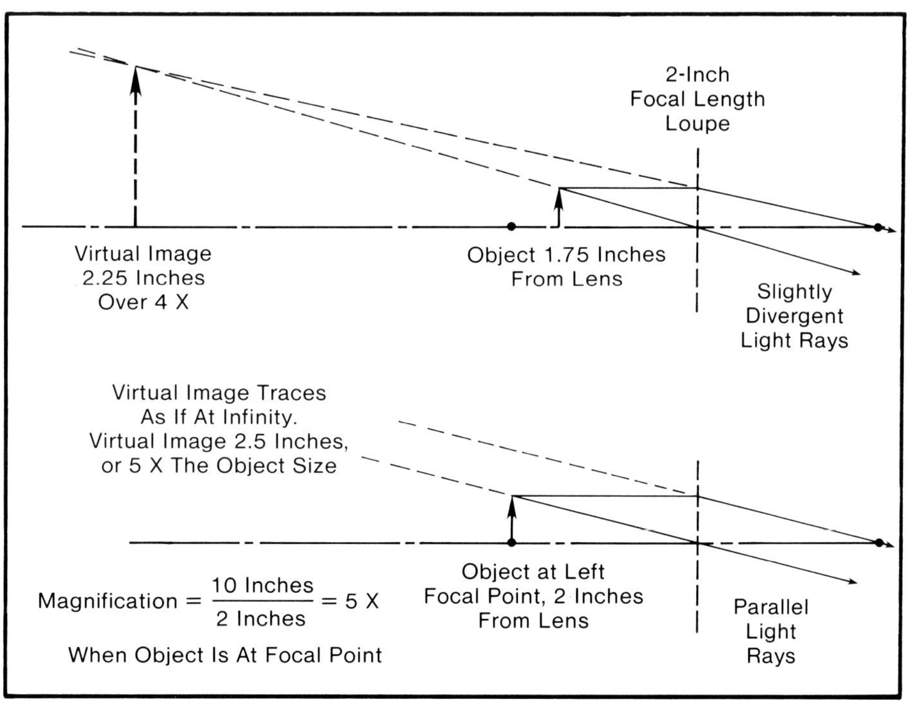

Figure 7-8. Virtual Image Size and Position with a Loupe.

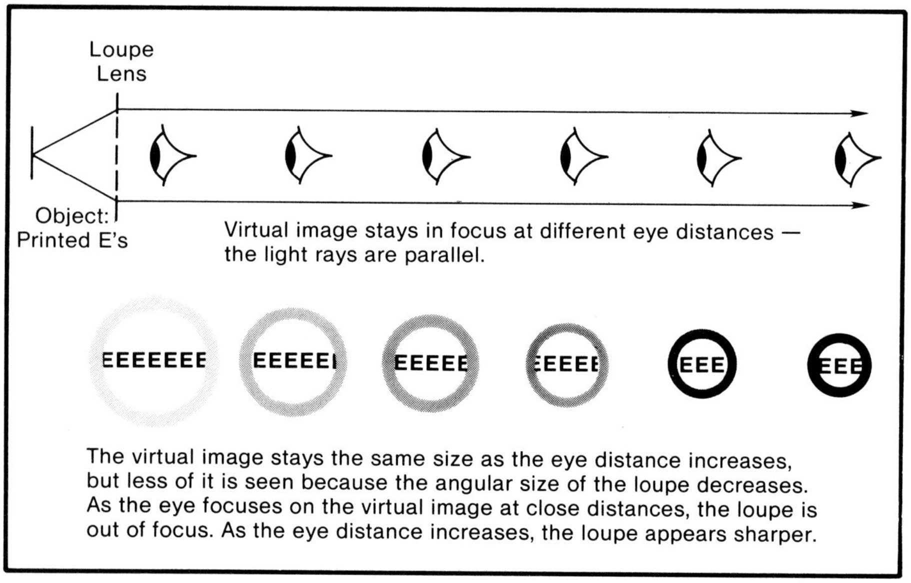

Figure 7-9. How the virtual image appears at different viewing distances.

of image light are parallel, the angular size of the magnified object stays the same. As a result, the magnified image appears relatively large compared to the magnifying glass until, at a relatively great distance, an extremely small portion of the object completely fills the aperture of the magnifying glass.

Checking for Parallel Light Rays

There are a number of readily available optical instruments that can easily be focused at infinity which means they create sharp images with parallel light rays. Binoculars, refracting or terrestrial telescopes or cameras that focus on a ground glass, such as view cameras or single-lens reflex cameras, are all instruments that can be used for this purpose.

When such an instrument is focused on a distant object and locked in this focus position, it can then be placed so that it views through a magnifying lens (Figure 7-10). The magnifier is moved back and forth in relation to the object until it appears sharp as seen through the eyepiece of the viewing instrument. The magnifier is then exactly its focal length distance away from the object. If the magnifying lens is fastened at this location, a test regarding field size can be performed.

The optical instrument is backed away from the magnifying lens along the optical axis of the magnifying lens. At first only a relatively large area of the object will be seen as sharp. The glass and its frame will be badly out of focus because the test instrument is focused at infinity. As the viewing instrument is moved farther away, less of the object field will be seen. The magnifying lens will gradually appear sharper as the magnifier-to-viewing-instrument distance is increased. The image of the object will remain sharp and its angular size will remain constant, proving that the light rays leaving the magnifier are parallel. At a considerable distance only a tiny fragment of the object will be visible. The angular size of the image of the magnifying glass becomes smaller and smaller as the distance is increased, while the magnified image size remains constant. Theoretically, as the viewing instrument's distance becomes infinite, the angular size of the magnifying lens will become zero and none of the magnified image will be seen.

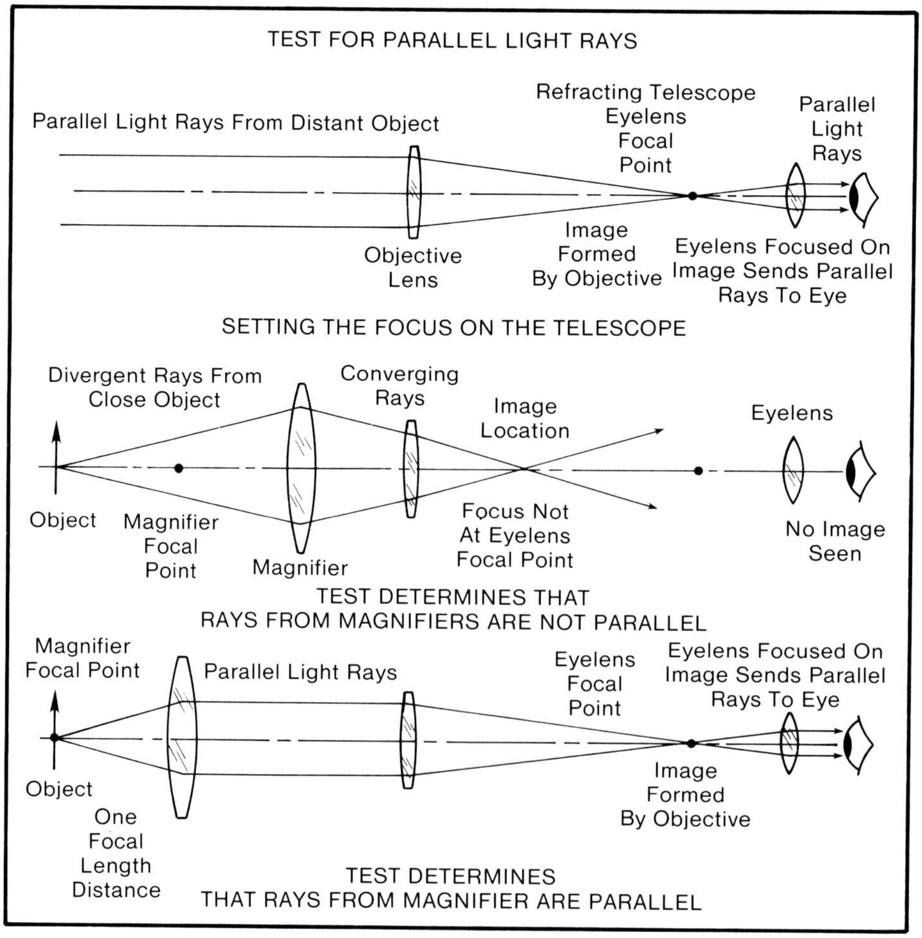

Figure 7-10. Test For Parallel Light Rays

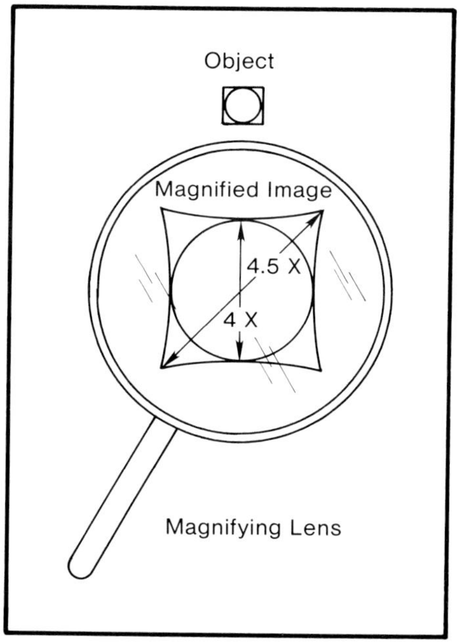

Figure 7-11. Pincushion Distortion

Pincushion Distortion by Magnifying Lenses

In the discussion of magnifying lenses, it has been assumed that the entire field of any lens is magnified the same amount. If it is a 5X magnifier, it is expected that the edges of the field would be magnified the same amount as the center of the field. Unfortunately, this is not the case. Only in special cases do the edges of a lens magnify the same as the center does. Because the edges usually magnify more than the center, a square that nearly fills the field will look like a pincushion. This is called *pincushion distortion* as seen in Figure 7-11. A circle centered in the field looks round because all parts of the circle are equally distant from its center. If a square is drawn enclosing the circle, the centers of its sides touch the circle and therefore are equally distant from the center of the circle. The corners of the square, however, are farther out in the field; they are magnified more, which makes them appear farther out than they really are. The sides of the square curve outward to the corners, each section receiving slightly more magnification than sections nearer the center.

A biconvex magnifying lens with curves of equal strength on both sides has pincushion distortion when the eye is close to the lens and also when the eye is farther away from the lens (Figure 7-12). But the distortion disappears when the eye-to-lens distance is twice the lens-to-object distance (focal length). This particular condition is an example of what is called the aplanatic condition; the same amount of bending takes place at each lens surface. When a plano-convex lens is used with the plano side toward the eye, and the eye is fairly close to the lens, the same condition occurs (Figure 7-13, Figure 7-14). However, when the eye is farther away, the light coming to the eye is hardly bent by the plano surface. Nearly all of the bending occurs at the convex surface. This violates the aplanatic condition so that the image distortion is great. If the convex surface of the plano-convex lens is toward the eye, the aplanatic condition results when the eye is at a great distance from the lens, and the distortion is least under these conditions. When the eye is close to the lens, the condition is not met and distortion occurs.

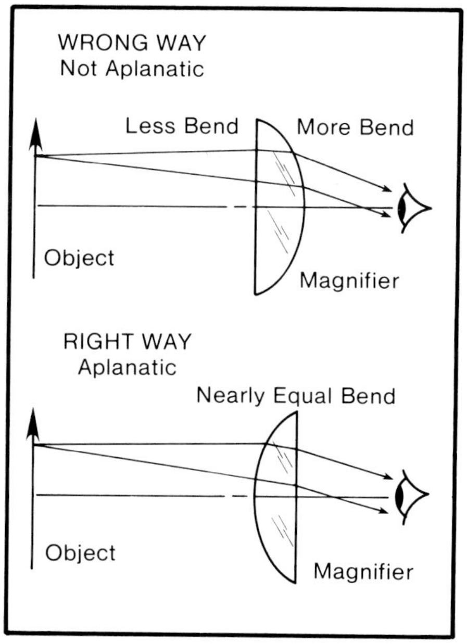

Figure 7-13. Plano-Convex Magnifier at close eye-to-lens distance.

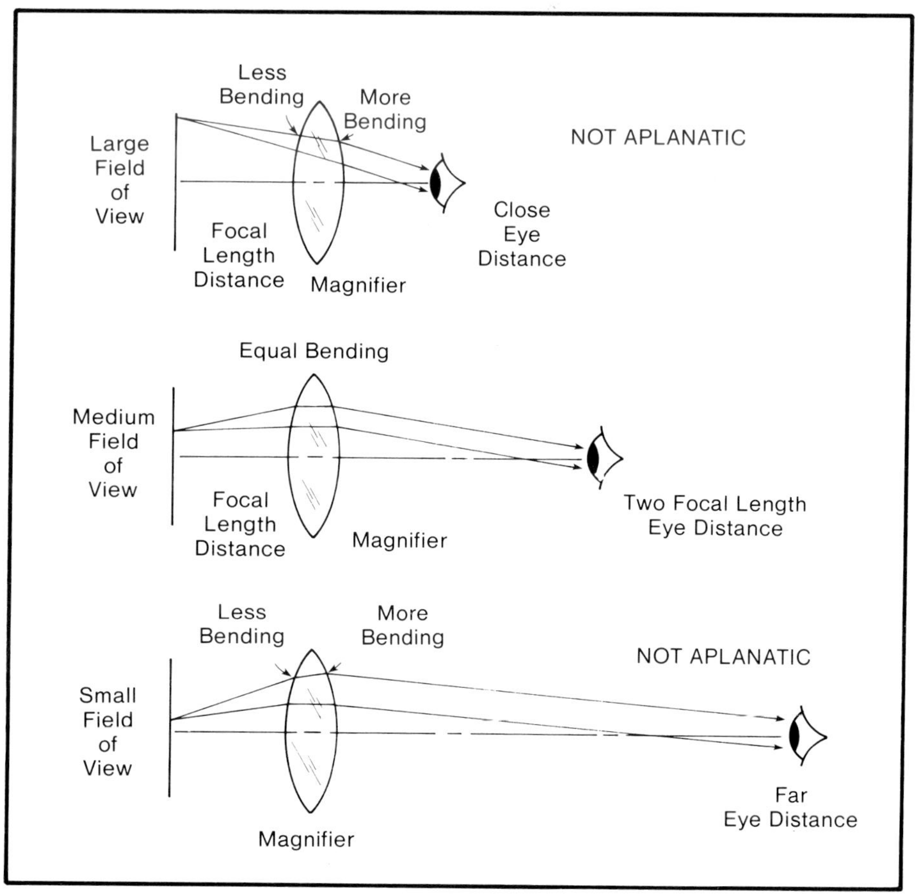

Figure 7-12. Biconvex Magnifying Lens

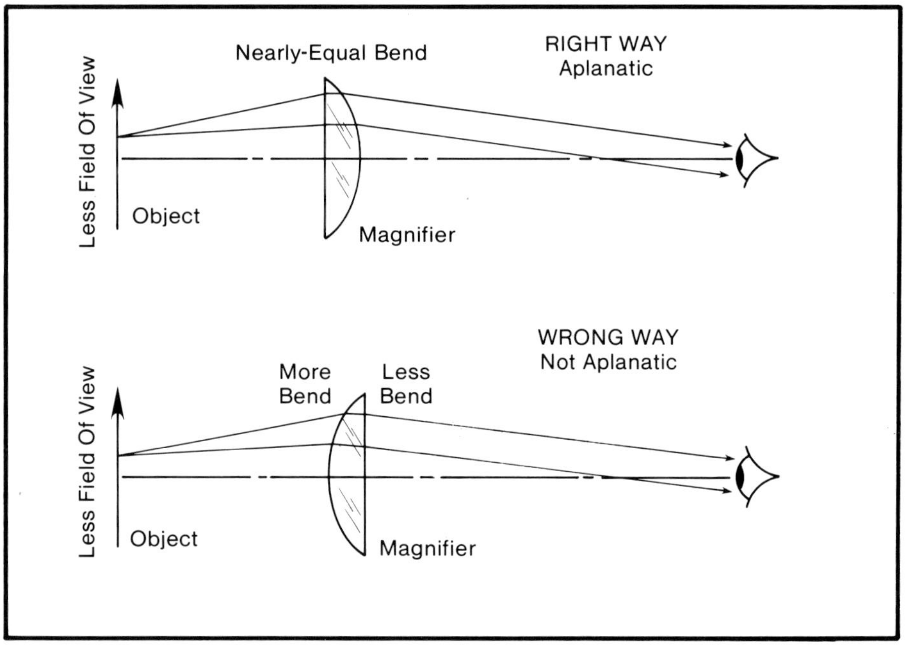

Figure 7-14. Plano-Convex Magnifier at great eye-to-lens distance.

Distortion is considered one of the major aberrations of lenses. Another is chromatic aberration caused by the dispersion of glass. This means that glass has different indices of refraction to different wavelengths of light. The spreading of white light into a spectrum is an example of dispersion.

A wedge is a prism with a relatively small angle (Figure 7-15). In some ways it is like a section of a positive lens. When the angle of the incident ray of light equals the angle of the ray leaving the other side of the prism, the same amount of bending occurs at each surface. The aplanatic condition is met, and the total angle of bending is the least. Also, the amount of spread in the spectrum is least. This indicates that chromatic aberration is least in a magnifying lens when the aplanatic condition is met.

Magnifiers Used as Eyelenses or Oculars

In many instruments such as microscopes, telescopes and the reflex finders on cameras, the eyelenses, often called oculars, are magnifying lenses that magnify the image created by objective lenses.

In a microscope the objective lens has a very short focal length and forms a sharp real image in the microscope tube. The object is close to the objective lens but still beyond the focal length so that a real image is formed. The eyelens is arranged so that it focuses on a real image, giving it magnification. Since the focal length of the eyelens is usually one inch or less, its magnification is usually 10X or greater.

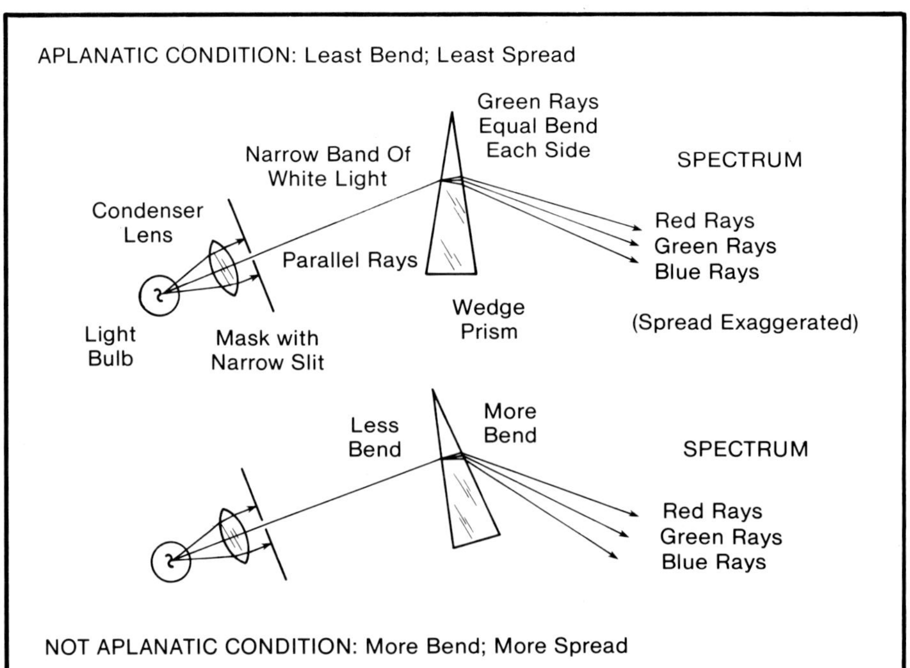

Figure 7-15. Effect of the aplanatic condition on chromatic aberration.

Telescopes are used to view distant objects, often millions of miles away. In order to get good visual separation of stars, a magnified image is required; so the telescope objective lens has a long focal length (Figure 7-16). The objective lens forms a real image, and the eyelens acts as a magnifying lens to increase the magnification.

Schematic Ray Tracing of a Refracting Telescope

Up to now schematic ray tracing has been done with single lenses. However, ray tracing is also very useful in the design of instruments with more than one lens. A telescope consisting of two thin lenses can serve as an example. As shown in Figure 7-17, the objective lens has a focal length of 4 inches and a diameter of 1 inch. The eyelens has a focal length of 1 inch and a diameter large enough to permit passage of all the necessary light rays. In this example, the lens is placed in two positions. The lens is first placed 0.75 inch from the focal point of the objective lens. Second, it is positioned 1 inch from the objective focal point.

The object consists of two stars, five degrees apart. One star is on the

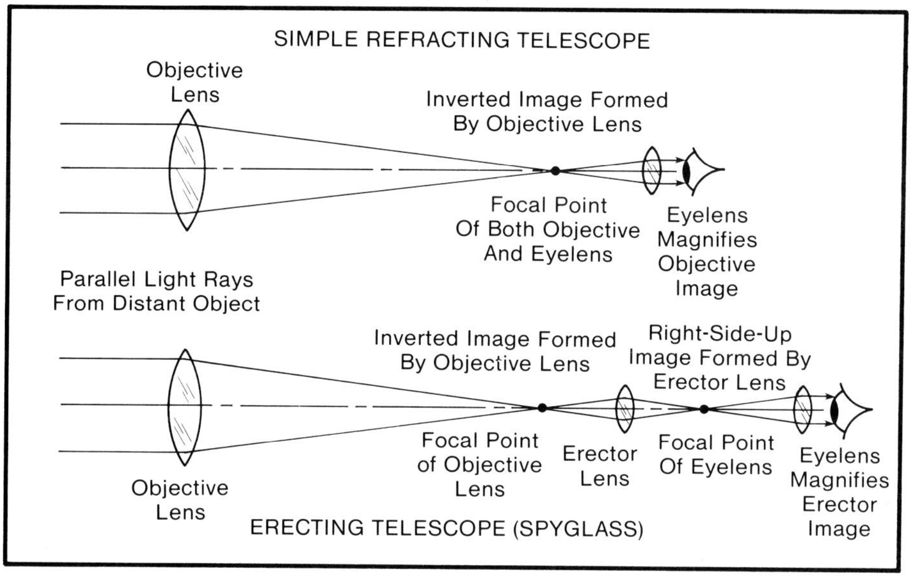

Figure 7-16. Two Simple Telescopes.

telescope axis; the other star is 5 degrees below the first star. The two stars form what could be called the "Arrow" constellation, and the lower star forms the point of the arrowhead. The schematic ray traces are shown in Figure 7-17 in three phases.

In Phase 1, the real image formed by the objective lens is found (both its location and size are required). In Phase 2, the virtual image is located, and magnification calculated. In Phase 3, the eyelens is moved to a position 1 inch from the objective focal point and the magnification is calculated.

Phase 1: Finding the real image formed by the objective lens.

Step 1: The lens axis (axial ray from the first star) is drawn.

Step 2: The objective lens is drawn as an equivalent refracting plane 1 inch in diameter. The focal point is drawn 4 inches to the right of the objective equivalent refracting plane. The eyelens equivalent refracting plane is drawn 0.75 inch to the right of the objective lens focal point.

Step 3: A ray from the first star is drawn parallel to the axis up to the objective lens. This ray is parallel to the axis, therefore it goes through the focal point of the lens. The point at which the ray passes through the focal point is the location of the real image.

Step 4: A ray is drawn from the second star to the objective lens. It is drawn 5 degrees to the axis so that it goes through the center of the objective lens. Rays that go through the center of a lens are not bent, therefore they continue on.

Step 5: These rays determine where the point of the arrowhead is in the real image formed by the objective lens. The location of the image is shown and the image can be drawn.

Phase 2: Finding the virtual image.

Step 1: The axis, lenses and focal points of both lenses are drawn. The eyelens focal point is 1 inch from the equivalent refracting plane.

Step 2: A ray is drawn from the real image arrow point parallel to the axis to the eyelens refracting plane. This ray bends to go through the focal point of the eyelens.

Step 3: A ray is drawn from the real image arrow point through the center of the eyelens. This ray is not

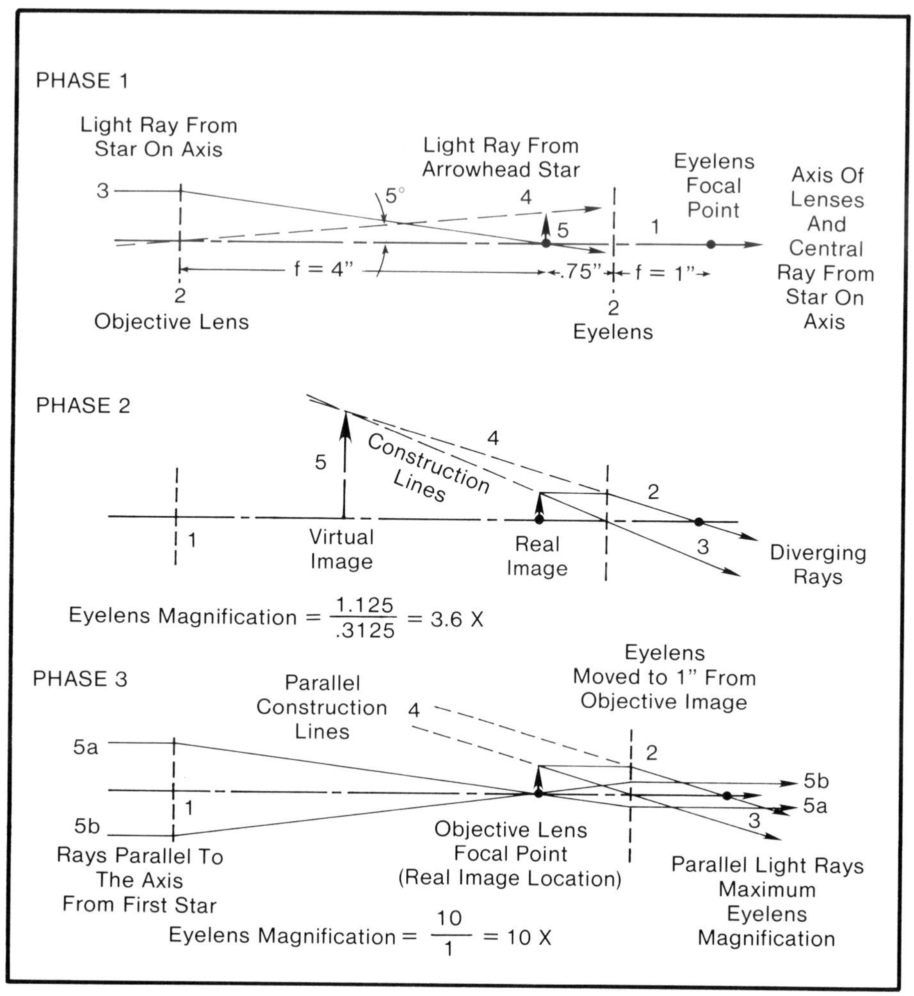

Figure 7-17. Refracting Telescope

bent. Because the rays drawn in steps 2 and 3 are diverging, they do not come together to form a real image. Therefore, they form a virtual image to the left of the eyelens.

Step 4: Construction lines that are continuations of the rays drawn in steps 2 and 3 are drawn to the left of the lens. The virtual image is located where these construction lines intersect.

Step 5: The virtual image is drawn and measured. It is about 1.125 inches long. The real image is 0.3125 of an inch long. The magnification of the eyelens in this position is 1.125″ ÷ 0.3125″ = 3.6X.

Phase 3: The eyelens is moved to a new position 1 inch from the real image. The position of the virtual image is found and the new magnification is calculated.

Step 1: The axis and the lens position are drawn. The left focal point of the eyelens and the focal point of the objective lens coincide.

Step 2: Rays are drawn as in steps 2 and 3 of Phase 2. It is found that these rays are parallel.

Step 3: The construction lines that are continuations of these rays are drawn (to the left). These are also parallel.

Step 4: It must be concluded that the virtual image is located at infinity because the construction lines are parallel. When this occurs, the magnification of the eyelens is calculated by the formula:

$$\text{Magnification} = \frac{\text{10 inches}}{\text{focal length of eyelens in inches}}$$

In this case 10 ÷ 1 = 10X, the magnification of the eyelens. Because the real image is 0.3125 of an inch, the virtual image will appear to be 10 x 0.3125 of an inch or 3.125 inches.

Step 5: Rays from the first star are drawn, resulting in rays 5a and 5b exiting the eyelens parallel to the axis. These rays show that the first star will be seen on the axis, while the second star will be seen by rays 2 and 3 at an angle off the axis.

Problem

The drawing shows a simple thin lens microscope with an objective lens that has a focal length of 1 inch and an eyelens with a focal length of 2 inches. The object is 0.5 of an inch long and the object distance is 1.5 inches. The distance between the two lenses is 4 inches. Trace the rays and find the location and size of the virtual image. Determine the magnification of the microscope in this position. To obtain maximum magnification, the eyelens must be moved. Indicate the eyelens location for maximum magnification. Plot the rays to show this.

Microscope

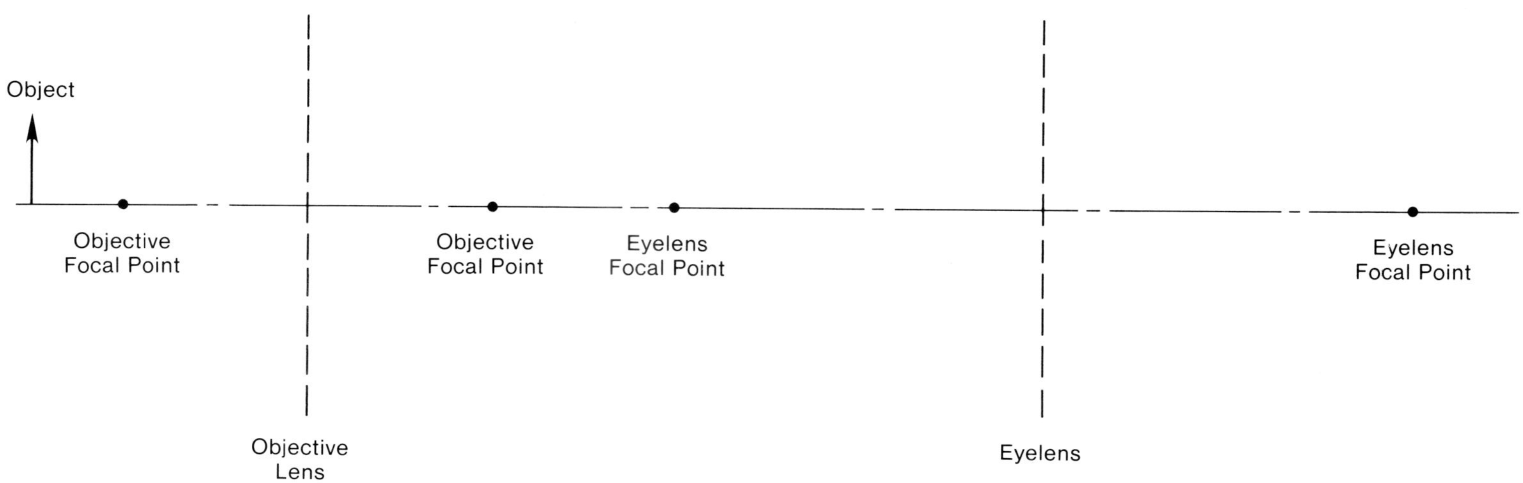
Object
Objective
Focal Point
Objective
Lens
Objective
Focal Point
Eyelens
Focal Point
Eyelens
Eyelens
Focal Point

Chapter 8

VIDEO LESSON 3

Diopters, Thick Lenses, and Forming Images with Negative Lenses

Summary

Lens power is often given in diopters which is advantageous because diopter powers can be added. Diopter power is commonly used when describing the strength of eyeglasses and binocular eyepieces. A long working-distance loupe is developed by using a magnifying glass and a monocular (half of a binocular). The principles of ray tracing are extended to include thick and compound lenses, and principle planes are introduced. Some uses of negative lenses and some methods of finding the focal lengths of negative lenses are given. Schematic tracing of virtual images formed by negative lenses is shown.

Lens Power

The focal length is a useful measure of the power of a lens. A short focal length results in a more powerful lens and a greater magnifying power if it is a positive lens.

Another measure of lens power frequently used is the diopter. The diopter power of a lens is the reciprocal of its focal length in metres:

$$\text{diopter} = \frac{1 \text{ metre}}{\text{focal length in metres}}.$$

It is usually easier to compute in millimetres:

$$\frac{1{,}000 \text{ mm}}{\text{f.l. in mm}}$$

or in inches: $$\frac{39.4 \text{ inches}}{\text{f.l. in inches}}.$$

A one-diopter lens has a focal length of one metre, while a two-diopter power lens has a focal length of 0.5 metre. A three-diopter power lens has a focal length of .333 metre, and so on. The reciprocal of the diopter power equals the focal length in metres. A 1Δ lens is a 1-diopter lens.

Diopter measure is used with both positive and negative lenses, therefore, + and − signs are used to distinguish them. A $+2\Delta$ lens is a positive lens of 2-diopter power which has a focal length of 500 millimetres (50% of 1,000 millimetres), or about 20 inches. A -3Δ lens is a negative lens with a -3 diopter power that has a focal length of 333 millimetres, or about 13 inches.

A great convenience of diopter power is that it can easily be used to find the combined focal length of two or three lenses placed close to each other. If a $+1\Delta$ lens and a $+2\Delta$ lens are used together, the combination has a power of $+3\Delta$. A $+2\Delta$ lens and a $+3\Delta$ lens gives a combination lens of $+5\Delta$. A $+3\Delta$ lens used with a -2Δ lens gives a $+1\Delta$ lens combination.

Common Uses of Diopter Power

The closest distance that many cameras will focus is about 3 feet. Positive supplementary lenses, sometimes called Portra* lenses or close-up lenses, can be placed in front of the camera lens to permit focusing at closer distances. The camera lens is set at the infinity setting, and a +2 supplementary lens (+2Δ) is placed in front of the camera lens as shown in Figure 8-1. An object placed 0.5 metre (20 inches) in front of the lens will be imaged sharply in the photograph. The drawing shows how this is possible. Light rays from an object point are parallel after they are refracted by the closeup lens. The camera lens focuses these parallel light rays in the same way it would if they came from a great distance (infinity).

When a camera lens is focused on infinity, it is set in such a way that parallel light rays will focus sharply on the film. When placed in the parallel light coming from the supplementary lens, it focuses that light on the image plane (film). The following table shows the focal lengths of various positive supplementary lenses in centimetres and inches. The focal length of the close-up lens equals the lens-to-object distance when the camera lens is focused at infinity.

*Trademark of Eastman Kodak Company

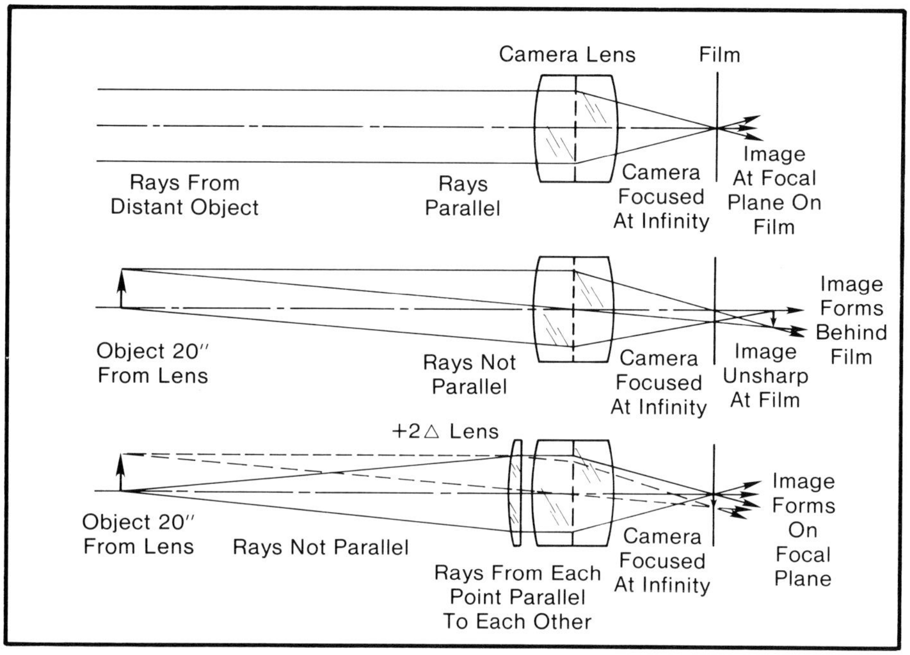

Figure 8-1. Plus 2 Diopter Lens used on camera focused at infinity to focus on object 20 inches from lens.

Lens	Focal Length Centimetres	Inches
+ 1Δ	100	39.4
+ 2Δ	50	19.7
+ 3Δ	33	13.1
+ 4Δ	25	9.8
+ 5Δ	20	7.9
+ 8Δ	12.5	4.9
+10Δ	10	3.9

Photographers often use supplementary lenses to "change the focal length" of their camera lenses. Positive supplementary lenses "shorten the focal length" (the combined focal length is less than that of the camera lens alone).

An approximate focal length of the combination supplementary + camera lens can be found using diopter addition. If a camera lens has a focal length of 8 inches and it is used with a +2Δ lens (Figure 8-2), the combined focal length becomes 5.7 inches.

The diopter power of the camera lens is found first.

$$\frac{39.4 \text{ inches (1 metre)}}{8 \text{ inches}} = 4.9\Delta$$

Then the diopter powers are added.

$$4.9\Delta + 2\Delta = 6.9\Delta.$$

The result is then converted to inches:

$$\frac{39.4 \text{ inches}}{6.9\Delta} = 5.7 \text{ inches}$$

In practice the two lenses cannot be placed very close to each other since their mounts interfere with each other. If the distance between the lenses is included in the calculations, the following equation is more accurate than using the diopter addition. This equation uses the focal lengths instead of the diopter power:

$$\text{f combination} = \frac{f_1 \times f_2}{f_1 + f_2 - d}$$

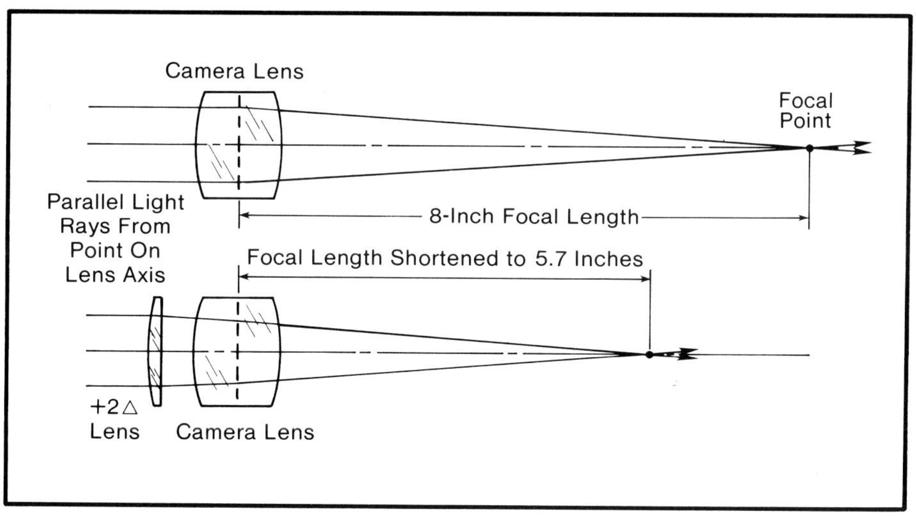

Figure 8-2. Plus 2 Diopter lens used to shorten focal length of 8-inch camera lens.

Assume that d, the distance between the lenses, is 0.5 inch.

$$f \text{ combination} = \frac{19.7'' \times 8''}{19.7'' + 8'' - 0.5''}$$

$$= \frac{157}{27.5} = 5.8 \text{ inches.}$$

This equation shows that using the diopter method was quite close. There is only about 0.1 inch difference between the two methods in this example.

If a -3Δ lens is used to lengthen the effective focal length of a 6-inch camera lens, with 0.5 inch spacing between the supplementary and camera lens, the diopter equation becomes:

$5.57\Delta - 3\Delta = 3.57\Delta = 11.0$ inches

The focal length equation becomes:

$$f \text{ combination} = \frac{6 \times (-13.1)}{6 - 13.1 - 0.5}$$

$$= \frac{-78.6}{-7.6} = +10.3 \text{ inches.}$$

In this example, there is a greater difference in the results obtained by the methods.

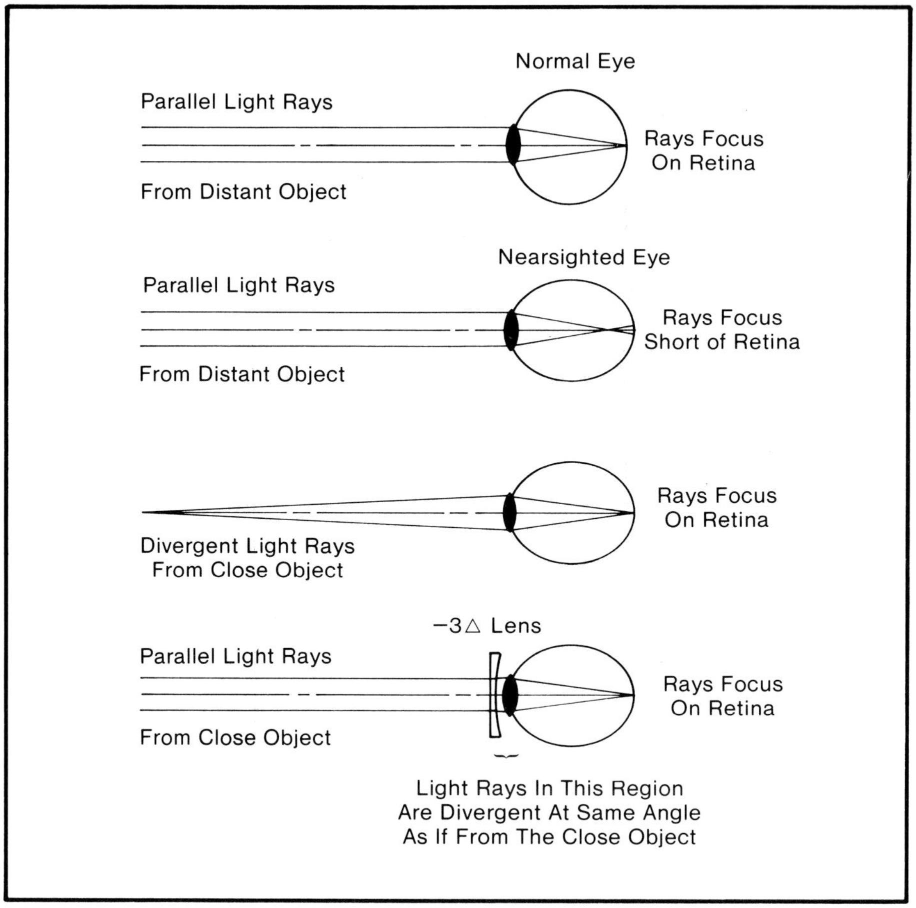

Figure 8-3. Using – Diopter Lens to correct nearsighted eye.

Eyeglasses

The power of eyeglass lenses is measured in diopters. Myopic (nearsighted) eyes focus easily on close objects but not on distant objects (Figure 8-3). Negative diopter lenses are required to permit clear distant viewing. Hyperopic (farsighted) eyes, on the other hand, require positive lenses to permit clear close-up viewing.

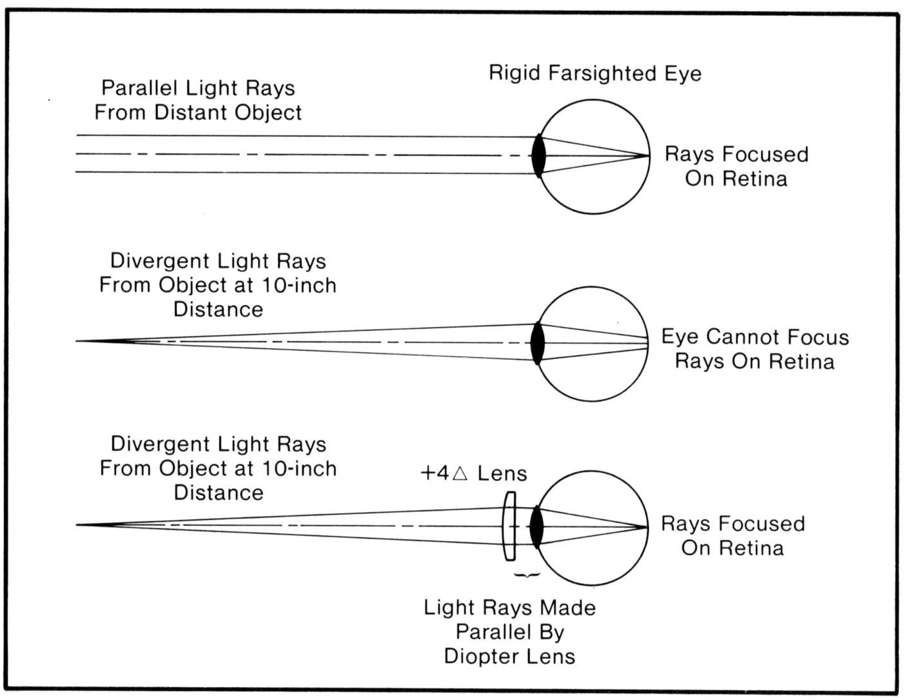

Figure 8-4. Using a +4 Diopter Lens as a Reading Glass.

For example, an older person who has lost the flexibility in his eyes, can see well at infinity, but needs reading glasses that permit him to read at a 10-inch distance. He needs positive reading glasses with a focal length of 10 inches or +4 diopter lenses (Figure 8-4). At a distance of 10 inches from the object, the lenses will refract the rays so that they come to the eyes in parallel paths. The eyes which can see well at infinity, focus the parallel light as though it were coming from infinity and the page will look sharp to the farsighted person. If he decides to get trifocal glasses, the upper portion would be without power (0Δ), the lower portion would have a strength of a $+4\Delta$, while the central portion would have an in-between power of about $+2\Delta$ to $+3\Delta$ to provide clear vision at intermediate distances.

Correction for nearsighted lenses is discussed later in this chapter.

Diopter Markings on Binoculars

Looking through binoculars requires the eyes to be quite close to the eyepieces in order to see the entire field covered by the binoculars. This usually prevents a viewer from wearing eyeglasses.

Many binoculars have center focusing (Figure 8-5). The center focus controls both binocular eyelenses. They move back or forth through the images formed by the objective lenses until the images appear sharp to the viewer. If a person had equal vision in both eyes, the central focusing mechanism would suffice, but one's left and right eyes usually differ and require different diopter correction. The binocular center-focus arrangement cannot adjust for this.

To compensate, most binoculars have an additional adjustment on one of the eyelenses which permits changing the focus for one eye. The eyelens mount is threaded and the focus change is accomplished by rotating the eyepiece. The outside of the focusing eyepiece is calibrated in diopters.

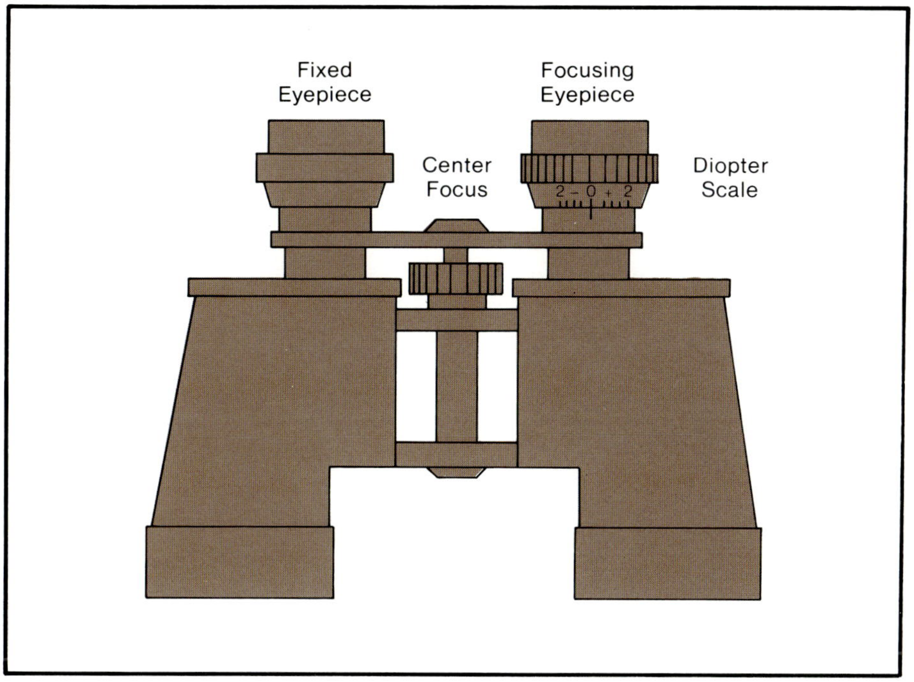

Figure 8-5. Binocular with Center Focusing and Correcting Eyepiece.

When the eyepiece focus is set at zero, both halves of the binocular have the same focus setting. Rotating the adjustable eyepiece one way results in a positive diopter correction; rotating it the other way results in a negative diopter change. The amount of change is indicated on a scale.

It is not necessary that the individual know the difference in vision between his/her eyes to use the adjustable eyepiece. Looking through the fixed eyepiece, a person can focus sharply on a detailed object at some distance using the center adjustment. The viewer can then rotate the focusing eyepiece until the image is sharp. Once this is done both images will appear sharp, using the center adjustment on objects at various distances.

The diopter setting on the focusing eyepiece is an important indicator. Set at zero, there is no difference between the diopter powers of one's eyes. If it is not at zero, the individual needs to remember the setting so as to avoid repeating the calibration procedure. Even those people who do not wear glasses may find that there is a slight difference of vision between their eyes. These people will be more comfortable using binoculars if they adjust the focusing eyepiece properly.

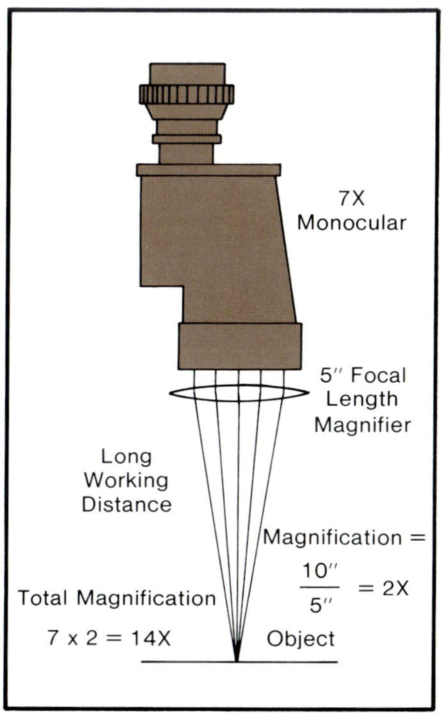

Figure 8-6. Long-Distance Working Magnifier.

A Long Working-Distance Loupe

Earlier it was shown that short focal length lenses have greater magnifying power. To obtain very great magnifications, it is necessary to use very short focal length lenses. In order to use such magnifiers (or loupes) it is necessary to hold the lens very close to the object to be magnified at a distance equal to its focal length.

This is quite satisfactory if the object being viewed has no potential danger to it. Sometimes it is necessary to have great magnification of moving cutters or of very hot objects. Other times it is necessary to magnify an object inside some device where it is impossible to get the magnifier, let alone the observer's eye and head. For these purposes, the long working-distance loupe is needed.

This device can be made with a magnifying lens and half of a binocular. The magnifying lens is placed in front of the binocular objective lens (Figure 8-6). The binocular half (monocular) is focused at infinity. When the magnifying lens is placed at an object distance equal to its focal length, the light is refracted so that the rays leaving the magnifier are parallel. These parallel light rays are focused by the monocular as though they had come from infinity.

The magnification of the combination of the magnifier and monocular is the product of magnifying the powers of the loupe lens and the monocular. If a 5-inch lens is used, its magnifying power is 10 inches (normal close-viewing distance) divided by 5 inches or 2X. It gives a 5-inch working distance from the lens to the object where it produces parallel emerging rays. If the monocular has a power of 7X, the long working-distance loupe has a total magnification of 2 X 7 or 14X, while maintaining the 5-inch working distance. Compare this to a simple loupe with a 14X magnification (focal length of 10 divided by 14 or .71 inches) and about .75 of an inch working distance. The actual working distance would be even less than this because in order to have this much power the monocle would have to be a very thick lens and its equivalent refracting plane would be located somewhere near the center of the lens.

Schematic Ray Tracing with Thick Lenses

In the schematic ray tracing discussed previously, lenses were assumed to have a "zero" thickness. A thin lens was represented as an equivalent refracting plane. All the refracting or bending of the light rays was shown to occur at this plane.

Since there is no such reality as a lens with zero thickness, the "thin lens" is a useful concept:

- It serves as a useful introduction to the reality of thick lens ray tracing.
- It is useful in the first steps of the design of advanced optical systems and is frequently used for this

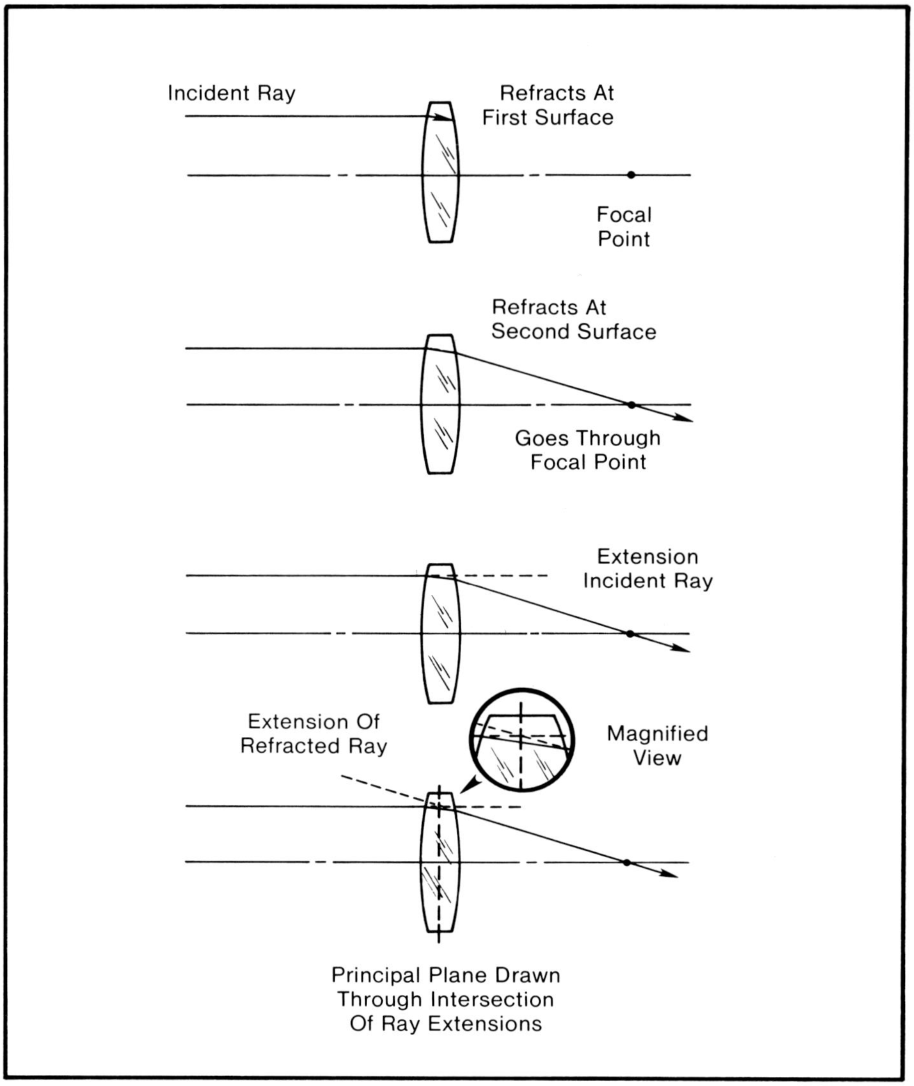

Figure 8-7. Locating the Principal Plane of a Thick Lens.

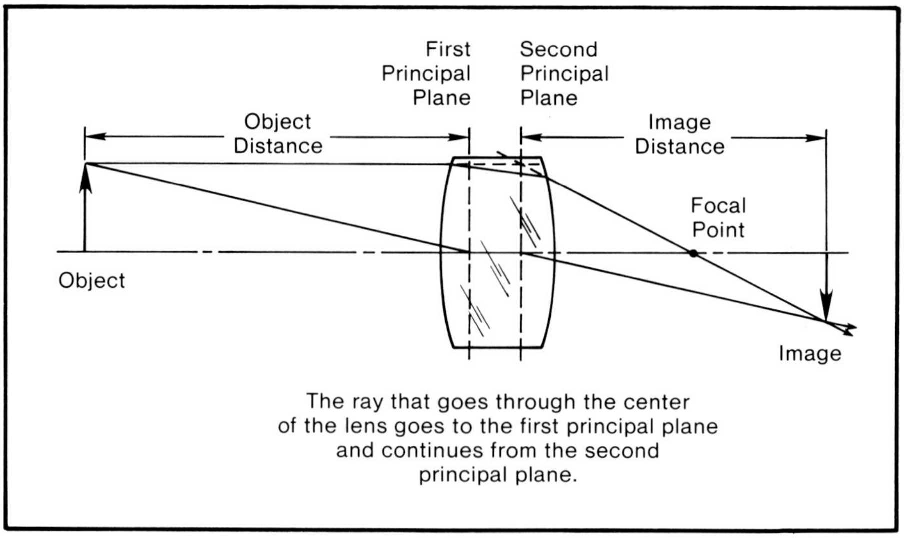

Figure 8-8. Simple schematic ray trace with a Thick Lens.

purpose by optical engineers.

- Most thick lens ray tracing concepts are but further elaborations of thin lens ray tracing principles.
- Some lenses, such as refracting-telescope objectives, so closely approach the thin lens in reality, that thin lens tracing is sufficient for the design of systems using such lenses.

However, real lenses are different from thin lenses. Figure 8-7 shows a thick lens with its optical axis. A light ray parallel to the axis is shown incident to the front surface of the lens. As the light ray strikes the surface of the lens, it is refracted or bent slightly downward as shown. The amount of bending is determined by the angle of incidence and the index of refraction of the lens. The method of finding the exact amount of bending will be discussed in a later chapter.

When the ray passes through the rear surface of the lens, it is again refracted toward the optical axis and crosses the optical axis at the focal point. Now construction lines are drawn, extending the entering ray to the right, parallel to the axis (as though undeviated by the lens) and extending the exiting ray to the left, undeviated, until the two extensions intersect. A line drawn through the intersection perpendicular to the optical axis locates the *equivalent refracting plane.* In a thick lens, this plane is called the *second principal plane.* The focal length is measured from the focal point to the intersection of the second principal plane and the axis.

If a light ray is drawn from the right side of the lens parallel to the axis, it is refracted in much the same way (exactly the same if the radii of curvature are the same) so that it crosses the axis at the left of the lens. When construction lines are drawn as before, they are found to intersect a small distance to the left of the second principal plane. This intersection determines the location of the *first principal plane.* There are also two principal planes within a thin lens, but they are co-planar meaning they occupy the same plane.

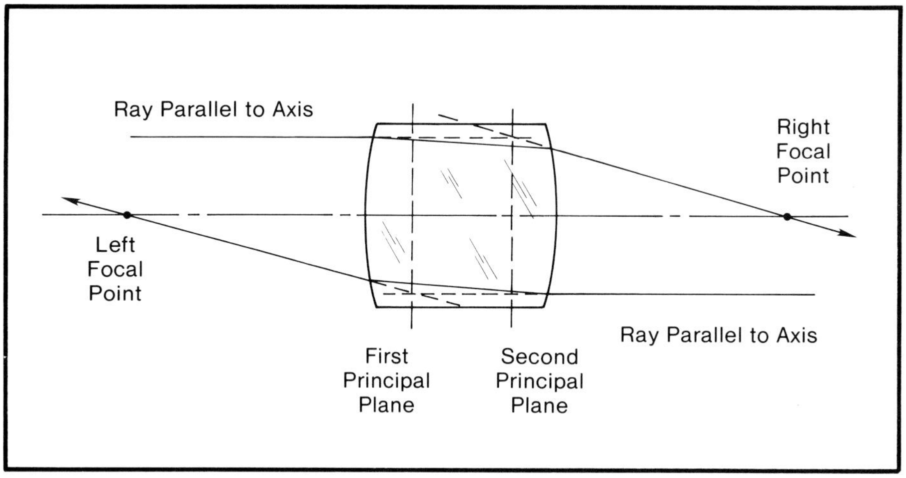

Figure 8-9. The two Principal Planes of a Thick Lens.

In real lenses, which have thickness, the two principal planes are important. The object distance is measured to the first principal plane at the left, while the image distance is measured to the second principal plane at the right. Figure 8-8 shows how a ray from the object traveling parallel to the axis is refracted at the second principal plane to pass through the focal point. A second ray directed to the center of the lens travels along the axis between the principal planes to exit at the same angle as the entrance angle.

In Figure 8-9, a single parallel ray from each direction is shown. In Figure 8-10 several rays are shown. Note how each ray is drawn as being refracted at the appropriate principal plane.

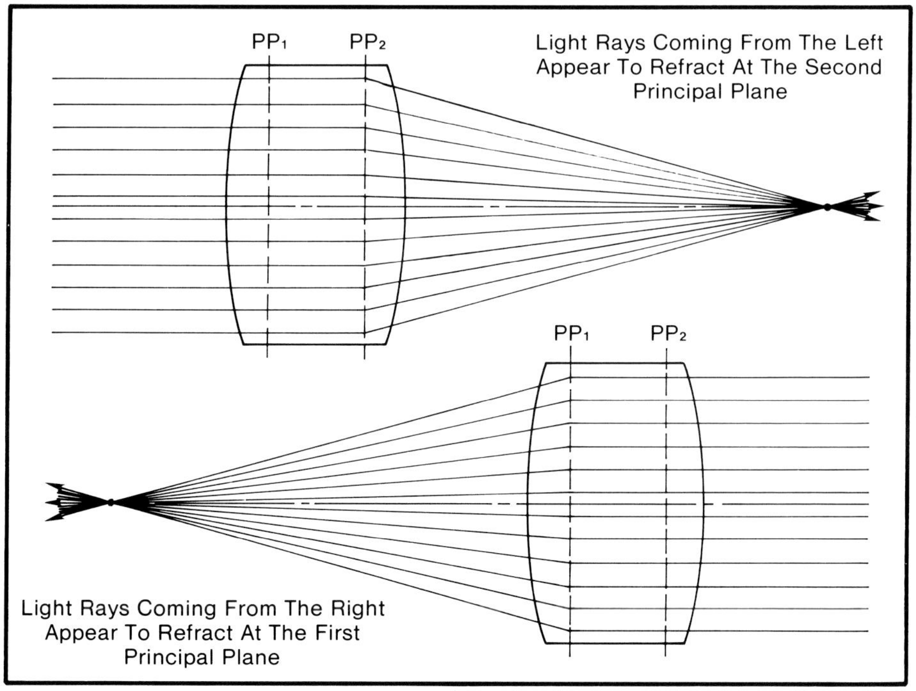

Figure 8-10. The effect of light ray direction on where refraction appears to take place.

Ray Tracing with Compound Lenses

In practice few single element thick lenses are used. Most real lenses are compound (multi-element lenses). A compound lens (like a thick lens) has two principal planes (Figure 8-11). The illustration shows a triplet (3 element) camera lens. The axis is drawn and a ray parallel to the axis is traced showing its refraction at each lens surface. The construction lines are added, and the location of the principal plane is found. The distance from the focal point to the principal plane is called the effective focal length and is often abbreviated "E.F.L." The effective focal length is the focal length used in optical equations such as those used to find image and object distances, magnification ratio and image size. When optical designers complete the design of a lens, they draw a formula sheet (Figure 8-12). The lenses are drawn to scale and many specific items of information are listed, enough so that a manufacturer can produce the lens.

Important values given on the formula sheet are the thickness of each element, its diameter and the glass from which it is made. Many times the index of refraction is given as well as the radius of the curvature of each lens surface. All dimensions are given in millimetres with parts of millimetres shown in decimal notation.

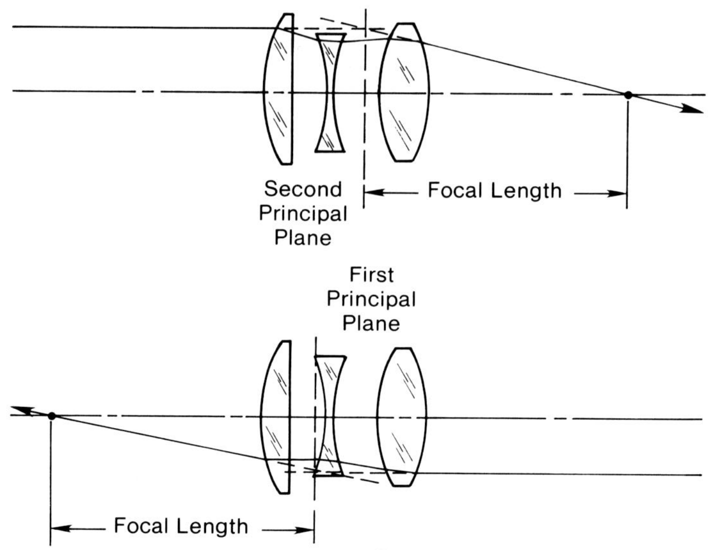

Figure 8-11. Finding the two principal planes.

When designing a camera and its lens mounts, the distance from the point at which the back lens surface intersects the axis to the focal point is a significant design dimension. This is called the back focus (abbreviated "B.F.L.") The principal plane can be drawn if the B.F.L. and the E.F.L. are known. Locate the focal point using the back focus, then measure the effective focal length along the axis from the focal point. This is the location of the principal plane.

When using a compound lens in a schematic ray trace of an optical system, it is often enough to indicate the lens by drawing only its principal plane. Other compound lenses, prisms, mirrors and beam splitters that may be in the system can be placed in relationship to the principal plane. The second principal plane is needed when designing the spacing of the optical elements in the system.

ARBIDI LENS FORMULA SHEET

FORMULA: 27-G-327 3-Inch f/2.0 Triplet Aperture = f/2.0
EFL = 74.2 for Impflex Camera BFL = 55.23
Field: 29° Prinĉipal Wavelength D FFL = 56.76
Semi-diagonal (42.4) F-C Achromatism

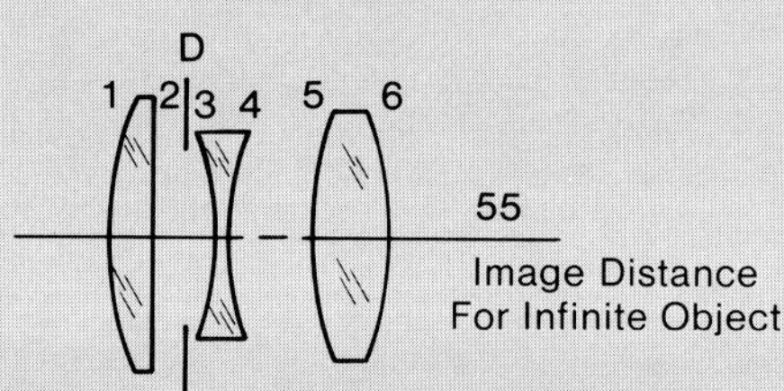

Surf.	Clear Aperture	Axial Beam	Radius	Thickness	Glass	N_D	V
1	36.21	36.21	44.47				
				8.03	ABC-21	1.6783	55.7
2	35.03	35.03	Plano				
				4.10			
D	29.21	39.20		4.75			
3	27.17	27.17	−56.33				
				2.00	EDF-19	1.6625	31.5
4	26.91	26.91	38.52				
A							
				13.05			
5	28.45	28.20	48.50				
				10.98	ABC-21	1.6783	55.7
6	32.68	32.40	−77.48				

Image Height	Vign.	% Illum.	% Field	Angular Field	Diaphragm Apertures	
0.0	1.00	100	0	0°	f/2.0	29.21
14.8	.87	83	35	11.3°	f/2.8	20.65
23.4	.75	72	55	17.5°	f/4.0	14.60
29.7	.67	65	70	21.8°	f/5.6	10.32
36.1	.58	53	85	25.9°	f/8.0	7.30
42.4	.55	49	100	29.0°	f/11	5.16
					f/16	3.65
					f/22	2.58

Figure 8-12. Arbidi Lens Formula Sheet

Some Uses of Negative Lenses

As has been indicated earlier, negative lenses have many uses. They are important elements of compound lenses and permit the corrections of lens aberrations. Without negative lenses, the achievement of sharp images in optical instruments and cameras would be almost impossible. Negative elements form the eyelenses of Galilean Telescopes, and are used as finder lenses in the optical viewfinders of cameras.

One very valuable use of negative lenses is in eyeglasses that correct nearsighted vision. An eye that is nearsighted is too long for the focal length of the eye's lens. The focal point falls in front of the retina (light-sensitive surface) so that a sharp image cannot be formed. By placing a negative lens in the eyeglass the positive power of the eye's lens is weakened. Its effective focal length is made longer so that the focal point falls directly on the retina forming a sharp image.

If you are nearsighted, you can find the diopter power required to correct your eyes in the following manner (Figure 8-13). With your glasses off, find the farthest distance you can focus sharply on a detailed object, such as a printed page. Start from a greater distance and gradually approach the object. The far distance is the greatest distance at which your eyes can sharply see the object. Try this one eye at a time since your eyes may be optically different. Once you know the farthest distance an eye can see the object sharply, that distance is the focal length of the negative lens required to correct your vision for infinitely distant objects.

This is shown in the drawing. The unaided eye can focus sharply at 10 inches. A 10-inch focal length negative lens is shown to diverge parallel light to the same degree as light coming from an object at 10 inches without the lens. Since a 10-inch focal length negative lens has a diopter power of 40 divided by −10, it is a −4Δ lens.

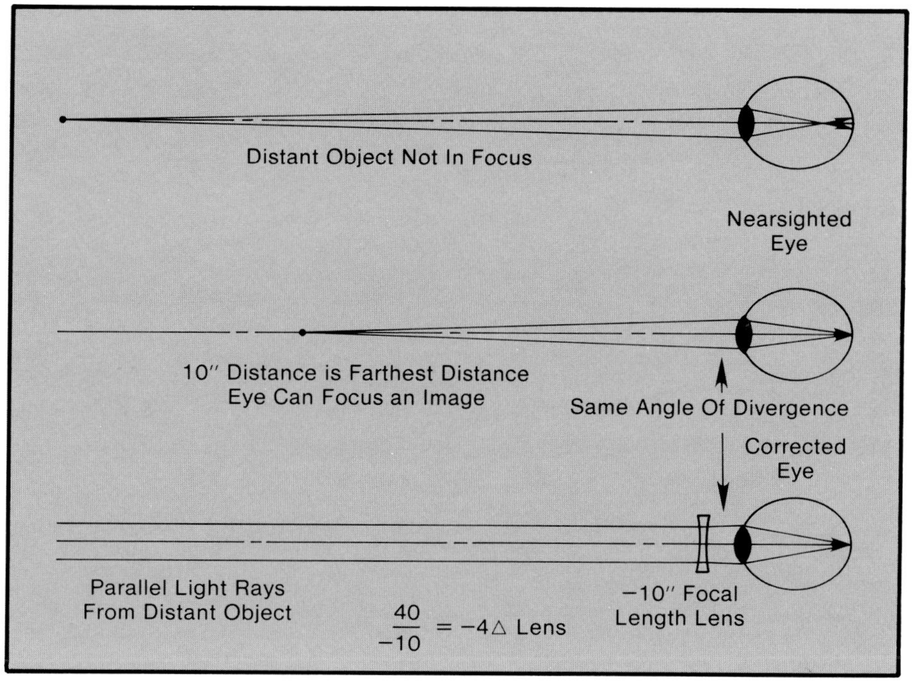

Figure 8-13. Test for negative eyeglass lens focal length.

The following table relates the nearsighted eye far distance with the diopter power of the lens required to correct it.

Far Distance	Lens Power
80 inches	−½Δ
40 inches	−1 Δ
20 inches	−2 Δ
13 inches	−3 Δ
10 inches	−4 Δ
8 inches	−5 Δ

Finding the Approximate Focal Length of Negative Lenses

It is relatively easy to find the focal lengths of positive lenses because they create real images whose sizes can be measured. However, without precise optical instruments, the focal length of negative lenses can only be approximated.

One way is by using a positive lens, whose focal length is known, as an optical probe. Mount the negative lens in such a way that one can see a distant object through it (Figure 8-14). Assume that the probe lens has a 12-inch focal length. Looking through the probe at the negative lens, move the probe gradually toward the negative lens until you can see its surface sharply. Measure the distance between the lenses. This should be about 12 inches (it may help to sprinkle some talcum powder on the negative lens so that you can locate the surface easily). Now move the positive lens probe closer to the negative lens until you can see the distant object sharply. Measure the distance between the negative and positive lenses and subtract it from 12 inches. This will give you the distance the positive lens was moved and that is the approximate focal length of the negative lens. This works only with negative lenses whose focal lengths are shorter than that of the positive lens being used. In Figure 8-14, the positive lens has moved seven inches which approximates the focal length of the negative lens.

Another method requires both eyes. Place some powder or crayon marks on a window through which you can see distant objects (Figure 8-15). Hold the negative lens between your eyes and the window, and looking through the lens with both eyes, change the lens-to-window distance until the distant objects

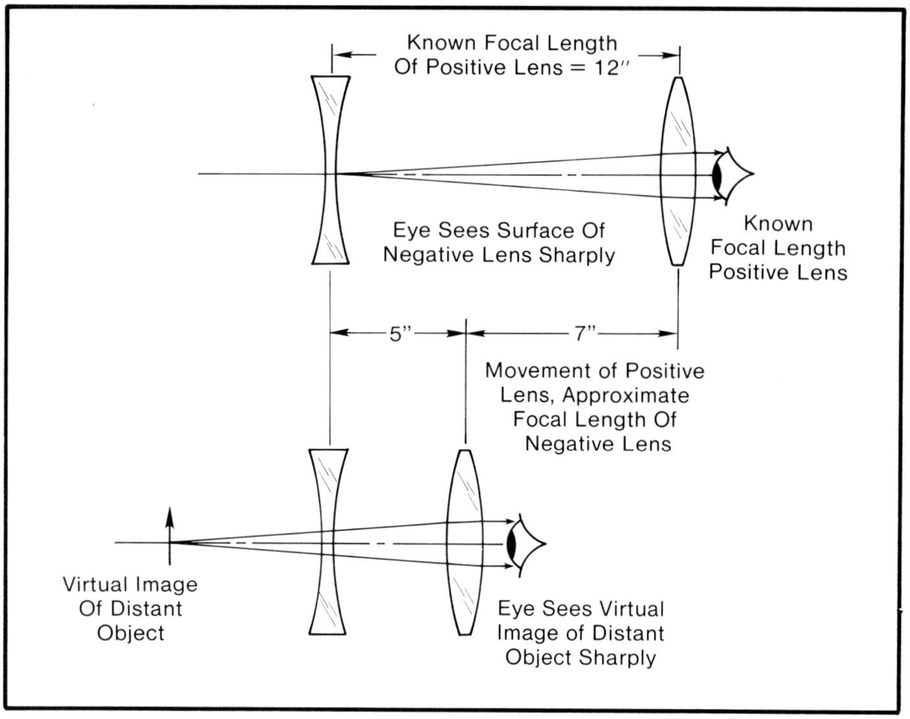

Figure 8-14. One method of finding approximate focal length of a negative lens.

appear to be in the plane of the window glass as seen around the lens. As you move the lens back and forth the distant objects will appear to move in space. Stop when they appear to have moved forward exactly to the window glass. Measure the distance between the lens and the glass. This distance approximates the focal length of the negative lens.

Schematic Ray Tracing— Negative Lenses

Negative lenses diverge parallel light rays and create virtual images rather than real images. Tracing the rays is similar to tracing the rays through a positive lens. However, the light is diverged by the negative lens and the ray from the arrow point parallel to the axis is drawn through the focal point on the *same side of the lens as the object.* This is a construction line and is drawn as a dashed line. The construction line is an extension of the exiting ray but in practice, the construction line is used to locate the exiting ray path.

A ray going through the center of a negative lens is undeviated, as in a positive lens. It is clear that these two rays, both from the arrow point, are diverging as they exit from the lens. The image is located where the construction lines extended from these two rays intersect. Because the rays intersect on the same side of the axis as the point at which they originated, the image is upright. It is also smaller than the object because the image point is closer to the axis than the object point.

In the first drawing (a) the object is located at the focal point (Figure 8-16). The image is located halfway between the object and the lens and is half the size of the object. In the second drawing (b) the object distance is greater than the focal length. The image is somewhat closer to the focal point and is smaller than when the object was at the focal point. In the third drawing (c) the object distance is less than the focal length. The image is larger than in the other two cases and is located closer to the lens. When the object is in contact with the lens, it should be obvious that the image and object are located in the same position, and the image has become the same size as the object.

Another Method for Finding the Focal Length of a Negative Lens

The focal length of a negative lens can also be found by using schematic ray tracing. If a light source is at infinity, its virtual image will be located at the focal point (Figure 8-17). The rays of light from the source will be diverged by the lens into the shape of a cone. The diameter of the lens will form the outer limit of the cone, while the focal length will determine the cone angle. The lines constructed as extensions of the outer limit diverged

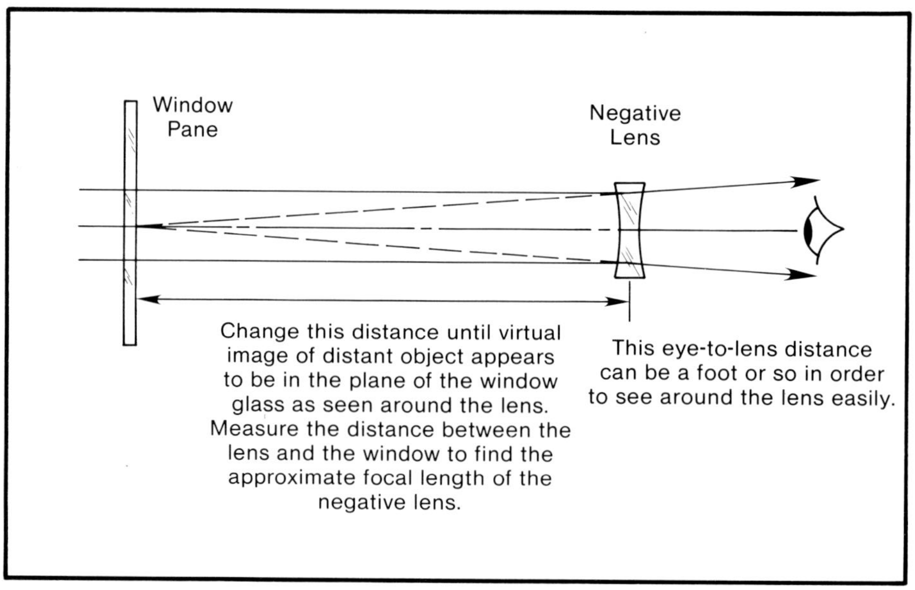

Figure 8-15. Another method of finding the approximate focal length of a negative lens.

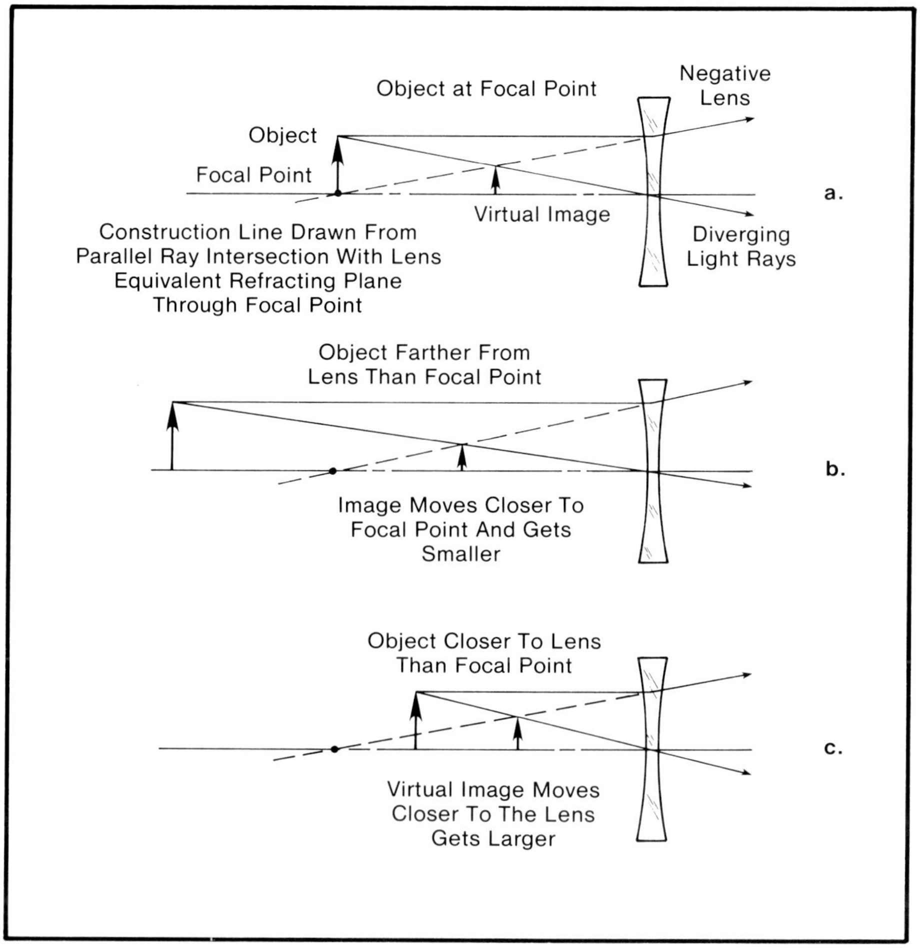

Figure 8-16. Schematic Ray Tracing with a Negative Lens.

Problem

Using schematic ray tracing, find the size and location of the virtual images for the three objects shown in the drawing. The lens has a focal length of −4 inches and the objects are all 1 inch high. Identify the images by number. Draw actual rays as solid lines and construction lines as dashed lines and show the emerging rays. Describe the size of the image of this object when the object distance is 100 feet. Show the approximate location of the image on the drawing.

rays, drawn toward the axis, form an apex of the cone at the focal point.

Therefore, the cone angle is fixed by the lens diameter and focal length, and if the focal length is unknown, it can be found in the following way. Measure the diameter of the lens and draw a circle on a piece of paper twice the diameter of the lens. Hold the lens in the rays of the sun (a convenient source of parallel light rays) so that the diverging rays form a circle of light on the paper. Change the lens-to-paper distance until the circle of light has a diameter equal to the circle drawn on the paper. Measure the distance between the lens and the paper to find the focal length of the lens. The schematic ray trace shows why this is true.

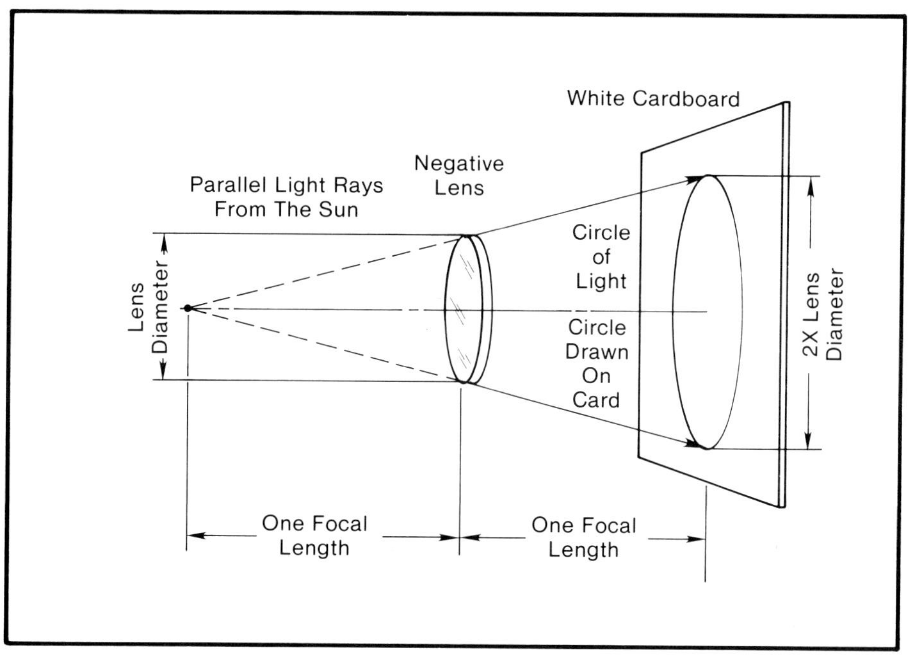

Figure 8-17. Using the sun's rays to find the focal length of a negative lens.

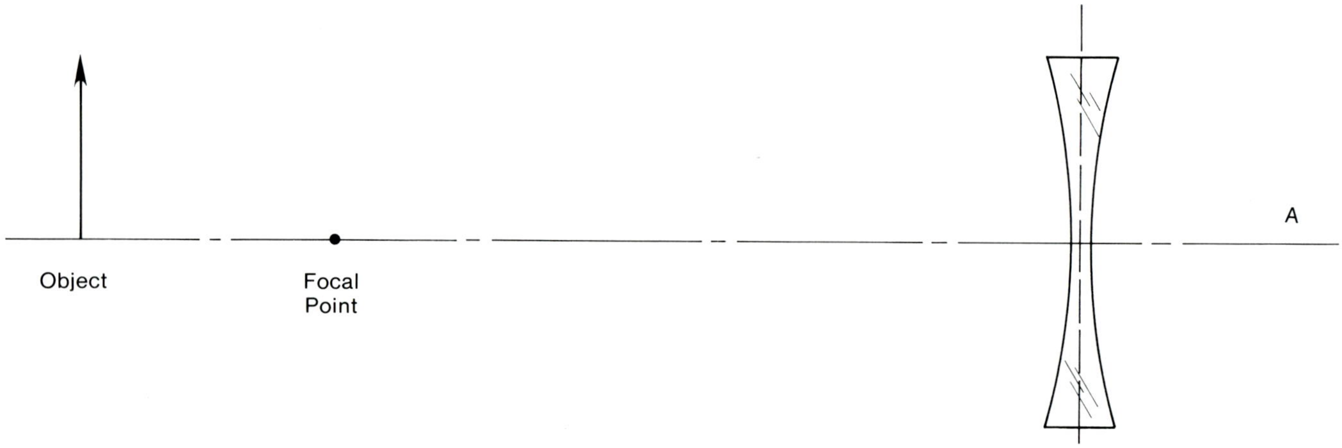

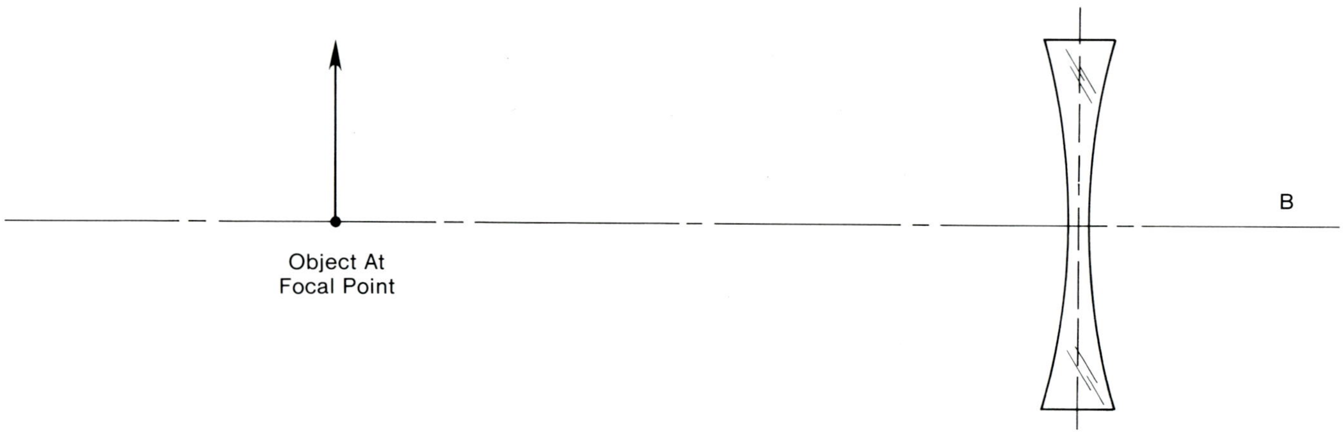

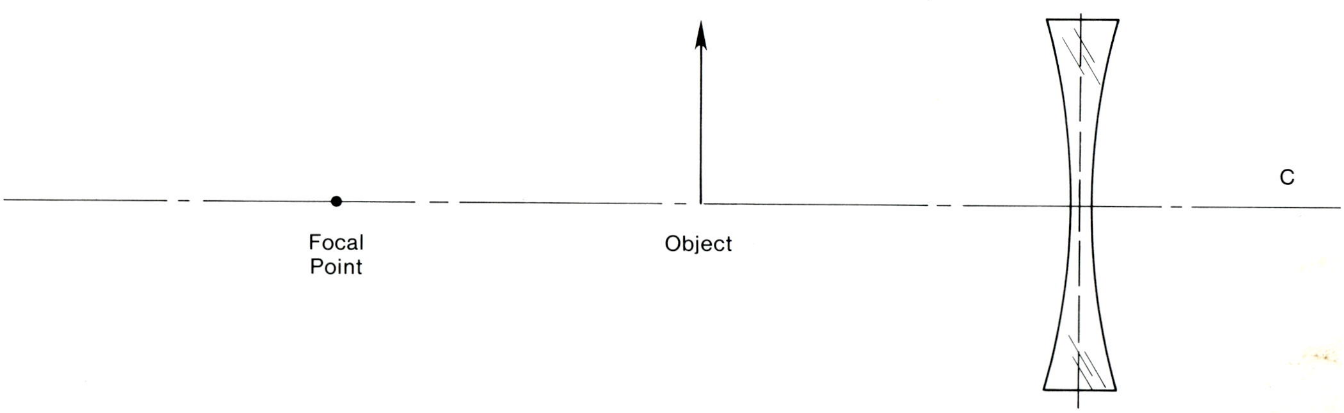

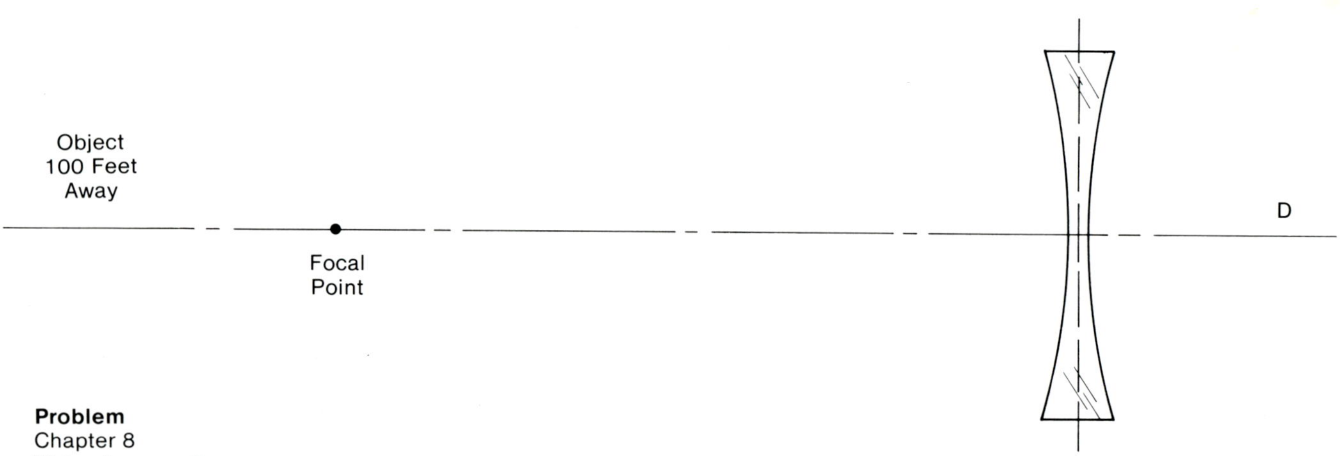

Problem
Chapter 8
Video Lesson 3

Chapter 9

VIDEO LESSON 4

Magnification, Parallax and Schematic Ray Tracing of Compound Lenses

Summary

When the object distance is infinitely large, the magnification is proportional to the focal length of the lens. When the object distance is finite, the image-distance to object-distance ratio is the same as the image-size to object-size ratio. One of the most important relationships in optics is the relationship between focal length, image distance, and object distance:

$$\frac{1}{f} = \frac{1}{u} + \frac{1}{v}$$

Optical scales are added to magnifiers and microscopes for measuring small dimensions. Parallax, which has to do with the apparent positions of objects at various distances and viewpoints, can be used to determine the distances of objects. Schematic ray tracing can be used to show parallax displacement. A collimator is a useful optical device that produces parallel light rays for test purposes. A projection lens bench is described as well as its use in finding the nodal points of projection lenses. The basic methods of schematic ray tracing of two-element compound lenses are introduced.

Relationship of Magnification to Focal Length

Object distance is the distance from the object to the principal plane. *Image distance* is the distance from

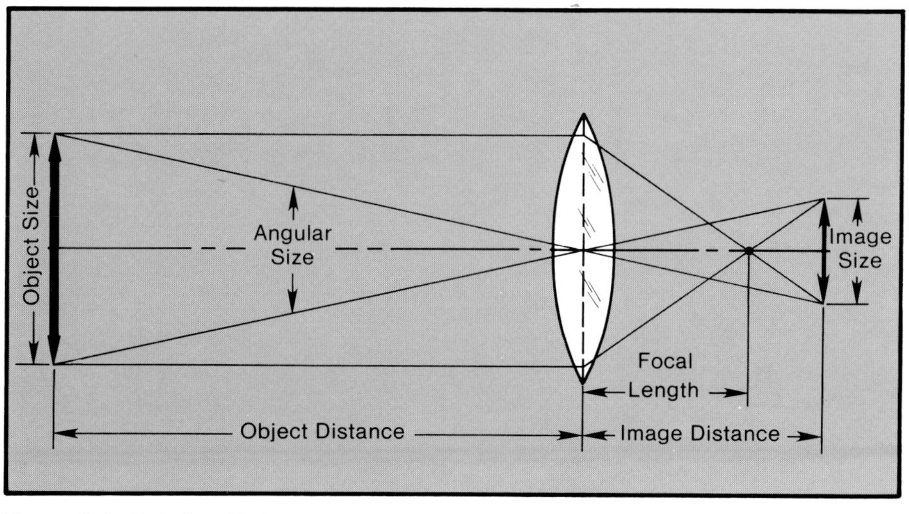

Figure 9-1. Relationship between object size, object distance, image size, image distance, and focal length.

the sharp image to the principal plane. When the object distance is infinitely large (object is at infinity), the image distance equals the effective focal length (Figure 9-1).

Object size is any linear dimension of the object measured perpendicular to the lens axis. For example, when a dimension on the object is at an angle other than perpendicular, it is its projection on a plane perpendicular to the axis that becomes the object size. *Image size* is the linear dimension of the image of the object dimension measured in the image plane perpendicular to the lens axis.

The *angular size* is the angle between the rays of light coming from each end of the object passing through the center of the lens and exiting from the lens to the ends of the image of the object. If the object is at a great distance, the image size is directly proportional to the focal length of the lens. For example, the image formed by a 12-inch focal length lens will be exactly twice as large as the image formed by a 6-inch focal length lens (Figure 9-2).

When the object is at infinity, the image distance is equal to the focal length. The object can be at a distance that is considered to be "optical infinity" (several hundred feet), but is still of course, at a finite distance.

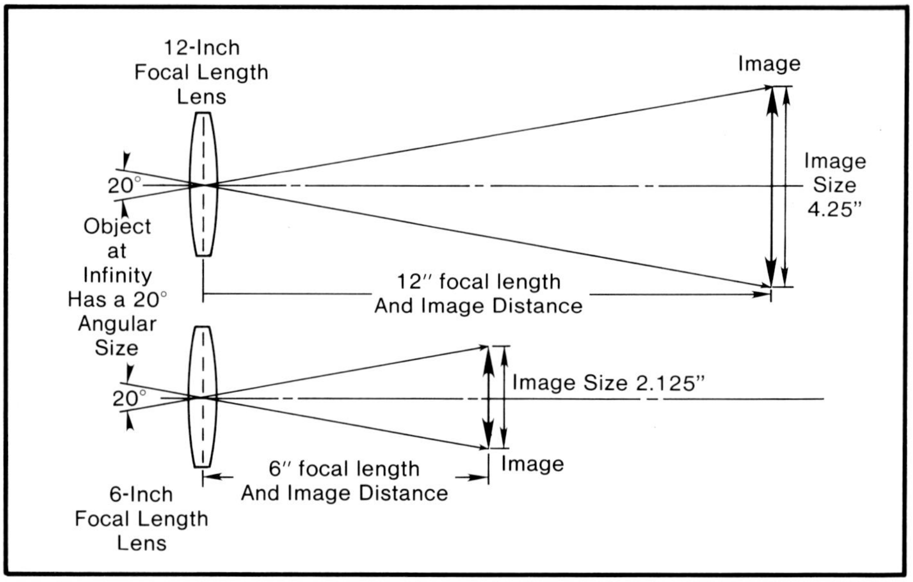

Figure 9-2. Angular Size and Effect of Focal Length on Image Size.

In Figure 9-3 the object is several hundred feet in distance and has an object size of 100 feet (1200 inches). It is imaged by lenses of 2-inch, 4-inch, 6-inch, and 8-inch focal lengths. The lenses make images of the object whose lengths are shown in the table. The magnification in each case is very small: the image size in inches is divided by the object size of 1200 inches.

Lens Focal Length	Image Size	Magnification
2	0.75 in.	÷ 1200 = .00063
4	1.50 in.	÷ 1200 = .00125
6	2.25 in.	÷ 1200 = .00188
8	3.00 in.	÷ 1200 = .00250

The image size is directly proportional to the focal length of the lens. Mathematically, this means that the ratio of the focal length divided by the magnification is a constant. In the above example, the constant is very large because the magnification is so small:

$$2 \div .00063 = 3200$$
$$4 \div .00125 = 3200$$
$$6 \div .00188 = 3200$$
$$8 \div .00250 = 3200$$

The general rule is shown by the equation:

$$\frac{\text{focal length}}{\text{magnification}} = C$$

Note that in this case, the object is at a constant optically-infinite distance so that the image distance is equal to the focal length.

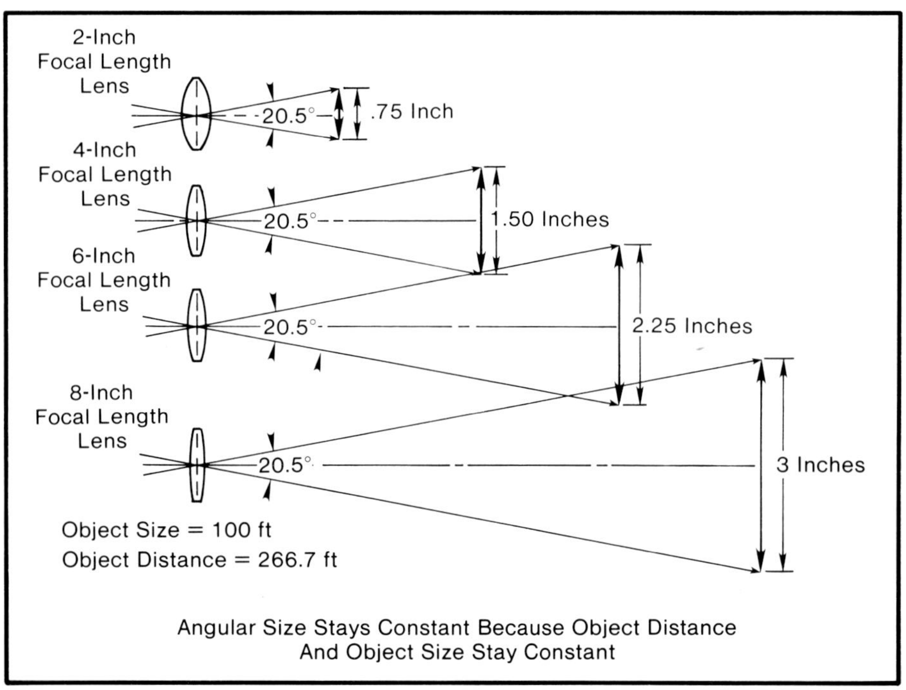

Figure 9-3. Effect of Focal Length on Magnification.

Object-Image Size Relationship at Finite Distances

When the object is at a close (finite)

distance, the image distance will be greater than the focal length. With a lens of a given focal length, the ratio of the image size to object size (the magnification) is also equal to the ratio of image distance to object distance:

$$m = \frac{\text{image size}}{\text{object size}} = \frac{\text{image distance}}{\text{object distance}}$$

There is another relationship that is very useful. In Figure 9-4 the object of a known size is at a measurable close distance from the lens. The image distance is greater than the focal length, which is unknown. In a real lens it is very difficult to measure the image distance because we do not know the location of the principal plane. However, we can measure the image size and then calculate the image distance quite accurately.

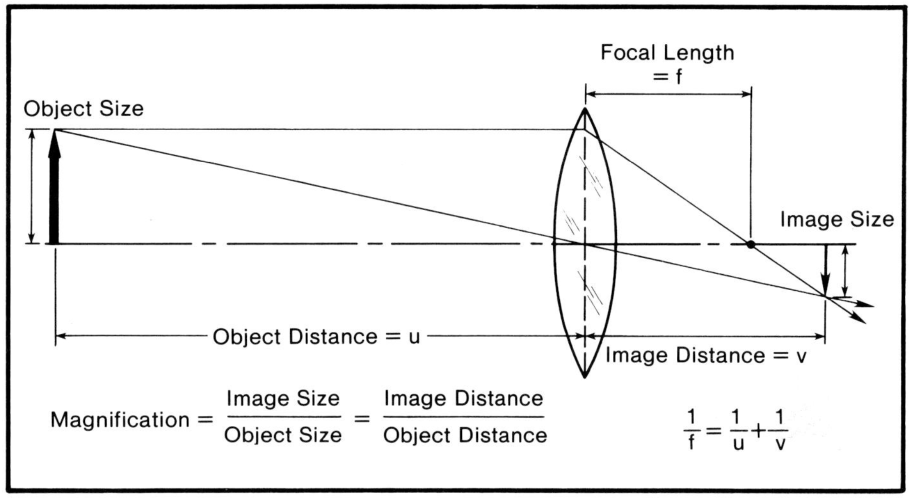

Figure 9-4. Close-Distance Object- Image Relationships.

Optical Scales

A magnifier, loupe, long working-distance magnifier, or microscope can be used as a measuring instrument. This is accomplished by putting a scale in either the object or image plane. For loupes or magnifiers, the scale is made at a magnification of 1X; but, because it is to be magnified, the lines are made very narrow and finely divided scales can be used. For simple scales, the pattern can be photographically recorded on a glass plate and cut to fit the mount. Transparent scales with opaque lines are called reticles. The finest reticle scales are made by engraving lines on wax coated glass. The thin wax is removed by a pantograph-guided needle to form the reticle pattern. The glass is then etched with hydrofluoric acid and filled with a white, black or colored fill. The finest fillers are made of a pigment dispersed in sodium silicate (waterglass).

In long working-distance loupes and microscopes, the scale is put in the image plane of the objective lens where it is magnified by the eyelens. Depending on the magnifying power of both the objective and the eyelens, the magnification of the scale may be of the order of .01X which means that the scale is 1/100 of the size of the object it measures.

The most common scales, linear-dimension scales, are very accurate rulers. Scales used for the determination of the measurement of various curves of various radii of curvature or protractors for measuring the angles of very small tapered objects are also put on transparent support scales for use with magnifying instruments (Figure 9-5).

A calibrated loupe with a transparent scale can be used to measure the image size quite accurately. The object size can be measured easily with a yardstick; but, because of the difficulty of locating the principal planes, it is difficult to accurately measure the object distance. However, if the object distance is made fairly large, around 20 feet, the error caused by not knowing the exact location of the secondary principal plane is minimized.

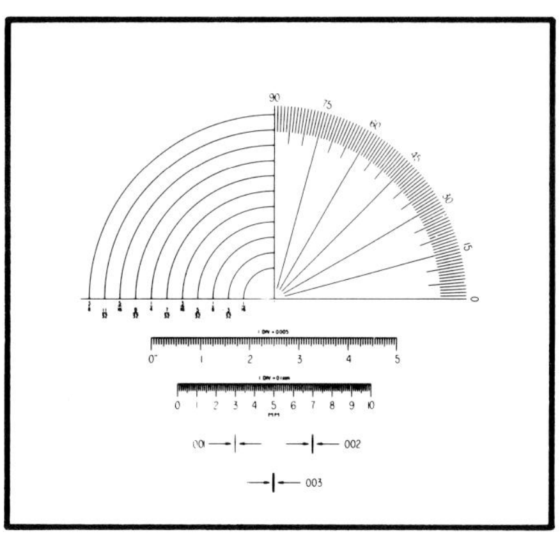

Figure 9-5. Loupe Scale.

If, for example, the object size is 6 feet (72 inches), the object distance (240 inches), and the image size is 1 inch (Figure 9-6), by the ratio equation:

$$\frac{72 \text{ inches}}{240 \text{ inches}} = \frac{1 \text{ inch}}{v}$$

$$v = \frac{240}{72} = 3.33 \text{ inches}$$

The image distance is found to be 3.33 inches (Figure 9-7). One of the most useful equations in optics is

$$\frac{1}{f} = \frac{1}{u} + \frac{1}{v}$$

where f is the focal length, u is the object distance and v is the image distance. We can use this equation to find the focal length in the example just given.

$$\frac{1}{f} = \frac{1}{240} + \frac{1}{3.33}$$

$$= 0.00417 + 0.3$$

$$= 0.30417$$

$$f = \frac{1}{0.30417} = 3.29 \text{ inches}$$

When the focal length is known, the location of the focal point can be established.

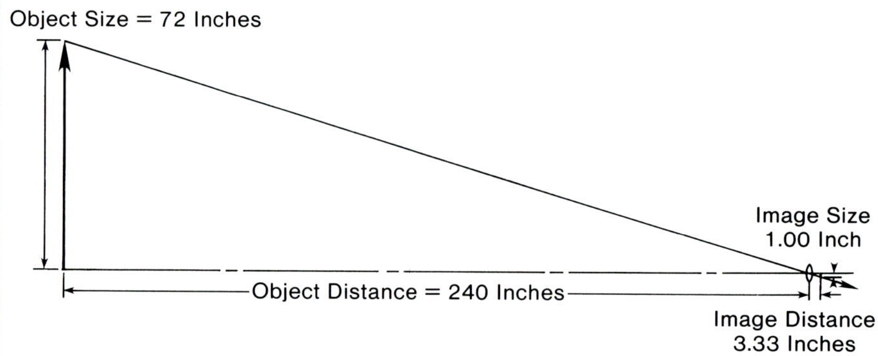

$$m = \frac{\text{Image Size}}{\text{Object Size}} = \frac{\text{Image Distance}}{\text{Object Distance}} = \frac{1 \text{ Inch}}{72 \text{ Inches}} = \frac{v \text{ (Image Distance)}}{240 \text{ Inches}}$$

$$V = \frac{240 \times 1}{72} = 3.33 \text{ Inches} \qquad \frac{1}{f} = \frac{1}{u} + \frac{1}{v} \qquad \frac{1}{f} = \frac{1}{240} + \frac{1}{3.33}$$

$$\frac{1}{f} = .00417 + .30000 = .30417 \qquad f = \frac{1}{.30417} = 3.288 \text{ Inches}$$

Figure 9-6. Example of Object-Image relationships.

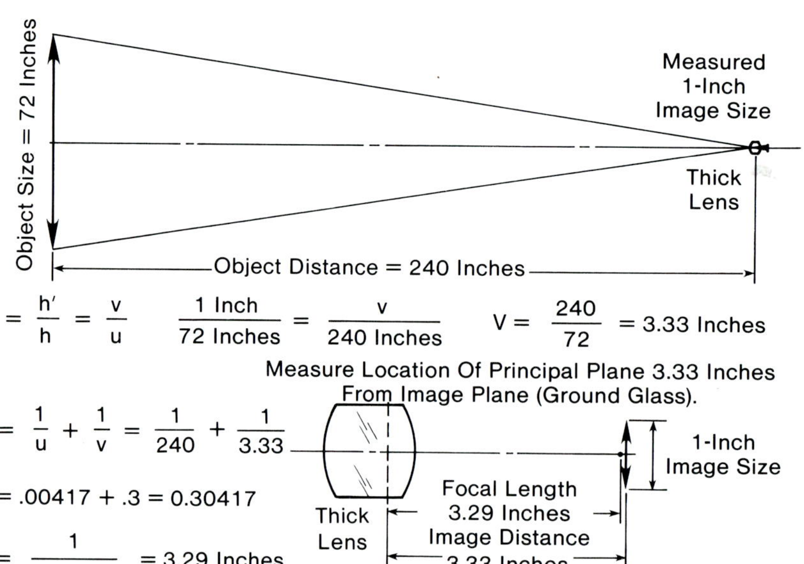

$$m = \frac{h'}{h} = \frac{v}{u} \qquad \frac{1 \text{ Inch}}{72 \text{ Inches}} = \frac{v}{240 \text{ Inches}} \qquad V = \frac{240}{72} = 3.33 \text{ Inches}$$

$$\frac{1}{f} = \frac{1}{u} + \frac{1}{v} = \frac{1}{240} + \frac{1}{3.33}$$

$$= .00417 + .3 = 0.30417$$

$$f = \frac{1}{0.30417} = 3.29 \text{ Inches}$$

Figure 9-7. Finding Location of Principal Plane in a thick lens.

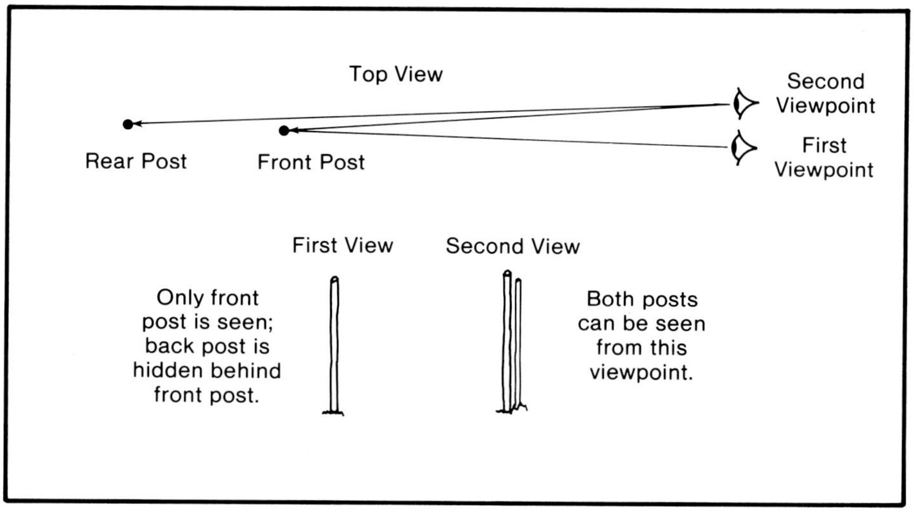

Figure 9-8. Parallax

Parallax

It is sometimes useful to locate the position of a virtual image quite accurately. In one of the methods for finding the focal length of a negative lens, it was necessary to place the virtual image of the distant object in a window glass. This was accomplished by stereo vision; the use of two eyes to see depth. Another method, the parallax method, can be used with one eye to verify the position found with two eyes. Parallax might be called the relative displacement of objects seen at different distances when the viewpoint changes. Let's explain this by using an example. There are two fence posts located at different distances from a viewer, one is fairly close and the other is farther away as shown in Figure 9-8. At the viewer's first position the two objects are lined up one behind the other. If the viewpoint is changed slightly to the right, the close object will appear to the left of the far object. If the viewpoint had been shifted slightly to the left, the close object would appear to be to the right of the far object. This shifting of the relative positions of objects at different distances with a change of viewpoint is called parallax.

If a window pane has a scratch or grease pencil mark on it and a virtual image of a distant object outside the window is viewed with one eye only, parallax can be used to assure that the virtual image is exactly in the window glass (Figure 9-9). The eye is shifted left and right. If the glass and virtual image are not in the same plane, there will be a shift of the image relative to the mark on the glass. When the virtual image is exactly in the same plane as the mark on the window, there will be no shift in the position of the mark in relation to the position of the image as the eye moves left and right.

The same principle can be used with a real image (Figure 9-10). A piece of glass or thin plastic with a scratch on it can be held behind a positive lens, roughly in position of the real image formed by the lens. When the image and the scratch on the glass are in the same plane (the same distance from the lens) they will not shift relative to each other as the eye moves left and right. The greater the displacement as the eye moves left and right the greater the displacement between the image and the scratched glass.

Schematic Ray Trace of Displacement

An optical block is a special kind of prism. It has at least two plane-parallel polished surfaces. When viewing a vertical line through an

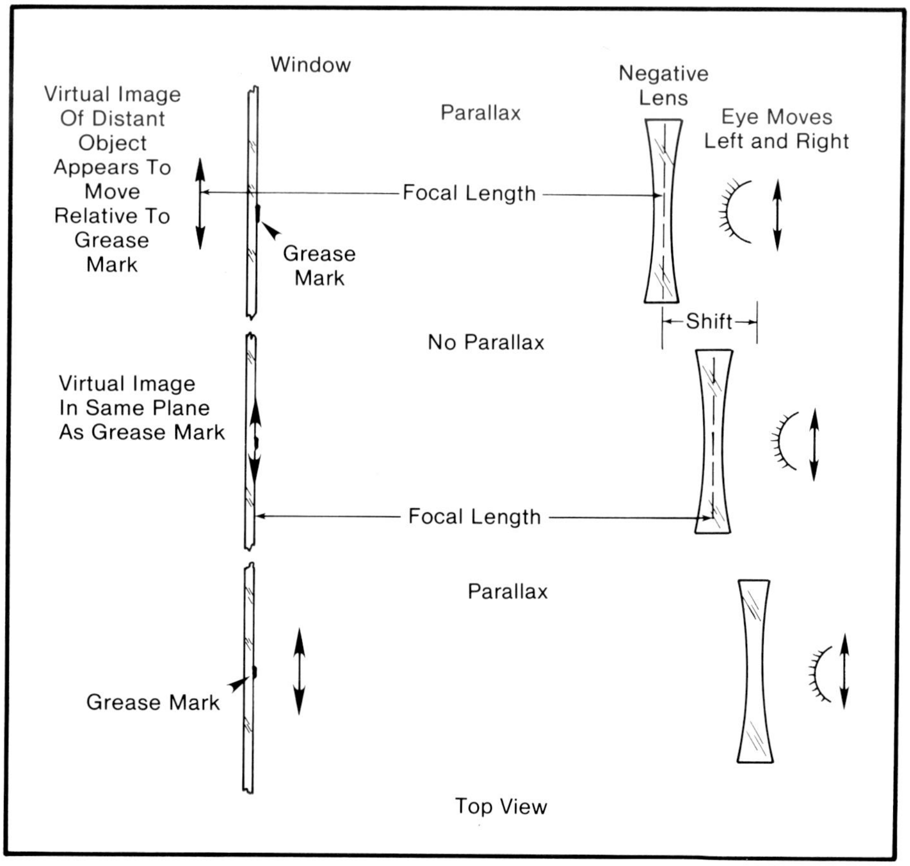

Figure 9-9. Using Parallax to find Focal Length of a negative lens.

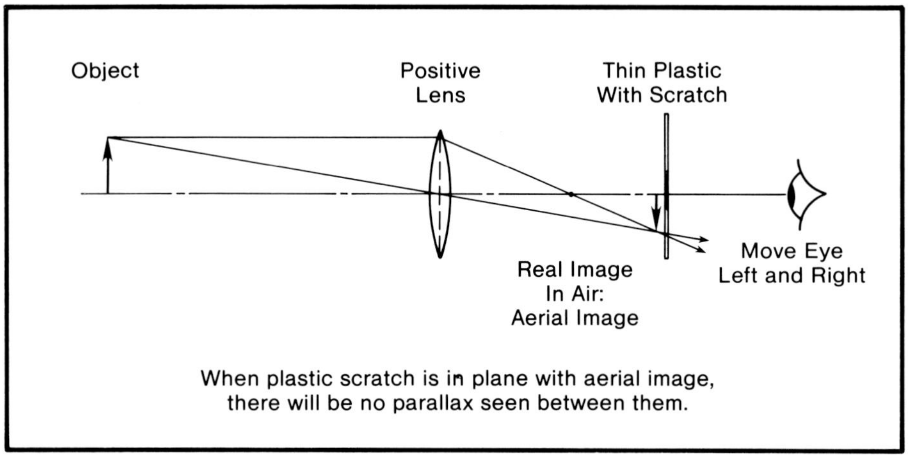

Figure 9-10. Using Parallax to find the position of an aerial image.

optical block, whose surfaces are perpendicular to the optical axis, the image of the line is aligned with the line itself by direct vision (Figure 9-11). If the block is rotated counterclockwise, about an axis perpendicular to the optical axis, the virtual image shifts to the right. If the block is rotated clockwise, the virtual image shifts to the left. In neither rotated case will the line, as viewed through the block, be aligned with the line as seen by direct vision.

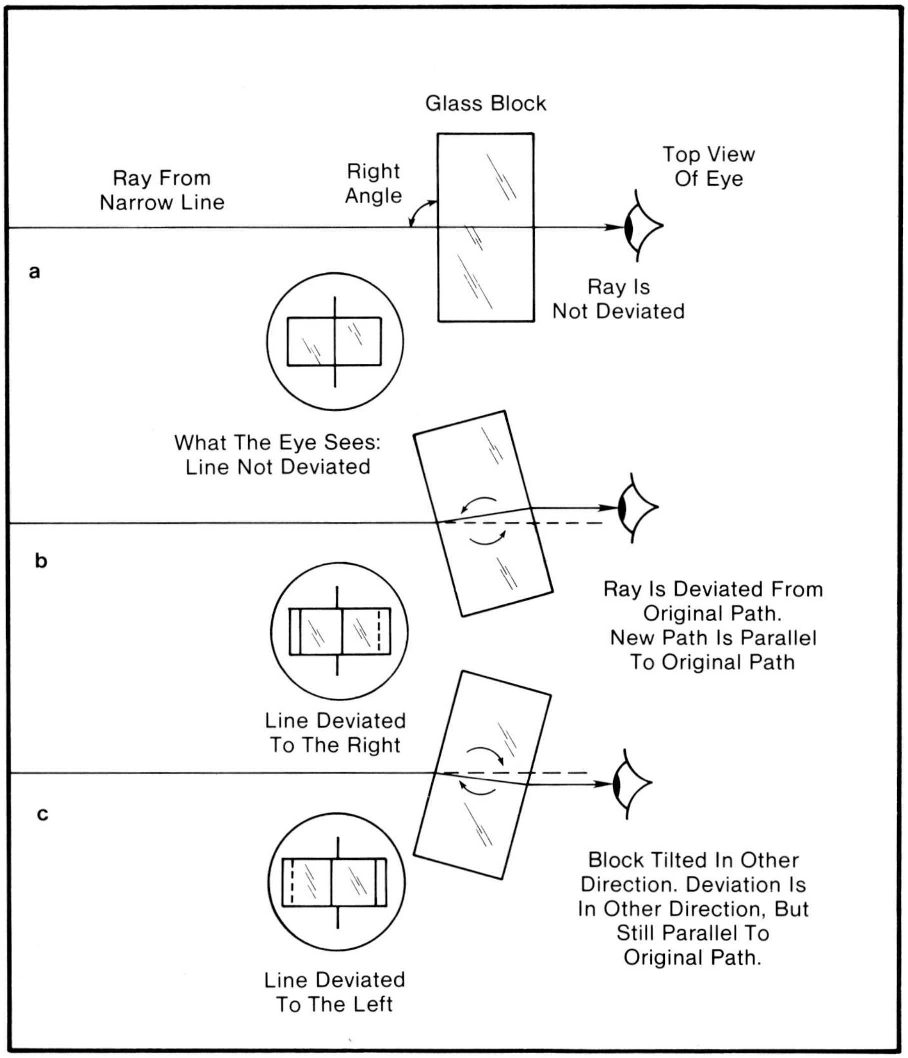

Figure 9-11. Light Ray Deviation by a Glass Block.

The first drawing (Figure 9-11a) shows a schematic ray trace when the block faces are perpendicular to the axial ray. Because both faces are perpendicular to the ray, there is no refraction. The ray goes straight through the block.

The second drawing (Figure 9-11b) shows the block rotated counterclockwise. Now the light ray coming from the vertical line is refracted by the glass. Inside the glass, the ray moves straight, at the new angle, until it exits from the rear surface of the block. Here it is refracted again. When both faces of a block are parallel, the path of the exiting ray is parallel to the path of the entering ray, but its position with respect to the optical axis has been displaced. In Figure 9-11c, the block is rotated clockwise and the image is displaced to the left.

As the rays pass through the tilted optical block, the amount of displacement is independent of object distance, and depends only on the thickness of the block, the refractive index of the glass and the angle of incidence of the ray away from the perpendicular to the block surface. Displacement increases with an increase of any of these last three factors.

The differing amounts of apparent displacement at different distances can be demonstrated by using the optical block. Two vertical-line objects are set up at different distances from the glass block. One is fairly close; the other is at quite a distance. The block is placed perpendicular to the line of sight established by the objects. A positive lens is placed in a way that its axis is on the line of sight and it images the objects through the glass block onto a ground glass (Figure 9-12).

Therefore, only the close line-object can be seen because the distant object is in a straight line behind the

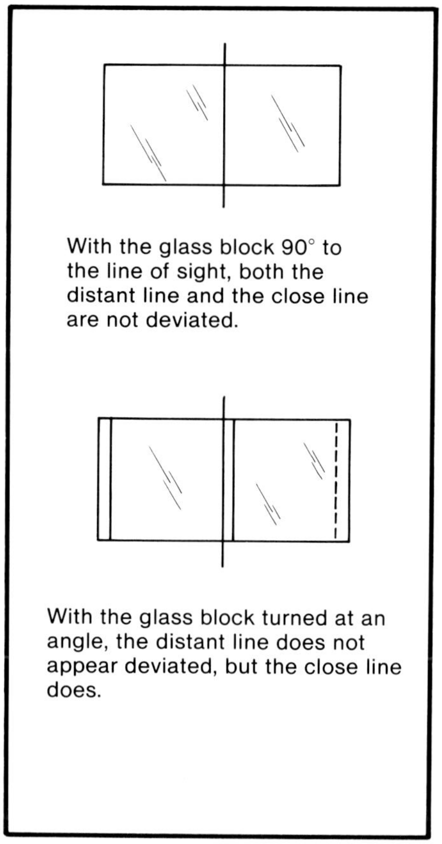

Figure 9-12. Parallax with a glass block.

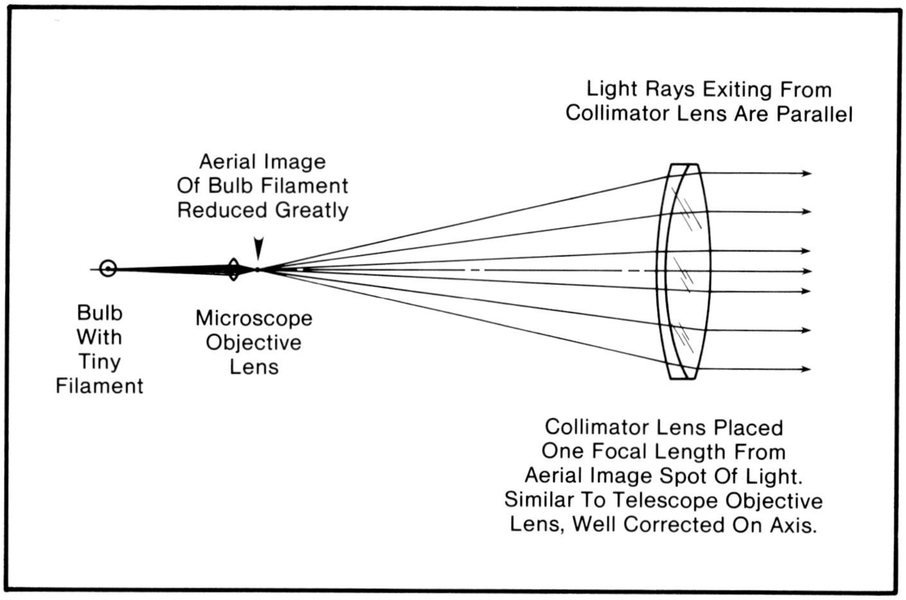

Figure 9-13. Collimator

close object. However, when the block is rotated, the viewpoint shifts, permitting one to see around the near line and thereby view the distant line. This is the parallax effect. As the block is rotated, the image of the near line moves faster than the image of the distant line because the incident angle to the block perpendicular changes faster for the near line than for the distant line.

A Collimator

A collimator is an optical device that produces parallel light rays. The image of a tiny light is placed exactly at the focal point of a collimating lens (a positive lens corrected for axial aberrations). As Figure 9-13 shows, the light source image is exactly one focal length away from the lens; the light rays exiting from the lens are parallel. If you place your eye anywhere in the beam of light directly in front of the collimating lens, it will appear to be filled with light. If you place a positive lens in the parallel beam of light, it can focus the parallel light rays to a single point of light just as if it were focusing parallel light rays from a star. In fact, collimator lenses are used as artificial stars to study the images of lenses (Figure 9-14). If the optical block is placed in the parallel beam between the two lenses and rotated, the position of the point image formed by the positive lens does not move. The image of the light source is not displaced. The position of the image is established by only two optical properties of the experiment, the position of the imaging lens, which establishes the optical axis and the angle of the rays incident on this lens. Rotation of the block while displacing the rays, did not change their angle provided they came from "infinity."

A Projection Lens Bench

The purpose of a lens bench is to test and measure optical characteristics of lenses. A projection lens bench can be used to measure the focal lengths of projection lenses and to locate their principal planes by lo-

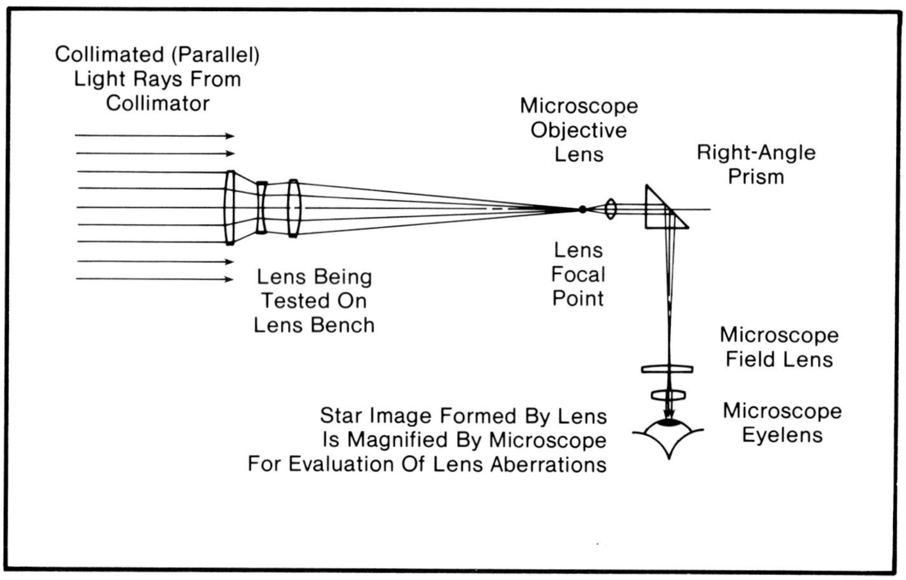

Figure 9-14. Collimator Lens Bench

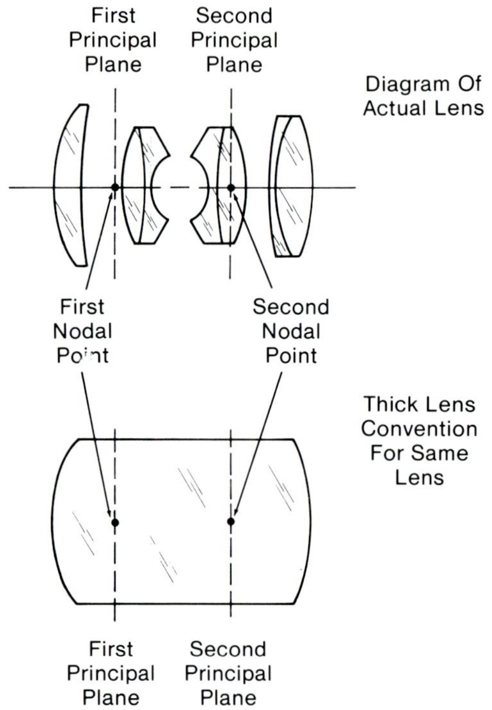

This thick lens convention is used when an existing lens of unknown configuration is to be part of a design, when a preliminary design is being made and the lens itself is not yet designed, or as a time saver in drawing the layout of an optical instrument where the lens design is known, but it is not necessary to the layout.

Figure 9-15. Nodal Points and Thick Lens Convention.

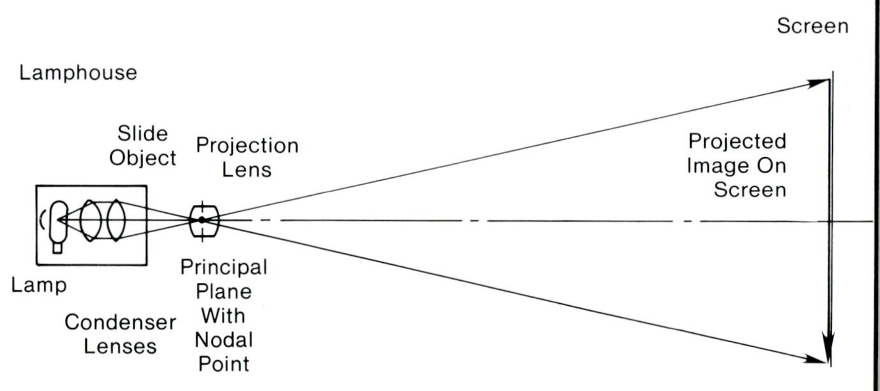

Figure 9-16. Projection Lens Bench

cating their *nodal points*. The nodal point is located at the intersection of the principal plane and the optical axis (Figure 9-15).

Figure 9-16 shows an illuminated object, a test slide, located at the front of the lamphouse of a projection-type lens bench. A single positive lens is located in front of the object and focuses the object on a screen placed several feet away from the lamphouse. The schematic ray trace shows how the rays form the image on the screen. Figure 9-17 illustrates the lens rotating about its nodal point. The principal plane is no longer perpendicular to the line of sight from the projector to the screen. The schematic ray trace shows that the image has not moved on the screen. This is an important fact in optics. *When a lens forming an image is rotated about its nodal point, the position of the image does not change.*

In Figure 9-18 the lens has been rotated around a point other than its nodal point. The physical center of the lens is no longer on the line of sight and the schematic ray trace shows that the image has moved from its original position. If the lens is rotated back and forth about the point shown (not the nodal point) the image will shift back and forth on the screen. In other words, if a lens is rotated about any point other than the nodal point, the image will move. This principle is used in the projection lens bench to find the nodal point.

In front of the lamphouse is a bench base similar to that used on a machinist's lathe (Figure 9-19). There is a support on the bed of this base that can slide closer to the test object or further away from it. In the center of this support is a vertical hole that contains a metal rod having a V-shaped trough on top. This rod can be raised and lowered and the trough can be pivoted or rotated on the rod.

Projection lenses almost always have thick, long barrels for two reasons. The first is to permit a fairly large amount of focusing movement. The second reason is that lenses are often compound lenses that are several centimetres in length. This type of mount allows the lens to be placed in the V-trough and raised or lowered until it is centered in front of the test slide in the lamphouse (Figure 9-20).

The entire support (vertical rod, trough, and lens) can be moved backwards and forwards to focus the image sharply on the screen.

The trough containing the lens is now rotated back and forth (Figure 9-21). If the image moves back and forth on the screen in the same direction as the *front* of the lens, the lens nodal point is in *front* of the pivot point provided by the vertical rod. If the movement of the image is in the same direction as the movement of the *back* of the lens, the nodal point is in *back* of the pivot point.

In the first example, the lens is moved backwards in the trough and refocused by sliding the support forward. In the second example, the lens is moved forward in the trough and refocused by sliding the support backward.

By process of trial and adjustment a point will be found where the image does not move as the lens and trough are rotated. When this occurs, the nodal point is directly over the pivot point. This location

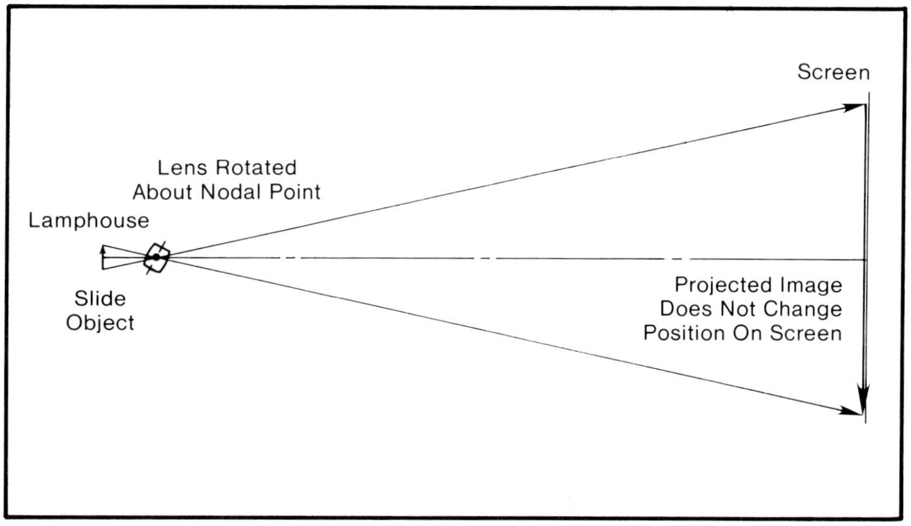

Figure 9-17. Rotating Lens About Its Nodal Point.

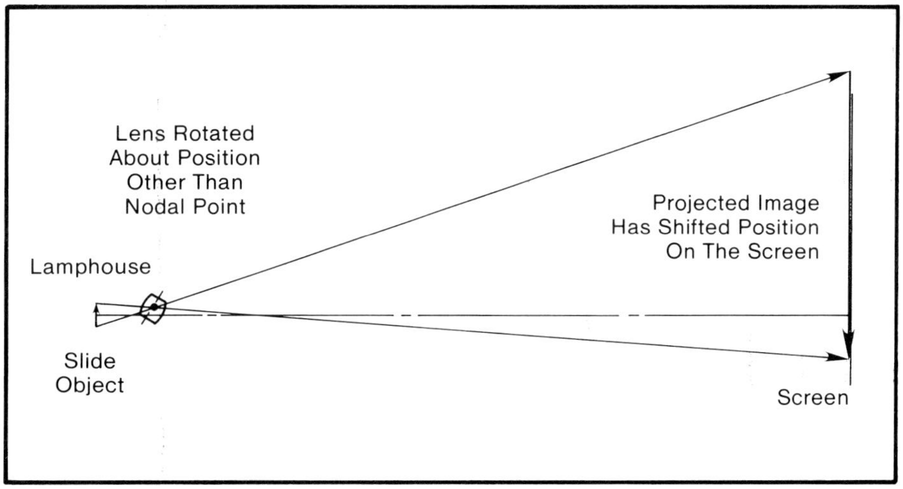

Figure 9-18. A lens, rotated at any point other than the nodal point, causes the image to shift position.

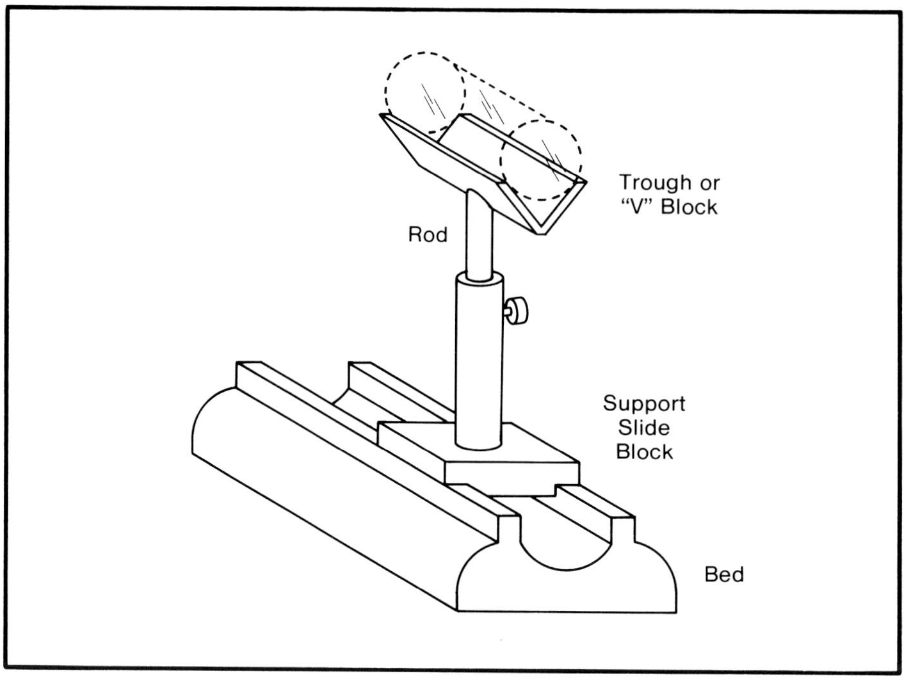

Figure 9-19. Projection Lens Bench Base

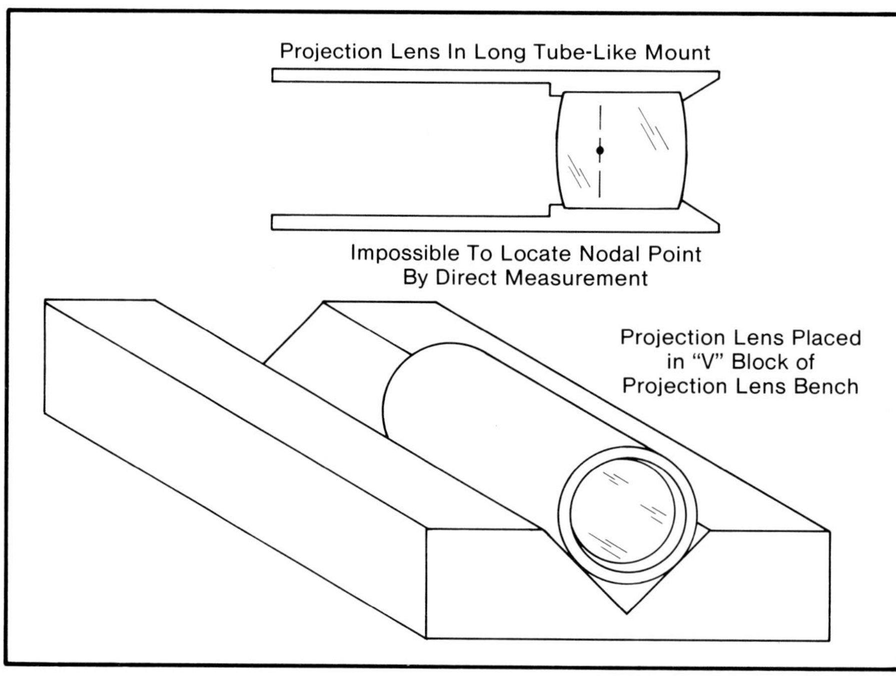

Figure 9-20. Slide Projector Lens in "V" Block.

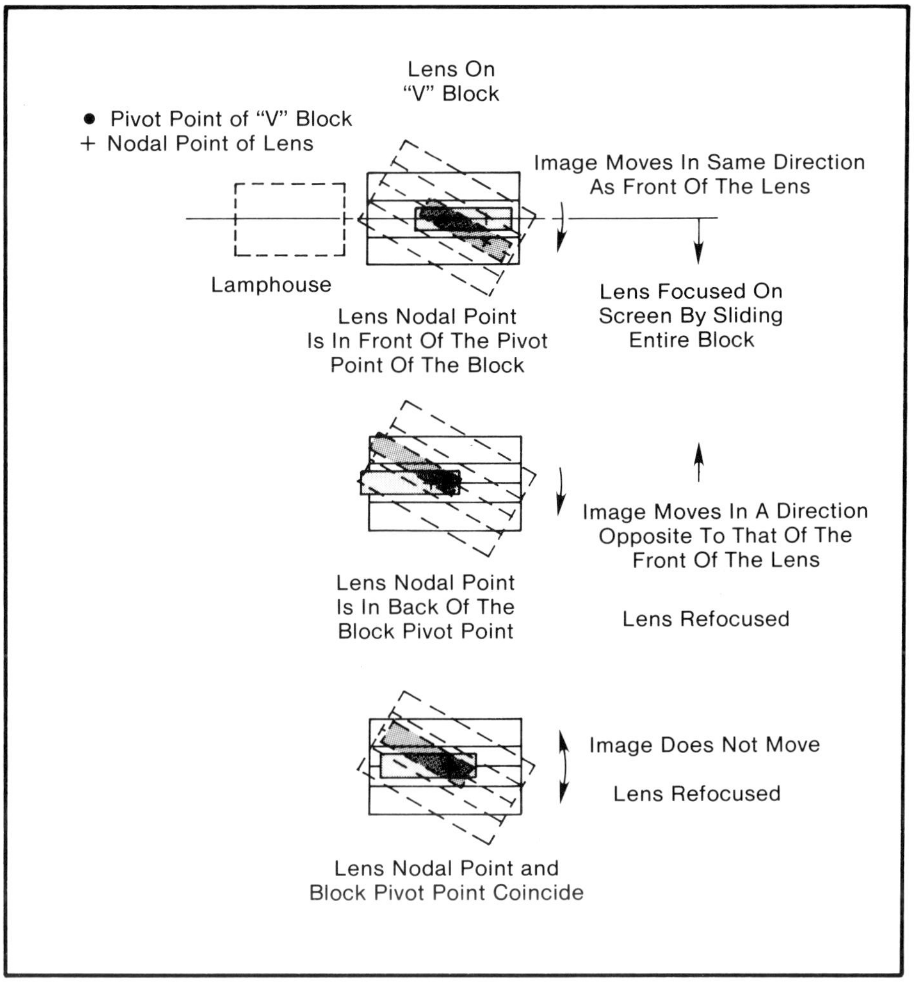

Figure 9-21. Finding the Nodal Point

can be marked on the lens and used for focal length measurements.

Some projection lenses are triplet construction; that is, they have three elements, usually two positive elements with a negative lens between them (Figure 9-22). In lenses of this type, the nodal point is near the center of the lens. Its exact location depends on the relative strengths of the front and rear positive lenses. Such lenses are typical of projection lenses with focal lengths of about four or five inches used for 35 mm slide projectors.

For long-throw projection in order to fill a screen at a great distance, a longer focal length lens is needed: perhaps as long as eight to eleven inches. Such a lens could be made with the same triplet design, but it would be huge. The barrel would be over a foot long and five or six inches in diameter, and it would be very heavy.

An alternative method is a telephoto design in which there is a positive lens element in front and, spaced at some distance behind, a negative lens element. In practice, both the positive and negative lenses are cemented assemblies to correct aberrations. In the next example, single lenses are used for simplification.

Schematic Ray Trace of a Simple Telephoto Projection Lens

In Figure 9-23 a screen with a projected image is shown at the left and the axis and two lenses are shown with their focal points. The positive lens has a focal length of one inch, and the negative lens has a focal length of two inches. The space between the equivalent refracting planes is 0.625 inch.

The projected image distance in the drawing is 12.5 inches from the positive and refracting plane, and the slide is located about 0.0625 inch beyond the positive lens focal point. The slide object size is slightly larger than 0.25 inch and the pro-

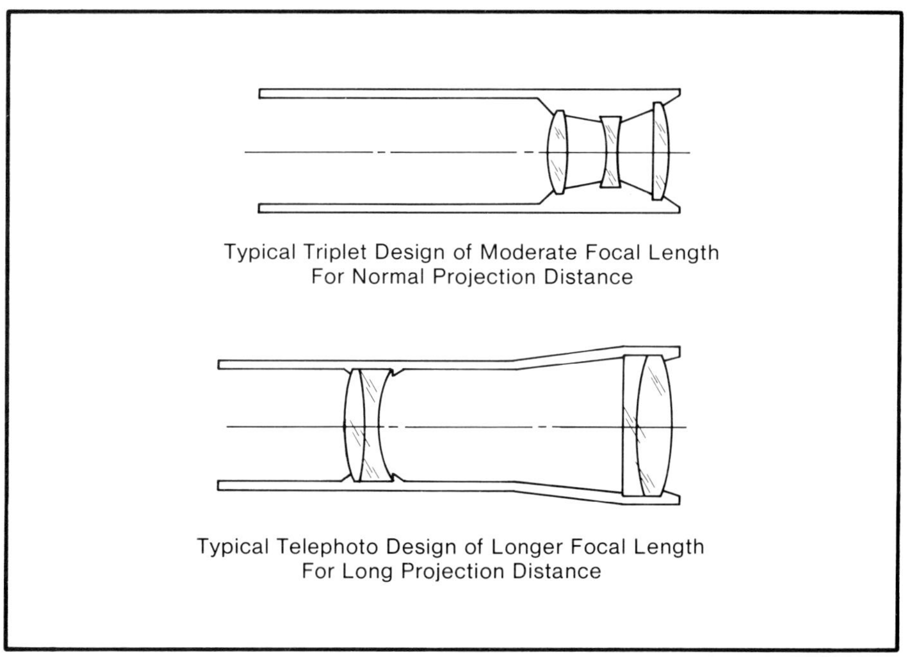

Figure 9-22. Typical Slide Projection Lenses.

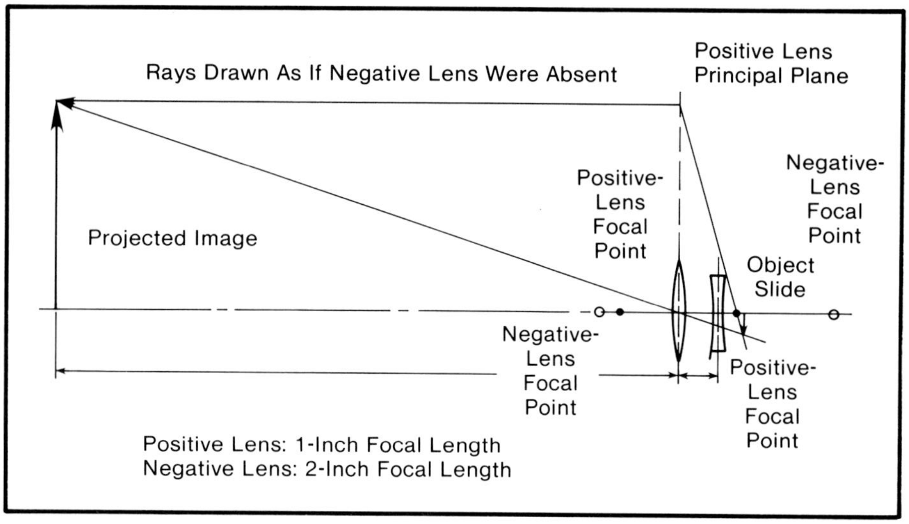

Figure 9-23. Schematic Ray Trace of a Telephoto Projection Lens.

jected image size is about 3.25 inches. The solid dots are the focal points of the positive lens.

Two rays locate the image of the positive lens, ignoring for the moment the presence of a negative lens. A ray from the arrow point parallel to the axis is drawn to the positive lens principal plane. As always, the ray which is parallel to the axis is refracted through the focal point. The axial ray goes from the object arrow point through the center of the positive lens without refraction. Because of the reversibility principle, we can think of the slide as an object and the projected image on the screen as an image; or we can think of the large arrow as the object and the small arrow as the image.

In Figure 9-24 the negative lens rays are shown as dashed lines and the negative lens focal points are dots with circles around them. Two rays are needed that will locate the final position and size of the object of both lenses. The image formed by the positive lens now serves as the object for the negative lens. A ray drawn parallel to the axis from the original object arrow point (the slide) to the negative lens will have a construction line that goes through the left focal point of the negative lens. At the refracting plane, this ray is extended. Therefore, a ray that comes from the projected-image arrow point on the screen is refracted at the positive lens so that it is parallel to the axis. At the negative lens, it is refracted away from the axis, and the direction is found by the construction line drawn through the left focal point.

The other ray goes through the center of the negative lens. It must pass through the arrow point of the original slide object. At the positive lens it is refracted to go through the screen arrow point.

The intersection of these two rays is the point of the final object arrow point of the compound lens. It is about 0.625 inches beyond the negative lens. The final object arrow is larger than the original object arrow (about 0.375 inch), and it is farther to

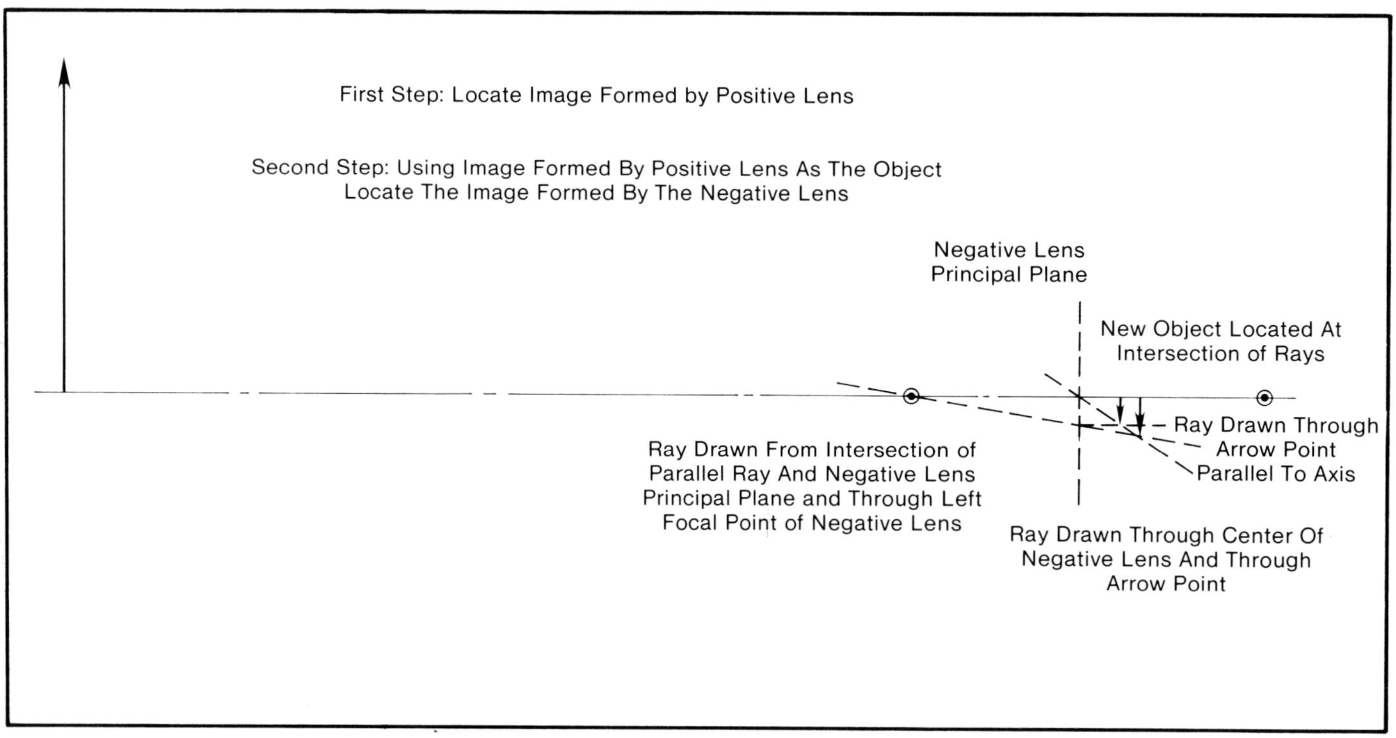

Figure 9-24. Schematic Ray Trace of Compound Projection Lens.

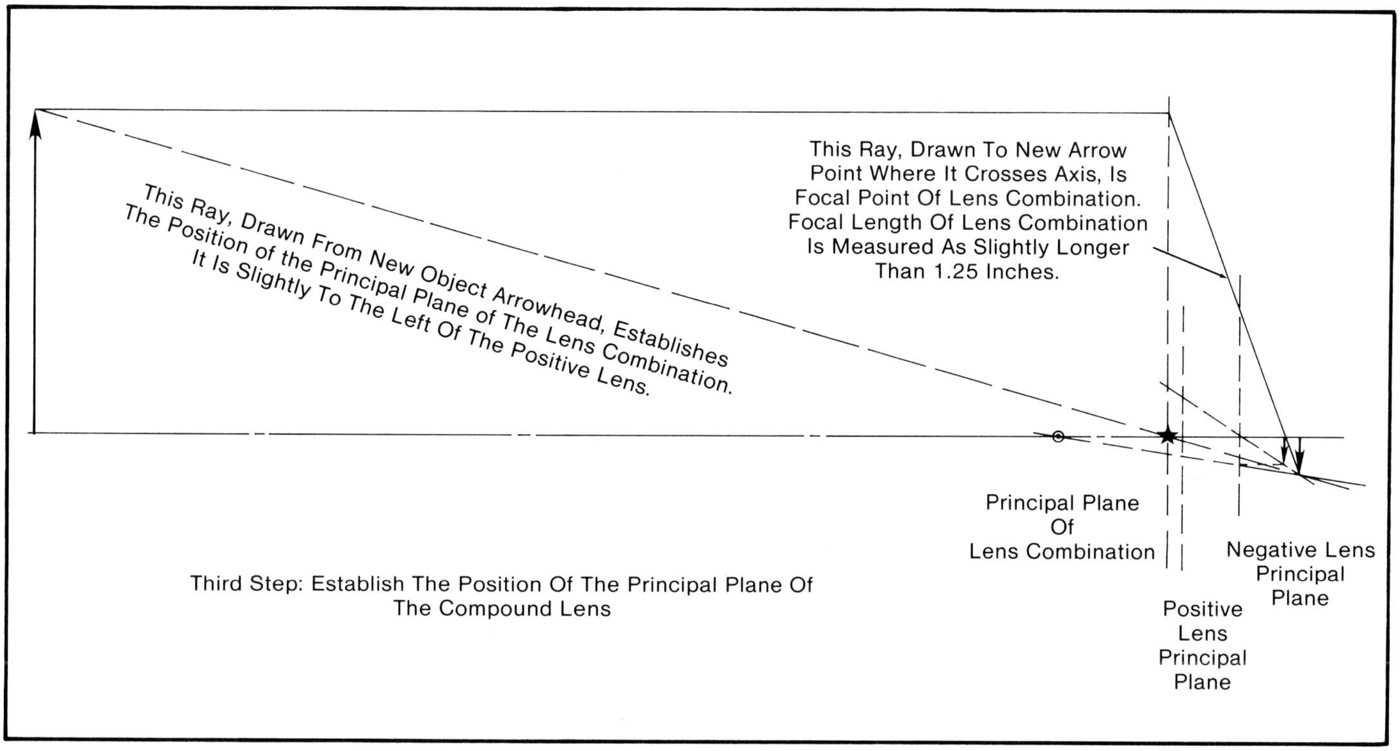

Figure 9-25. Schematic Ray Trace of a Compound Projection Lens.

the right. This is as expected, because the focal length of the compound lens must be greater than that of the positive lens alone. In practice this means the lens would have to be moved farther from the new object in order to focus its image on the screen. In order to keep the screen image the same size, the slide image would have to be increased.

Now it is necessary to find the location of the principal plane of the compound lens and the focal length of the compound lens. Figure 9-25 shows how this is done. A ray is drawn from the new object arrow point to the projected image arrow point. This ray crosses the axis at the nodal point of the compound lens as shown by the star. The principal plane of the compound lens is drawn through this nodal point.

Now a ray is drawn parallel to the projected arrow point to the compound lens principal plane. This ray is refracted to go to the new object arrow point. The lens focal point is at the intersection of this ray and the lens axis. Measuring the distance between the compound lens nodal point and the compound lens focal point gives the focal length of the compound lens, which is about 1.25 inches.

An equation that gives the focal length of two thin lenses verifies the schematic ray trace solution:

$$\frac{1}{f_c} = \frac{f_1 \times f_2}{f_1 + f_2 - d} \text{ where}$$

f_c = focal length of combination
f_1 = focal length of first lens
f_2 = focal length of second lens
d = distance between the lens

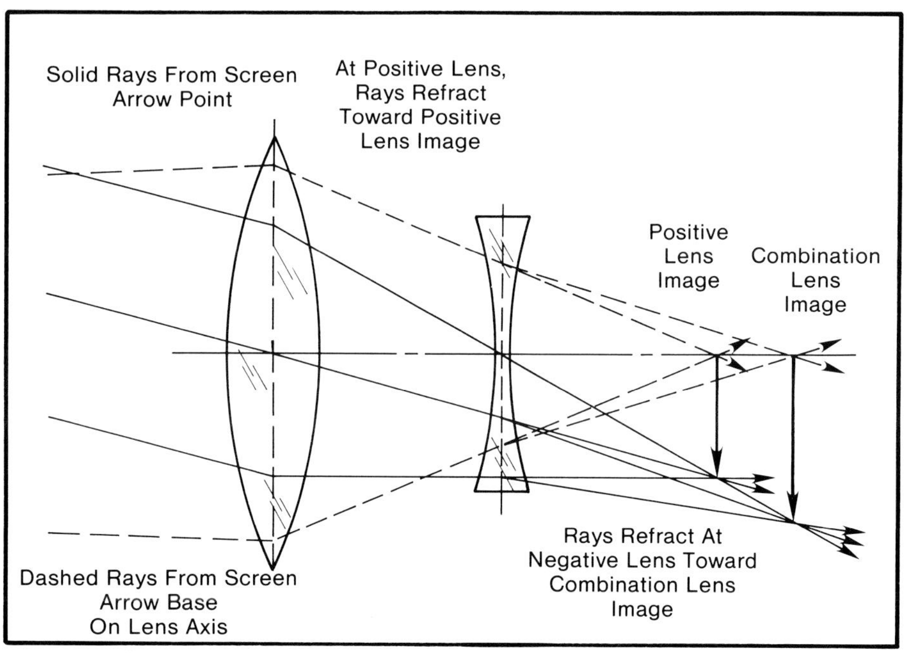

Figure 9-26. Schematic Ray Trace of Compound Projection Lens: Close-Up of Final Ray Paths.

$$\frac{1}{f_c} = \frac{1 \times (-2)}{1 + (-2) - 0.625}$$

$$\frac{1}{f_c} = \frac{-2}{-1.625} = 1.231 \text{ inches}$$

which is close to the measured value.

Figure 9-26 shows an enlarged view of the lens and images, and how the rays would actually go through the two lenses.

Reversed Telephoto Lens

The lens design just shown as a projection lens with a positive front element and negative rear element, is also used as a telephoto lens on cameras. It makes possible the construction of compact lenses with relatively long focal lengths.

For wide-angle lenses, a reverse telephoto lens is sometimes used. This type of lens has a negative front element combined with a positive rear element (actually groups of elements for correction of aberrations). The reverse telephoto lens (Figure 9-27) moves the equivalent refracting plane of the compound lens to the rear which makes a long back focus with a relatively short effective focal length.

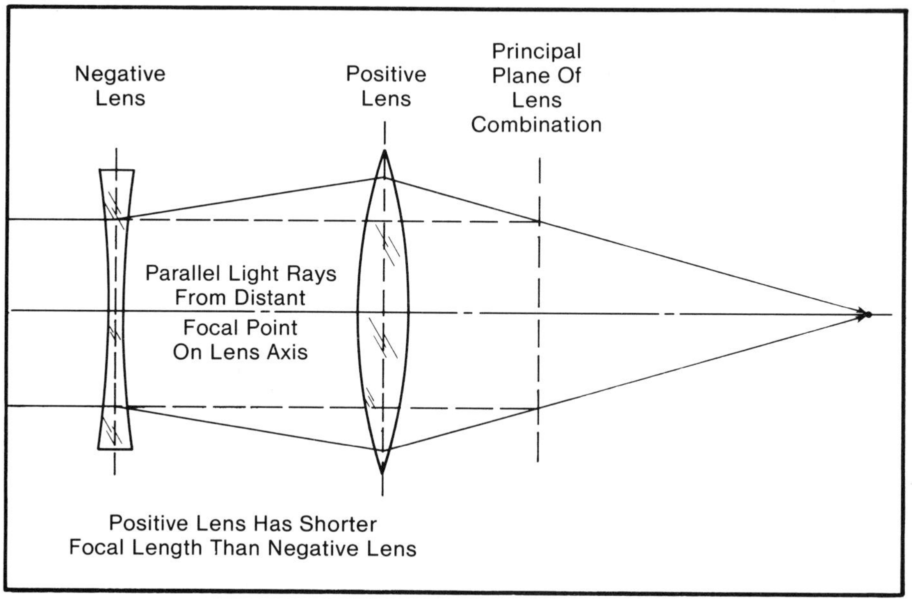

Figure 9-27. Reversed Telephoto Wide- Angle Lens.

The long back focus is useful in single-lens reflex cameras where the moving mirror requires clearance space. The problem for this chapter involves a lens of this type with a very special placement of the negative lens.

Problem

A positive lens with a 1-inch focal length is used as a projection lens. The screen is 7 inches from the lens, and the projected image size is 1.5 inches. Locate the slide position and determine the object size by schematic ray tracing.

Now a negative lens with a 1.25-inch focal length is placed in the left hand focal point of the positive lens between the lens and the screen. Locate and determine the size of the screen image as seen by the positive lens. Determine the size and location of an object required to produce the same image of the screen with the compound lens.

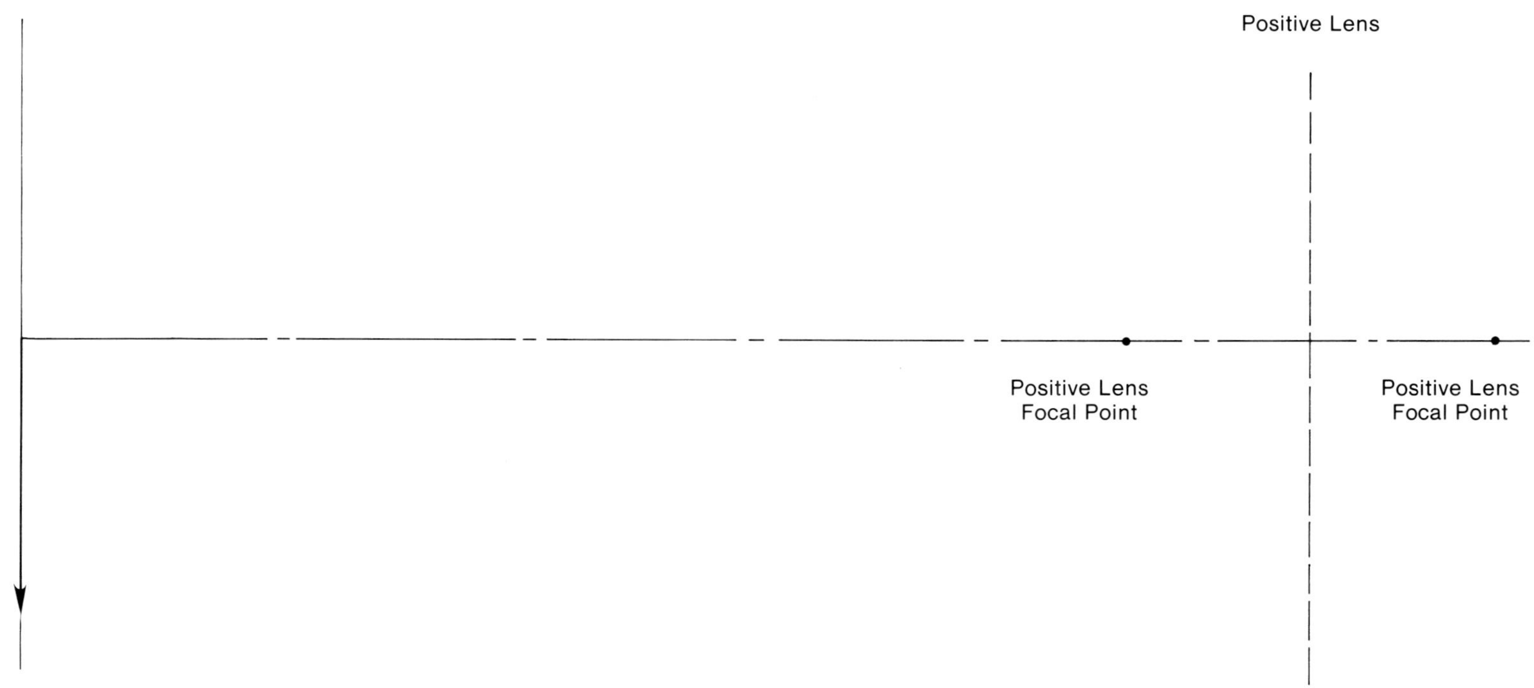

Problem
Chapter 9
Video Lesson 4

Chapter 10

VIDEO LESSON 5

Compound Lenses, Variable Focal Length Projection Lenses, Conjugate Nomograph, and Tele-extenders

Summary

When making a schematic ray trace of a compound lens formed of two positive thin lenses, each lens is treated separately. The image formed by the first lens becomes the object for the second lens. A variable focal length projection lens is designed using schematic ray tracing. Conjugate distances are defined as the object and image distances. A nomograph is developed which relates the two conjugate distances with focal length. When any two of these factors are known, the third is easily found. A tele-extender which lengthens the focal length of the lens is defined and designed.

Schematic Ray Tracing with Two Positive Thin Lenses

In the last chapter, the method of tracing rays through a compound lens formed of a negative and a positive thin lens was shown. The same principles are used when tracing rays through a compound lens formed of two positive lenses.

In Figure 10-1, the principal planes (equivalent refracting planes) of two positive thin lenses are shown. The focal length of the first lens is one inch and the focal length of the second lens is 0.75 inch. The lenses are placed 0.5 inch apart. The object is three inches from the first lens.

As previously shown, the two rays required to find the image of the first lens alone are drawn. One is parallel to the axis from the object-arrow's point to the lens. This ray is refracted and passes through the focal point. The other ray passes through the center of the lens without bending. The image is formed 1.5 inches behind the first lens and is about 0.38 of an inch long. This image is now treated as if it were the object of the second lens.

In Figure 10-2, a construction line is drawn parallel to the axis from the arrow point of this new object to the second lens where it becomes a ray. Note that this ray passes through the left focal point of the first lens as it travels to the original object point. This ray is parallel to the axis between the lenses and appears to pass through the second lens without bending because we are using it to locate the real path before it reaches the second lens. The dashed ray to the right of the second lens is the path the ray would use if the second lens were not there. It becomes a

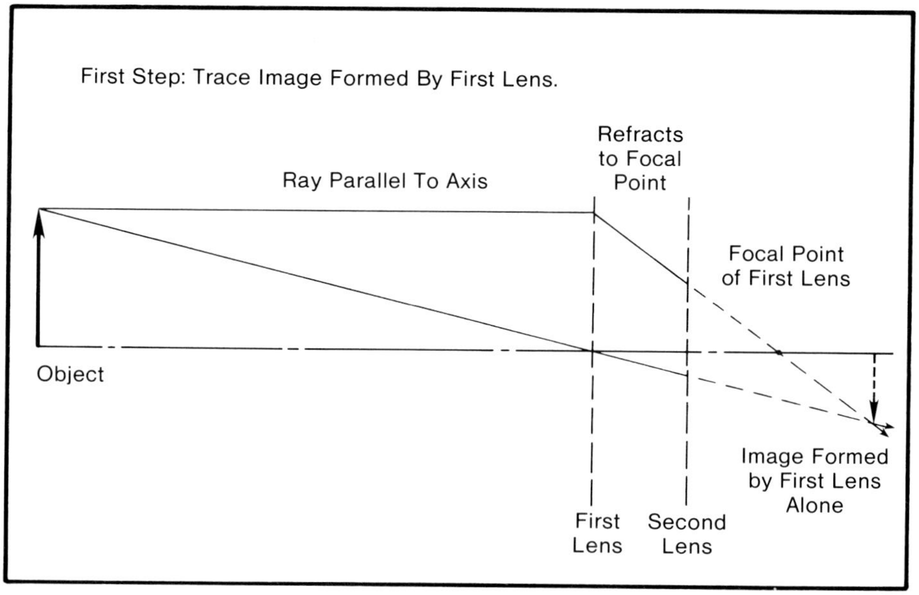

Figure 10-1. Schematic Ray Trace of a Compound Lens formed of two thin Positive Lenses.

construction line when the second lens is in place. This ray actually refracts to go through the focal point of the second lens because it is parallel to the axis as it reaches the second lens. This is shown as a solid line because it is an actual ray path.

A ray is now drawn through the center of the second lens to the arrow point of the second object (image of the first lens). The intersection of these two rays determines the location and size of the real image of the compound lens.

In Figure 10-3 the construction lines and rays forming the image of the first lens are eliminated, and the rays are shown as refracted by the second lens to form the real image. The upper refracted ray is continued backwards through the second lens and the parallel ray from the object is continued. The intersection of these two rays determines the location of the principal plane of the compound lens. The focal point of the compound lens is located at the intersection of the upper refracted ray and the axial ray. Measurement of the illustration reveals that the focal length of the compound lens is about 0.563 inch. The tracing can be checked with the equation given in the previous chapter. The effective combination focal length would be:

$$f_c = \frac{f_1 \text{ x } f_2}{f_1 + f_2 - d}$$

$$f_c = \frac{1 \text{ x } .75}{1 + .75 - .50} = \frac{.75}{1.25}$$

$$= 0.600 \text{ inch}$$

When using schematic ray tracing to find the images of compound lenses formed by two thin lenses, we must first assume that the rays pass through one of the lenses without refracting to provide the construction lines used to determine the actual location of the image. In Figure 10-3, the actual paths of the rays are shown as they refract at both lenses.

The Term "Conjugate"

The object distance and image distance are the two *conjugates* of an image-forming optical arrangement. The object distance is usually the long conjugate, and the image distance is the short conjugate (Figure 10-4).

However, with projectors, microscopes and enlargers, the object is close to the lens and the enlarged image is formed at a fairly great distance from the lens. Thus the object distance becomes the short conjugate, and the image distance becomes the long conjugate.

The following are descriptions of several combinations:

Camera lenses, telescope objectives
$u = f + x =$ long conjugate
copying objectives etc.
$v = f + x' =$ short conjugate
Microscope objectives, enlarger
$u = f + x =$ short conjugate
Projection lenses:
$v = f + x' =$ long conjugate

A Variable Focal-Length Projection Lens

The videotape session for this lesson illustrates a variable focal-length

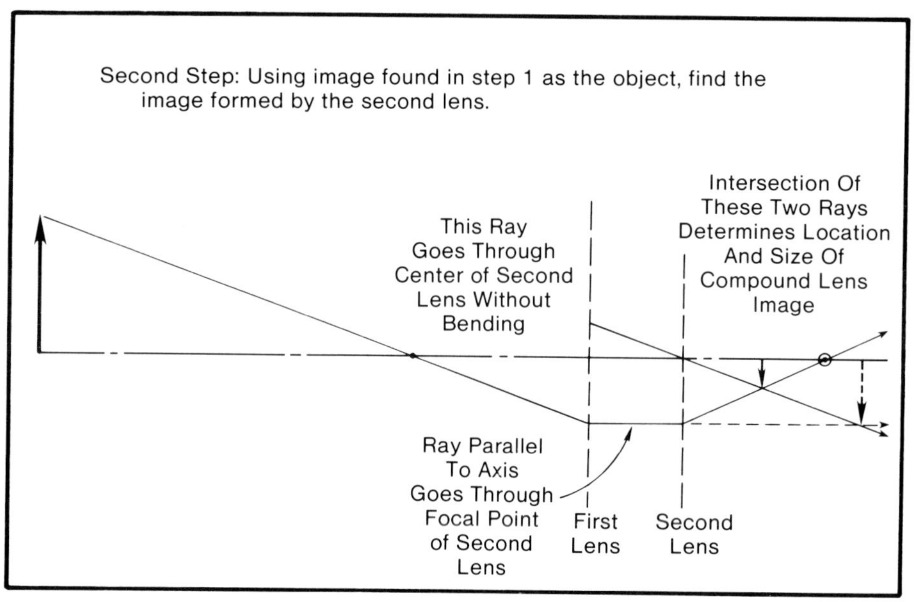

Figure 10-2. Schematic Ray Trace of Compound Lens.

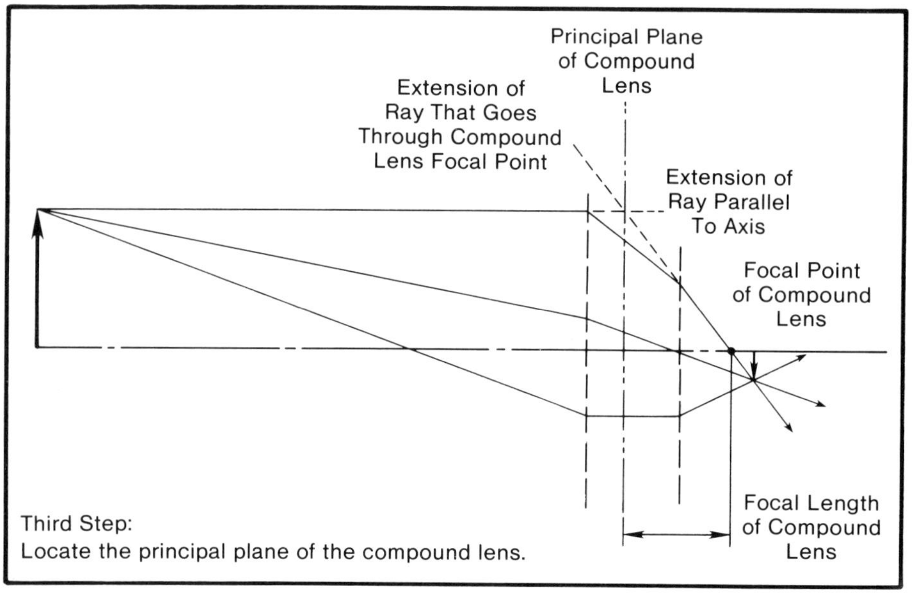

Figure 10-3. Schematic Ray Trace of Compound Lens.

projection lens. A positive projection lens is used to project the image of a slide on the screen at a given distance.

A negative lens is placed at the focal point of the projection lens and the image goes out of focus (Figure 10-5). As the slide is moved farther away from the projection lens (the short conjugate is increased) the image comes into focus. The new image on the screen is the same size as the original image using only the positive lens. This is always true regardless of the focal length of the negative lens.

When constructing the schematic ray trace that shows how this is accomplished, imagine the positive lens to be looking through the negative lens toward the screen. The negative lens makes the screen look smaller and closer than it actually is. Looking from the positive lens viewpoint, a virtual image of the screen can be seen.

In Figure 10-6, the screen with the arrow image is positioned at your left. Using only the positive lens, two rays locate the object size and position. In this case, the object is a slide in the projector.

In the second diagram (Figure 10-7), the negative lens has been added. The focal points of the negative lens are shown as are two rays that locate the size and position of the virtual image (the nomograph, described later in this chapter, could be used to locate the position of the virtual image).

In the third diagram (Figure 10-8), the positive lens is imaging the virtual image as though the negative lens were not there. The construction line is drawn from the image toward the object. This is possible because optical systems are reversible. Note that the object is the same size, but that it is farther from the positive lens. The slide in the projector had to be repositioned to reestablish sharp focus. This optical combination creates the same size image on the screen as was imaged by the positive projection lens alone. Figure 10-9 illustrates the actual ray paths with both lenses in place.

If the position of the negative lens is changed, the image size on the screen changes, and the slide must be refocused. When the negative lens is moved farther from the positive lens, the image size on the screen is increased. When the negative lens is moved closer to the projection lens, the image size is decreased.

Thus, a wide range of image sizes can be produced by placing a negative lens at different distances from the projection lens (Figures 10-10, 10-11). A single negative lens, however, introduces aberrations that reduce the quality of the projected image.

A simple zoom projection lens is designed to have positive and negative components with variable spacing between the two components. By making the negative and positive components of several elements, the aberrations can be corrected. Zoom projection lenses with this basic design are manufactured for both slide and motion picture projectors. A well designed zoom lens changes the image size without changing the focus, so that the example lenses are

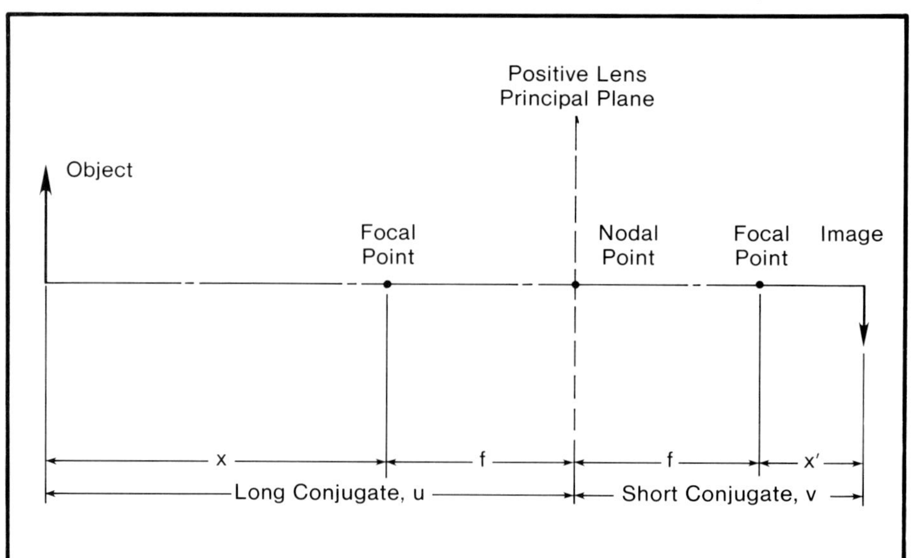

Figure 10-4. Conjugate Definition

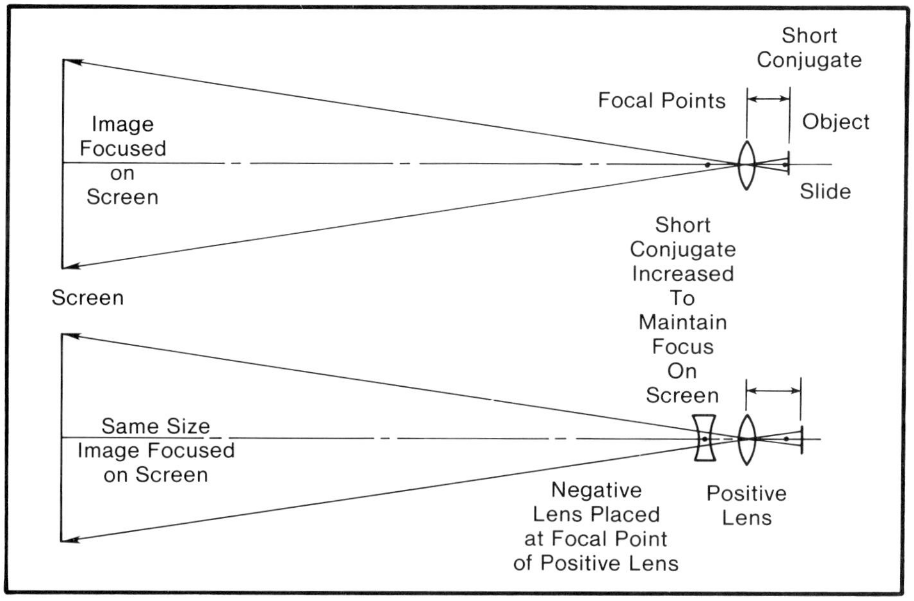

Figure 10-5. Variable Focal Length Lens

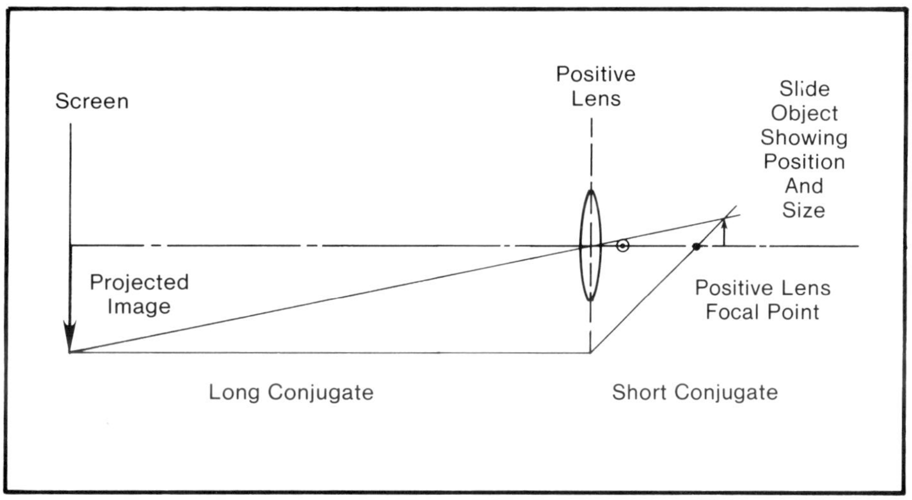

Figure 10-6. Zoom Projection Lens: finding location and size of object (slide) with positive lens only.

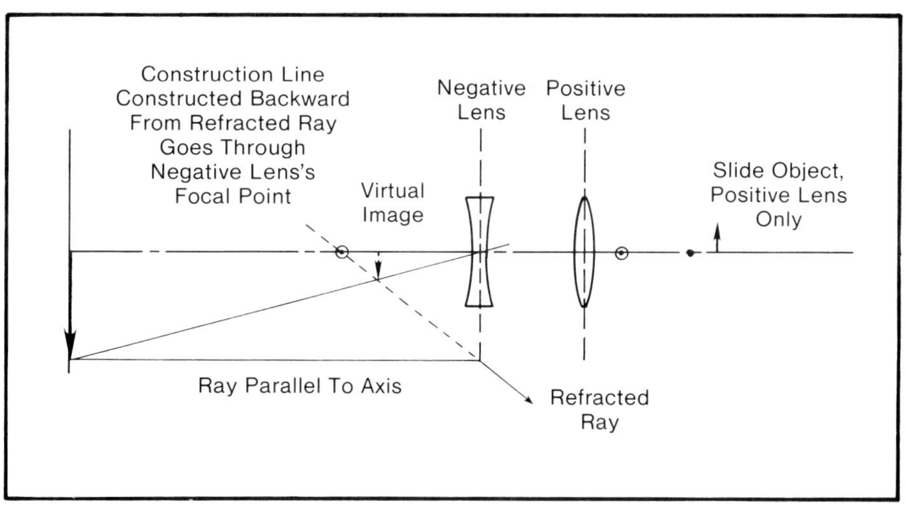

Figure 10-7. Finding Virtual Image Created By Negative Lens

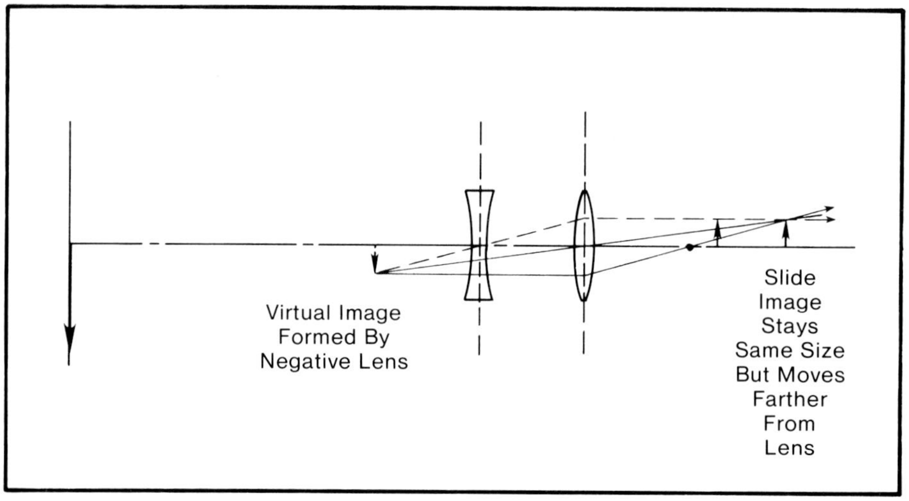

Figure 10-8. Positive Lens Imaging the Virtual Image formed of the Screen Image by the Negative Lens.

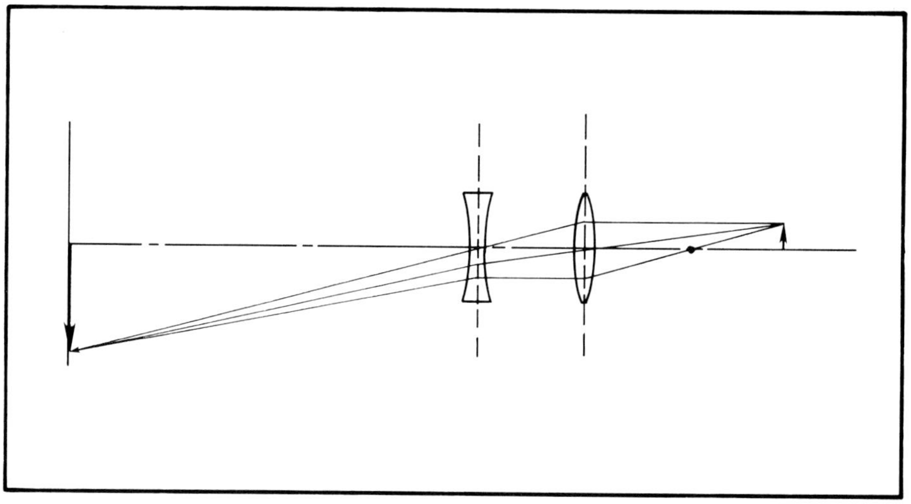

Figure 10-9. Actual ray paths with both lenses.

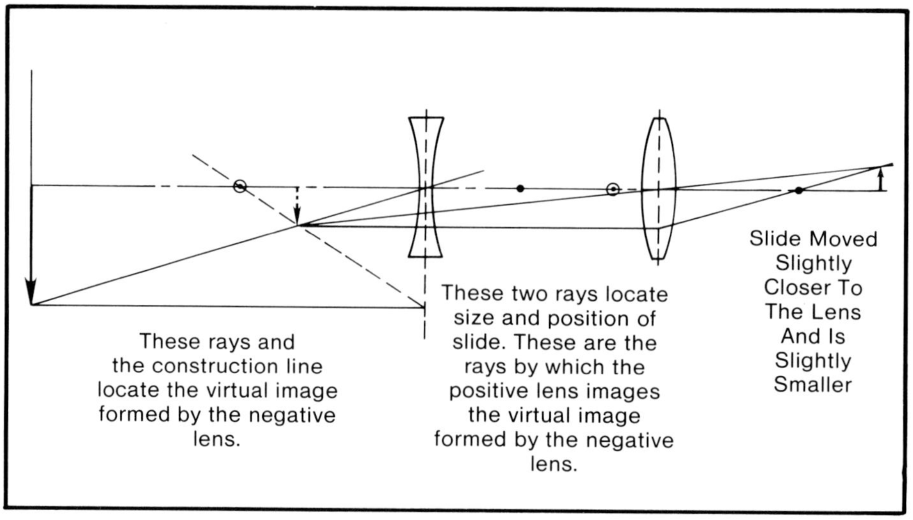

Figure 10-10. Negative Lens Moved Farther from the Positive Lens.

more properly called "variable focal-length" projection lenses. Zoom lenses for cameras have complicated spacing mechanisms that maintain lens focus when image size is changed.

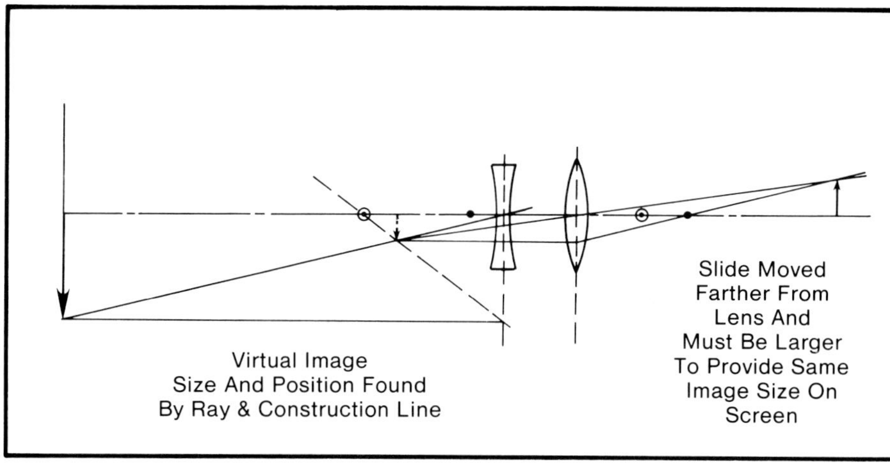

Figure 10-11. Negative Lens Moved Closer to the Positive Lens.

Development of an Object-Distance, Image-Distance Nomograph

Figure 10-12 illustrates the typical image formation diagram. The focal length of the lens is f, the object-distance is u and the image-distance is v. The distance x is the object extension and the distance x′ is called the image extension. The drawing shows that:

$$u = x + f, \text{ and}$$
$$v = x' + f$$

Two methods have been given for relating these distances. One is using schematic ray tracing, and the other is using the mathematical equation:

$$\frac{1}{f} = \frac{1}{u} + \frac{1}{v}$$

A third method is the nomograph method. A nomograph is a scaled diagram that relates three variable mathematical relationships. When two of the variables are known, the third is found by the use of a straightedge. In this nomograph, the variables are f, u, and v, and they are related mathematically as shown in the equation above. The nomograph can be made to any scale and with any units of measurement as long as the scale and the units are consistent. Millimetres are commonly used when describing optical elements, however this course has been using inches because most American students think in inches more easily than they think in metric units.

The First Quadrant

The first quadrant of the nomograph is the upper right quadrant (Figure 10-13). This quadrant is used to find the distances of f, u, and v when **positive lenses are used to form real images.** If the focal length f is known, the extensions are found by subtracting the focal length from u or v. If a straightedge is placed so that it intersects the known distance (u or v) and the focal length, it will then intersect the unknown distance on the other axis. Note that u = f + x and that v = f + x′ on these axes, as defined earlier. For example, when u (object distance) is 15 inches and the focal length is 5 inches, the straightedge intersects the horizontal scale at about 7.5 inches so that v is 7.5 inches, or x′ is 2.5 inches (Figure 10-14). By the equation:

$$\frac{1}{f} = \frac{1}{u} + \frac{1}{v}$$

$$\frac{1}{5} = \frac{1}{15} + \frac{1}{v}$$

$$\frac{1}{v} = \frac{1}{5} - \frac{1}{15} = \frac{2}{15}$$

$$v = \frac{15}{2} = 7.5 \text{ inches}$$

When the straightedge is placed at 45°, u = v, both u and v are twice f, and u + v is equal to 4 x f. For example, if f = 5, both u and v are found to be 10, and u + v = 20 (Figure 10-15). This means that when the magnification is 1, both the object distance and image distance are twice the focal length, and the object-to-image distance is 4 x f, or 20. By rotating the straightedge around the point, f = 5, it will be found that 20 is the smallest value

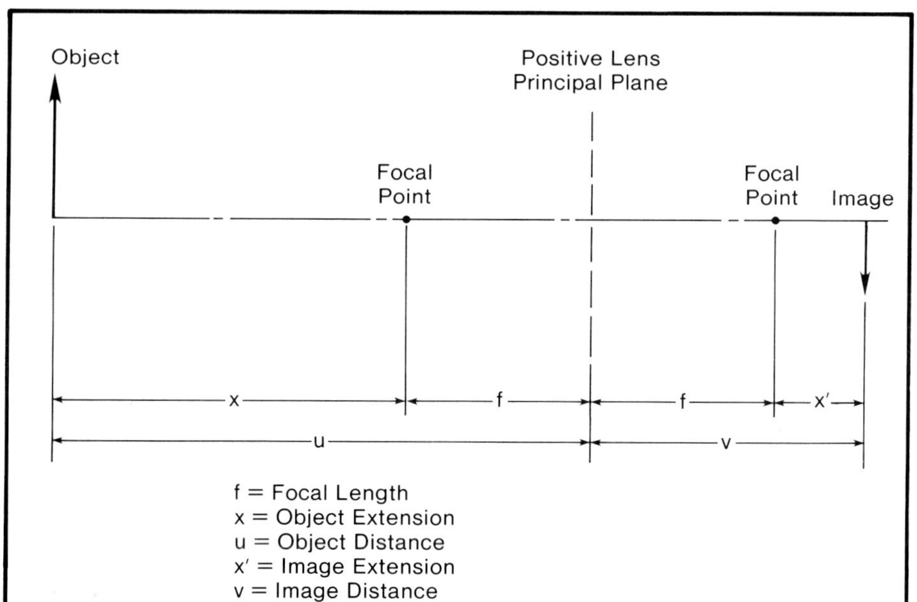

Figure 10-12. Typical Image-Formation Diagram.

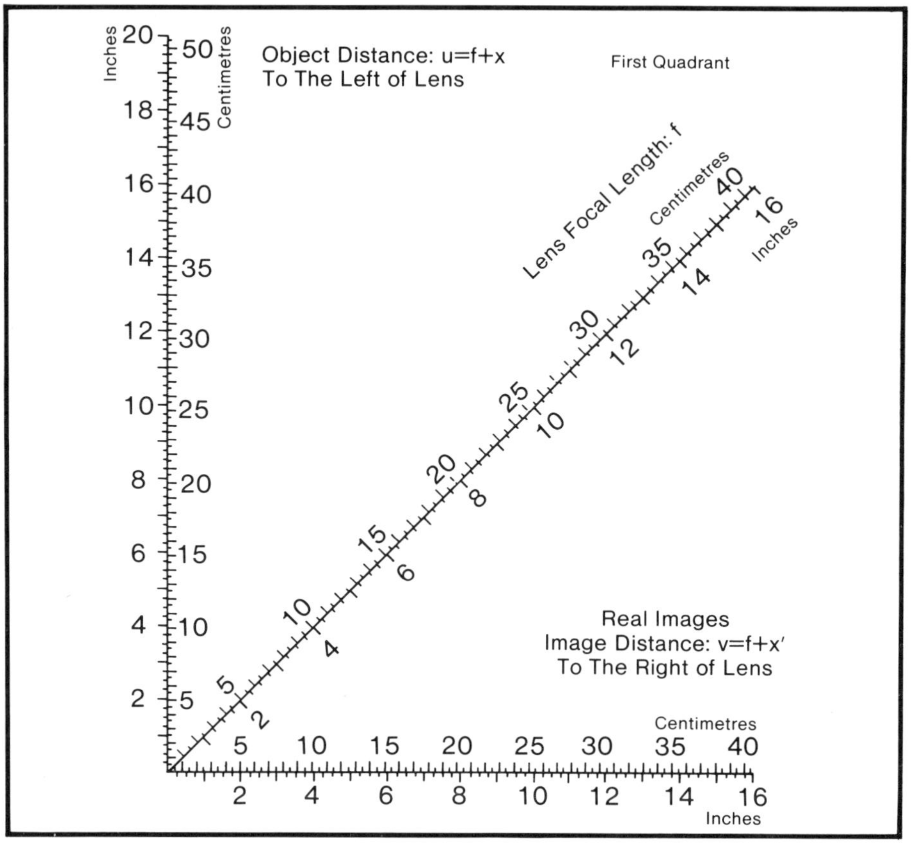

Figure 10-13. Image Formation Nomograph: First Quadrant, Upper Right.

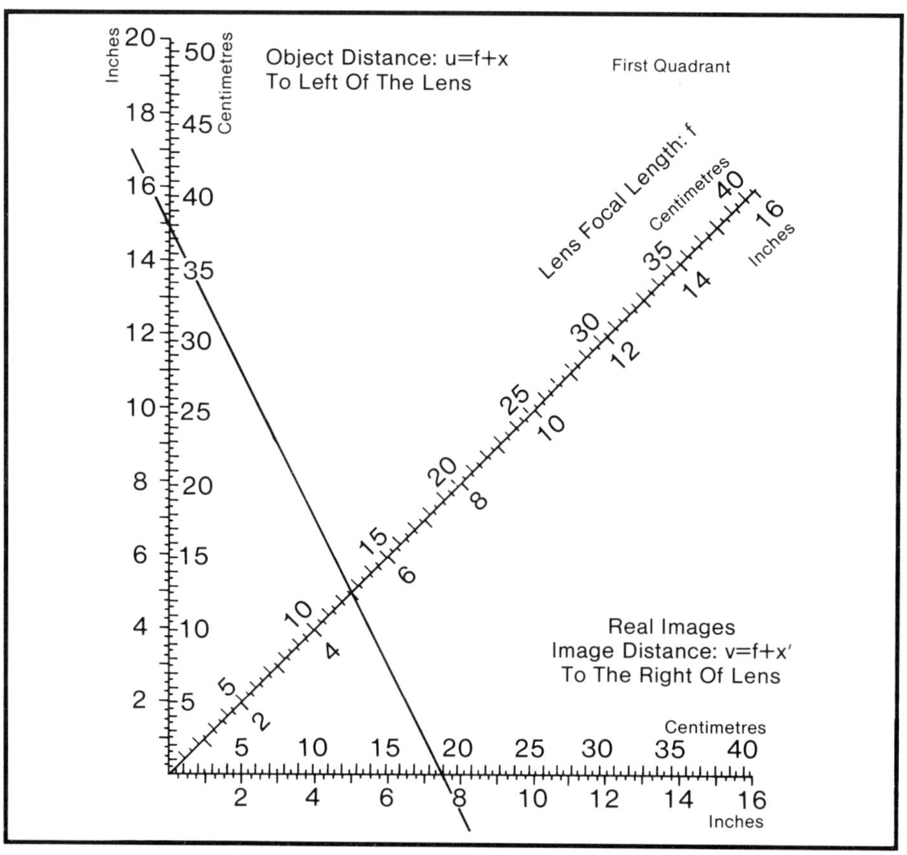

Figure 10-14. Image Formation Nomograph: First Quadrant, upper right.

that can be obtained for u + v; so that the smallest object to image distance obtainable with a positive lens is 4x its focal length, and this occurs when the magnification is 1.0.

The Second Quadrant

The upper-left quadrant is the second quadrant (Figure 10-16). **Images to the left of the vertical scale are virtual images,** and this quadrant is used to find the **location of virtual images of positive lenses.** While the first quadrant can be used by itself, the second quadrant must be used in conjunction with the first quadrant.

Positive lenses can create virtual images. When the object distance is less than the focal length, and when the object is to the left of the lens, the virtual image is also to the left of the lens. Figure 10-17 shows an object placed 3 inches to the left of a 5-inch focal length lens. The virtual image is found to be about 7.5 inches from the lens by the equation:

$$\frac{1}{f} = \frac{1}{u} + \frac{1}{v}$$

$$\frac{1}{5} = \frac{1}{3} + \frac{1}{v}$$

$$\frac{1}{v} = \frac{1}{5} - \frac{1}{3} = \frac{3}{15} - \frac{5}{15} = \frac{-2}{15}$$

$$v = \frac{-15}{2} = -7.5 \text{ inches}$$

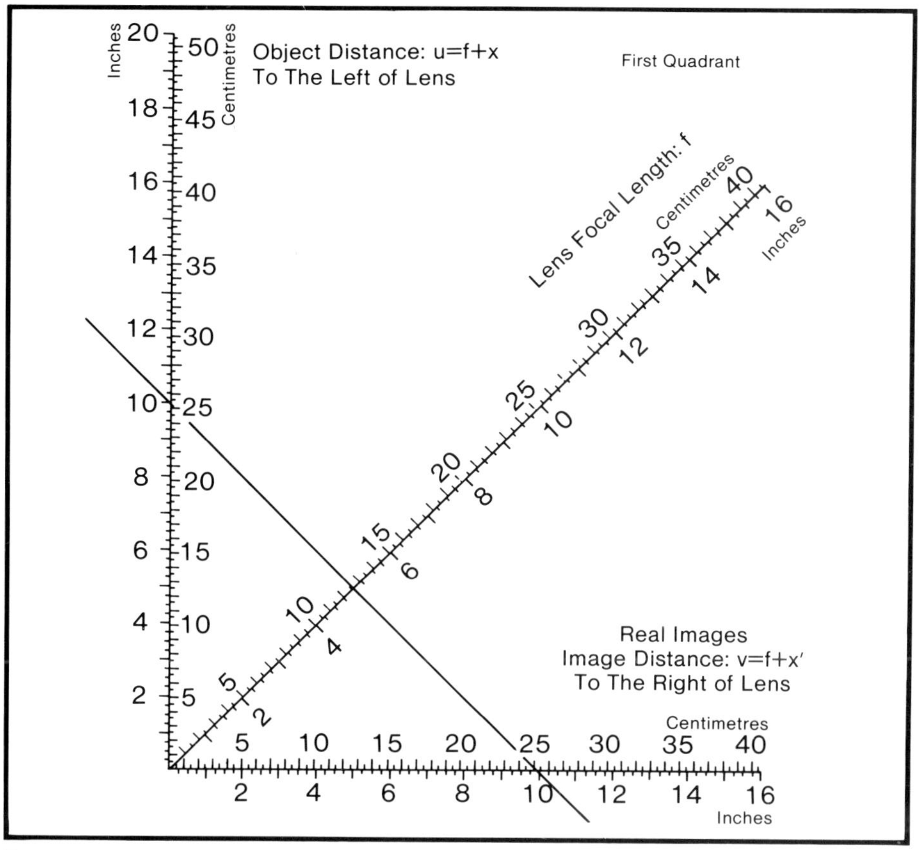

Figure 10-15. Image Formation Nomograph: First Quadrant, Upper Right.

The Third Quadrant

The third quadrant is the lower left quadrant and **is used for negative lenses** as shown in Figure 10-18. The 45° focal length fulcrum line is extended from the first quadrant through the zero point, and *gives the focal lengths of negative lenses*. The horizontal axis indicates distances to the left of the lens and only for virtual images.

In Figure 10-19, the negative lens has a focal length of −5 inches, and the object distance to the left of the lens is 10 inches (Figure 10-11). A straightedge placed on the −5-inch fulcrum, and rotated until it intersects the 10-inch mark on the vertical scale of the first quadrant (the distance to the left of the lens) intersects the left horizontal axis just beyond the −3-inch mark. This means that the virtual image is located slightly more than 3 inches to the left of the lens. Checking with the equation, we find:

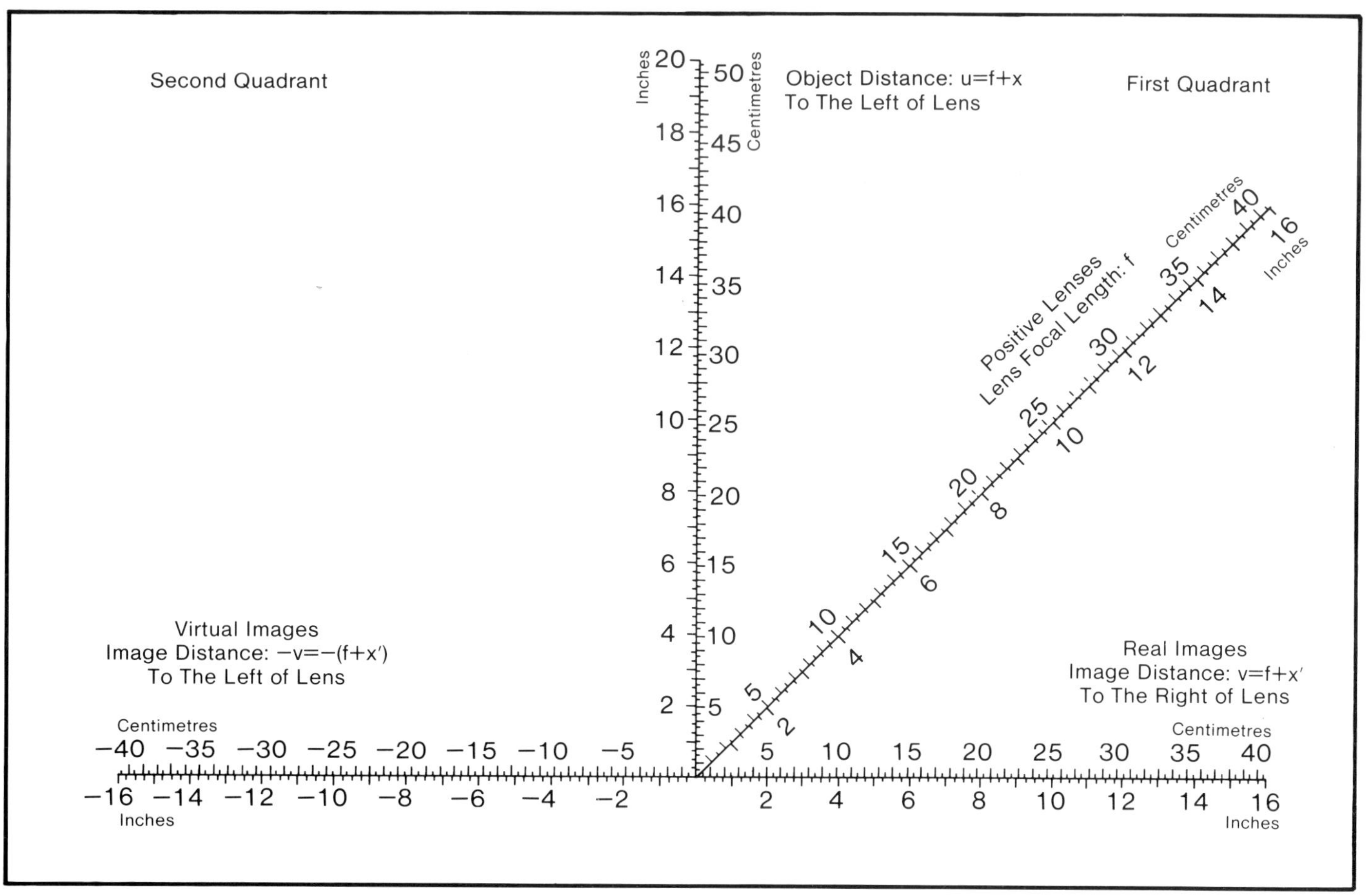

Figure 10-16. Image-Formation Nomograph.

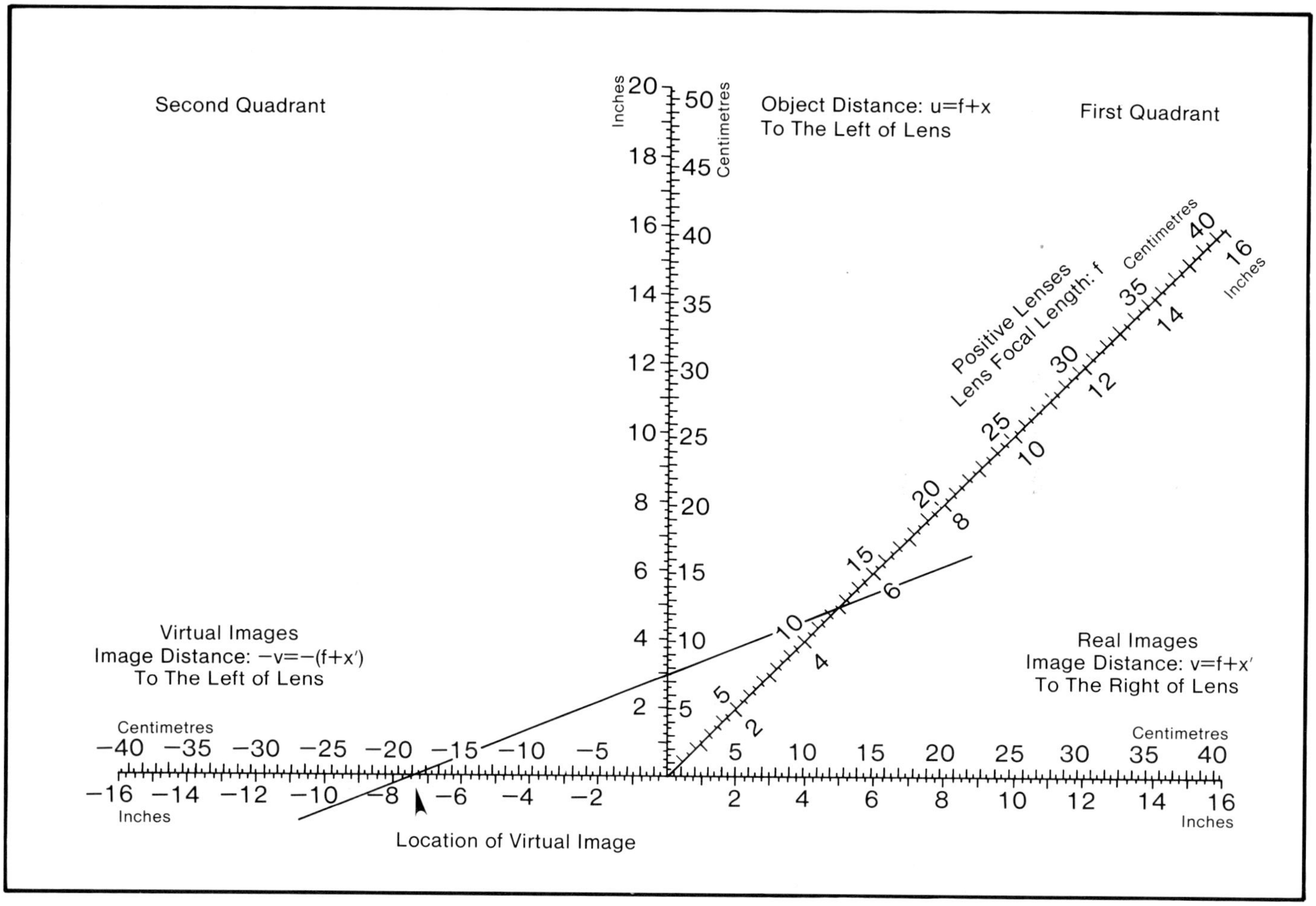

Figure 10-17. Image Formation Nomograph.

$$\frac{1}{f} = \frac{1}{u} + \frac{1}{v}$$

$$\frac{-1}{5} = \frac{1}{10} + \frac{1}{v}$$

$$\frac{1}{v} = \frac{-1}{5} - \frac{1}{10} = \frac{-2}{10} - \frac{1}{10} = \frac{-3}{10}$$

$$v = \frac{-10}{3} = -3.33 \text{ inches}$$

With the full page nomograph the distance is found to be between −3.25 inches and −3.5 inches—close enough for many first layout approximations by an optical engineer.

The Fourth Quadrant

The fourth quadrant (Figure 10-20) is the lower right hand quadrant and **is used when the object and the image are at the right of a negative lens.**

In this example, the object is 2.5 inches to the right of the −5-inch focal length of the lens. The straightedge shows that the image is about 5 inches from the lens. Checking, we find:

$$\frac{1}{f} = \frac{1}{u} + \frac{1}{v}$$

$$\frac{1}{v} = \frac{1}{u} - \frac{1}{f}$$

$$= \frac{1}{2.5} - \frac{1}{(-5)}$$

$$= 0.4 + 0.2$$

$$v = \frac{1}{0.6} = 1.67 \text{ inches}$$

Limits to the Use of the Nomograph

The nomograph is useful with positive and negative single-element lenses and compound lenses when the focal lengths and object distances are not too great. However, when the object distance is very great compared to the focal length, the value of x′ gets very small, and even with the required nomograph scale, its measurement is relatively inaccurate. The locations of the two principal planes of compound or thick lenses must be known if the nomograph is to be used successfully (Figure 10-21). The value of v is measured from the second principal plane, while the value of u is measured from the first principal plane. Figure 10-22 shows the approximate locations of the first two principle planes in a variety of thick lens shapes.

The nomograph can also be used for the approximate solutions of problems in which a negative lens is combined with a positive lens such as the variable focal-length projec-

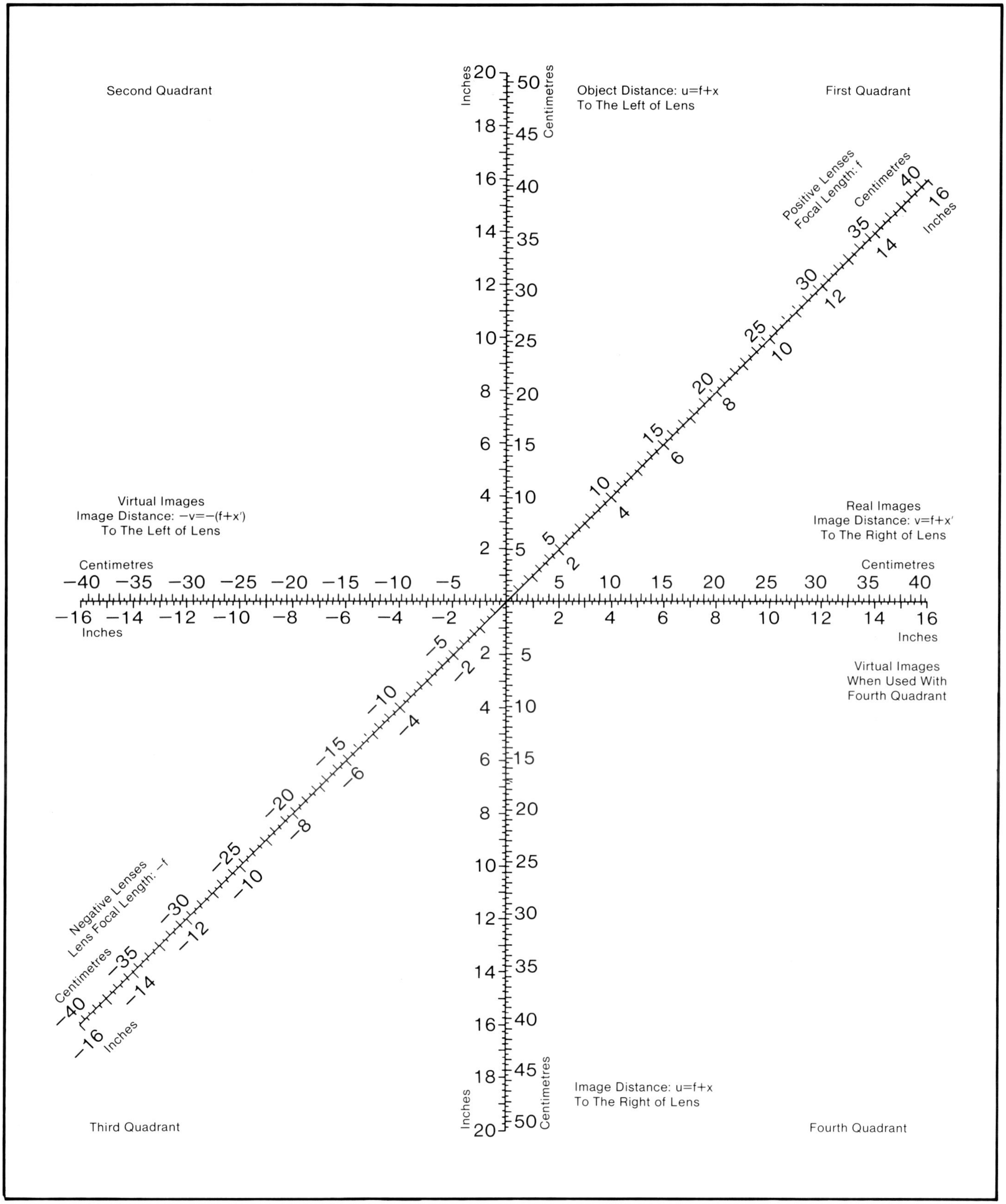

Figure 10-18. Image Formation Nomograph.

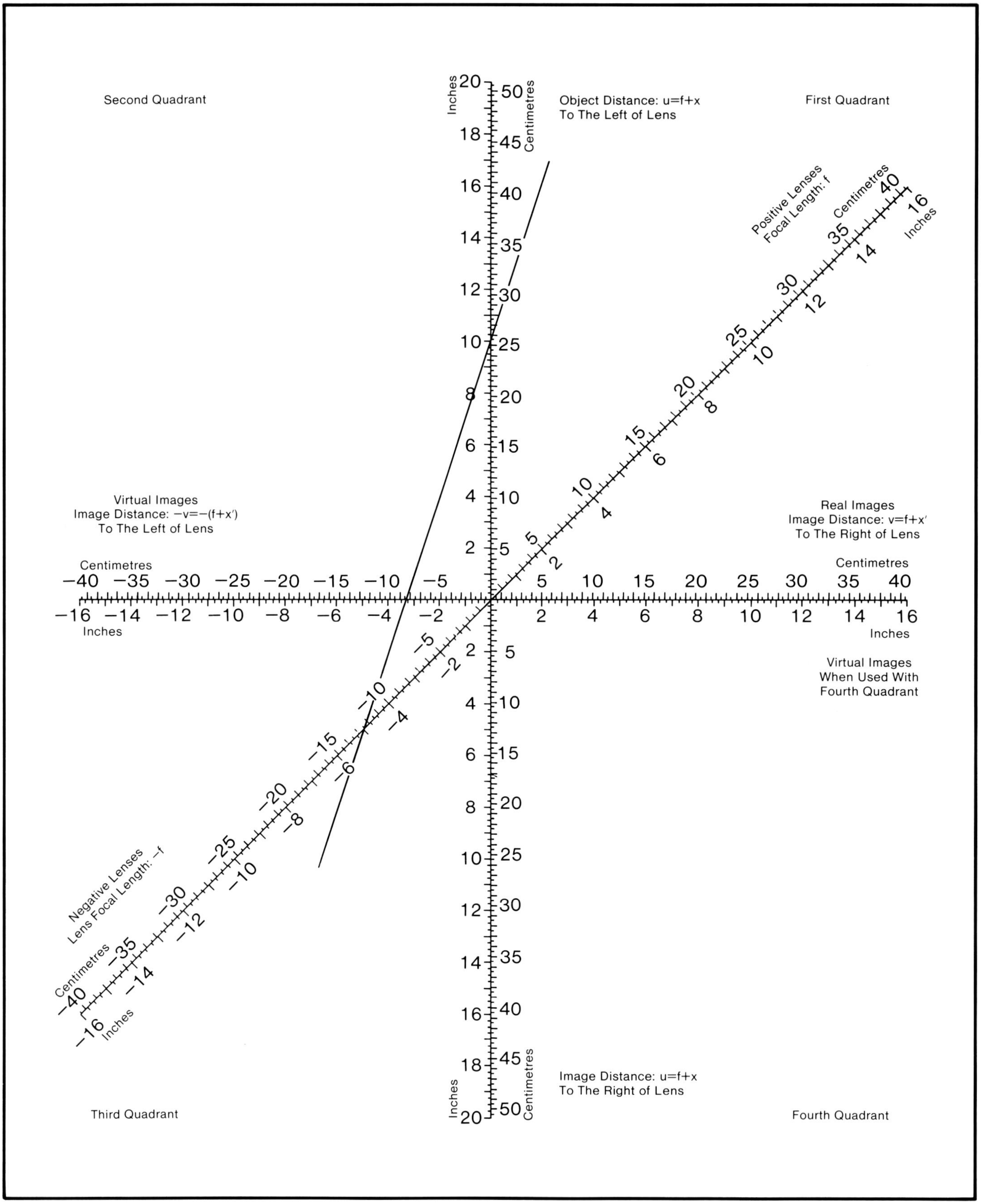

Figure 10-19. Image Formation Nomograph.

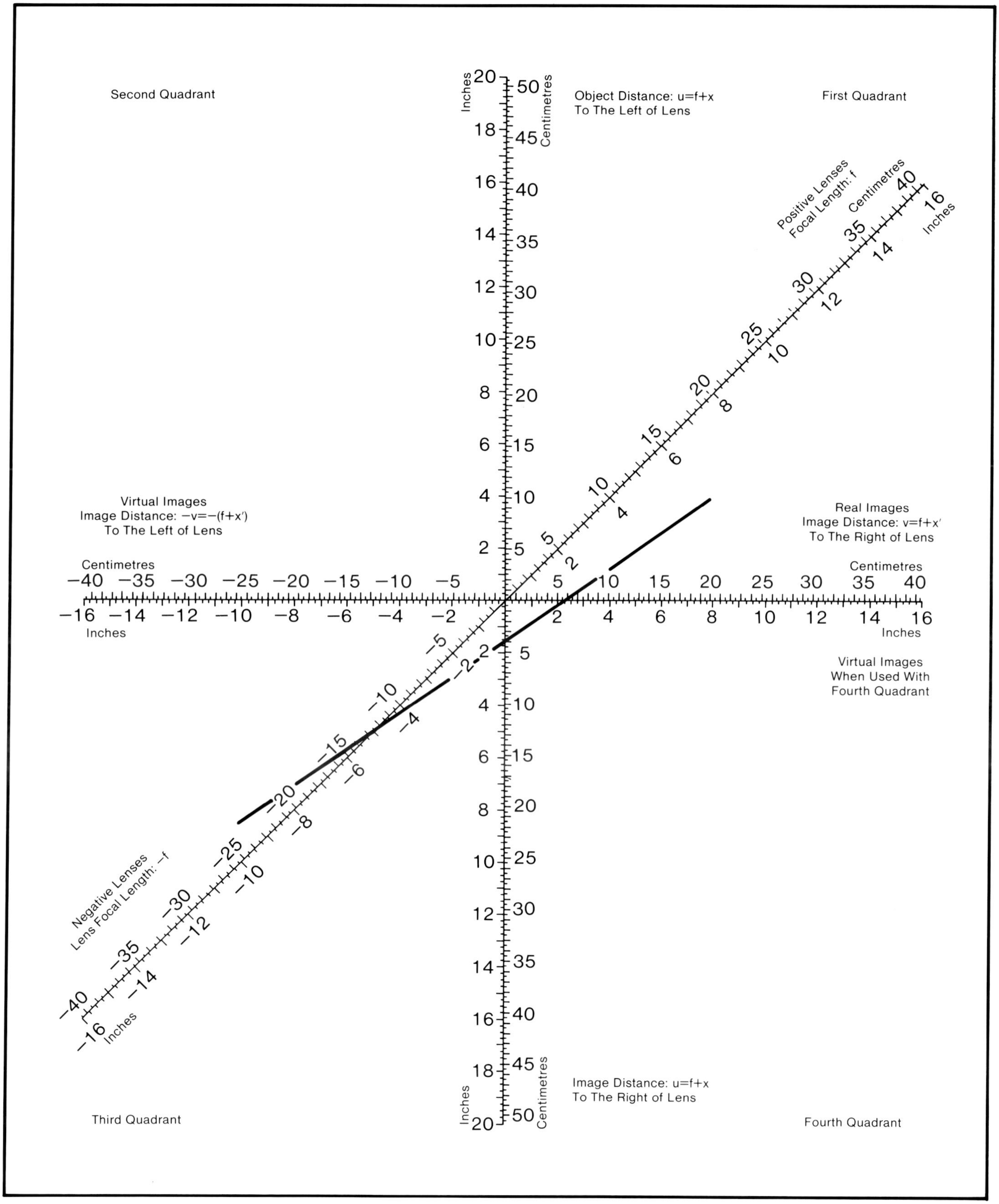

Figure 10-20. Image Formation Nomograph.

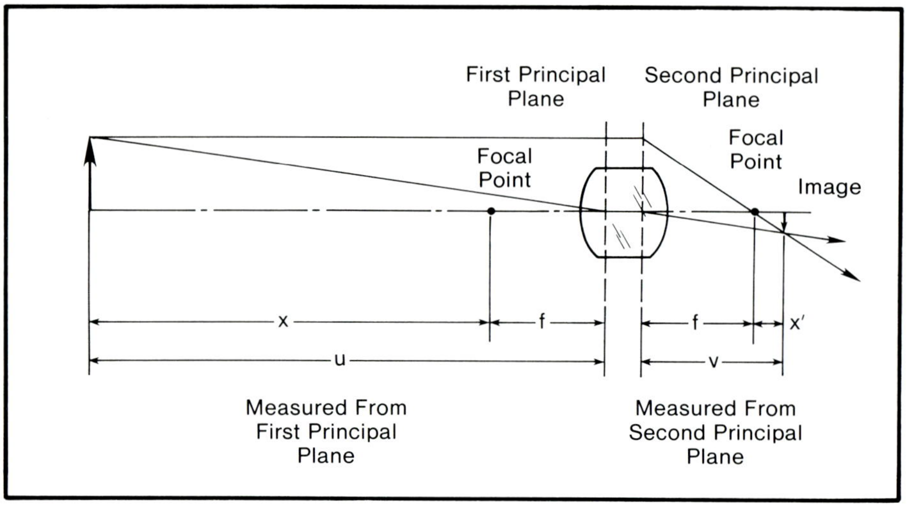

Figure 10-21. Image and Object Distances with a Thick Lens.

tion lens and tele-extender lens described later in this chapter. In both cases, locating the virtual image formed by a negative lens is the first step, and this can be accomplished with the nomograph.

Tele-extenders

The variable focal-length lens described earlier is made by placing a movable negative lens on the long conjugate side of a positive lens. A tele-extender is made by placing a fixed position negative lens on the short conjugate side of a positive lens.

Photographers who use 35 mm single lens reflex cameras with interchangeable lenses, find many uses for lenses having long focal lengths. Such lenses are common, but often are quite costly. A relatively inexpensive way to achieve the same results is to use a tele-extender which is an optical device that lengthens the focal length of lenses the photographer already has on hand. Tele-extenders are usually made in two powers: a 2X tele-extender doubles the focal length of any lens used with it, while a 3X tele-extender triples the focal length.

Cameras which use interchangeable lenses are made so that the lenses can be easily removed from the camera. The most common device for attaching the lenses is called the bayonet mount (Figure 10-23). The camera has a lens mount seat and a receptacle for the bayonet mount on the lens. Each lens has a matching lens seat which positions the lens accurately on the camera, and a bayonet ring which, with a partial turn, anchors to the receptacle and holds the lens firmly in place. Each lens is made so that its optical image is in focus on the film plane when the lens is seated. The seat is called the reference seat because it accurately positions the lens at a reference distance from the film plane. Each lens places its focused image the same distance behind the reference seat.

The tele-extender is placed in a mount that fits the camera's lens seat on one end, and accepts the bayonet of the camera lens on the other end.

The negative lens (usually a compound lens) is placed so that its focal point is on the film plane. In this position, it enlarges the image by a factor of 2 or 3, depending on whether it is a 2X or 3X extender.

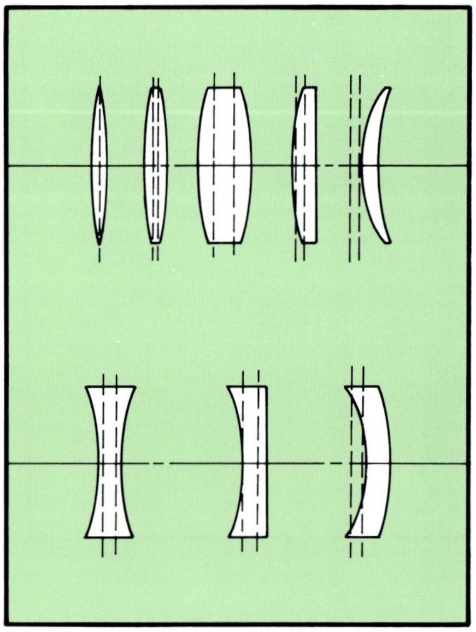

Figure 10-22. Approximate location of the Principal Planes in some Single Element lenses.

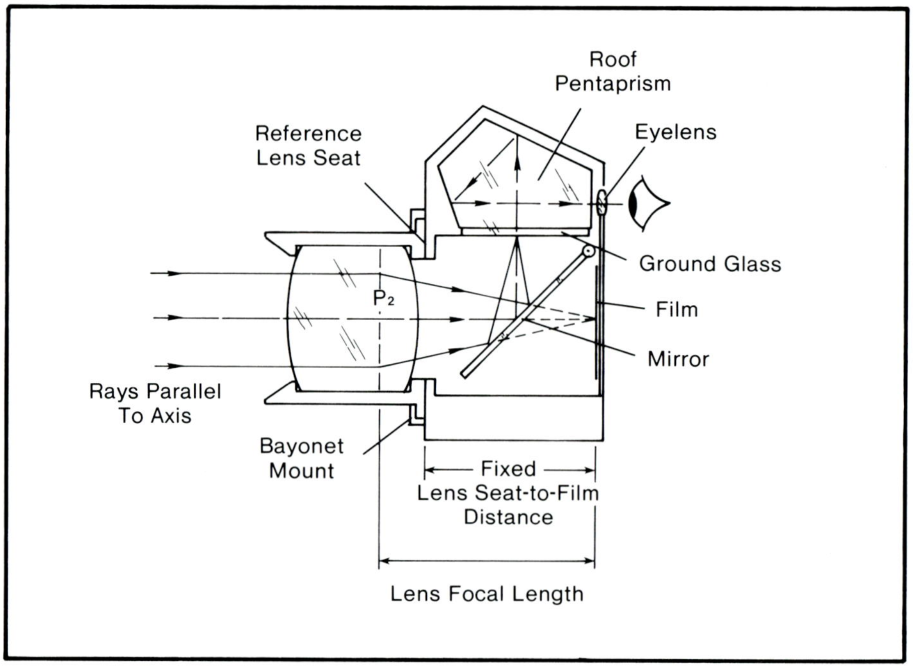

Figure 10-23. 35 mm Single Lens Reflex Camera Showing Lens Seat & Bayonet Mount.

Figure 10-24 shows where the virtual image of the negative lens is formed—a distance *d* in front of the film plane. The ray trace shows the two rays used to locate the virtual image of the negative lens. One ray is drawn from the camera image-arrow through the center of the negative lens. The other ray is drawn parallel to the axis from the arrow point. The working line, which is an extension of the ray, must pass through the focal point of the lens, which is on the film plane. The virtual image is located where these two rays intersect, equidistant between the optical center of the negative lens and the negative lens focal point. It is one-half the size of the real image. This, of course, is with a 2X tele-extender.

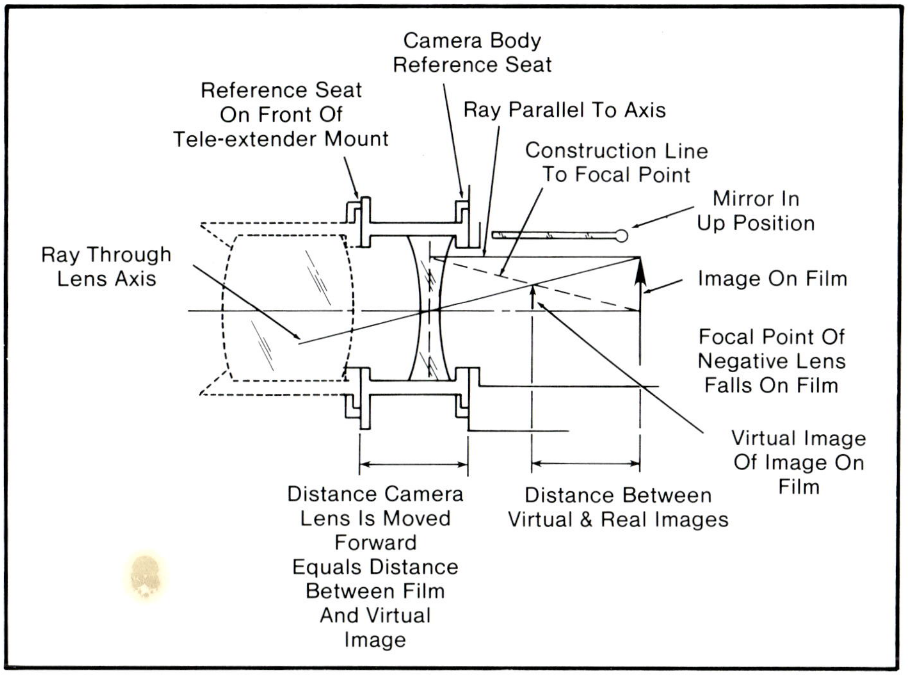

Figure 10-24. Schematic Ray Trace of Tele-extender In Place On Camera

The image formed by the camera lens must focus on this virtual image location. Therefore, it must be placed at a distance *d* in front of the negative lens. The dimension between the two seats on the tele-extender mount must be equal to the distance.

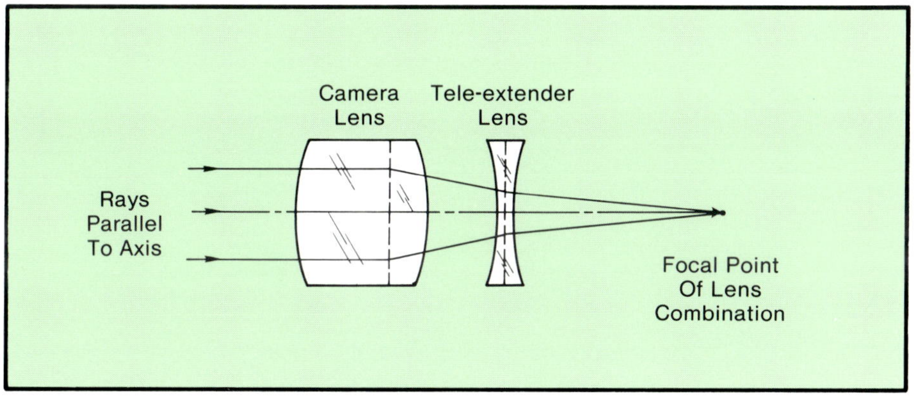

Figure 10-25. Schematic Ray Trace Showing Paths Of Rays Parallel To Lens Axis Going Through Camera Lens and Tele-extender.

The camera lens "thinks" it is placing the image in this location, but the tele-extender lens lengthens the focus and places the image on the film (Figure 10-25).

If you were the camera lens looking at the film through the tele-extender, the film would look half-size to you because you were looking through a negative lens (Figure 10-26). However, the image you make would look the same size to you as if the tele-extender were not there, so that the image looks as though it were on a smaller film. In reality, the image has become twice-size on the film.

There are two effects to consider when using a tele-extender. First, the aberrations are increased slightly, because additional lenses have been added. Therefore the use of a tele-extender may reduce the definition of the image slightly. Second, the effective aperture of the lens is reduced.

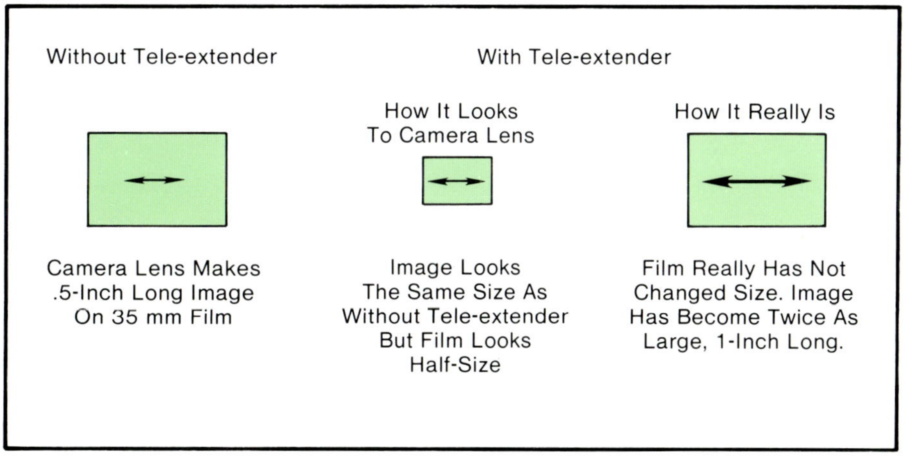

Figure 10-26. How Image Looks to Camera Lens, and How It Really Is.

The Effect of Tele-extenders on *f*-numbers

The *f*-number is a term used to define the aperture and it is determined by dividing the focal length by the diameter. When tele-extenders are used they alter the *f*-number. For example, if an *f*-2 lens is used in combination with a 2X tele-extender, the effective *f*-number becomes *f*-4. If we had an *f*-4 setting on the lens and used it with a 2X tele-extender, it would really be operating at *f*-8. Thus a 2X tele-extender reduces the amount of light reaching the film by factor of four (two *f*-stops). This can be seen when you cut the diameter of a circle to one-half; the area of the circle becomes one quarter. A detailed discussion of *f*-numbers is covered in the next chapter.

Problems

1. The drawing represents the upper right quadrant of the nomograph presented in this lesson with generalized values of f, x and x′. There is a mathematical equation that is often used in solving optical problems:

$$x \cdot x' = f^2$$

which says that the distance x times the distance x′ equals the focal length squared. Using the geometric rule that states that similar sides of similar triangles are proportional, derive the above equation.

2. Magnification is a relationship between image size and object size:

$$m = \frac{\text{image size}}{\text{object size}}$$

It is optically true that magnification is equal to image distance divided by object distance.

$$m = \frac{x' + f}{x + f} = \frac{v}{u}$$

Using this relationship and the similar triangle relationship, derive the following equations to find the two conjugates; that is, object distance and image distance.

Object conjugate

$$u = x + f = f + \frac{f}{m}$$

Image conjugate

$$v = x' + f = f + mf$$

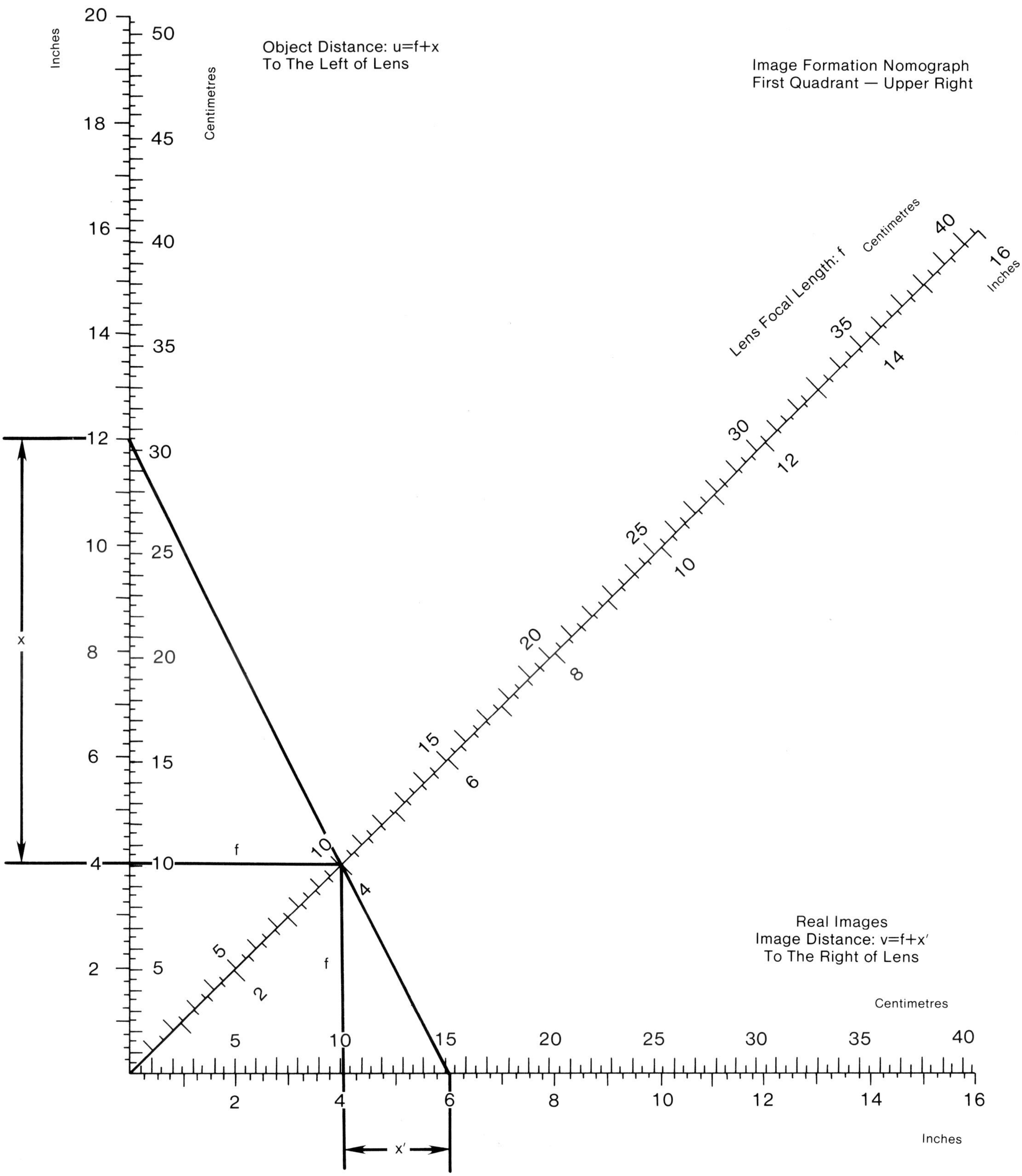

Problem
Chapter 10
Video Lesson 5

Chapter 11

VIDEO LESSON 6

f-numbers, f-numbers with Tele-extenders, Types of Diaphragms, and Diaphragm Shapes

Summary

f-numbers are defined as the ratio of the lens focal length divided by the diameter of the lens opening. Smaller *f*-numbers mean larger lens openings and greater image illuminance. *f*-numbers are commonly used on camera lenses along with exposure times to establish photographic exposures. Use of a 2X tele-extender is found to reduce the illuminance of the image by a factor of 4X at the same marked *f*-number. Diaphragms are used to change lens openings. Diaphragm openings may be round, square or oddly shaped. Single or two-blade diaphragms for automatic cameras are usually oddly shaped. An oddly shaped diaphragm may cause slightly less sharp images at a given aperture, and may show up in out of focus images of small bright object areas.

f-numbers

The speed of camera lenses is calibrated in *f*-numbers or relative apertures. The *f*-number is the focal length of the lens divided by the diameter of the lens. If a lens has a focal length of 2 inches and a diameter of 1 inch, its *f*-number is:

$$f\text{-number} = \frac{f}{d} = \frac{2}{1} = f/2$$

As shown in Figure 11-1, light parallel to the axis coming to the lens, forms a cone behind the lens, the apex of which is located at the focal point. The cone angle of all *f*/2 lenses is about 28°.

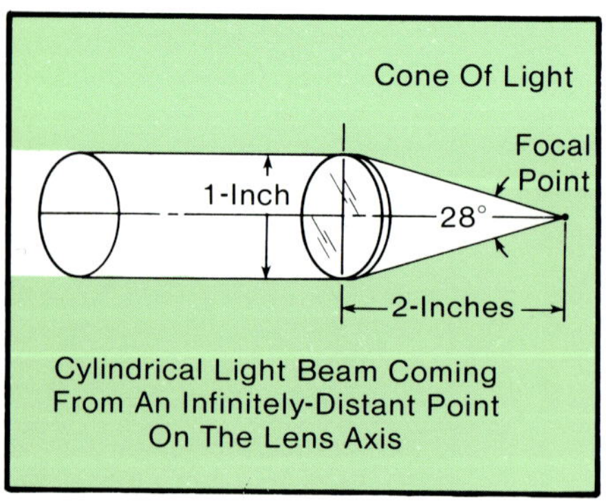

Figure 11-1. f/2.0 Lens.

A lens with a diameter of 0.5 of an inch and a focal length of 1 inch is also an *f*/2 lens and produces a cone angle of 28° (Figure 11-2). Since the cone angle is the same as with the *f*/2 two-inch lens, the illuminance at the focal point is the same. Illuminance is a measure of the brightness of light falling on a surface (in this case it is the film in the camera on which the lens is placed). A lens with a diameter of 2 inches and a focal length of 4 inches is also an *f*/2 lens, and has a cone angle of 28°. The illuminance at the focal point is the same as for the other two.

If this is not apparent, perhaps looking at the brightness of images another way may help. In Figure 11-3, a gray card 24 inches square is placed 10 feet away from two lenses.

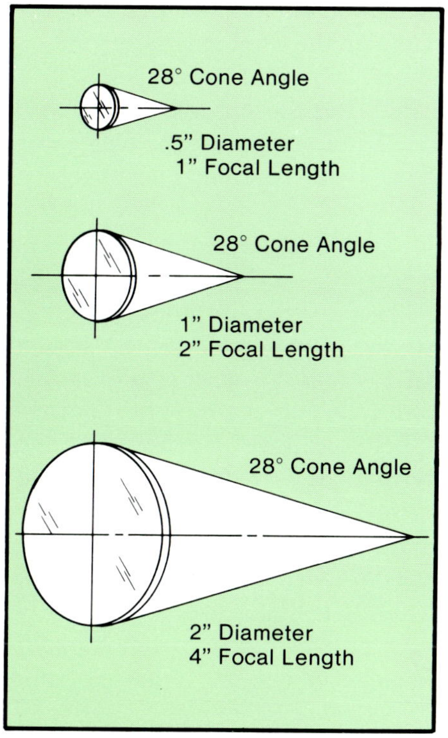

Figure 11-2. All f/2.0 lenses.

One is an $f/2$, 4-inch focal length lens, while the other is an $f/2$, 2-inch focal length lens. The magnification with the 4-inch lens is:

$$4 \div 120 = 0.333X$$

while the magnification for the 2-inch focal length is:

$$2 \div 120 = 0.0167X$$

The image of the gray card formed by the 4-inch lens is:

$$0.333 \times 24 = 0.800 \text{ in. sq.}$$

The area of the image over which the light from the object must be spread is:

$$0.800 \times 0.800 = 0.640 \text{ sq. in.}$$

The image formed by the 2-inch focal length lens is:

$$0.0167 \times 24 = 0.400 \text{ in. sq.}$$

Its area is:

$$0.400 \times 0.400 = 0.160 \text{ sq. in.}$$

The area of the image formed by the 4-inch lens (0.640 sq. in.) divided by the area of the image formed by the 2-inch lens (0.160 sq. in.) is 4. The light from the object square has to be spread over an area 4 times as large as the area with the 4-inch lens as with the 2-inch lens.

However, the lens opening of the 4-inch lens is larger, so it lets in more light. The amount of light is directly proportional to the area of the lens opening. The 4-inch, $f/2$ lens has a lens opening diameter of 2 inches (the focal length divided by the f-number).

The area of the aperture of the 4-inch focal length lens is:

$$A = \pi \left(\frac{d}{2}\right)^2$$

$$= 3.142 \times 1 \times 1$$

$$A = 3.142 \text{ square inches}$$

The area of the aperture of the 4-inch focal length lens is:

$$A = \pi \left(\frac{d}{2}\right)^2$$

$$= 3.142 \times 0.5 \times 0.5$$

$$= 0.785 \text{ square inches}$$

Since 4 times .785 equals 3.142, the lens opening of the 4-inch lens has 4 times the area of the lens opening of the 2-inch lens. The 4-inch lens, therefore, allows 4 times as much light to enter as the 2-inch lens.

However, since the size of the image formed by the 4-inch lens has 4 times the area of the image formed by the 2-inch lens, the light must be spread over an area 4 times as great. Thus, the amount of light *per unit area* is exactly the same with each lens.

This is the reason an $f/2$ lens always has the same image illuminance regardless of its focal length. This is a great convenience to photographers. The amount of exposure on a film is:

$$\text{Exposure} = \text{Illuminance} \times \text{Time}$$

or

$$E = i \times t$$

The photographer calculates the exposure by using the f-number and shutter speed. A given exposure might be a camera setting of $f/2$ at 1/100 second. The f-number represents the illuminance on the film, while the shutter speed represents the time. The photographer knows that the illuminance with all $f/2$ settings on various focal length lenses will be the same with the same subject.

The above discussion assumes that lenses transmit light without any loss which is not quite true. In Chapter two, the RAT formula

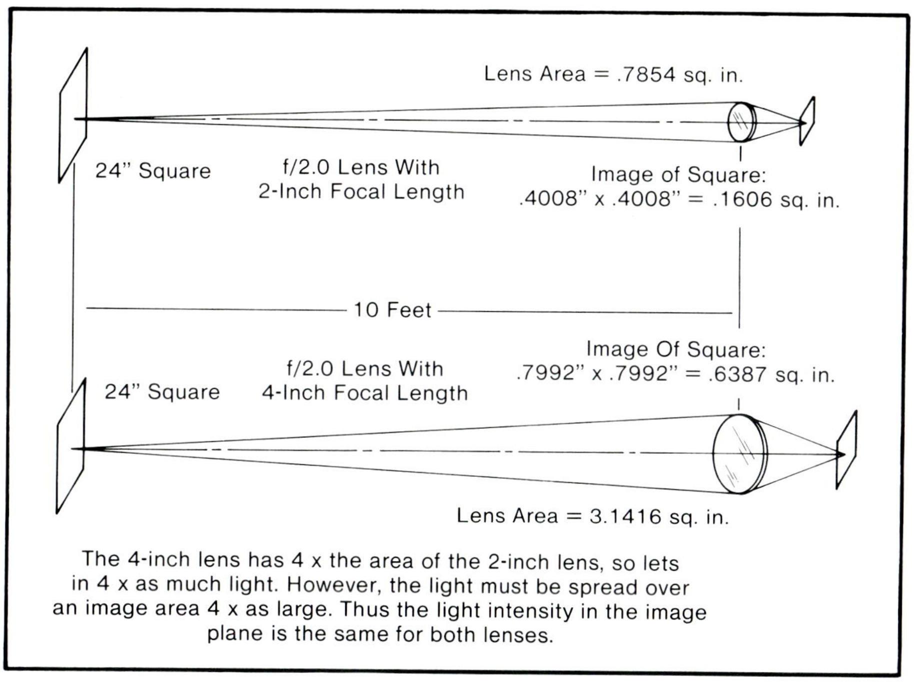

Figure 11-3. Demonstration of Equal Image Illuminance with Equal f/ Numbers.

Figure 11-4. 5-Bladed Iris Diaphragm.

states that all of the incident light falling on transparent material can be accounted for by adding the amounts of reflected, absorbed and transmitted light. The *f*-number discussion above makes the assumption that all of the light entering a lens is transmitted.

In practice, this is nearly true. The loss due to absorption by the glass in the lenses usually is less than one-half percent.

In lenses made before World War II, the big loss of light was due to reflection from the glass-air surfaces—about 5 percent for each surface, which amounts to about 40 percent for a camera lens with eight glass-air surfaces.

In modern coated lenses, however, even those with many glass-air surfaces, losses due to reflection are minimized, so that the transmission is about 98% or more of the light. The 2% loss is negligible.

In most camera lenses there is a variable opening or iris diaphragm made of thin metal leaves arranged to form a nearly circular aperture (Figure 11-4). Moving a lever, or rotating a ring, changes the size of the aperture. The apparent diameter of this aperture as seen through the front lens, is the diameter used in calculating the *f*-number (Figure 11-5).

By simply adjusting the diaphragm and using different *f*-numbers, a photographer can change the aperture for many reasons (Figure 11-6). When the subject is brightly illuminated, smaller lens openings (large *f*-numbers) are used to get the right exposure. In dim light, the lens must be opened; that is, large lens openings (small *f*-numbers) are used to get enough light to the film for exposure. Lens openings are also used to control the depth of field.

As long as the camera lens is used by itself, the *f*-numbers marked on the diaphragm control remain valid. This is true for most photographs. However, when the lens is extended for extreme close-ups, additional exposure must be given and, as mentioned in Chapter 10, when a tele-extender is used, the *f*-number markings no longer represent true lens openings (Figure 11-7). When a 2-inch lens is used with a 2X tele-extender, the focal length becomes 4 inches. The lens opening is still 1 inch. When the lens is set at *f*/2, with the 2X tele-extender, the true value of the 1-inch opening becomes:

$$f\text{-number} = \frac{4}{1} = f/4.$$

The *f*/4 is two stops less than *f*/2, and the illuminance at the image plane is only ¼ of what it is at *f*/2. With a 2X tele-extender the photographer must open up two stops above the marked *f*-value to get the true *f*-number. Another alternative would be to increase the exposure time by 4X (E = I x T) rather than changing the aperture.

As indicated in an earlier chapter, a change of one stop, that is, from

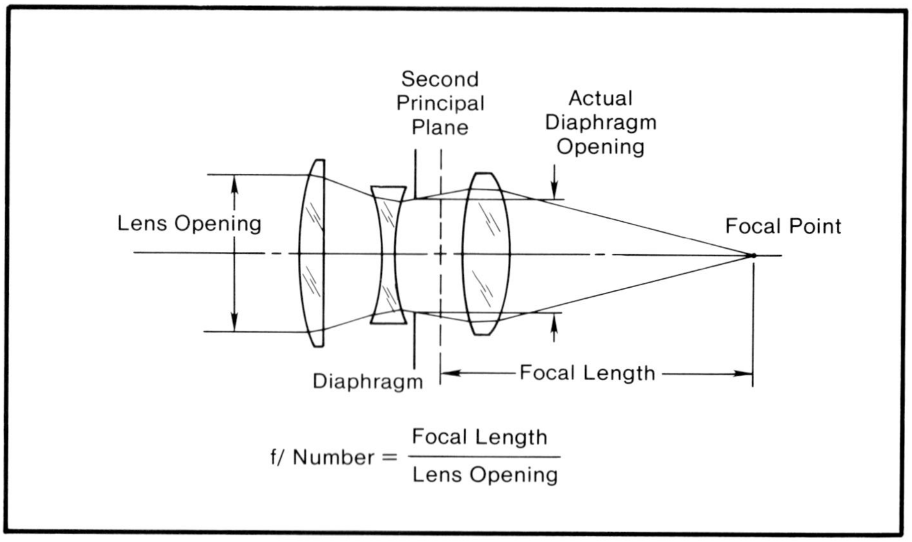

Figure 11-5. Effective Lens Opening compared to Diaphragm Opening.

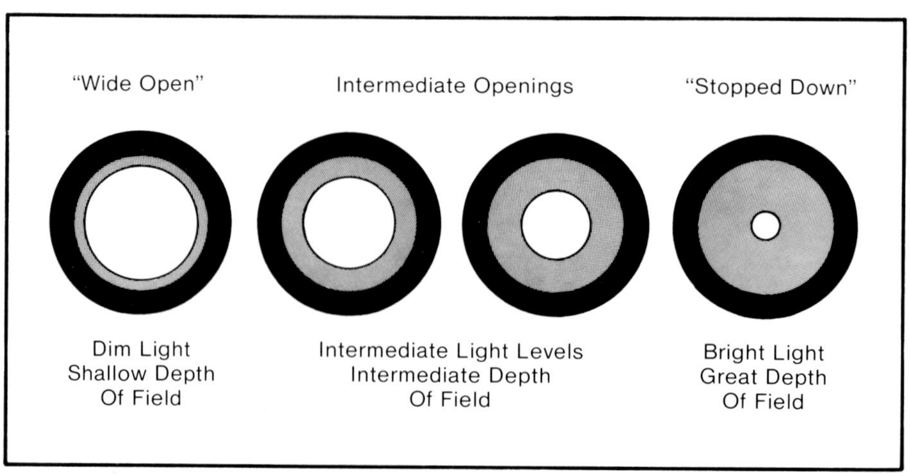

Figure 11-6. Lens Openings.

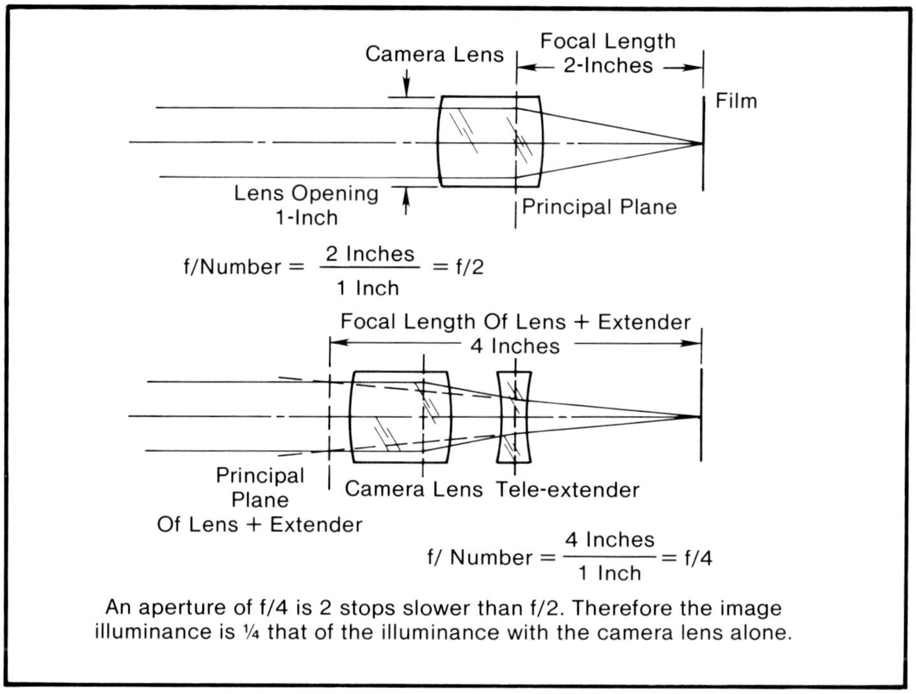

Figure 11-7. Effect on f/Number of a Tele-extender.

one *f*-number to the next either doubles the amount of light that enters the camera, or cuts it in half. This means that it changes the illuminance on the film plane by a factor of 2X.

The *f*-number series is given in the left column of the following table. Going down the column, each *f*-number allows one stop less light to enter, that is, one-half the amount of light. The second column shows the amount of light that enters. If the amount for *f*/1 is arbitrarily set at a value of 1, then for *f*/1.4, one stop less, the value is $\frac{1}{2}$, and so on.

Each successive *f*-number is 1.414 times the *f*-number above it. The *f*-number *f*/1.4 is a rounded value of *f*/1.414 (1.414 is the square root of two). The comparative amount of light is the reciprocal of the square of the *f*-number. Because 1.414 squared is 2, the comparative image illuminance is the reciprocal of 2 or $\frac{1}{2}$.

The amount of light is controlled by the area of the lens opening, and the area is a square function. The area of a circle is πr^2. The square of the radius is the variable. The *f*-number is a reciprocal linear function:

$$f\text{-number} = \frac{f}{D} \text{ or } \frac{f}{2r}$$

The *f*-number series changes by the square root of 2, while the illuminance changes by a factor of $\frac{1}{2}$:

$$1 \times \tfrac{1}{2} = \tfrac{1}{2},$$
$$\tfrac{1}{2} \times \tfrac{1}{2} = \tfrac{1}{4},$$
$$\tfrac{1}{4} \times \tfrac{1}{2} = \tfrac{1}{8}, \text{ and so on.}$$

The smallest lens opening available on most camera lenses is usually *f*/16 or *f*/22. Smaller openings are not provided because diffraction causes a decided loss in definition when the openings are very small.

When a camera has a through-the-lens exposure meter, the photo cell measures the illuminance at the film plane and does not rely on marked *f*-numbers. Therefore, no additional allowance for the use of the tele-extender is necessary. However, if a manual exposure meter is used, the lens must be opened two stops, as shown in the table, or the exposure time lengthened by a factor of four.

Types of Diaphragms

The leaf-type diaphragm, discussed in the previous section, is the most commonly used diaphragm in more expensive cameras. Simple cameras may have only one lens opening, usually about *f*/11.

A series of differently sized apertures can be punched in a circular metal plate to control the amount of light reaching the film. The plate can be pivoted so that when rotated, a hole centers on the lens axis (Figure 11-8). Such apertures were common in cameras in the 1800's and were called "Waterhouse Stops." Usually each aperture had half the area of the next larger aperture and twice the area of the preceeding smaller aperture, so that the aperture stops represented one-stop difference in exposure. They usually had a series of click stops so that each aperture was

Full *f*-Number Series	Comparative Image Illuminance	*f*-Number with 2X Extender	Comparative Illuminance With 2X Extender
f/1	1	*f*/2	1/4
f/1.4	1/2	*f*/2.8	1/8
f/2	1/4	*f*/4	1/16
f/2.8	1/8	*f*/5.6	1/32
f/4	1/16	*f*/8	1/64
f/5.6	1/32	*f*/11	1/128
f/8	1/64	*f*/16	1/256
f/11	1/128	*f*/22	1/512
f/16	1/256	*f*/32	1/1024
f/22	1/512	*f*/45	1/2048
f/32	1/1024	*f*/64	1/4096
f/45	1/2048	*f*/90	1/8192
f/64	1/4096	*f*/128	1/16384

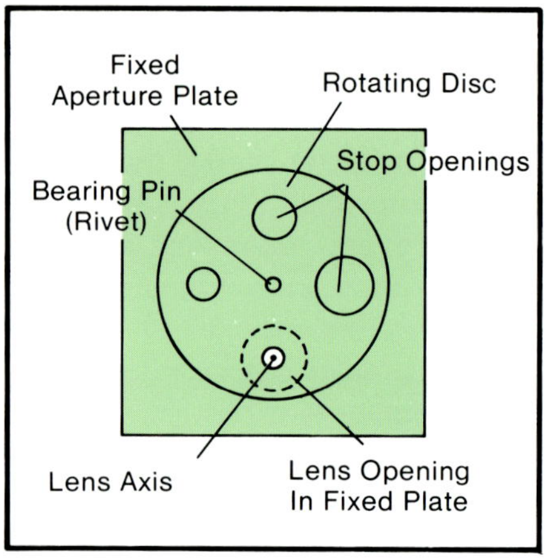

Figure 11-8. Rotating Waterhouse Stops.

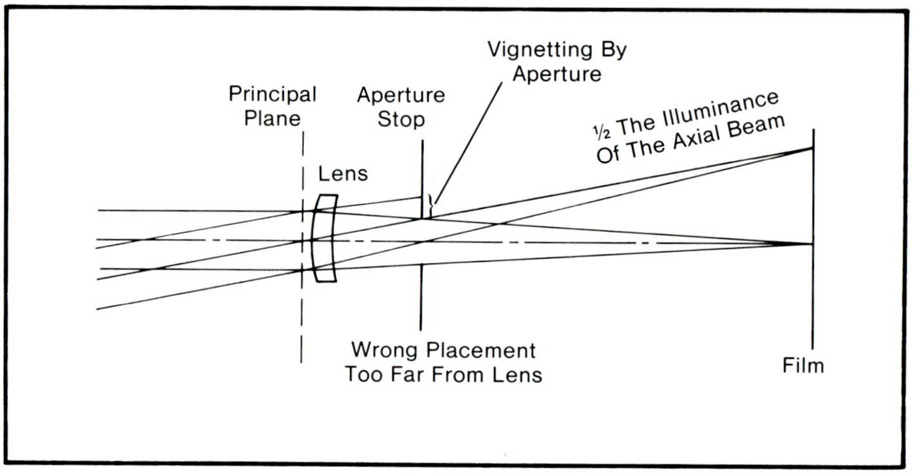

Figure 11-9. Vignetting By Aperture Stop.

accurately centered with the lens when the click stop was in place. These stops, being one *f*-number apart, led to the term "Stop" meaning twice or half the exposure of the next setting.

The location of the diaphragm is important. Usually the stop is located at or near the principal plane of the lens. If the opening is too far behind or in front of the lens, a shadowing effect called vignetting takes place (Figure 11-9). Vignetting causes a loss of light in part of the image and this, of course, is undesirable.

Location of Diaphragm

There is usually just one location where the lens designer can place the diaphragm. However, several factors must be considered:

- It must be in a location where reducing the area by half will reduce the film plane illuminance by half.
- It must not vignette, that is, it must not lower the illuminance more at the edges than in the center.
- It must be placed in a location to remove certain skew rays (rays other than those parallel to the axis) in order to minimize the effects of certain aberrations.

In symmetrical lenses, the location is in the exact center of the lens. With the meniscus lenses, the diaphragm is placed on the concave side of the lens. If it is a single meniscus, it is at some distance from the lens. The diaphragm is placed close to a chromatically corrected meniscus lens. In asymmetrical lenses, it is placed where the designer calculates that it will satisfy the above conditions. Several examples are shown in Figure 11-10.

Diaphragm Shape

So far, only round, or nearly round apertures have been discussed. However, the lens opening does not have to be round. It can be triangular, square, or any other shape if there were a good reason to have it that shape. For example, in a catadioptric mirror lens, the lens opening

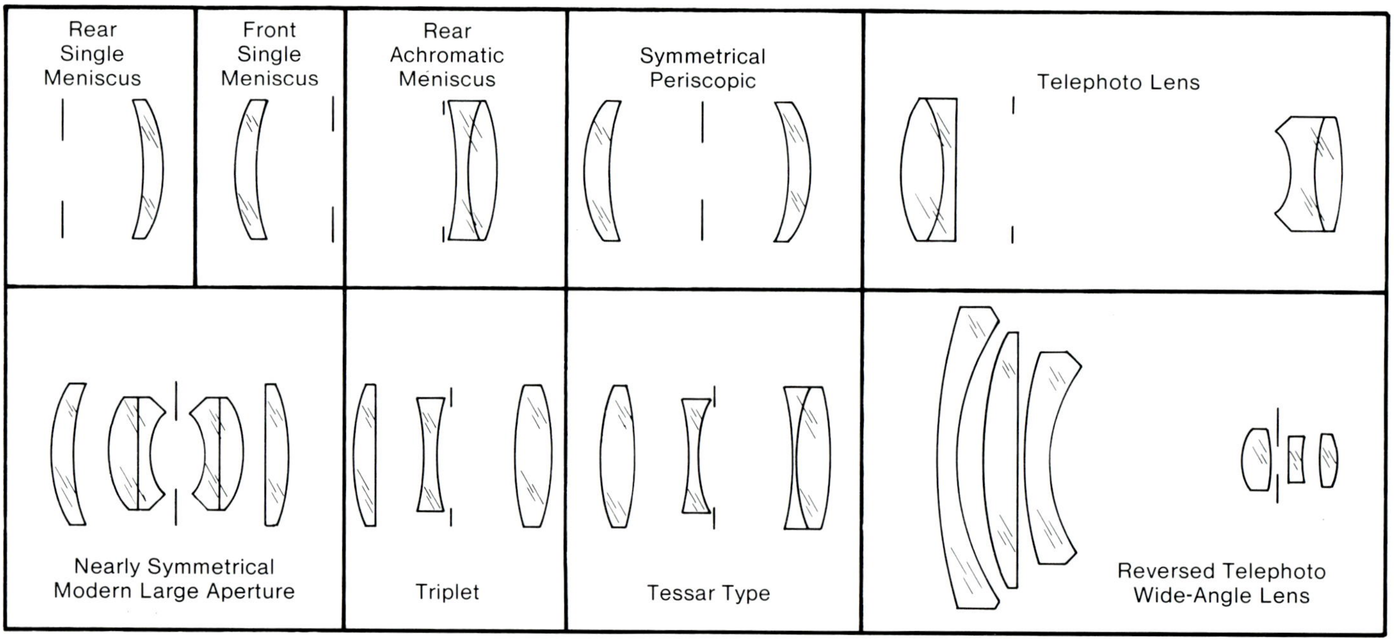

Figure 11-10. Diaphragm Locations With Different Types Of Lenses.

is donut-shaped, because it is necessary to have a small mirror in the center of the "lens" opening (Figure 11-11).

The videotape session shows the effect the shape of the diaphragm has on the out-of-focus images of points of light. Figure 11-12 shows how a round diaphragm creates "circles of confusion," (round circles of light) when the film is placed in front of or behind, the image plane. When the image is in focus, the light forms a point image. Figure 11-13 shows how a star-shaped diaphragm also forms a point image of the point-of-light subject when it is in focus, but now the out-of-focus images are star shaped.

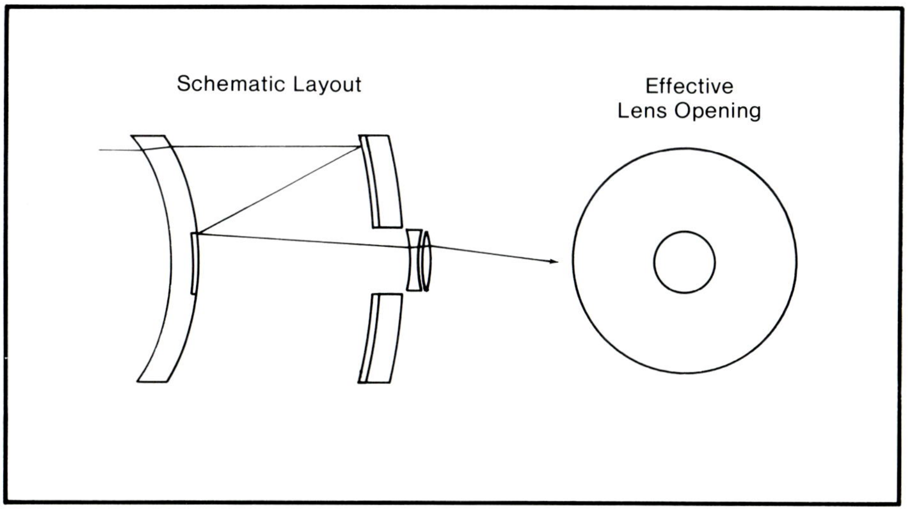

Figure 11-11. Mirror Telescopic Lens.

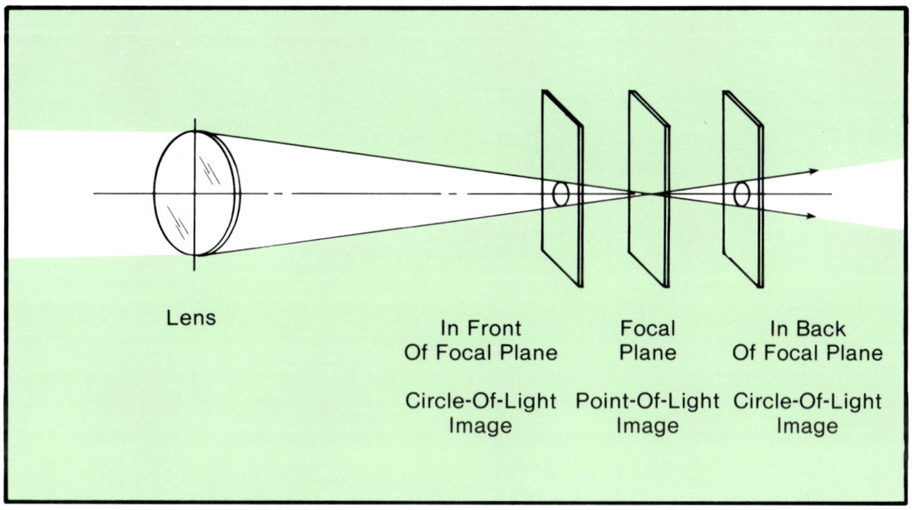

Figure 11-12. Circles Of Light.

The out-of-focus images of stars, formed by the mirror lens described above, are little donuts of light, technically called annular rings.

One advantage that round lens openings have over oddly shaped openings is improved image quality. The image formed by the central portion of a lens has fewer aberrations than the image formed by the outer portion of a lens. In the case of the star image, the part of the light passing through the tips of the points of the star goes through the outer portion of the lens. A round opening, with the same area as the star, uses only the central portion of the lens and therefore produces an image with fewer aberrations.

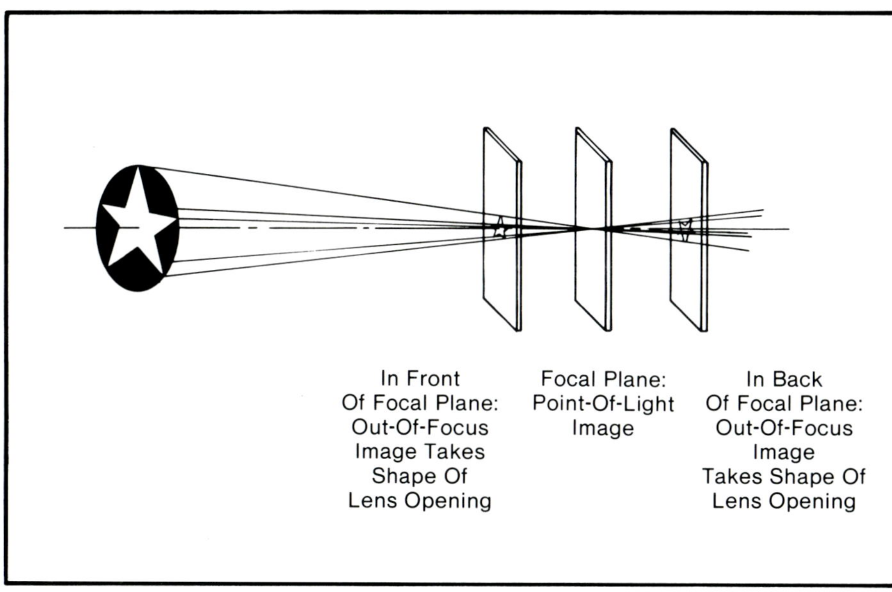

Figure 11-13. Shaped Lens Opening.

Diaphragms for Exposure Control Systems

Most modern exposure meters use cadmium sulfide, lead arsenide, or silicon-blue cells to measure light intensities. Such cells are variable resistance cells—their electrical resistance varies with the amount of light falling upon them. A very small battery is used to supply the electricity (Figure 11-14).

Earlier exposure meters used selenium cells and did not require

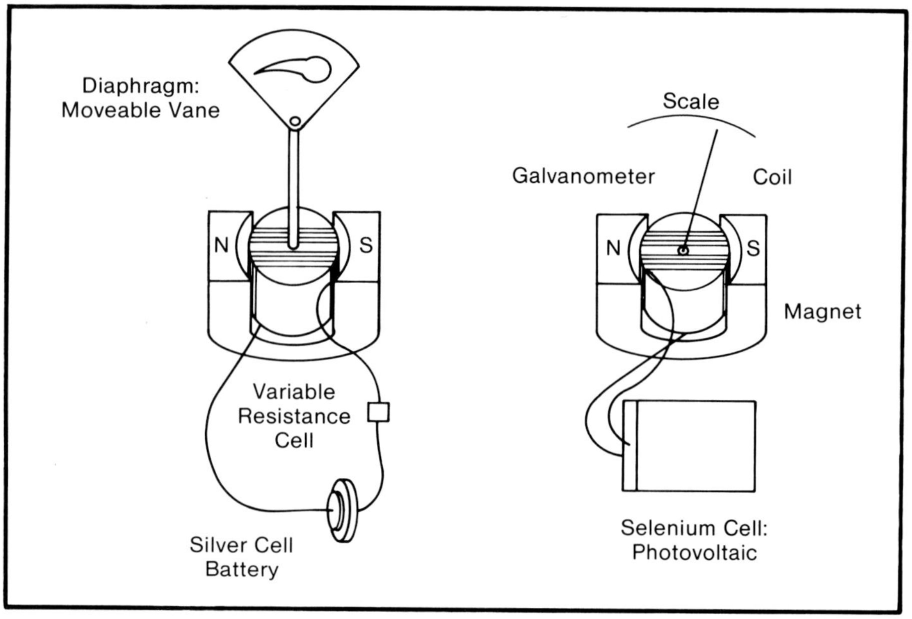

Figure 11-14. How A Light-Sensitive Cell Moves A Single-Blade Diaphragm.

batteries. Selenium cells are photovoltaic, that is, they generate electricity when light falls on them. The voltage at which the current is generated increases with the amount of light. A voltmeter attached to the cell varies its needle position as the light falling on the cell changes. Instead of calibrating the voltmeter scale in volts, the scale is calibrated in light values, often in candles per square foot. When aimed at a subject, this type of meter indicates the luminance of the light reflected by the subject. A calculator dial on the meter is used by the photographer to convert the luminance value into exposure in terms of *f*-number or exposure time based on the sensitivity of the film.

Several early cameras were provided with automatic exposure devices. These use the power of selenium cells to rotate a single diaphragm blade that automatically changed the *f*-number of the lens opening with changing light conditions. Designing the shape of the opening presented a challenge.

The selenium cell does not produce enough energy to operate a multiblade, round-shaped diaphragm. It generates just enough power to rotate a light, single-pivoted, thin metal blade. The moving coil of the galvanometer is attached to the rotating blade so that the blade moves through a total angle of less than 90°. The problem was to design an opening shape in the single blade which would give a continuously variable aperture area.

Figure 11-15 illustrates the solution. An aperture in the fixed lens opening plate was like an open figure eight. The aperture in the moving blade is a curved comet shape. When the light level is low the head of the comet is over the fixed opening, allowing the full amount of light to pass. The little projections in the fixed opening reduce the amount of light by less than a third of a stop. However, when the light level is high, the rotating blade swings over so that the narrow tail of the comet aperture is over the fixed opening. As the drawing indicates, this produces a small, nearly-square shaped opening. At in-between light levels the shape is peculiar, but as long as the area allows the correct amount of light to enter, the exposure is correct. On the few occasions when such cameras are used to take photographs in which there are small out-of-focus light sources, these are imaged as various shapes, depending on the aperture shape at the time of exposure.

Because of the continuously variable opening and the fast response of the cell and blade to scene lighting conditions, this type of diaphragm is often used on automatic-exposure movie cameras. If the light changes while the photographer is taking movies, the blade moves to keep the exposure correct. It also corrects the exposure when the camera's aim is moved from a sunny to a shady area.

Another solution to the same problem consists of a two-blade diaphragm (Figure 11-16). The two blades are hooked together so that when one blade goes down, the other goes up. The ends of the blades are V-shaped, with the angle of the V

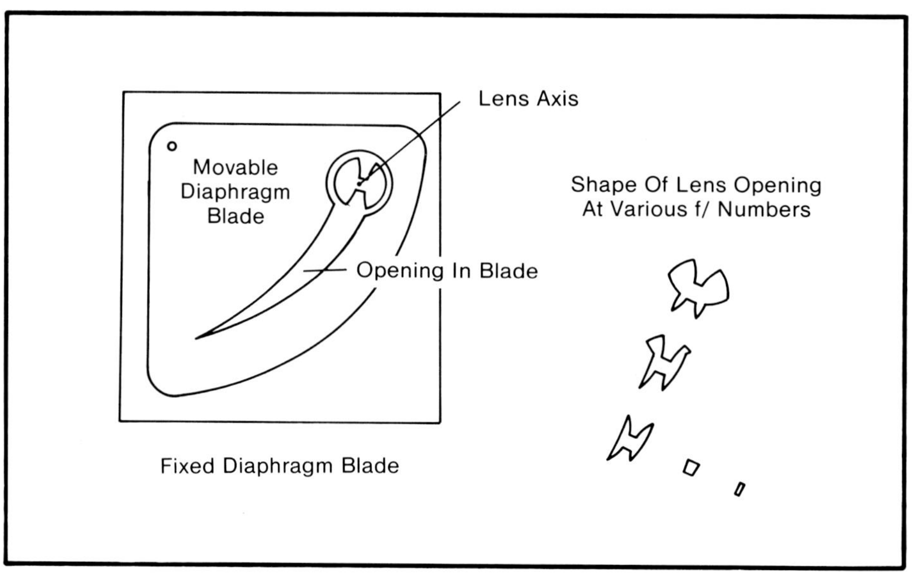

Figure 11-15. Single-Vane Variable Diaphragm.

being 90°. This makes a square-shaped opening. The selenium cell does not have enough power to activate a two bladed diaphragm, so the mechanism is attached to the camera release button, and the photographer's finger powers the movement. The upper blade has a cam slot cut in it near the top. The galvanometer needle sticks through the slot and moves back and forth in the slot, varying its position with the amount of light. When the release button is pushed, the needle stops the blades as the cam touches the needle. When the light level is low, the needle is at the narrow end of the cam so that the diaphragm remains fully open. When the light level is high, the needle is on the wide end of the cam, so that the blades move only enough to provide a very small lens opening. All that the cell has to do is move the needle to the position called for by the amount of light.

The square shape of the opening is perfectly satisfactory for a diaphragm, and has been used in many successful automatic-exposure cameras.

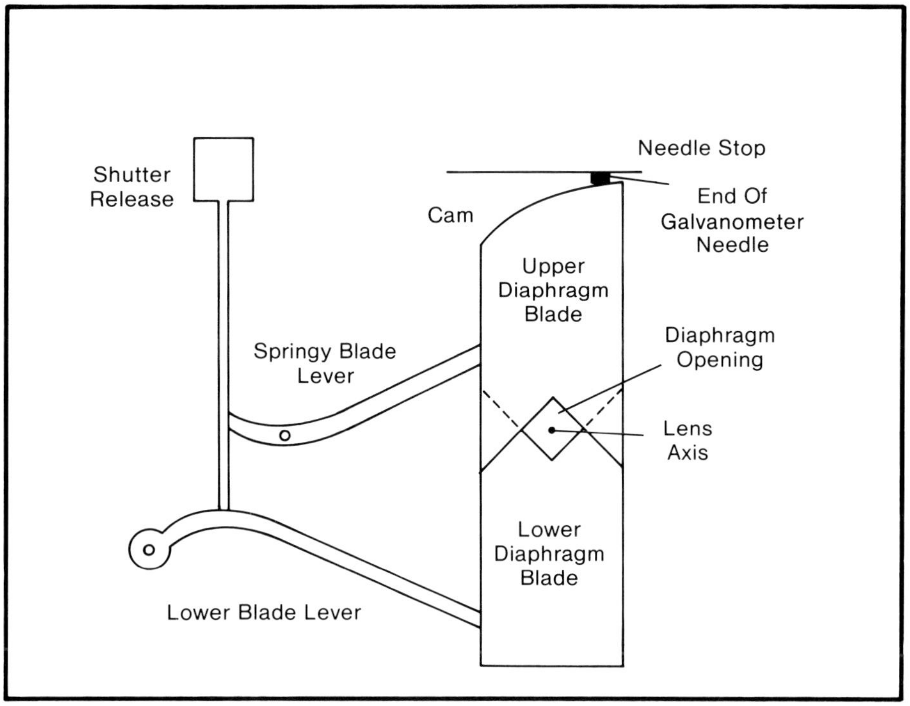

Figure 11-16. Two-Bladed Diaphragm for Automatic Camera.

Problem

Using thin lens theory, make a schematic ray trace of a 3X tele-extender. You will need to review the trace of the 2X extender and use ray tracing to find the size and position of the virtual image of a thin negative lens. Assuming the original lens has an $f/2$ opening, calculate the f-number when the lens is used with the extender.

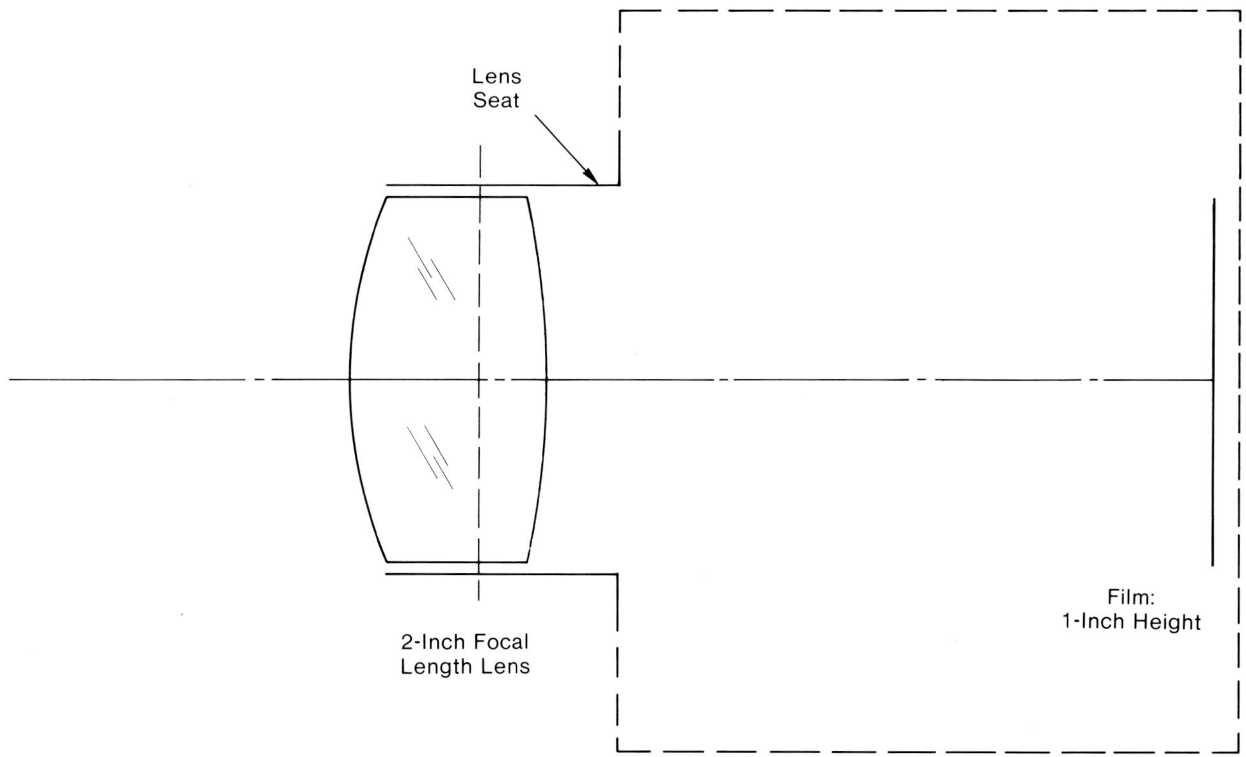

Problem
Chapter 11
Video Lesson 6

Chapter 12

VIDEO LESSON 7

Diaphragm Shapes, Schematic Ray Trace of a Telescope

Summary

The shape of the diaphragm has no effect on the out-of-focus images of most photographic subjects. However, the images of small, bright points of light in the subjects tend to be imaged with the shape of the diaphragm. A simple refracting telescope consists of an objective lens and an eyelens with a common focal point. A schematic ray trace of this type of telescope shows how the telescope magnifies images at a distance. The exit pupil of a telescope is the position where the eye is placed for best viewing. Its location and size are found by a schematic ray trace. One method of finding the power of a telescope is to divide the focal length of the objective lens by the focal length of the eyelens. A field lens is a positive lens placed in the common focal point of the objective lens and eyelens to increase the field angle of the telescope.

Circles of Confusion

A positive lens bends rays of light to form real optical images. Light coming from a single object point radiates in all directions. If some of that light strikes a positive lens, the light rays are then refracted to form a cone of light. The apex of the cone of light is where light rays merge and this apex is the focused image of the object point (Figure 12-1). Rays of light form circles of light both slightly in front and just behind the image point. These circles form fuzzy, out-of-focus images and they are called "circles of confusion" (Figure 12-2).

If a round diaphragm is placed near the lens (where the diameter is

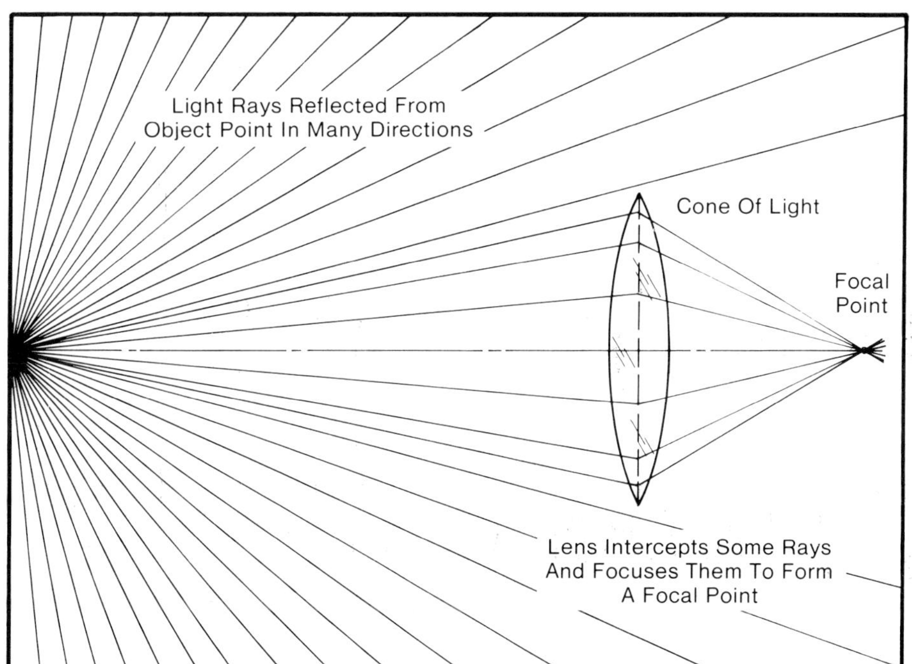

Figure 12-1. A lens images only those rays of light that come to it.

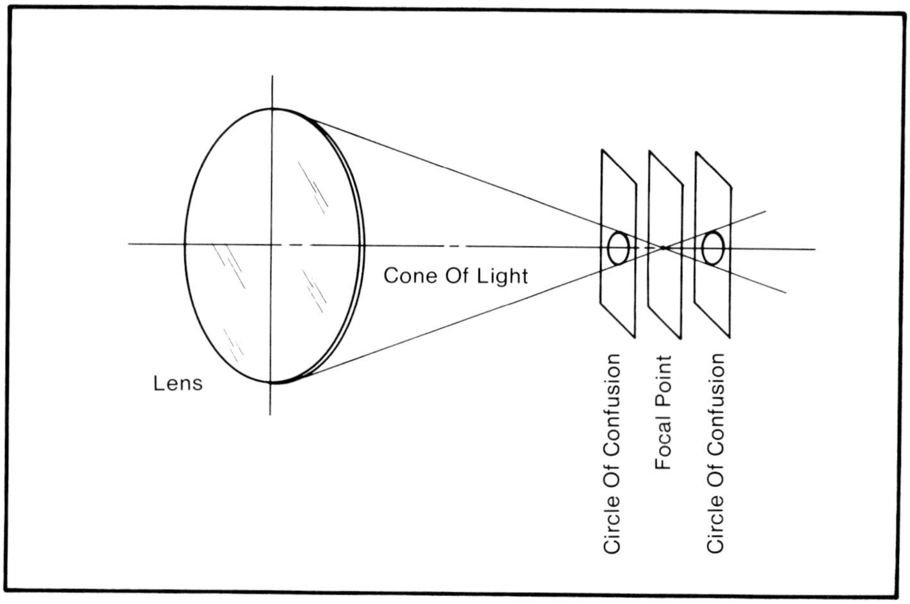

Figure 12-2. Circles Of Confusion.

smaller than that of the lens) the effective diameter of the lens and the cone angle of the cone of light is decreased. At the same distances from the lens, the circles of confusion are smaller, but they are still circular in shape (Figure 12-3).

If the round diaphragm opening is replaced with a diaphragm opening of a different shape, the shape of the cone and that of the out-of-focus images tends to take the shape of the opening. In this drawing, the shape is hexagonal (Figure 12-4). The overall shape is no longer a cone; it now approximates a hexagonal pyramid. If a ground glass is placed in front of the film plane, the out-of-focus image of the object point is no longer imaged as a circle of confusion. It might be called a "hexagon of confusion."

It is apparent that the out-of-focus image of an object point takes the approximate shape of the diaphragm opening, while the sharply focused image is a point.

When a camera lens has other than a round diaphragm opening, sharply focused image points are points. The entire sharp optical image, and later the picture, will be composed of sharply focused image points. This image cannot be distinguished from a sharply focused image made with a round lens opening.

When real objects are photographed, they are often located at different distances from the camera. Only objects at one distance can be sharply focused at one time in the image plane, while the images of objects at other distances are imaged out-of-focus (Figures 12-5 , 12-6).

The images of objects are composed of an infinite number of imaged object points. When the light from all of the object points is focused sharply, the composite image resembles the object in appearance. It is usually smaller and rotated 180°, but it has the visual aspects of the object. When objects are imaged slightly out-of-focus, the image blurs. The shape of the diaphragm cannot be distinguished from the blur. Only when there are object points that are small and very bright as compared to the surrounding image points, can the diaphragm shape be distinguished in the image.

Referring back to the images of stars in the night sky, we find they are ideal images with which to distinguish the diaphragm shape. In normal photographic subjects, reflections of the sun in rippled water form a number of bright object points that will reveal the diaphragm shape in the out-of-focus images. The spaces between leaves silhouetted against a bright sky also act as tiny bright object points that reveal the diaphragm shape in out-of-focus images.

Such scenes are relatively rare as photographic subjects, so that most people who use cameras with non-circular diaphragm opening shapes rarely become aware of that fact.

Lens Openings in Projection Lenses

Projection lenses are always used at maximum aperture to create bright

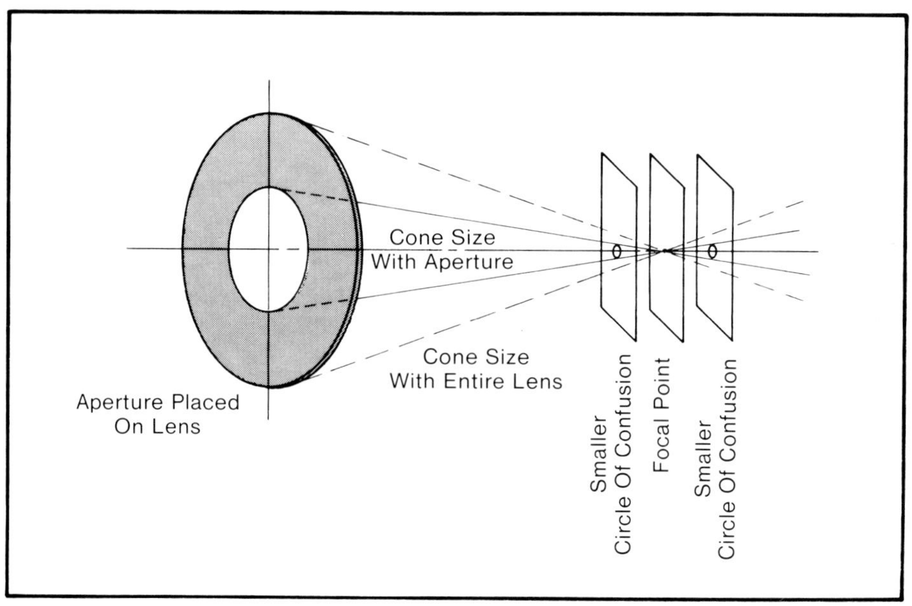

Figure 12-3. Smaller Circles Of Confusion With Smaller Lens Openings.

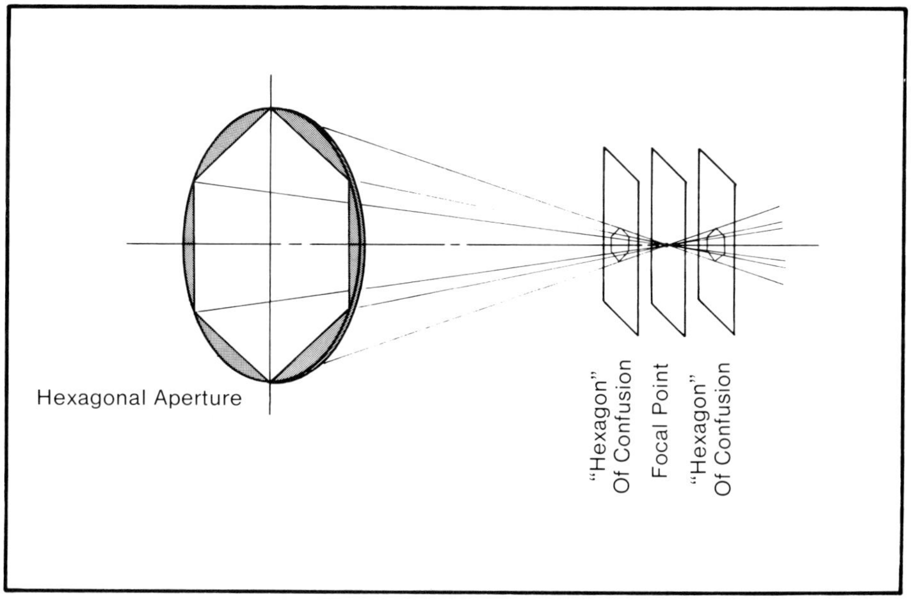

Figure 12-4. "Hexagons" Of Confusion Created By Hexagonal Lens Opening.

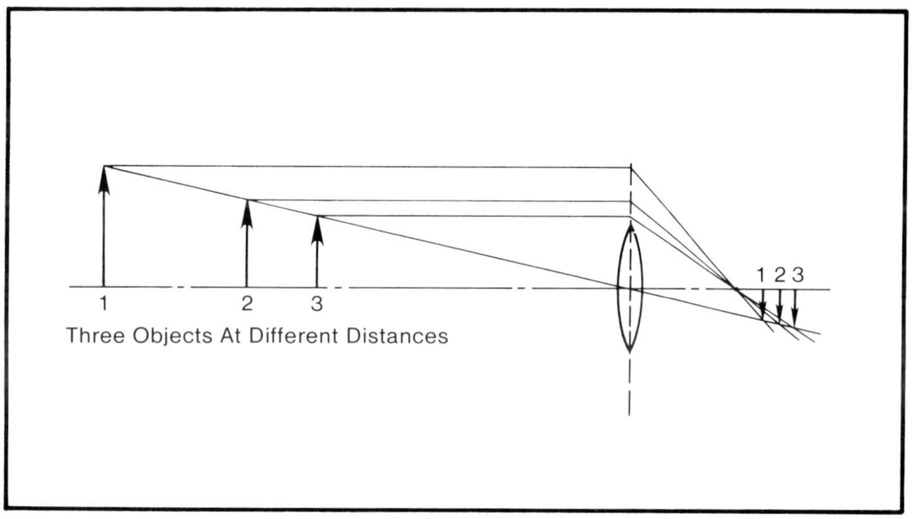

Figure 12-5. Locating The Image Positions Of Three Objects At Different Distances From A Lens.

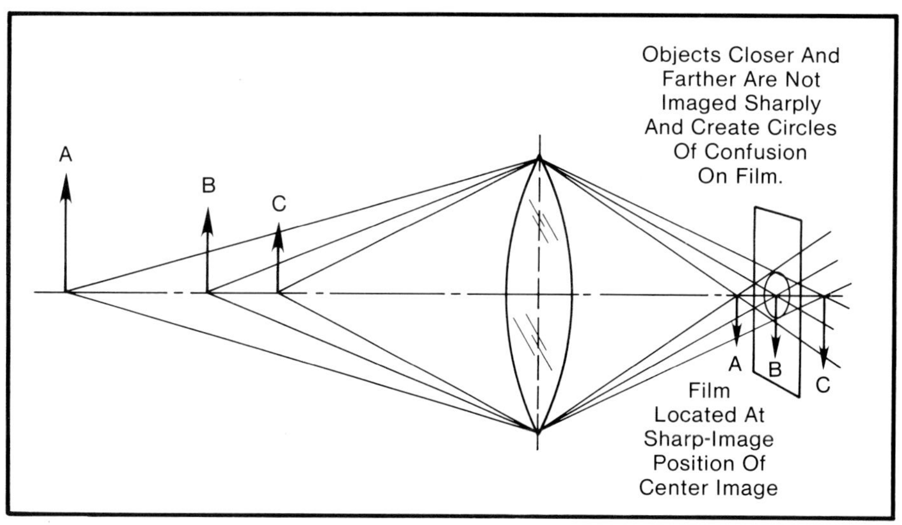

Figure 12-6. A lens focuses sharply at only a single distance at a given time.

screen images. Their aperture shape is round. The demonstrations on the video tape, where shaped lens openings are shown, are illustrations of what happens as images are formed.

If the hexagonal lens opening is placed on the projection lens, and a metal plate with rows of circular openings is placed in the normal slide gate, out-of-focus projected images are easily seen to be hexagonal in shape (Figure 12-7).

If both a slide and the metal plate are placed in the gate with a slight space between them, either the plate or the slide can be focused on the screen. When the slide is in focus, (Figure 12-8), the fact that the metal plate is there can only be detected by a diminished brightness in the projected image. As the focus is shifted to bring the circular holes in the metal plate into focus, the slide image breaks up into hexagonal areas, one for each hole in the perforated metal (Figure 12-9). As the perforated plate comes into focus, the hexagonal shapes change into circles. When the metal plate's holes are in focus, on the screen, the shape of the lens opening and the image on the slide no longer show (Figure 12-10).

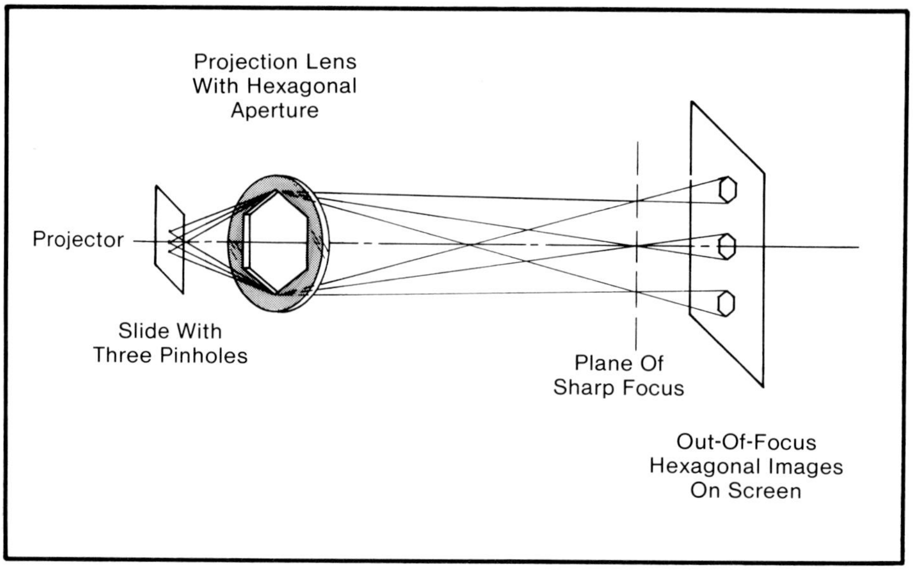

Figure 12-7. Out-of-focus images of points tend to take the shape of the lens opening in a projection system.

Periphery and Inner Portions of the Cone of Light

The rays of light on the image side of a lens are very orderly. In imaging a point of light on the axis, a round lens opening creates an almost perfect cone. Only the aberrations slightly mar the perfection.

The orderly peripheral rays create the "skin" of the cone of light. The internal rays are also very orderly (Figure 12-2). This can be seen by stopping down (reducing the aperture size) a lens with successively smaller lens openings (Figure 12-3). The height of the cone remains constant and is equal to the focal distance of the lens. However, the diameter of the base of the cone is reduced by stopping down. By eliminating the original peripheral rays, it can be seen that the remaining rays form a nearly perfect cone.

If a tiny pinhole is made in a slide, placed in a projector and sharply focused on the screen, a cone is formed whose base diameter is equal to the lens opening. The height of the cone is now the lens-to-screen distance or the long conjugate (Figure 12-11). If the pinhole is thrown slightly out-of-focus, the image becomes a small circle of confusion on the screen. Actually, it becomes a section of the cone near the apex (Figure 12-12). The sharp image of the pinhole has been shifted to a plane in front of or behind the screen. If the hexagonal mask is placed over the projection lens, a long hexagonal pyramid of light is formed (Figure 12-7). The light rays within this pyramid are, of course, orderly.

If bits of red, green and blue filter material are cut and assembled as in Figure 12-13, a test of the orderliness of the light can be made. If fashioned in the right proportions, the mix of red, green and blue light should produce a white image when focused on the screen. If the mix is not exact, the small spot of light on

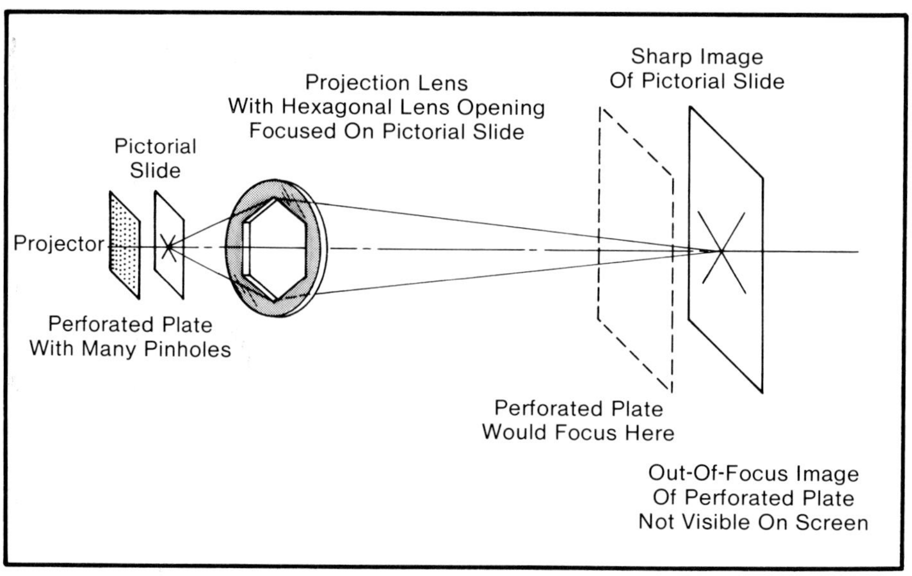

Figure 12-8. Focus Demonstration

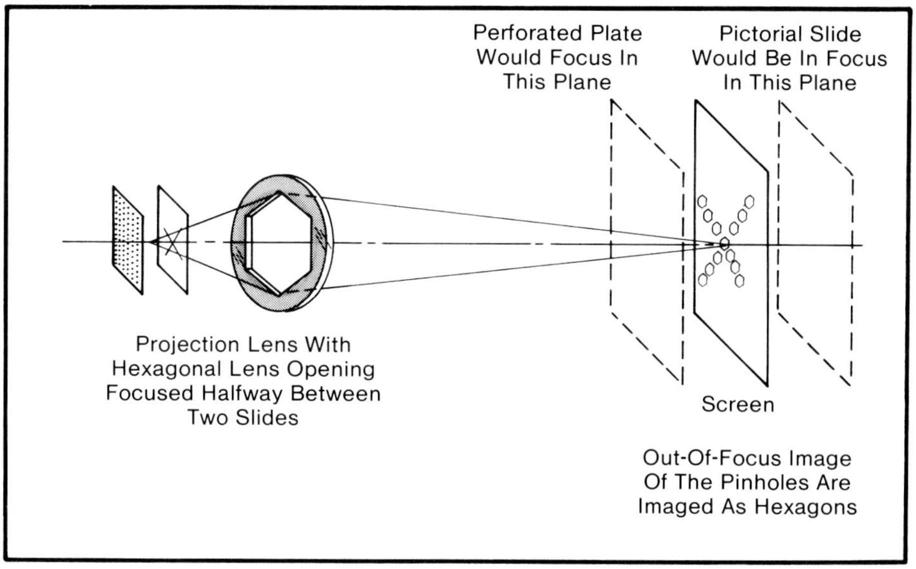

Figure 12-9. Focus Demonstration (Continued).

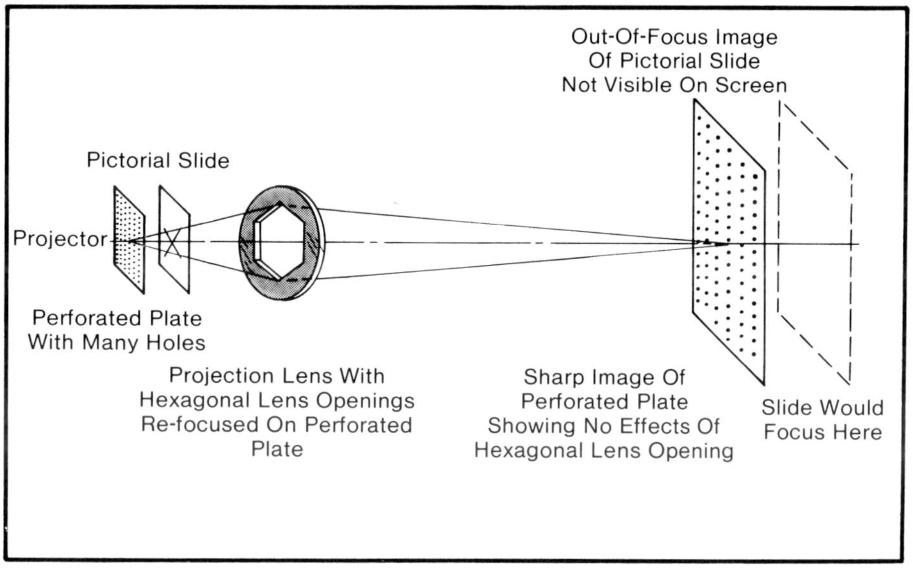

Figure 12-10. Focus Demonstration (Continued).

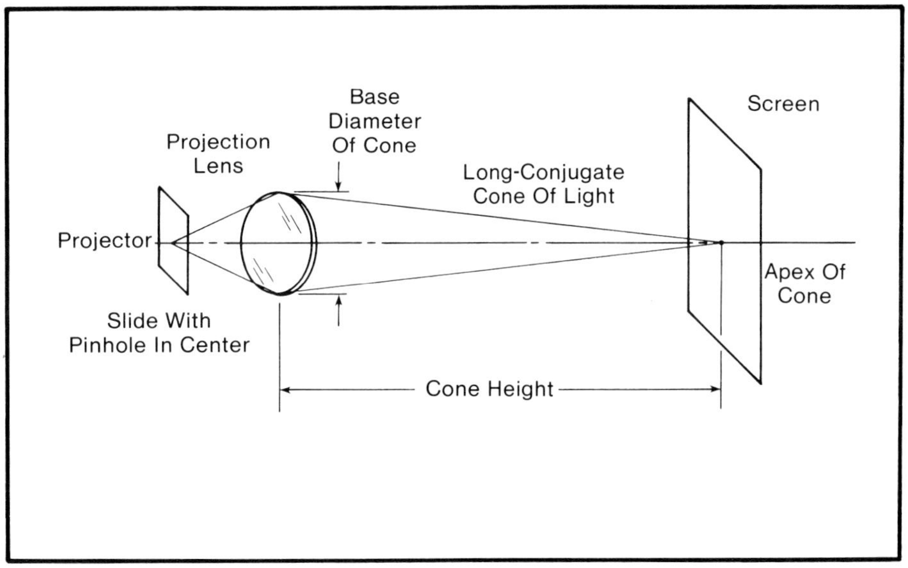

Figure 12-11. Long-Conjugate Cone Of Light Formed By Projection Lens.

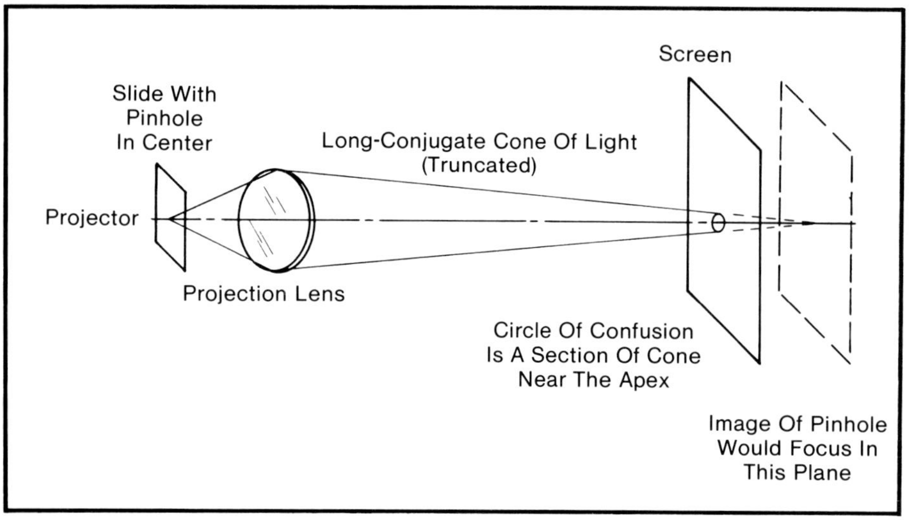

Figure 12-12. The circle of confusion is a conic section near the cone apex.

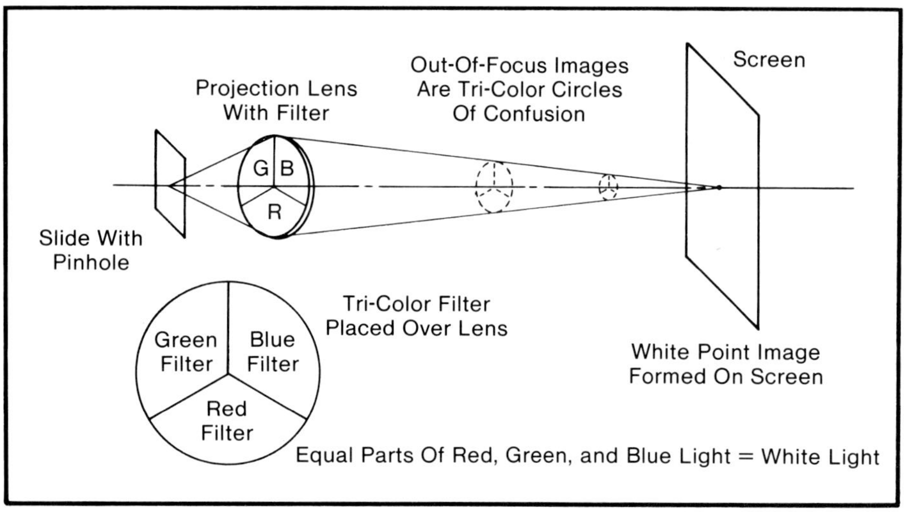

Figure 12-13. Tri-Color Filter Demonstration

the screen may have a slight tinge of color.

However, when the pinhole is thrown out-of-focus, not only is the out-of-focus image seen as a circle of confusion, but the colors and shapes of the filters can also be seen. This is because the paths of light within the cone of light are extremely orderly.

If the single pinhole is replaced with a slide into which a number of pinholes have been punched, a number of "stars" will be seen when the slide is sharply focused on the screen. These change gradually to multiple images of the colored circle as the slide is thrown out-of-focus (when the sharp image is moved in front of or behind the screen). This serves as a reminder that the light rays coming to an image-forming lens from an individual point in an object are alike—in this case, white light (Figure 12-14). As each ray strikes the lens, it is refracted so that it arrives at the same position in the image plane as do all the other rays from that object point. Putting the filter over the lens identifies which rays are going through each area of the lens, and indicates the orderly paths of the light rays as they converge to form the image of the point.

The out-of-focus image of the diaphragm is erect in front of the focal plane, while the out-of-focus images are inverted behind the focal plane. The ray trace shows why this occurs (Figure 12-15). In the videotape a yellow filter is used to dramatize the direction of the image when it is right side up and when it is inverted.

Using the Outer Rays as a Focusing Aid

The sizes of the circles of confusion determine the degree of fuzziness or blur in an image. When a lens is used with a small diaphragm opening, the angle of the cone of light is smaller. Therefore, at a given distance from the focal plane, the circles of confusion are smaller than with the lens diaphragm fully open. Closing the diaphragm opening of a camera lens

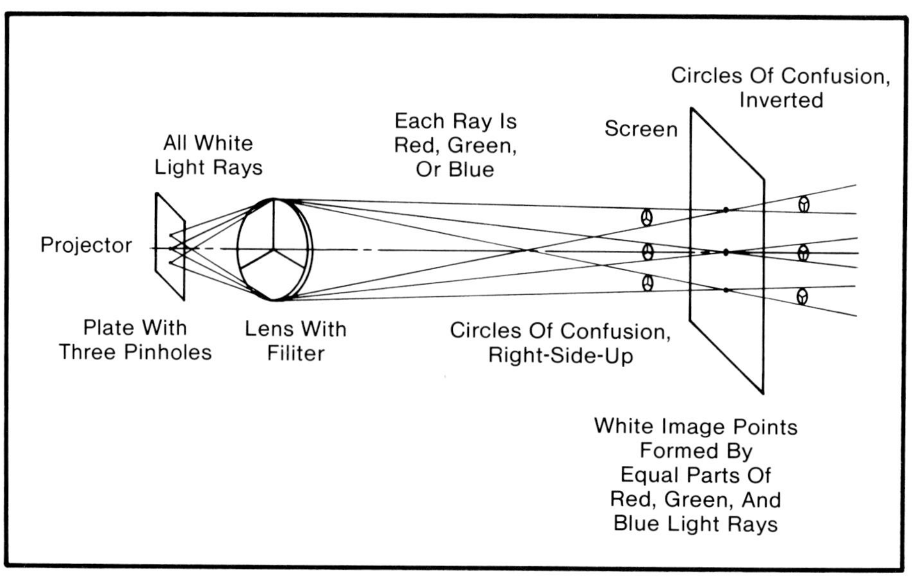

Figure 12-14. Inverted Circles Of Confusion Beyond The Image Points.

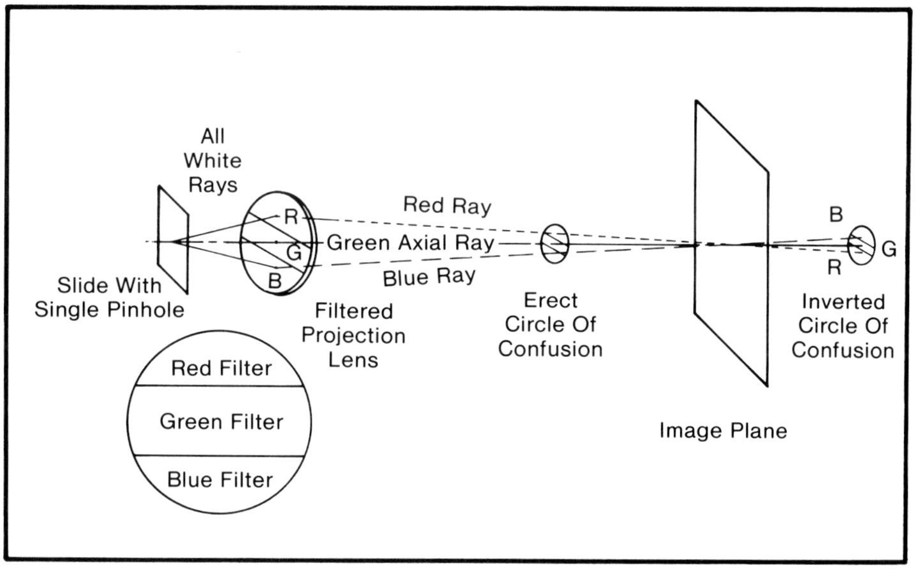

Figure 12-15. Schematic Ray Trace Showing Orientation Of Circles Of Confusion.

one or two stops below its maximum aperture generally improves its performance by increasing the depth of field (Figure 12-16). This also reduces residual aberrations and increases the tolerance to lack of film flatness. The effect of outer rays on depth of field can be used in making a focusing aid for a projector.

A piece of opaque tape can be placed over a projector lens so that only two small portions of the outer lens are left uncovered (Figure 12-17). If the perforated metal slide (used previously) is placed in the projection gate, the perforations can be imaged sharply on the screen as circles of light. However, when the image is gradually thrown out-of-focus, each circle of light is imaged as two circles of light. The amount of separation increases between them in proportion to the displacement of the plane of sharp focus from the screen. If a pictorial slide is projected with this split lens, it will appear as a double image when out-of-focus; the two images come together as it reaches focus positions (Figures 12-18, 12-19). This principle is used in cameras. Two optical wedges are molded side by side in the center of the ground glass, with their apexes opposite each other (Figure 12-20). The angle of the wedges is such that it accepts only light rays from the outer edges of the lens. As the image is focused, the eye sees two images of a subject when it is out of focus, but they come together and are seen as a continuous, unbroken image when the image is sharply focused. It is easy to understand that this system has limitations. When the lens is stopped down, the prism area goes black because it accepts light only from the periphery of the lens. Furthermore, with long focal-length lenses, especially those that have small maximum apertures, this system does not work.

A Long, Narrow Diaphragm Opening

The smaller the diaphragm opening, the greater the depth of field. Long, vertical, narrow lens openings have a greater horizontal depth of field than vertical depth of field. Figure 12-16 shows the circle of confusion size resulting from the full lens diameter in the upper drawing. The stopped down lens in the lower drawing shows a smaller circle of confusion. This can also be interpreted as showing a different depth of field from the long and narrow dimensions of the aperture just described. This is demonstrated in the videotape using a projection system with the perforated metal slide placed in the projector (Figure 12-21). Tape is placed over the projection lens to form a long, narrow vertical opening.

When the slide is in focus, we see rows of bright circles of light, separated by the dark image of the metal. As the projector lens is moved slowly, throwing the slide out-of-focus, the vertical dark areas stay

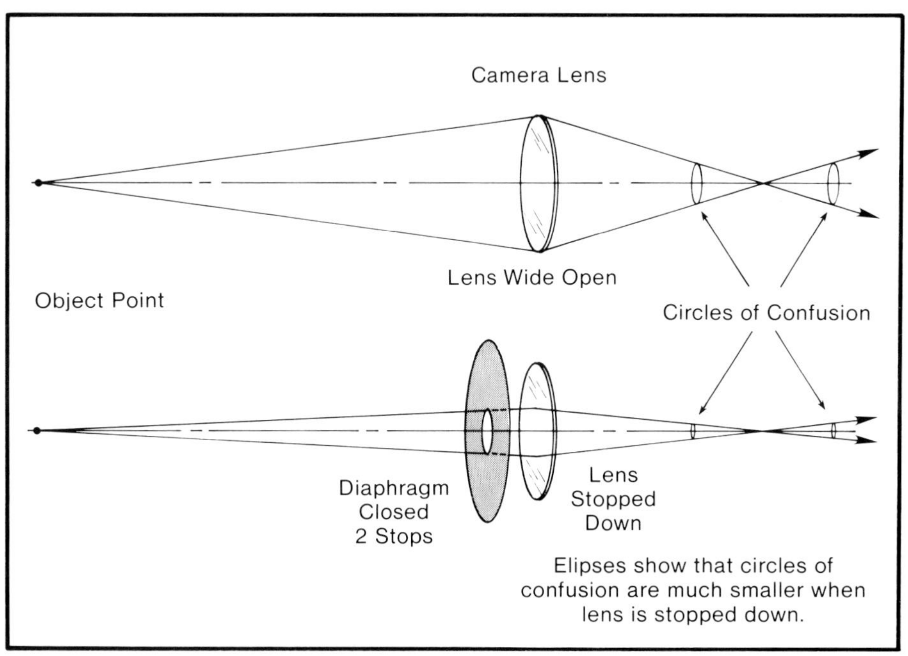

Figure 12-16. A smaller diaphragm opening produces smaller circles of confusion.

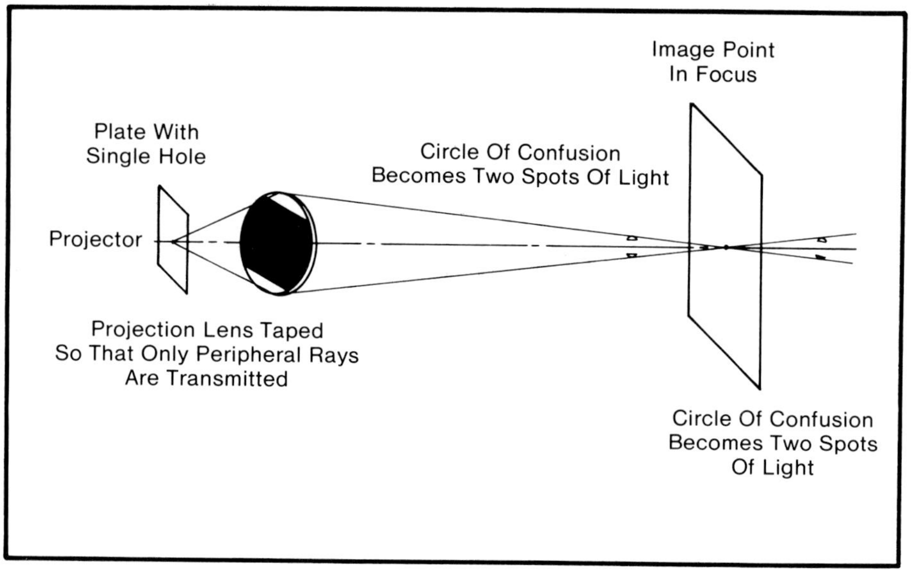

Figure 12-17. Using Peripheral Rays For Focusing.

Figure 12-18. In Focus.

Figure 12-19. Out-Of-Focus.

relatively in focus, while the horizontal areas go rapidly out-of-focus. Soon, all that can be seen is a row of vertical dark lines separated by bars of light.

If the tape is repositioned so that the lens opening is horizontal, the out-of-focus image shows horizontal alternate dark and light lines.

If a camera lens is equipped with a long, narrow lens opening, its out-of-focus images will show these same characteristics (detail that is parallel to the lens opening and is clearly imaged, while detail perpendicular to the lens opening is lost in blur). This indicates that, while odd-shaped lens openings can be used, there still is a limit beyond which the images formed are too greatly degraded.

Schematic Ray Tracing of a Telescope

Tracing the rays schematically helps to form an understanding of how a telescope forms images. A thin-lens design serves the purpose well. The objective lens of a telescope has a relatively long focal length in order to provide magnification of distant objects. A long focal length objective lens forms a large real image.

In Figure 12-22, the light rays coming to the lens are parallel because they are coming from a distant object point. They are shown parallel to the lens axis because the telescope is imaging a point that is on the lens axis. These two rays are bent by the lens so that they converge at the axis, showing the location of the focal point.

A simple astronomical telescope uses an eyelens with a short focal length. The eyelens is used as a magnifying lens to enlarge the image formed by the objective lens. In order for the light rays to be parallel as they leave the eyelens, the eyelens is placed at a distance equal to its focal length from the real image formed by the objective lens. All direct-viewing optical instruments are designed so that light rays leaving the eyelens are parallel, or

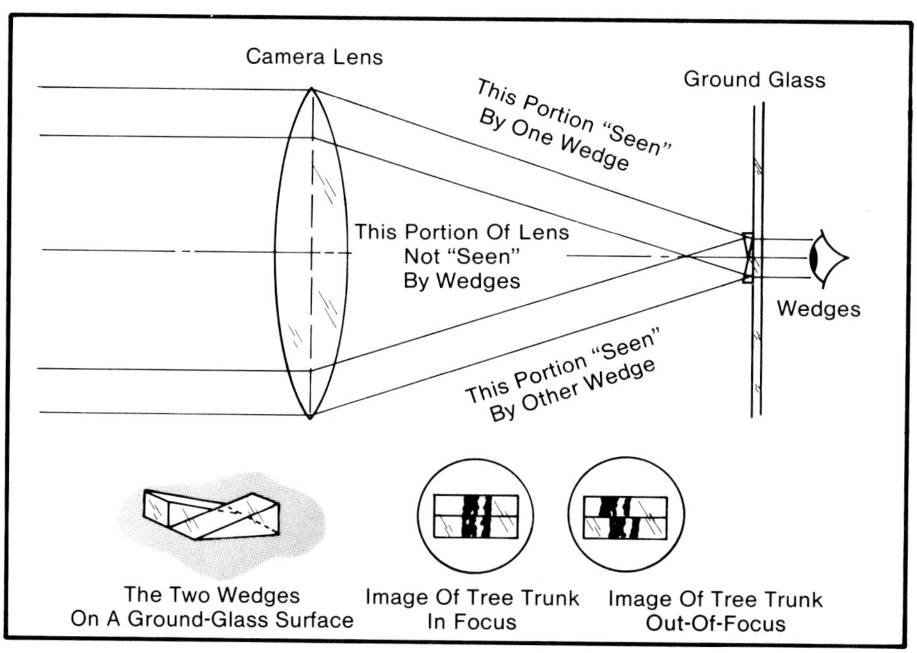

Figure 12-20. Opposed-Wedges Rangefinder For Single-Lens-Reflex Camera.

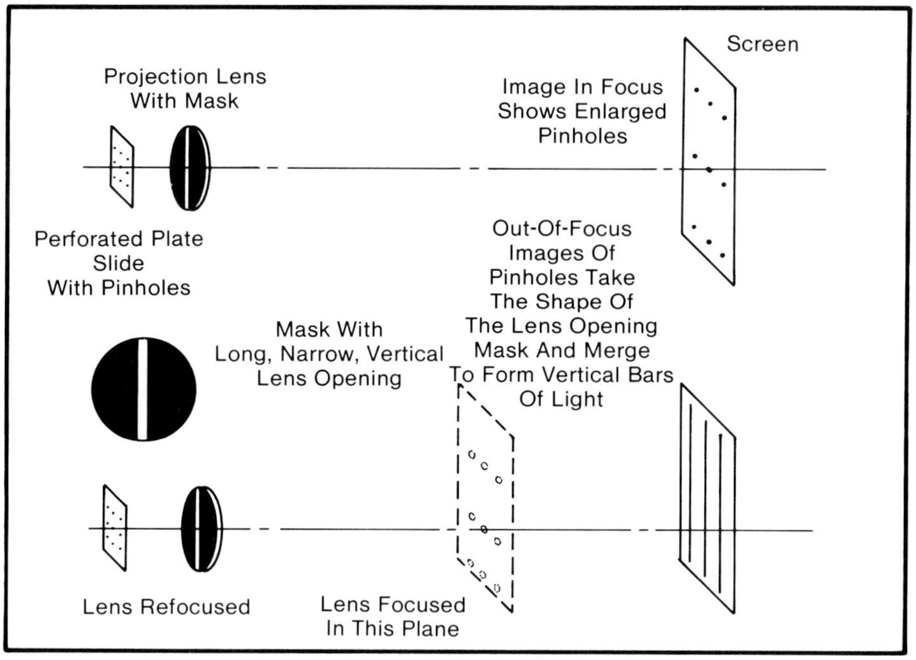

Figure 12-21. Demonstration: Out-of-focus images of points tend to take the shape of the lens opening.

nearly parallel, in order that the human eye can focus them. The eye muscles are more relaxed in this viewing condition.

Magnification

In the drawing, the focal length of the objective lens is shown as 7 inches, while the eyelens has a focal length of 1 inch. The magnification of a telescope is found by dividing the focal length of the objective lens by the focal length of the eyelens:

$$M = \frac{\text{f objective}}{\text{f eyelens}}$$

$$M = \frac{7\text{ inches}}{1\text{ inch}} = 7X$$

In this example, the magnification is 7X.

Image Size and Field Curvature

A telescope is aimed at an object arrow 1,000 feet away that is 87.5 feet high (Figure 12-23). The base of the arrow is on axis and the arrowhead is below the base. Light rays from the tip of the arrowhead will form a 5° angle with the axis. Three parallel rays from the arrowhead point are imaged by the objective lens. The arrow image is above the axis, with the arrowhead also above the axis. The size of the image is found to be 0.61 inches. This real image is an aerial image because there is no film or screen to intercept it.

The focal plane is shown to be flat. Actually, with most telescope objectives, the field is curved with the center of curvature being approximately equal to the focal length of the objective lens. However, since the field of view of telescopes is relatively small, the discrepancy between the curved field and a plane is small. In our example, assuming a 5° half-field, the arrowhead focuses only about 0.026 inches (1/40 of an inch) closer to the lens than does the base of the arrow. The eye easily accommodates this difference.

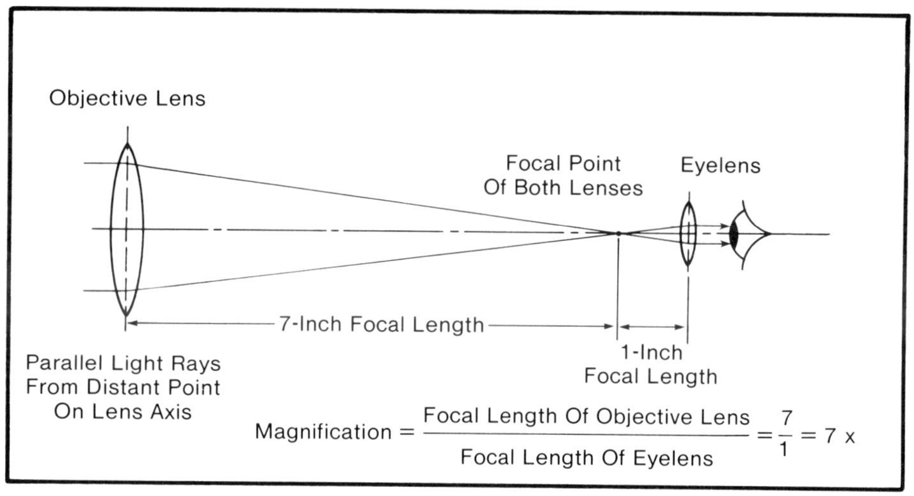

Figure 12-22. Ray Trace Of Simple Telescope Showing Magnification.

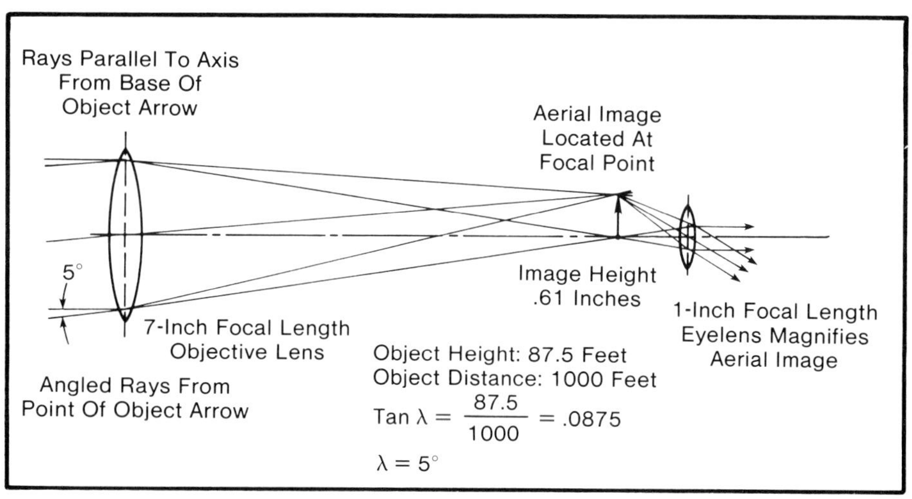

Figure 12-23. Simple Telescope Imaging Distant Object.

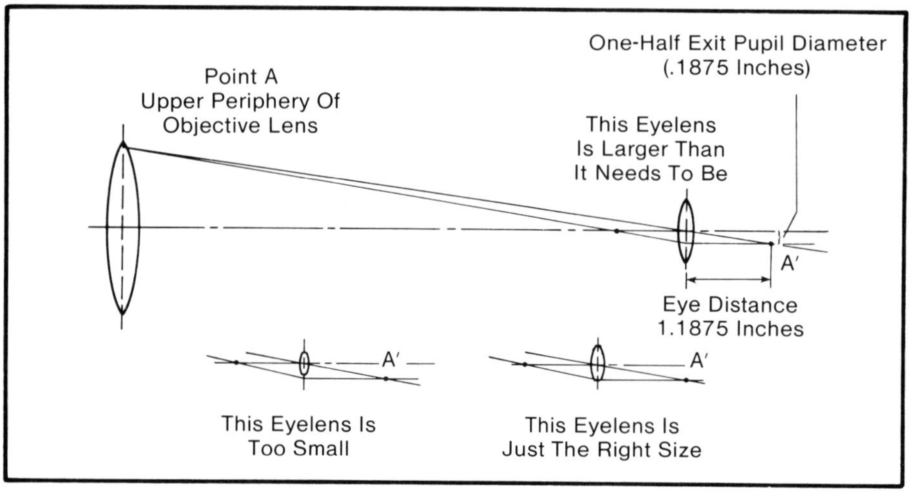

Figure 12-24. Finding Position And Size Of Exit Pupil.

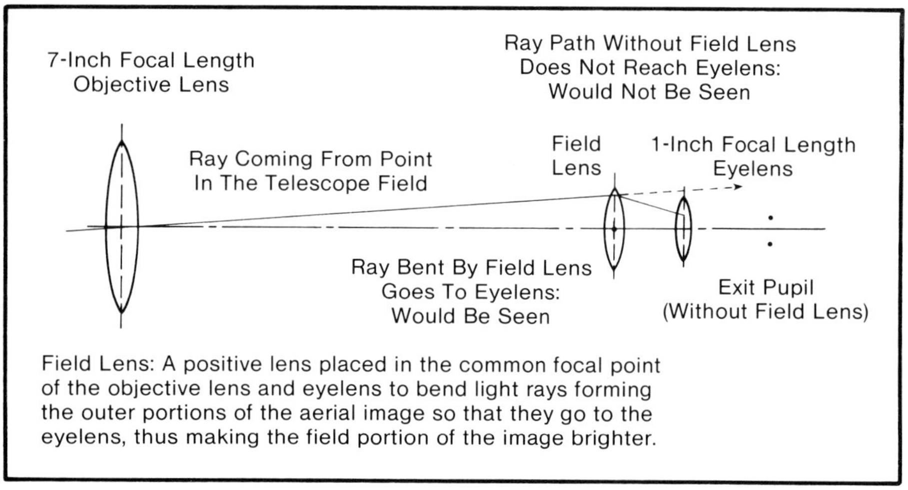

Figure 12-25. Location And Purpose Of A Field Lens.

Exit Pupils

It is convenient to have a position behind the eyelens where the eye can see the entire image formed by the telescope. It is desirable to have this position located at least 0.75 of an inch behind the eyelens, especially if one wears glasses. This position is called the "exit pupil" of the telescope. The distance between the eyelens and the exit pupil is called the "eye distance." The portion of the exit pupil (the eye distance) and its diameter can be found by schematic ray tracing.

The eyelens is placed with its focal point located at the image plane of the objective lens and forms an image of the objective lens itself. In Figure 12-24, Point A is one point on the periphery of the objective lens. When the image of Point A formed by the eyelens is located, both the diameter of the exit pupil and the eye distance are determined.

One ray passes through the optical center of the eyelens and goes through the lens without being refracted. The other ray passes through the front focal point of the eyelens where it strikes the principal plane of the eyelens and is refracted parallel to the axis. Where the two rays intersect, Point A′ is the location of the periphery of the exit pupil. The distance from the lens axis to Point A′ is one-half the diameter of the exit pupil.

Three eyelenses are shown in Figure 12-24. One is smaller than the exit pupil diameter and if used, some light will be lost. Another eyelens is larger than the exit pupil. The part of the lens that falls beyond the exit pupil is not used, so this lens' diameter is larger than it needs to be. The third eyelens is the same size as the exit pupil. This size permits viewing the entire image, yet no part of it is unused. By finding the diameter of the exit pupil, the proper diameter of the eyelens is also determined.

Field Lenses

If you were to look through the simple telescope described above, you would find that it had a very narrow field of view. You would see a bright central image, but as you look out into the field, the image gets dim and fades away completely.

Another lens is inserted into the telescope (Figure 12-25) to make the field of view brighter over a larger field angle. This lens is called a "field" lens because it broadens the field of view.

A field lens is a positive lens placed where the focal points of the objective lens and the eyelens coincide. It does not change the magnification of the telescope. The eyelens and the field lens work together to form the image of the objective lens and therefore determine the size and position of the exit pupil. This is discussed further in the following problems, and in the next lesson.

Problems

1. On the adjacent drawing use additional ray tracing to locate the position and determine the size of the exit pupil with the field lens in place. The exit pupil without the field lens is shown. State what has happened to the size and position of the exit pupil by the addition of the field lens. Hint: when tracing the rays, use the front focal point of the field lens.
2. Using the magnification equation given in the lesson, determine the magnification (power) of the telescope. Use the diameter of the objective lens and the diameter of the exit pupil to derive a new equation for magnification.

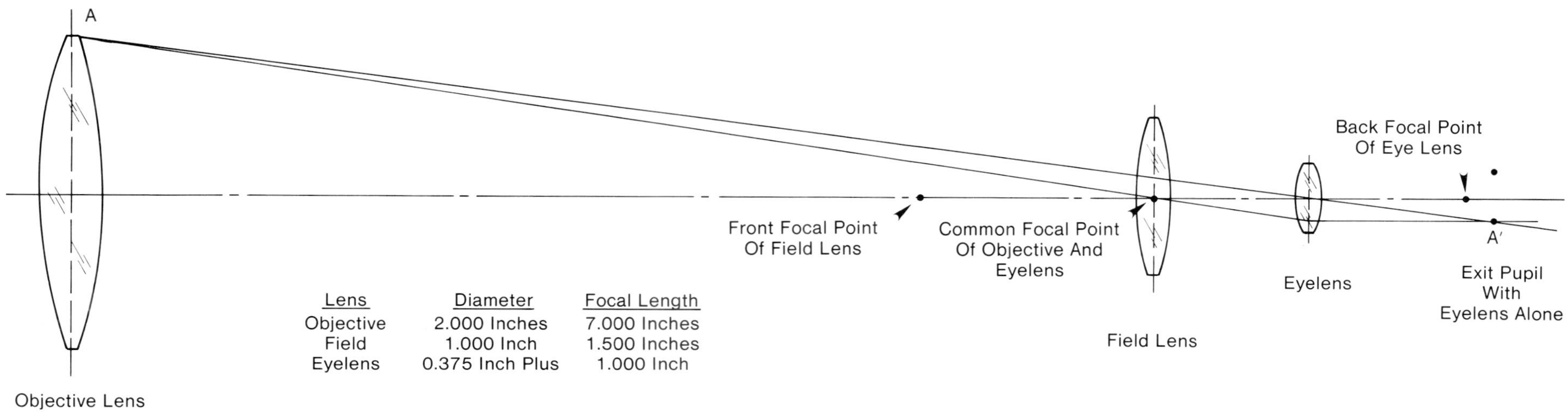

Lens	Diameter	Focal Length
Objective	2.000 Inches	7.000 Inches
Field	1.000 Inch	1.500 Inches
Eyelens	0.375 Inch Plus	1.000 Inch

Problem
Chapter 12
Video Lesson 7

Chapter 13

VIDEO LESSON 8

Field Lenses and Fresnel Lenses

Summary

Field lenses are added to telescopes to increase the field angle. Schematic ray tracing can establish the size and position of the exit pupil of the telescope with the field lens in place. The magnification of a telescope can be found by dividing the diameter of the objective lens by the diameter of the exit pupil. Another method is to divide the apparent angle by the actual angle; these two terms are defined. A fresnel lens is a true thin lens and is, in essence, a segmented plano-convex lens. It is often used as a field lens in projection-type equipment and in cameras. The diffusing characteristics of ground glass are shown diagrammatically as well as methods of changing these characteristics using liquids.

Determining the Size and Location of the Exit Pupil When a Field Lens is Added

The first part of this lesson is a step-by-step answer to the problems given at the end of the last lesson.

The diagram given in the problem is reproduced here half size (Figures 13-1 and 13-2). The objective lens has a focal length of 7 inches, the field lens has a focal length of 1.5 inches, and the eyelens has a focal length of 1 inch. The field lens is located in the aerial image formed by the objective lens at its focal point and the eyelens is located at a distance equal to its focal length from the field lens. Figure 13-1 shows the location of the exit pupil of the telescope without the field lens. Since the exit pupil is the image of the periphery of the objective lens formed by the eyelens, its location can be found by locating the image of point A on the objective lens. One ray is drawn from point A through the center of the eyelens. It goes through the center without bending.

The second ray is drawn through the focal point of the eyelens. (This is also the center of the field lens). Any ray going through the focal point of the eyelens will emerge from the lens parallel to the lens axis. The point at which this parallel ray intersects the first ray determines the position of the image of point A. This image point is called A′. Similarly, the point B on the objective lens is imaged at B′; the same distance from the eyelens as point A′ and the same distance from the axis

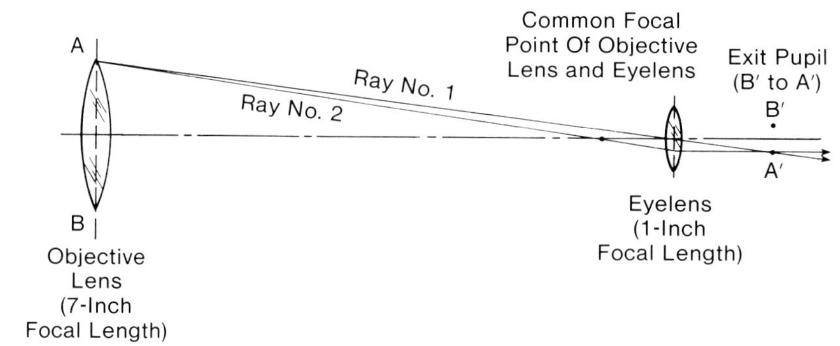

Figure 13-1 Locating The Exit Pupil Without The Field Lens.

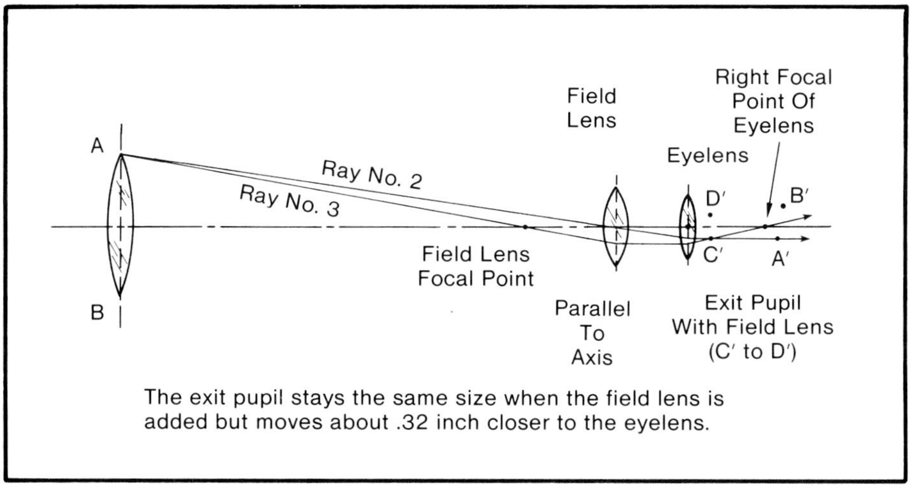

Figure 13-2. Locating the exit pupil with a 1.25-inch field lens.

as A′. These two points locate the image of the periphery of the objective lens and therefore determine the size and location of the exit pupil without the field lens in position. In the drawing we see that the exit pupil has a diameter of about 0.32 of an inch and is located at a distance of about 1.2 inches from the principal plane of the eyelens.

This establishes the size and position of the exit pupil without the field lens and is necessary as a reference point when determining the amount of change caused by the addition of the field lens.

The second ray that passes through the focal point of the eyelens also goes through the center of the field lens and is valid for tracing both with and without the field lens in position (Figure 13-2). The size of the new exit pupil will be determined by this ray. Since this ray determines the size of the exit pupil without the field lens, it will also determine the size *with* the field lens. The exit pupil remains the same size when the field lens is added. This solves the first part of the problem.

The second part of the problem asks: What other ray can be used to determine the location of the exit pupil? The hint given in the problem was to use a ray that goes through the front focal point of the field lens. Other rays can be used, but this is the easiest one. A third ray is drawn from point A through the front focal point of the field lens. As this ray strikes the field lens, it is bent so that it is parallel to the axis. All rays that go through either focal point of a lens emerge parallel to the axis. The opposite is also true; all rays parallel to the axis incident on a lens are refracted so that they pass through the focal point (this is an example of the reversability of the paths of light rays).

The third ray refracts at the principal plane of the eyelens and passes through its focal point. Since the eyelens is too small for this ray to go through, the principal plane must be extended beyond the diameter of the eyelens. It intersects the path of Ray 2 at the new location of the exit pupil and this is marked C′. This exit pupil is the image of the objective lens formed by the combined refractions of the field lens and the eyelens.

The position of the exit pupil is controlled by the focal length of the field lens as well as that of the eyelens. If the field lens focal length is longer, the third ray will cross the axis farther away from the field lens. After passing through the eyelens, it will intersect with the second ray farther behind the eyelens—locating the new exit pupil closer to the original exit pupil (Figure 13-3).

If, however, the focal length of the field lens is shorter, the third ray will locate the exit pupil closer to the eyelens. In all cases the diameter of the exit pupil remains the same because the path of the second ray is parallel to the axis.

In the example in which the focal length of the field lens is 1.5 inches, the new exit pupil is located about 0.32 of an inch behind the principal plane of the eyelens and is about 0.32 of an inch from the exit pupil formed by the telescope with the 1.5-inch field lens.

Using Another Ray

The best way of locating the new exit pupil is to use the ray that passes

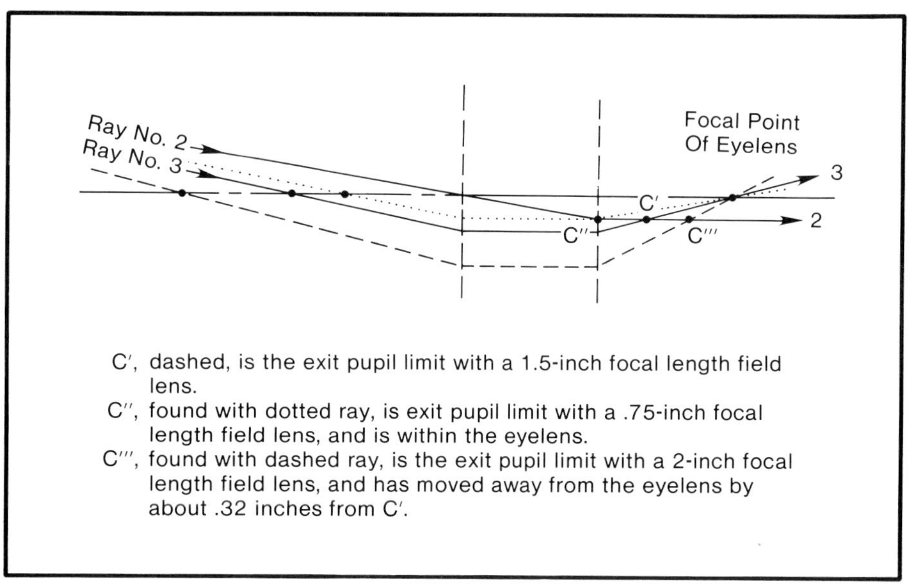

Figure 13-3. The Effect Of Field Lens Focal Length On Exit Pupil Position.

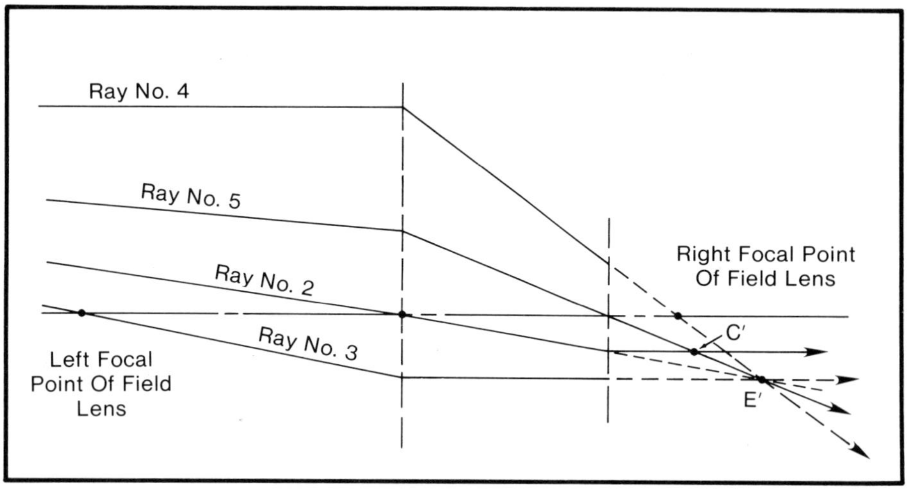

Figure 13-4. Using Another Ray To Locate The Exit Pupil.

through the first focal point of the field lens. There are other rays that can be used; however, their use introduces complications. In order to provide a good example of the orderliness of ray paths, an alternate will be traced.

In Figure 13-4, a ray (number 4) is drawn from point A parallel to the axis. The field lens refracts the ray so that it passes through the right focal point of the field lens and strikes the eyelens. Since the ray is no longer parallel to the axis and does not pass through the center of the eyelens, no method has been presented so far to determine the direction in which the ray will be bent. Therefore some other method must be tried.

If the eyelens were not there, the ray would go to point E′. This ray is one of the rays that determines where the field lens would image point A if the eyelens were not there. Ray number 3 is another ray that forms the image point E′. Ray number 2 is another ray that goes from point A through the center of the field lens without bending and intersects the other two rays at E′ (when the eyelens is not there).

These are just three of the infinite number of rays that form the image point E′. If you could put your eye at E′ and look toward point A (without the eyelens), the pupil of your eye would see the field lens filled with light, all coming from point A. Any ray coming to point E′ comes from point A. Point E′ can be used to find the other ray needed. When the eyelens is put back in the system, ray number 5 is drawn from point E′ through the center of the eyelens. It passes through point C′, straight through the eyelens and strikes the field lens. At the field lens it must be bent to go to point A, because all light rays from A go to point C′. This ray also locates C′ and hence the new exit pupil, but it is more difficult to use than ray number 2 which passes through the front focal point of the field lens.

Field Size of a Telescope

Inserting a thin field lens into a telescope will not change the power of the telescope. This is because the exit pupil has not changed size. This may be difficult to visualize. What *does* change, however, is the field size—the angular coverage image through the telescope.

In Figure 13-5, the telescope is shown without the field lens. A number of parallel rays are shown coming to the top of the objective lens from a distant point about 3° off the telescope axis. As they are bent by the objective lens, only one ray is bent so that it reaches the eyelens. The rays intersect in the aerial image plane. The eye receives only a small part of the light from the object point.

In Figure 13-6, the rays incident on the object lens are parallel to the axis. The light from all these rays goes through the eyelens and reaches the eye. The entire bundle of light from the object point reaches the eye, so the point is very bright. Without the field lens, this telescope covers about 6° actual angle (twice the 3° half angle) with the image becoming very dim at 3° off the axis.

Figure 13-7 shows how the field lens enlarges the field and makes the field completely bright. A 1.5 inch focal length field lens with a 1 inch diameter is placed in the telescope. Three parallel rays are incident to the objective lens and they are at an angle of 4° to the telescope axis.

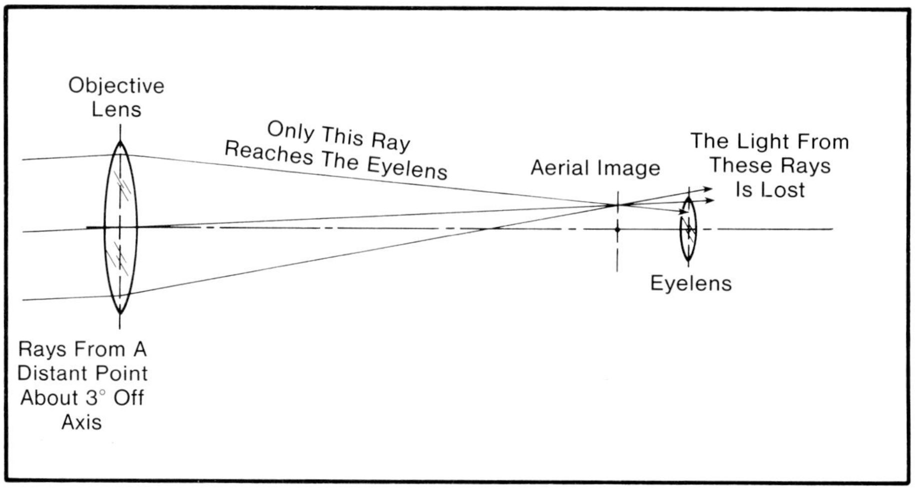

Figure 13-5. Without a field lens, many light rays slightly off axis are not used to form a visible image through the eyelens.

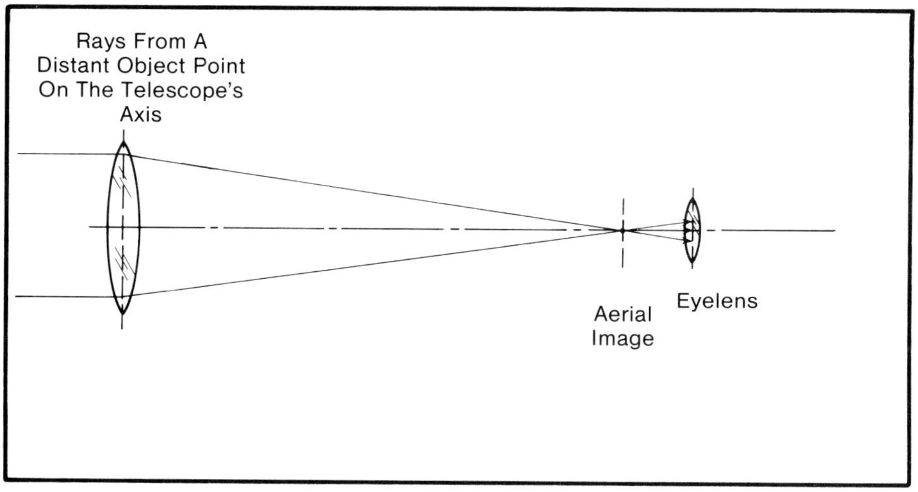

Figure 13-6. Without a field lens, all the light rays parallel to the axis reach the eyelens and form a bright axial image.

These three rays converge at point F on the periphery of the field lens. Ray number 2 determines the angle of coverage of the telescope because it crosses the axis in the center of the objective lens (Note that if the diameter of the field lens is increased, the angle of coverage of the telescope is increased).

Where do these three rays go after they are refracted by the field lens? Without the eyelens in place, the field lens would image the objective lens at an exit pupil farther to the right than C′D′. However, with the eyelens in place, ray number 1 goes through the center of the eyelens and through C′, which is the image of point A formed by both lenses.

The plane C′D′ is the image plane of the objective lens formed by both lenses. Since ray number 2 comes through the center of the objective lens, it must cross the axis halfway between C′ and D′ in the image plane. Ray number 3 would go to point B′ (Figure 13-1) if the eyelens were not there. This determines its direction between the field lens and the eyelens. It must emerge parallel from the eyelens to rays numbers 1 and 2. It is ray number 3 that determines how large the eyelens must be. In this instance, it must be about 0.32 of an inch in diameter to transmit ray number 3.

Since these rays are parallel and image the entire objective lens, the field lens has refracted the rays that would never come to the eyelens (without the field lens). The image of the object point at 4° to the axis will be bright when viewed through the eyelens.

Other Methods of Determining Telescope Power

Earlier in this chapter it was stated that field lenses increase the angular field coverage of the telescope, but do not change the magnification or power of the telescope.

The magnification equation for telescopes given earlier was:

$$\text{Magnification} = \frac{\text{focal length of object lens}}{\text{focal length of eyelens}}$$

It has now been determined that the power of a telescope is not dependent on the field lens, therefore this equation is still valid.

Based on the measurements of the telescope that has been developed in this chapter, it has been established that the diameter of the objective lens is 2.00 inches and the diameter of the exit pupil is about .3125 inch.

If the diameter of the objective lens (2.00 inches) is divided by the diameter of the exit pupil (.3125 of an inch), the result is found to be 6.4. This is almost the same number as the power found by the focal length equation, the difference being in the accuracy of our drawings. Although this is not theoretical proof, it is by example, an indication that the power (magnification) of a telescope can also be found by dividing the diameter of the objective lens by the diameter of the exit pupil.

$$\text{Magnification} = \frac{\text{Diameter of Objective Lens}}{\text{Diameter of Exit Pupil}}$$

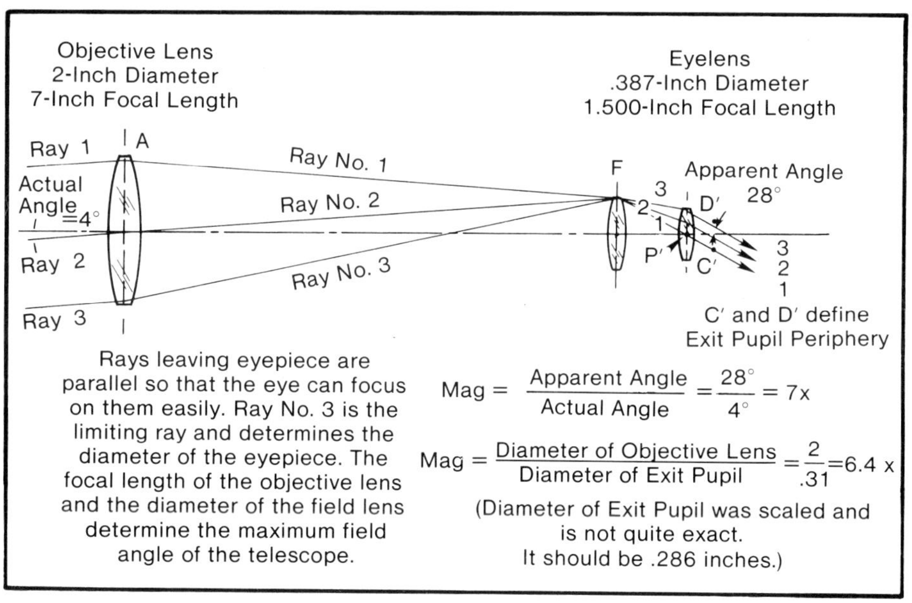

Figure 13-7. The Completed Telescope.

Binoculars and telescopes are often identified by the diameter of the objective lens and the power of the instrument. From these figures, the diameter of the exit pupil can easily be determined mathematically from the above equation:

$$\text{Exit Pupil Diameter} = \frac{\text{Diameter of the Objective Lens}}{\text{Magnification}}$$

Using this equation and 7X for the magnification, the diameter of the exit pupil is calculated to be .2857 inches. We were off about 0.032 of an inch. If a pair of binoculars is classified as being a 6 x 30 binocular, it means that the diameter of the objective lens is 30 mm and the power is 6X. Therefore, the diameter of the exit pupil is 30 ÷ 6 = 5 mm. The exit pupil of a 7 x 50 binocular is 50 ÷ 7 = 7+ mm diameter.

A third method of determining the magnification of a telescope is found by dividing the apparent angle by the actual angle of an object point at a distance from the telescope. In Figure 13-7 three parallel rays are drawn coming to the objective lens from a distant object point at an angle of 4° to the axis. The central ray goes through the center of the objective lens without bending and hits the field lens at point F. Since the field lens is in the focal plane, the other two rays are bent by the objective lens and meet at point F.

Without the eyelens, ray number two, which passes through the center of the objective lens, would be imaged by the field lens on the axis in the exit pupil plane of the field lens. However, it goes in this direction only until it is refracted by the eyelens. The new exit pupil is the image of the objective lens formed by both the field lens and the eyelens. Therefore, the ray is bent to go through point P′ on the axis. This is because the ray was located at the center of the objective lens as it entered the system.

The upper ray has gone through point A on the periphery of the objective lens. Therefore, it leaves the field lens headed toward the lower limit of the exit pupil found by the field lens alone. As it passes through the eyelens, it is bent toward the lower limit of the exit pupil of the complete telescope at point C′.

Similarly, the lower ray is bent toward the upper limit of the field lens exit pupil and then it is bent by the eyelens to go to the upper limit of the exit pupil of the complete telescope (D′).

All three rays are parallel as they leave the eyelens. The telescope is said to be afocal which means that the light rays leaving the eyepiece are parallel and the eye can focus them easily on the retina.

The apparent angle is 28° which is the angle at which the eye will see the original object point. The actual angle measures 4°. The magnification is 28° ÷ 4° = 7X, which is the same magnification found by the other two methods.

Again, this is not proof of the third magnification law, rather it is a demonstration of it.

Demonstrating Field Size

In the videotape of this lesson, two lenses are placed in front of an

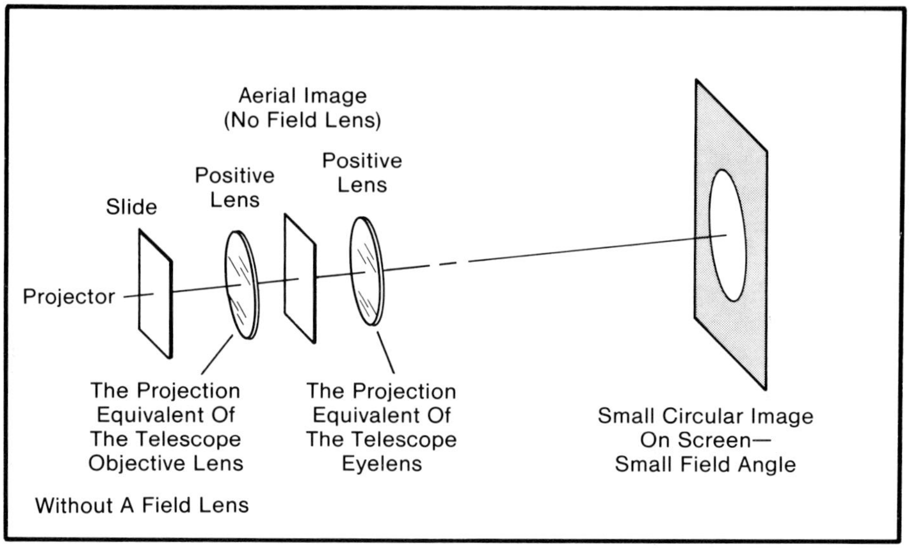

Figure 13-8 Projection Demonstration Of The Effectiveness Of A Field Lens.

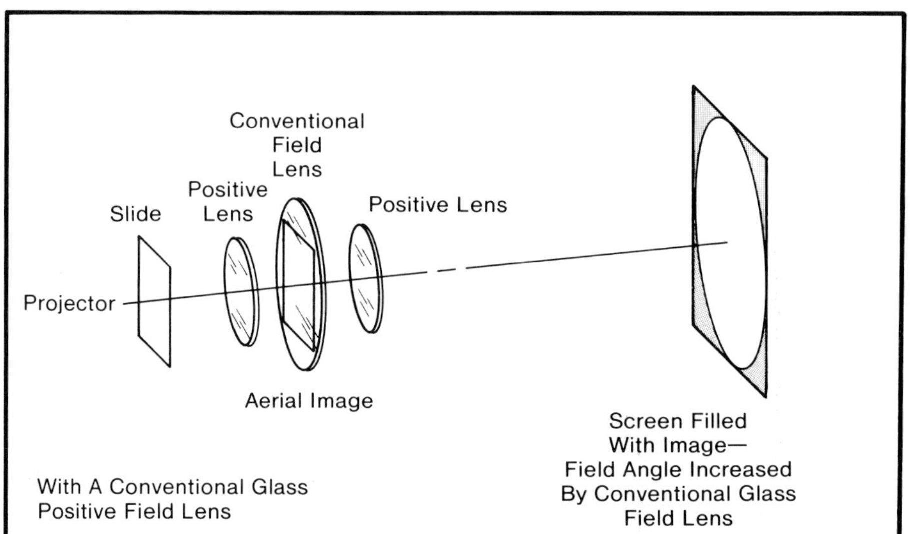

Figure 13-9. Projection Demonstration Of The Effectiveness Of A Field Lens.

illuminated slide. These simulate the objective and eyelenses of a telescope (Figure 13-8). The lens nearest the slide is the objective lens and forms an aerial image of the slide between the two lenses. The lens farthest from the slide acts as the eyelens and focuses the aerial image on a screen. Only a small, circular area of the slide can be seen on the screen.

A field lens is added between the two lenses in the position of the aerial image (Figure 13-9). Now the entire slide area can be seen on the screen. The field lens has increased the field angle of this optical system, just as a field lens in a telescope increases the field angle of the telescope.

If the image size is measured on the screen before and after insertion of the field lens, we will see that no change in image size takes place by the addition of the field lens. The addition of the field lens does not change the magnification of a telescope—it merely increases the field angle.

If the positive glass field lens is removed from the optical system of the projector and replaced with a thin, molded-plastic Fresnel lens which serves the same function as the glass field lens, the entire slide image will be seen on the screen (Figure 13-10).

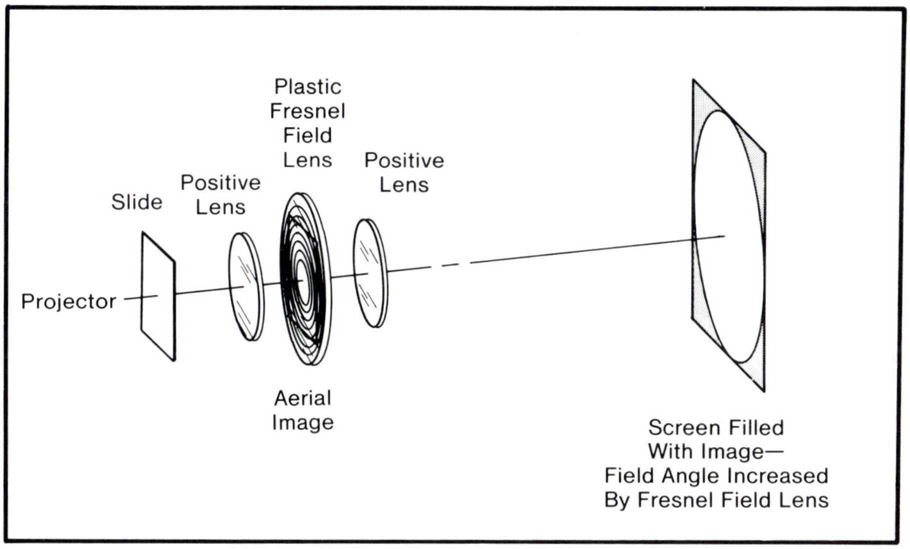

Figure 13-10. Projection Demonstration Of The Effectiveness Of A Field Lens With A Molded Plastic Positive Fresnel Field Lens

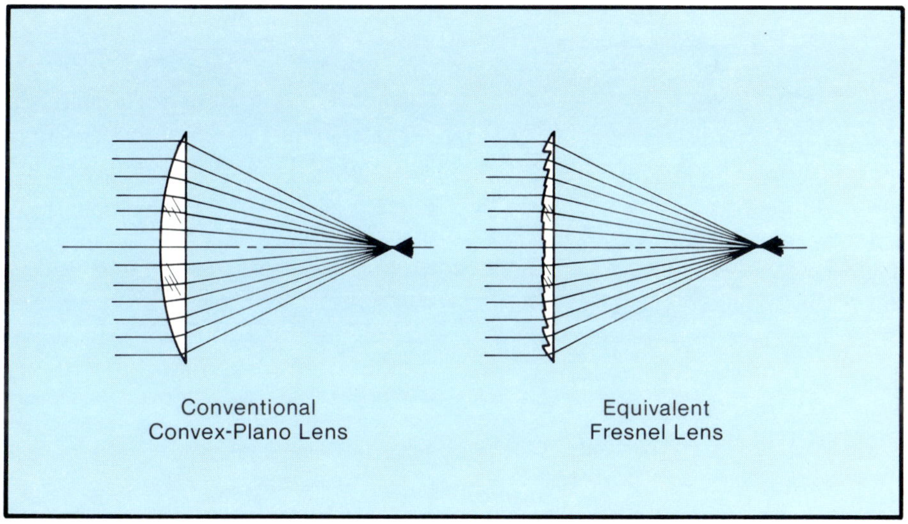

Figure 13-11. Fresnel Lens.

Fresnel Lenses

Figure 13-11 shows a number of light rays parallel to the axis coming to a plano-convex lens. Each ray is bent by the lens so that the rays come to a common point, the focal point, behind the lens. The axial ray is not bent at all because it goes through the center of the lens where it is perpendicular to the glass surfaces. The farther a ray is from the axis, the greater is the angle between the normal to the glass and the path of the ray. Therefore, the bending of the ray is greater. It is this angle that determines the amount of bending of a given ray.

In situations where a high degree of image quality is not required, a thick positive lens can be replaced by a Fresnel lens. In a Fresnel lens, the convex curve is broken into concentric sections which are located so that each section is equidistant from the plano side. The surface angle of each segment increases as it gets farther away from the axis. The surface angle is the same angle as the corresponding segment of the convex curve. In *most* ways, the Fresnel lens refracts light rays in the same way as a planoconvex lens having the same focal length. The drawings illustrate this.

Fresnel lenses cannot be ground and polished like normal lenses; they must be cast or molded. The lenses of theatrical spotlights are cast or molded glass. In large fresnel lenses the section widths are fairly large—0.50 to 0.75 of an inch or more for the larger lenses. Such spotlight lenses are commonly 12 inches or larger in diameter in the lights used at sports arenas and in professional motion picture and television studios. A conventional planoconvex lens with a 12 inch diameter will weigh many times more than its Fresnel counterpart.

View cameras and reflex cameras

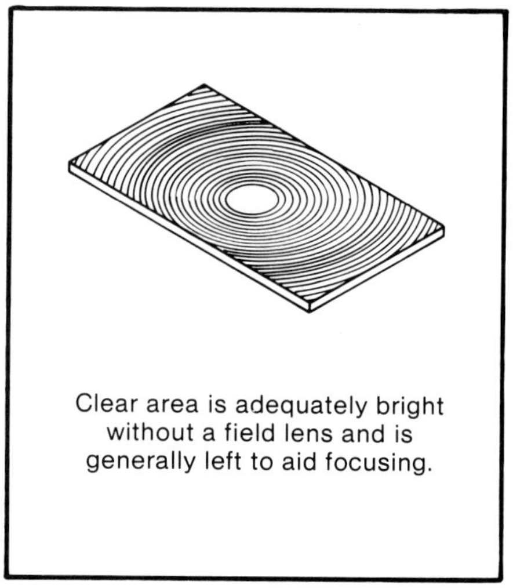

Figure 13-12. Fresnel Lens For Camera.

utilize a ground glass screen to help focus and frame the image. The ground glass viewing screen is placed in the focal plane of the camera and the camera lens forms an image on the ground glass. The photographer uses this screen image to determine the focus, depth of field, and frame of the scene.

The granular surface of the ground glass scatters the light which makes the image visible. However, it transmits more light along a line which is a radius originating at the center of the lens. When the eye is centered behind the ground glass, the central area of the ground glass is bright, while the corners are so dark the image can scarcely be seen.

A Fresnel lens in contact with the ground glass refracts the peripheral rays toward the axis (the usual eye position) so that the corners of the image appear brighter. The focal length of the Fresnel lens must be approximately equal to the focal length of the camera lens for best image illumination. In cameras with interchangeable lenses, a compromise Fresnel lens focal length must be chosen.

The Fresnel lenses used in cameras are made of molded, transparent plastic (Figure 13-12). Plastic sheets are heated until soft and are then pressed against the surface of a mold which has the Fresnel pattern cut into its surface. The plastic is forced into the "grooves" which form the Fresnel pattern on the surface of the plastic sheet. When cooled, the plastic retains the Fresnel pattern. The width of the "sections" of such Fresnel lenses are made relatively small, perhaps .030 of an inch, so that the depth of the grooves is slight and the plastic can be pressed into them. The grooves are usually visible through the ground glass viewing screen when viewed with a magnifier.

The molds are made of a special bronze alloy, and the Fresnel pattern is cut on a lathe with a specially shaped diamond tool. The Fresnel pattern is actually a series of concentric circles with a continuously changing angle—from a steep angle at the edges, to practically no angle at the center. The diamond tool cuts the groove so smoothly that the surface is equivalent to a polished surface. The bronze alloy is fairly soft and will not mold many Fresnel lenses without the tops of the grooves wearing away. Therefore, the bronze molds are plated with a hard metal such as chromium to make a more durable surface for molding, thus extending the life of the mold.

Large sized Fresnel lenses are used with rear-projection screens to provide even illumination across the viewing area. Rear-projection screens made of ground glass, matte acetate, or other diffusing materials, are used on table-model slide viewers, motion picture film editors, motion picture viewers, and in instruments used to make precise measurements, such as contour projectors.

The Characteristics of a Ground Glass Viewing Screen

As stated above, a ground glass screen can be placed in a camera lens image so that the image can be seen. It makes the image visible by scattering the light. It does not scatter the light evenly, however, and this is why the center seems bright while the corners seem dim when viewed on axis (Figure 13-13).

A method used to show the brightness of light sources at various angles can also be used to show the effect of the ground glass screen. In the first diagram a narrow beam of light is aimed at a ground glass perpendicular to its path (Figure

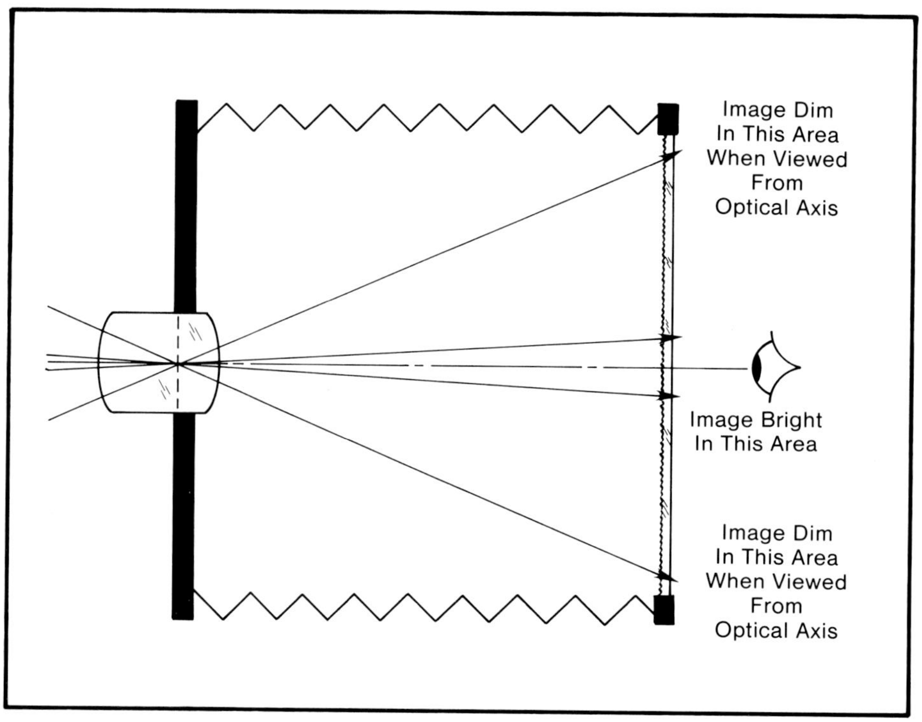

Figure 13-13. Camera Ground-Glass Viewing Screen.

13-14). In this diagram, it is assumed that the ground glass scatters the light evenly in all directions. A semi-circle is drawn with the intersection of the lightbeam and the ground glass surface as its center. Several eyes are drawn around the semi-circle. The even distance from the intersection point to the semi-circle is a graphical way of saying that the brightness of the light is the same in all directions.

Figure 13-14 shows a theoretical situation and compares it to Figure 13-15 which shows a real case. The oddly shaped diagram represents the relative brightness of the light in any direction by the distance along any line of sight to the intersection point. Obviously, the light is brightest along the axis. The light diminishes gradually, until at angle A (approximately 90°), there is no light seen at all. At the other eye positions shown, the relative brightness of the ground glass is shown by the distance from the intersection point to the curved line. If the axial brightness is 1, the relative brightness at A is 0, at B 0.60, and at C it is 0.86.

If water is put on the ground glass surface, the scattering pattern is changed. The axial brightness is increased, while the brightness off axis is decreased. The light coming through the screen is more specular because there are more orderly rays with very little spreading, and the rays scattered at great angles are considerably decreased. This is because the liquid fills the pits in the ground surface, making it similar to a polished surface. If there is a need to increase the specularity of a ground glass in a more permanent way, the ground surface can be rubbed with castor oil (Figure 13-16). This treatment makes the bright cone angle about 6°, with very little light being spread beyond the 6° cone angle. The castor oil hardens like a varnish and the treatment lasts for years.

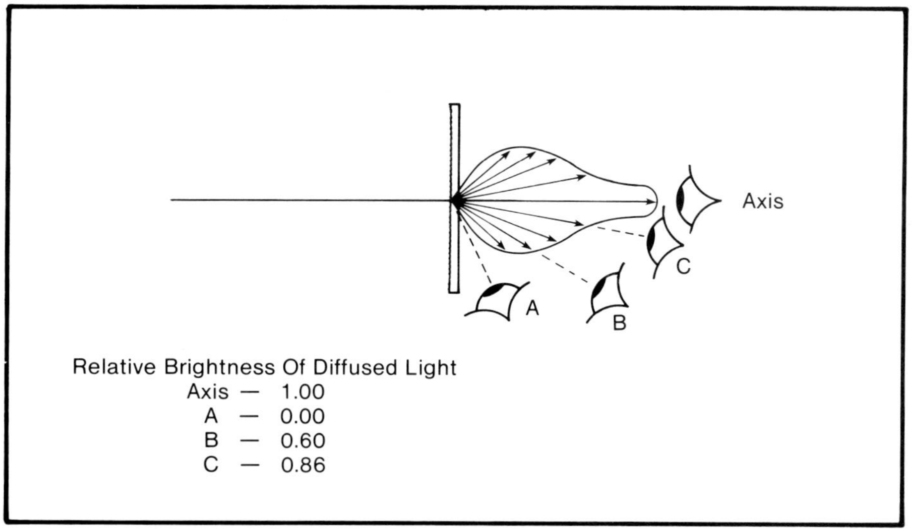

Figure 13-15. Typical Ground Glass, Polar Plot Of Light Distribution.

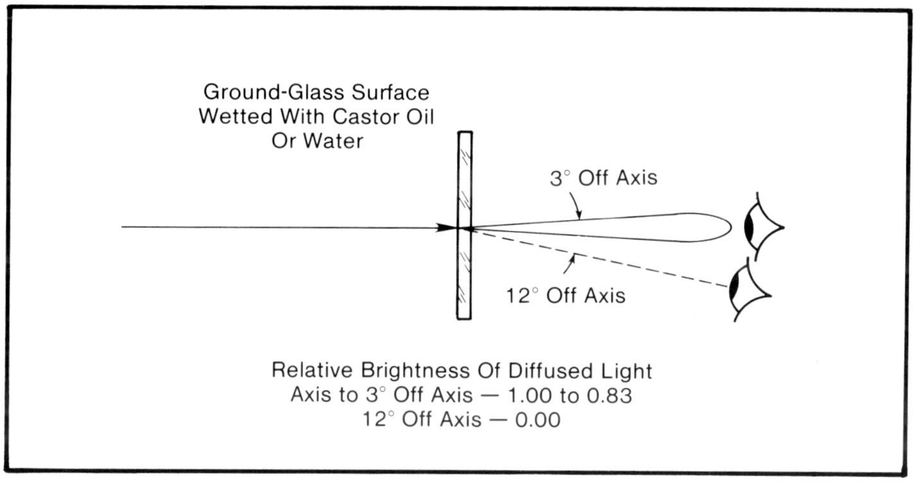

Figure 13-16. Ground Glass Wetted With Castor Oil Or Water.

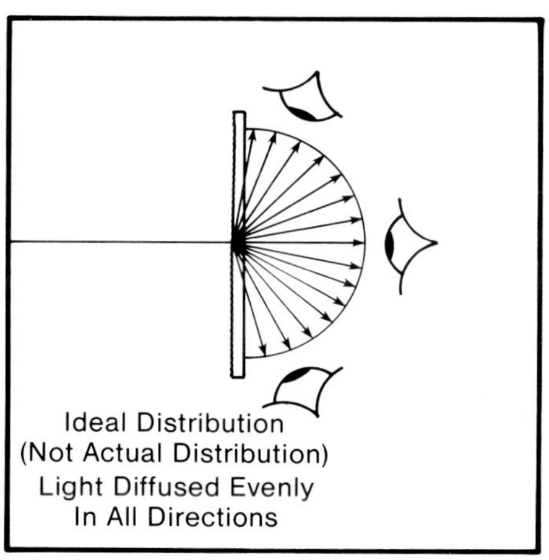

Figure 13-14. Theoretical Ground Glass, Polar Plot Of Light Distribution.

Problems

On the attached drawing, using schematic ray tracing, find the answers to the questions below. The objective lens has a diameter of 35 mm and a 7-inch focal length. The field lens has a 25.4 mm diameter and a 1.5-inch focal length. The eyelens has a 1-inch focal length.

1. What is the magnification of the telescope?
2. What is the exit pupil diameter?
3. What is the approximate eye distance? (This assumes thin lenses and requires a very accurate schematic ray trace).
4. What will the actual maximum field angle be?
5. What will the apparent maximum field angle be?
6. What would be an acceptable diameter for the eyelens?

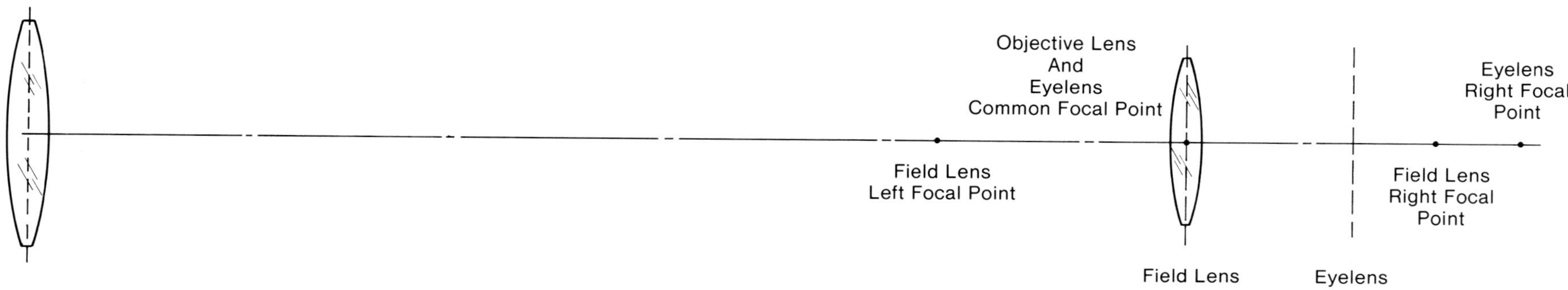

Problem
Chapter 13
Video Lesson 8

Chapter 14

VIDEO LESSON 9

Eyelens Size, Terrestrial Telescopes, Critical Angle and Snell's Law

Summary

Using the schematic ray tracing, the chapter 13 problem is solved. The exit pupil location and size are determined. Using the angular ray that goes through the center of the objective lens to the periphery of the field lens determines the actual angle. This same ray, after being refracted by the field lens and eyelens, determines the apparent angle. Two rays parallel to this ray determine the minimum size of the eyelens. A new angle magnification equation is given that is slightly more accurate than the one given earlier. Two methods are shown for converting the refracting telescope into a terrestrial telescope with an upright image. The first is to add a relay or inverting lens—the second is to add two Porro prisms. The concept of total reflection is introduced by taking a fish-eye view of the outside world. The critical angle is the angle at which total reflection first occurs. This angle is determined by the indices of refraction of the two optical media—air and water or air and glass. Snell's law of refraction is developed geometrically. Graphic ray tracing is introduced. This is a method of finding a ray path through a plane refracting surface using circles and straight lines.

Solving the Problem

The telescope optical system consists of an objective lens (7-inch focal length, 35 mm diameter), a field lens (1.5-inch focal length, 25.4 mm diameter), and an eyelens

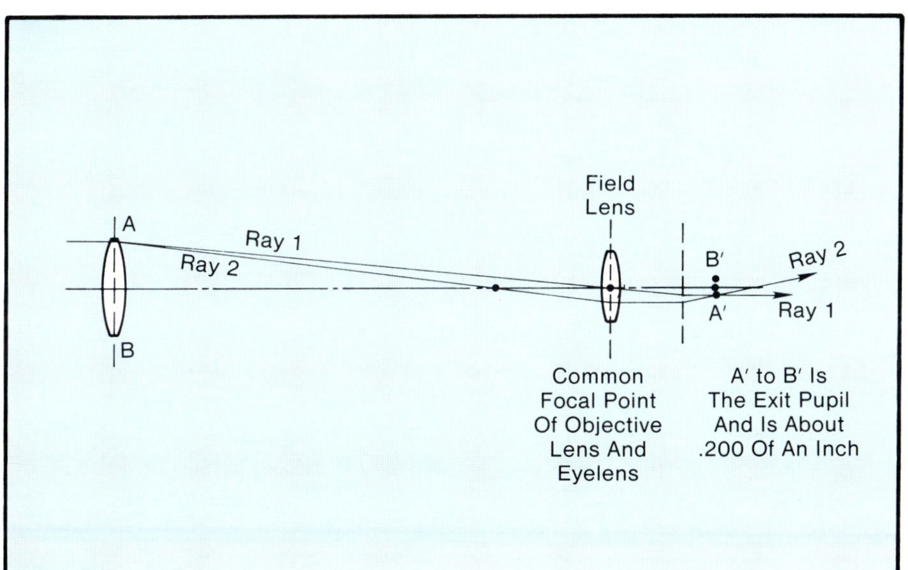

Figure 14-1. Locating The Telescope Exit Pupil With Both The Field Lens And The Eyelens.

(1-inch focal length, diameter to be found).

These lenses are located as shown in Figure 14-1 and the drawings have been reduced to half size.

The Exit Pupil

Ray number 2 is parallel to the axis and is drawn to Point A on the periphery of the objective lens. This ray is refracted to the focal point. Because it passes through the center of the field lens, it is not bent. As it strikes the eyelens, it is refracted and emerges parallel to the axis. This is because it goes through the focal point of the eyelens, which is positioned on the focal point of the objective lens.

Ray number 1 is drawn from Point A through the left focal point of the field lens. The field lens refracts Ray number 1 so that it emerges parallel to the axis. The eyelens refracts this ray through its focal point because it is parallel with the axis as it enters the objective lens. The two rays intersect at Point A′.

The exit pupil is defined as the image of the objective lens formed by the field lens and eyelens working together. A′ is the image of Point A and defines the location of the exit pupil. If rays are traced from Point B, as from Point A, its image would be found at Point B′, the same distance above the axis as A′ is below, and the same distance behind the eyelens. The distance between A′ and B′ is the exit pupil size. It is 0.200 inch in diameter and located about 0.5 of an inch behind the eyelens.

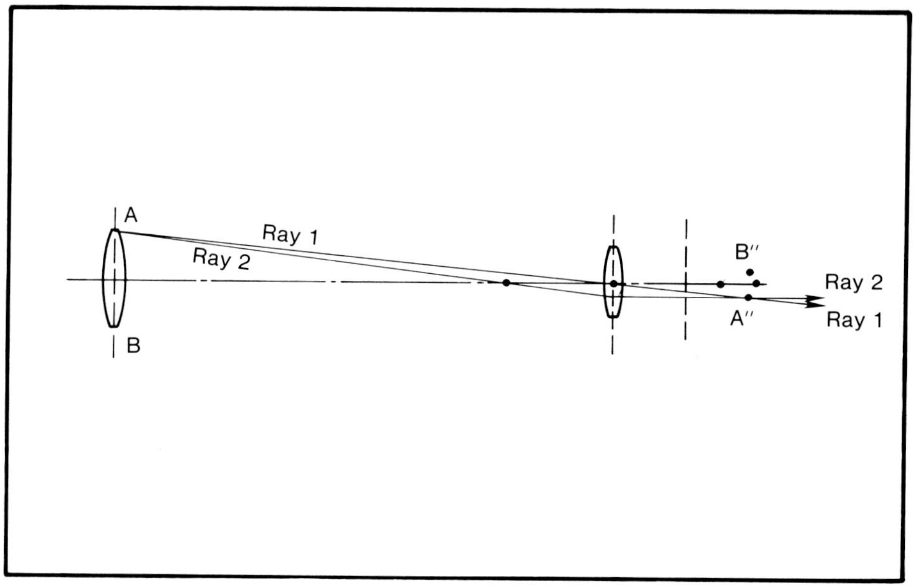

Figure 14-2. Locating The Exit Pupil With The Field Lens And Not The Eyelens.

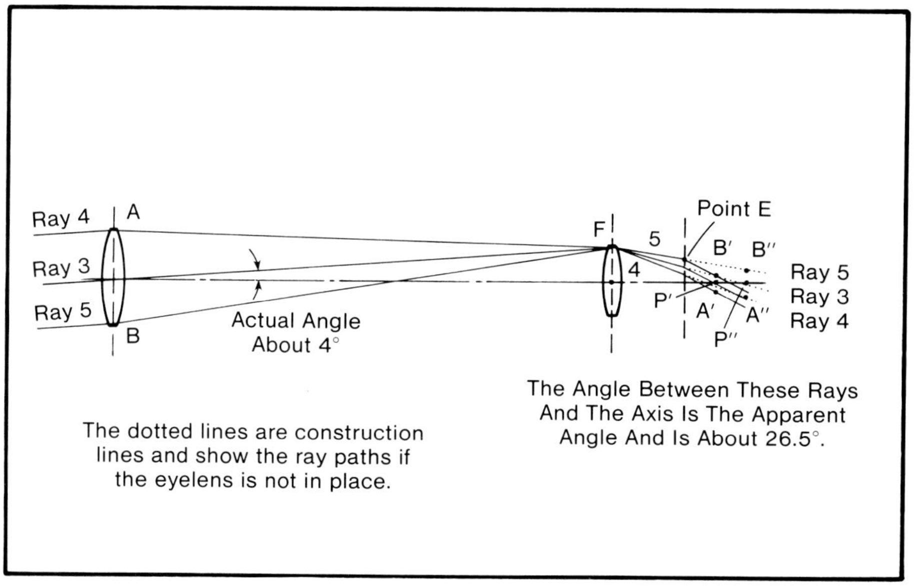

Figure 14-3. Finding The Apparent And Actual Angles And The Minimum Diameter Of The Eyelens.

The Actual and Apparent Angles

To find the actual and the apparent angles it is necessary to know the location of the exit pupil when the eyelens is not in the system. This is accomplished by locating the image of the objective lens formed by the field lens. In Figure 14-2, Ray number 2, which goes through the left focal point of the field lens and is refracted parallel to the axis by the field lens, is continued on as if the eyelens were not there. Ray number 1, which goes through the center of the field lens, is not bent and continues without bending because the eyelens has been removed. These two rays intersect at Point A″, and establish the image location of Point A without the eyelens. The location of Point B″ is found the same way. Point B″ is the same distance behind the field lens as Point A″ and the same distance above the axis as Point A″ is below.

In Figure 14-3, Rays number 1 and 2 have been removed for clarity, but both exit pupils with and without the eyelens are shown.

The actual angle is found by drawing a ray through the center of the objective lens to Point F on the periphery of the field lens. The *actual angle* is the angle between this ray and the axis, which in this case is about 4°. Rays at lens angles less than 4° are imaged, while rays more than 4° to the axis miss the field lens and are not imaged. The actual angle is the maximum angle at which object points are imaged. This ray is bent by the field lens. Since the exit pupil is the image of the objective lens formed by the field lens, this ray is bent by the field lens toward Point P″. This is the image of the center of the objective lens formed by the field lens alone.

However, as this ray strikes the eyelens it is refracted to Point P′, which is the center of the exit pupil with both lenses in place. The angle that this ray makes with the axis is the *apparent angle*, and is approximately 26.5°.

Finding the Minimum Diameter of the Eyelens

There is an entire bundle of parallel rays entering the objective lens at 4° to the axis. The limit rays, numbers 4 and 5, are shown coming to Points A and B, parallel to Ray number 3 and to each other. These rays are all focused at Point F, so that Rays 4 and 5 are drawn to Point F, which is the periphery of the field lens.

Ray number 3 is refracted by the field lens to cross the axis in the plane A″, B″ at Point P″ because that is where the field lens would image the center of the objective lens. However, at the eyelens, Ray number 3 is refracted to cross the axis in the plane A′, B′ because that is the exit pupil with both lenses in place. The dotted lines are construction lines that show where the rays would go if they were not refracted by the eyelens.

Ray number 4 would be bent toward A″ because this is the place the field lens images Point A. However, when Ray number 4 reaches the eyelens, it is bent toward A′, where both field lens and eyelens are working together to form the image of Point A. Note that this part of the ray is parallel to the equivalent segment of Ray number 3.

Ray number 5 is refracted by the field lens toward Point B″, but the eyelens refracts it toward B′. Since this ray is the farthest from the axis at the eyelens, it is the one that determines the minimum diameter of the eyelens. When measured, it is slightly over 0.38 of an inch from the axis. This is a radius, and thus the minimum diameter of the eyelens is slightly larger than twice this value which is *about 0.8 of an inch.*

A Revision of the Angle Magnification Equation

In the previous lesson, the angle magnification equation was given as:

$$\text{Magnification} = \frac{\text{apparent angle}}{\text{actual angle}}$$

If we divide the 26.565° by the 4° found in the ray tracing, the magnification is found to be 6.6X. However, by using the focal length equation, we find:

$$\text{Magnification} = \frac{\text{focal length of objective}}{\text{focal length of eyelens}} = \frac{7}{1} = 7\text{X}.$$

The exit pupil was found to be slightly less than 0.200 of an inch.

$$\text{Magnification} = \frac{\text{diameter of objective}}{\text{diameter of exit pupil}} = \frac{1.378}{0.200} = 6.89\text{X}$$

which is as close as can be achieved by the type of ray tracing used here.

The angle magnification law was given as:

$$\text{Magnification} = \frac{\text{apparent angle}}{\text{actual angle}}$$

This rule is only approximately true; it needs some revision.

The revision uses the tangents of angles. The tangent of an angle is a trigonometric relationship based on right triangles. The standard triangle is shown in Figure 14-4. The angle to the left is the angle on which the relationship is based. The right angle (90°) is at the bottom right. The three sides are labeled. The tangent of angle A is the ratio of the length of the opposite side divided by the length of the adjacent side. In the example shown, the opposite side is 1, the adjacent size is 2, therefore:

$$\tan A = \tfrac{1}{2} = .5$$

The angle whose tangent is .5 is 26.565°.

Two other triangles are shown; one smaller and one larger. It is apparent that as long as the angle is the same, the ratio between the two sides will be the same although the triangles are of different sizes. The tangents of the angles are used in the revised magnification equation.

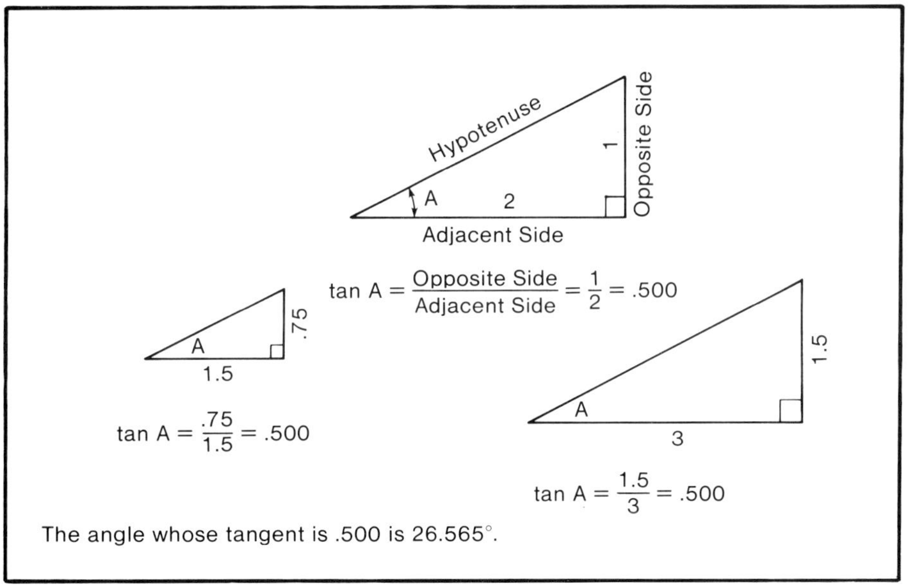

Figure 14-4. Tangents: A Trigonometric Relationship.

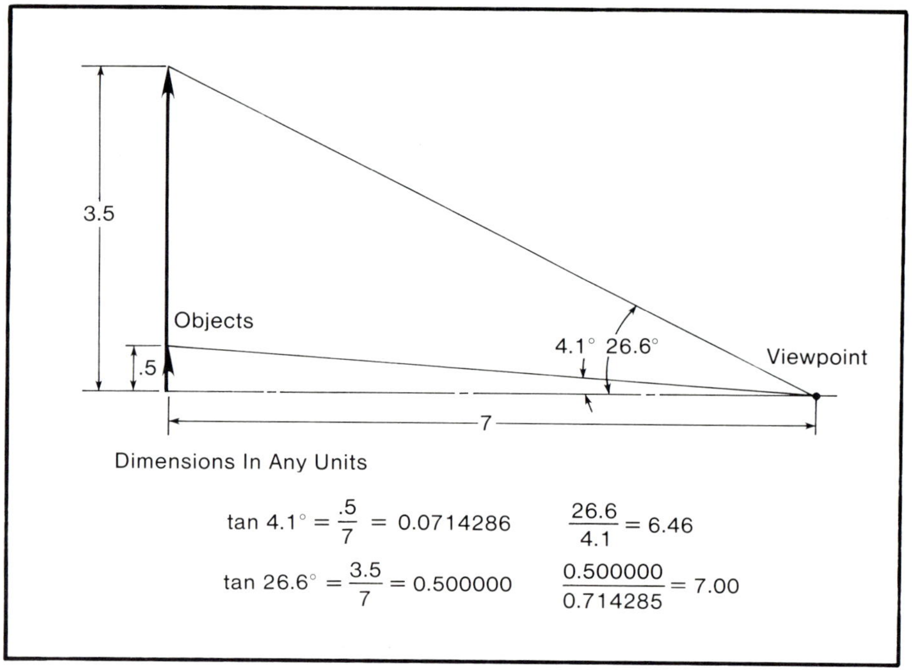

Figure 14-5. Example Of The Exactness Of Using Tangents For Magnification.

To demonstrate the need for this revision, we will use the following example. In Figure 14-5 there are two objects. One is 0.500 inch in height and is 7 inches from the point of view. The angle from the bottom to the top of this object is 4.1°. The other object is in the same position and it is 7x as large as the first object: 7 x 0.500 inches = 3.500 inches. We could say that the second object is the first object magnified 7 times. The angle from the bottom to the top of the second object is 26.565°.

The actual angle of a telescope is equivalent to a viewing angle without the telescope, hence a magnification of 1X. With the telescope, the apparent angle is equivalent to viewing the object with a magnification of 7X. If the apparent angle is divided by the actual angle we get 26.565° divided by 4.1° = 6.5, which is almost, but not quite, the magnification. Therefore the angle magnification equation gives only a close approximation of the magnification. A more precise equation is:

$$\text{Magnification} = \frac{\text{tangent of apparent angle}}{\text{tangent of actual angle}}$$

In the example, this works out as follows:

$$\text{Magnification} = \frac{\tan 26.565°}{\tan 4.1°}$$

$$= \frac{0.501}{0.072} = 6.99X$$

This is very close to 7X.

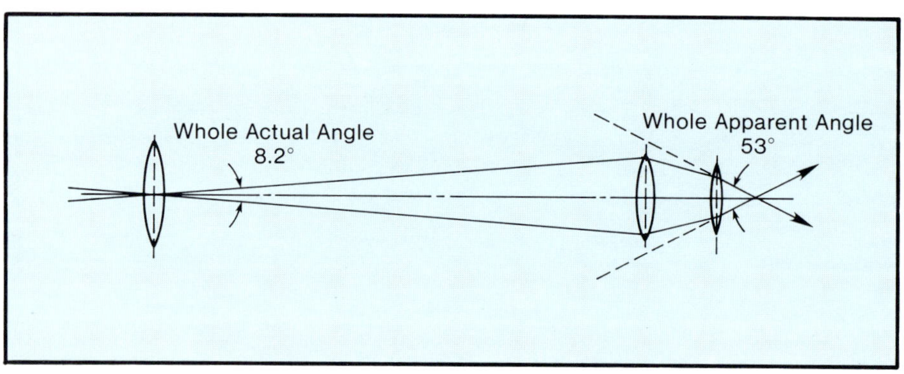

Figure 14-6. Telescope Whole Angles.

If we calculate the tangents exactly and carry out the math to more decimal places, we get even closer to 7X.

$$\text{tangent actual angle} = \frac{0.500}{7}$$
$$= 0.0714286$$

$$\text{tangent apparent angle} = \frac{3.500}{7}$$
$$= 0.500000$$

$$\text{magnification} = \frac{0.500000}{0.0714286}$$
$$= 7.00000X$$

Thus, we see that the tangent equation is the most accurate one.

Half Angles and Whole Angles

The actual and apparent angles shown in the drawings of the telescopes are really half angles. It is customary when tracing rays to do one half of the system. Assuming that the system is symmetrical (top to bottom) the other half works the same way. When the actual and apparent angles were measured in Figure 14-5, from the ray to the axis, it was clear that this represented the upper half of the system.

For the total angle, the 4.1° figure would be doubled: 8.2°. The total apparent angle is 53° (Figure 14-6).

At 1,000 yards, this telescope field width is 140 yards (twice the tangent of 4.1° x 1000) (Figure 14-7). The apparent object distance is 140 yards; that is, at 1000 yards it appears in the telescope the same size as it would appear to the naked eye at 140 yards.

$$\tan 26.565° = \frac{140 \div 2}{x}$$

$$.5 = \frac{70}{x}$$

$$x = \frac{70}{.5} = 140$$

The Inverted Image

Figure 14-8 shows the same telescope with the object arrow drawn above the axis. The image arrow is

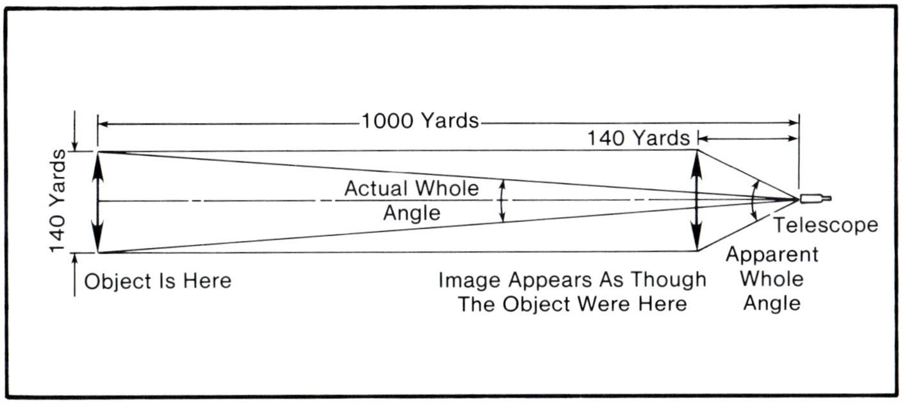

Figure 14-7. Locating The Apparent Position Of The Object.

shown at the exit pupil, but the arrowhead is below the axis. The image is inverted, or rotated 180°.

Looking at stars or the moon with this telescope would be satisfactory as it does not matter if the stars appear upside down. We call this a celestial or astronomical telescope.

However, looking at sporting events or ships at sea, it would be difficult to use this telescope. There are several ways to turn the image right-side-up, and two of these methods will be discussed in the following sections.

Use of an Erector Lens

An erector lens is, as its name suggests, a lens that rotates the image 180° so that, in a telescope, it is seen erect (not inverted). The image has been formed inverted by the objective lens and is erected by the erector lens.

Figure 14-9 shows how a simple lens inverts the image in a one-to-one relationship. The image is the same size as the object—which in this case is the image formed by the objective lens. Figure 14-8 shows this simple erector lens inverting the image in the telescope. In practice, this simple arrangement would require another field lens in the common image point of the objective and erector lenses. So instead, the erector lens is made up of two lenses, and moved closer to the objective image so that the left erector lens serves as a field lens as well as being part of the erector lens system.

Figure 14-10 shows that when the erector lens is used, the field lens is moved to the left slightly, so that the sharp image falls slightly to its right. The eyelens focuses on this sharp image. Because the objective lens image is inverted, the image seen by the eyelens is erect.

Erector lenses are mounted with the field lens and eyelens, and the four lenses move together to focus on the objective image. In terrestrial telescopes, in which erect images are important, relatively close objects form images farther back from the objective lens than its focal length.

A Porro-Prism Erecting System

Another method of erecting an image is to use a pair of Porro prisms. A Porro prism is a right angle prism with rounded ends to minimize breakage. It may have a groove cut

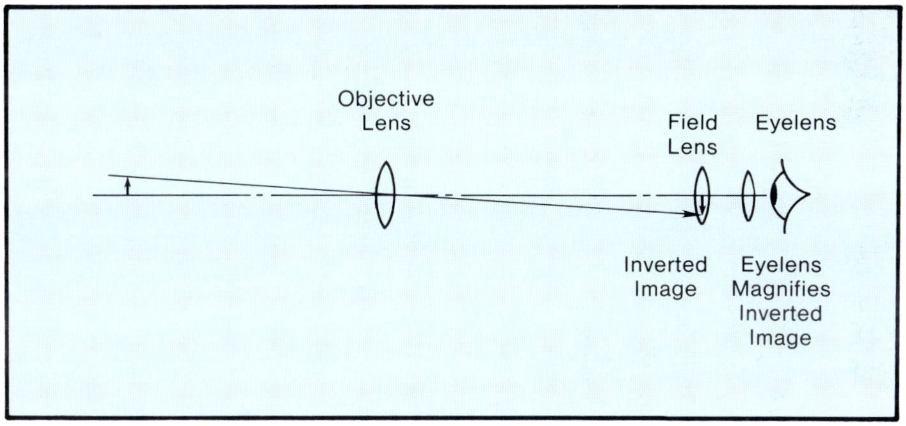

Figure 14-8. Image Inversion By Simple Refracting Telescope.

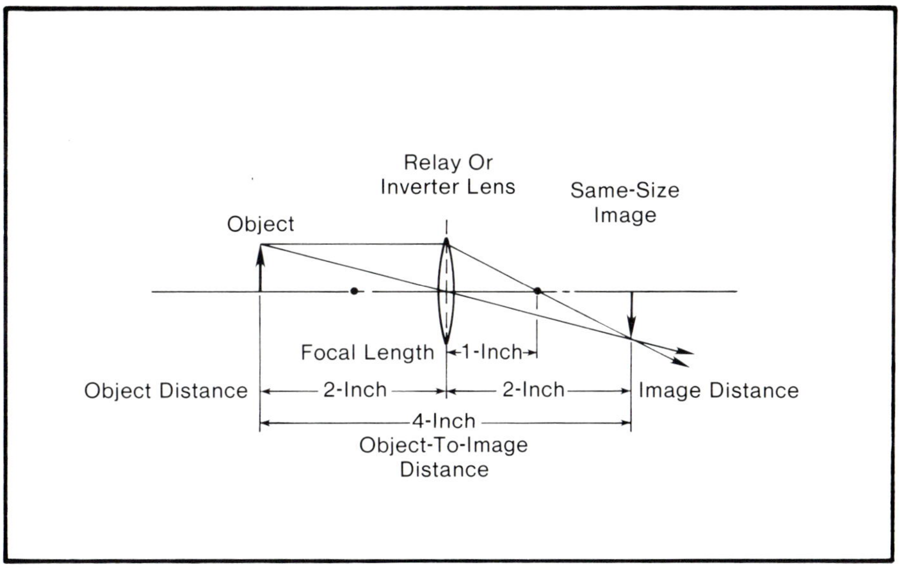

Figure 14-9. Erector Or Simplified Relay-Lens Function.

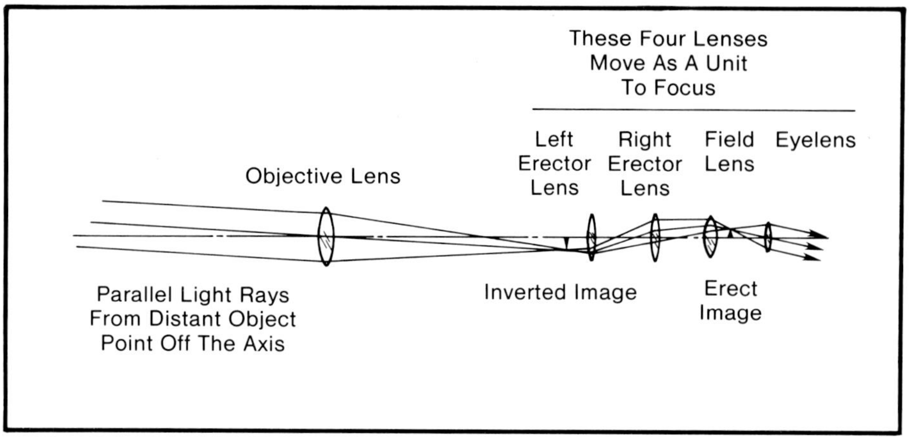

Figure 14-10. Terrestrial Erecting Telescope.

across the face of the hypotenuse (large face) for mounting purposes.

Figure 14-11 shows the light path of a monocular, which is a terrestrial telescope, using Porro prisms to erect the image. A single Porro prism would erect the image of the telescope, but would leave it reversed left to right. In addition, the image would be directed back toward the objective lens and would be inconvenient for viewing.

To correct for this a second Porro prism is placed hypotenuse to hypotenuse with the first prism, but at right angles to it. The placement has to be at exactly 90° or the image will be rotated slightly.

Figure 14-12 shows the prism assembly. This method of erecting the image makes a much shorter telescope than would be possible through the use of a relay or erector lens. A pair of monoculars mounted together is called a "binocular" or a "pair" of binoculars.

This drawing follows the path of the letter F through the mounted pair of Porro prisms and shows the letter F both "erected" and left-to-right reversed.

Refraction

In the first chapter, refraction was defined as the bending of light rays as they leave one optical medium and enter another at an angle. As rays go from a less-dense to a denser medium, they bend toward the normal (which is a line constructed perpendicular to the optical surface). As they go from a denser to a less-dense medium, rays bend away from the normal (Figure 14-13).

Refraction is demonstrated in the videotape lesson by placing a coin in the bottom of an opaque cup (Figure 14-14). The TV camera was aimed into the cup at such an angle that the coin could not be seen. When water is poured into the cup, the coin becomes visible. The drawings illustrate this effect and the cross-section diagram shows how the refraction of light rays at the air-water surface permits the coin to be seen.

The index of refraction is a measure of the density of a material such as glass or water. Technically, the index of refraction of a material is the ratio of the speed of light in air (or vacuum) to the speed of light in the material. The approximate speeds of light and the indexes of refraction of several optical materials are shown in the table below.

The speed of light in space or air is constant for all wavelengths. However, in other optical materials, the speed varies with the wavelength. This causes dispersion and the separation of light's wavelengths into a visible spectrum. It is common to give the index of refraction of a glass a wavelength of 589.6 nanometers, which is the wavelength of the D line in the sodium spectrum. The index of refraction of this wavelength is shown as n_D.

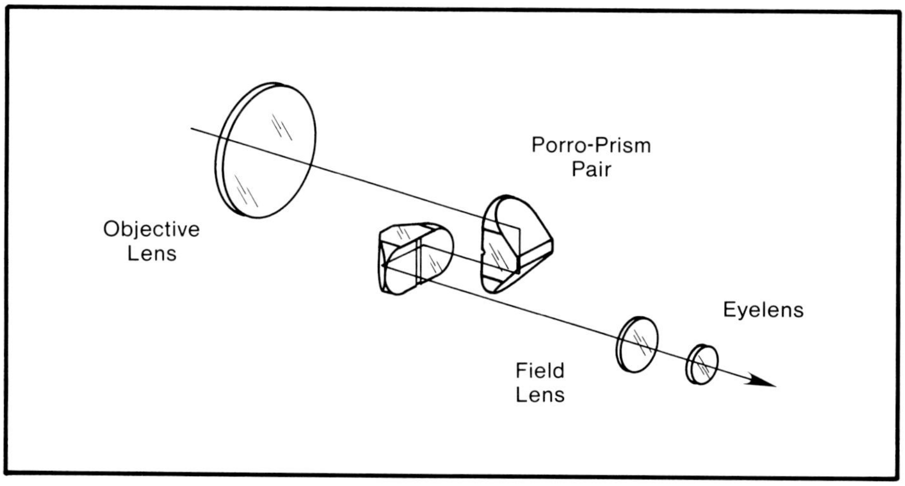

Figure 14-11. Monocular (Erecting Telescope).

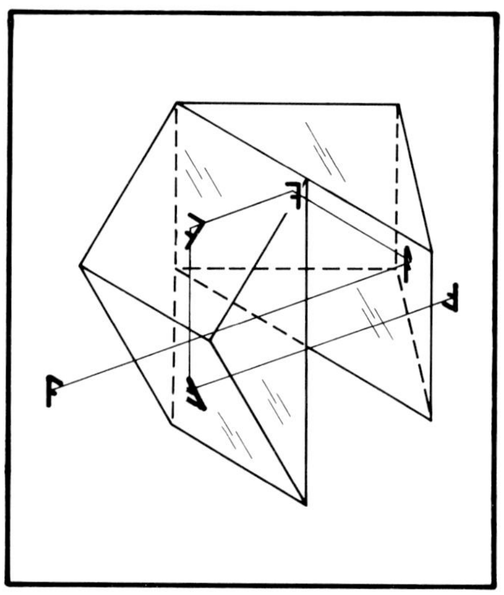

Figure 14-12. How A Porro-Prism Pair "Rotates" An Image 180°.

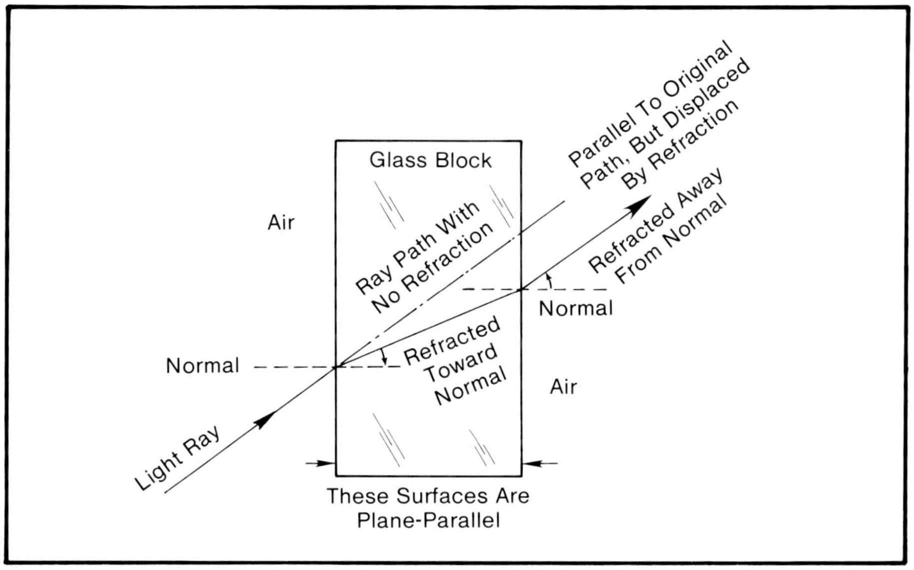

Figure 14-13. Refraction.

Optical Material	Speed of Light (Sodium D line)	Index of Refraction (n_D)
Space	186,300 mps	(Air) 1.0000
Water	139,760 mps	1.3330
Common Crown Glass	124,200 mps	1.5000
Dense Flint Glass	112,916 mps	1.6499
Rare Earth Glass (Flint)	95,051 mps	1.9600
Diamond	77,079 mps	2.4170

Snell's Law

Prior to 1600 A.D., optical scientists were unable to find an equation that would describe the amount of bending that takes place when a light ray enters or leaves glass. In the early 1600's, a Dutch mathematician and scientist named Snell solved the problem. He formulated what is known as Snell's Law of Refraction.

This series of drawings gives a geometric display of Snell's law. Figure 14-15a shows a horizontal surface of glass with air above it. A light ray enters the glass at Point O, at an angle I, the angle of incidence. The angle of incidence is measured from the normal (a perpendicular to the surface). The dot-dash line shows the path the ray would have followed if it had not been refracted. The solid line in the glass shows the refracted ray. The refracted ray is bent toward the normal as the ray goes from air (a less dense medium) into glass (a denser medium).

In Figure 14-15b, a construction line has been drawn perpendicular to the glass-air surface, and parallel to the normal. The intersection of the construction line with the dot-dashed line is shown as Point P. The intersection of the construction line with the refracted ray is shown as Point P′.

The exact location of the construction line is not critical. What is important to note is that the ratio of the lengths of the lines OP′ and OP is constant.

$$\frac{OP'}{OP} = C$$

When OP′ changes by a move of the construction line, OP changes a proportional amount.

In this drawing, the angle I is shown in the glass. This is the same as angle I in air because both the normal and the unrefracted ray path in air and glass are straight lines. In geometry, these are known as opposite angles, and they are equal. This is visually evident. I′ is shown in the glass as the angle of the refracted ray.

In Figure 14-15c, lines are constructed perpendicular to the normal to intersect points P and P′. Since the normal and the construction line are parallel, the lengths of both h lines are equal.

There are two important triangles formed, Triangle OAP and Triangle OA′P′. OAP represents a triangle formed by the unrefracted ray, while OA′P′ represents a triangle formed

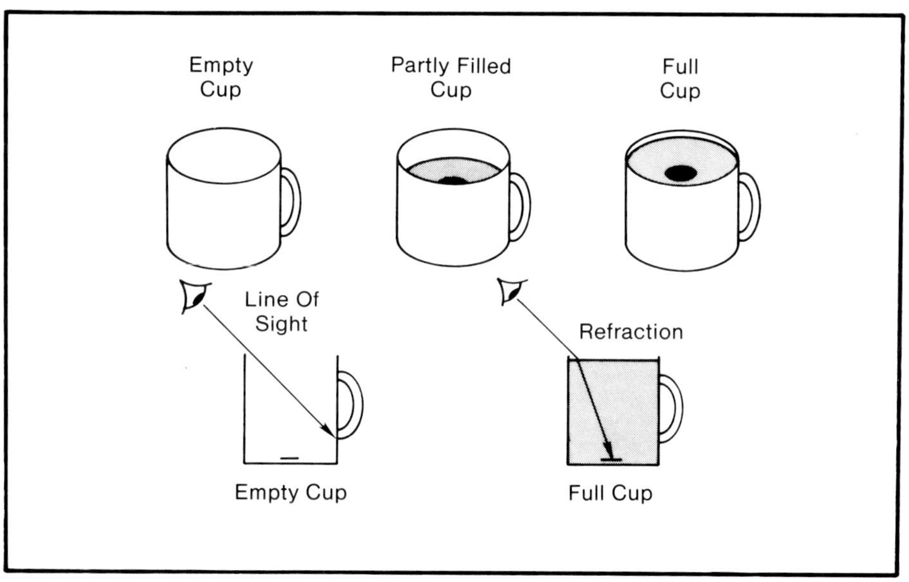

Figure 14-14. Cup-And-Coin Demonstration Of Refraction.

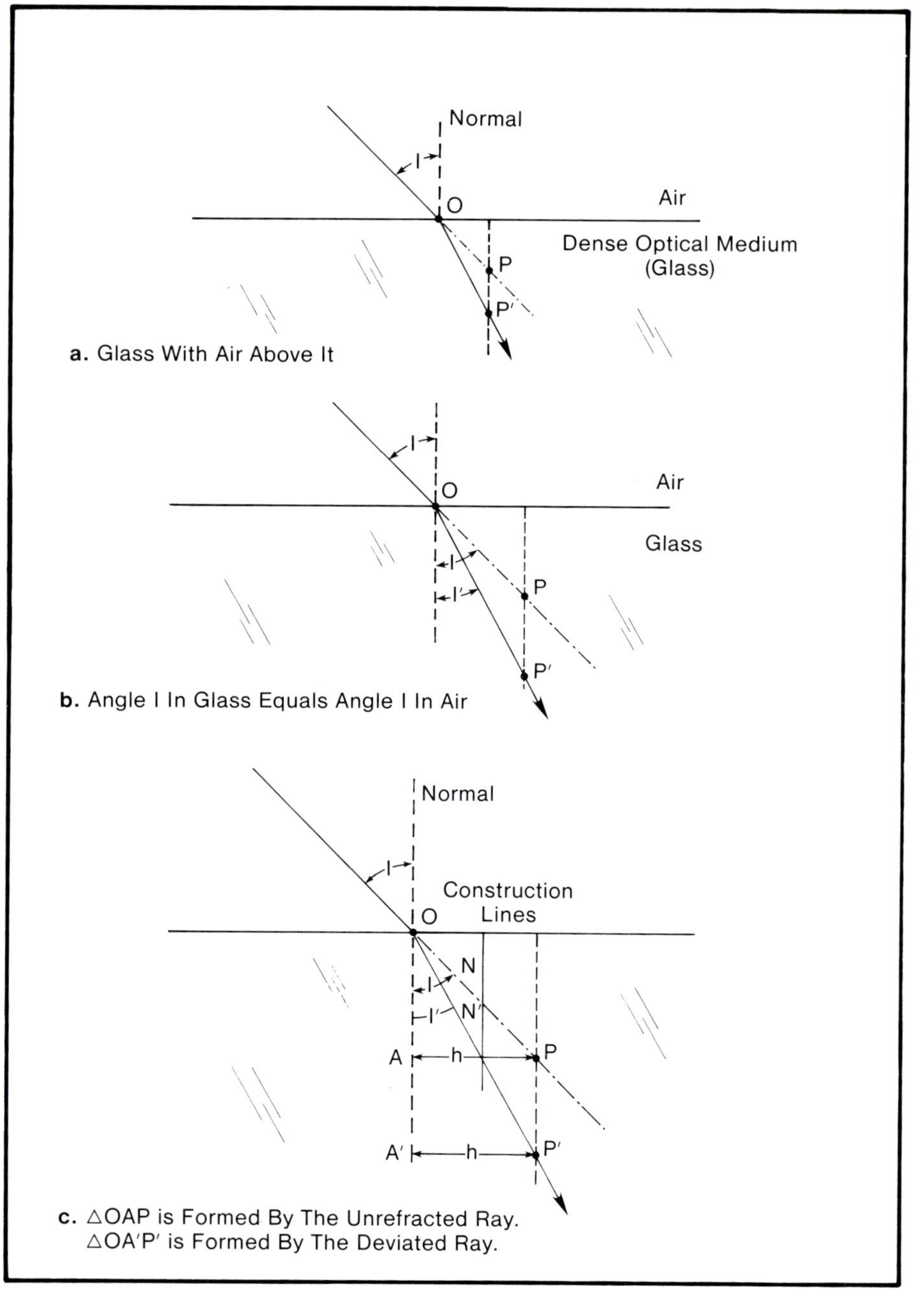

Figure 14-15. Geometric Proof Of Snell's Law.

Line N represents the index of refraction in air, and N′ represents the index of refraction in the glass. In this example N, the index of refraction of air, is 1.000. The index of refraction of the glass can be found by the ratio:

$$C = \frac{OP'}{OP} = \frac{N'}{N} = \frac{N'}{1} = N'$$

The constant is the index of refraction of the glass. If the value 1.000 is substituted for N in the equation, then:

$$\sin I = N' \sin I', \text{ and}$$

$$N' = \frac{\sin I}{\sin I'}$$

In this example, N′ is 1.500.

Total Reflection—Critical Angles

In the earlier discussions of Porro prisms, the total reflection of light off the small faces was assumed, although no mention was made of applying a reflecting metallic material such as silver or aluminum to the surface of the glass.

No metallic coating is needed because the glass surfaces reflect nearly all the light without any coating. This is because most of the light rays are beyond the critical angle.

If a fish (or a diver) in a pond of clear water with a still surface looks straight upward, he sees out of the water and can see the sky. However, all the view above the water from

by the deviated ray.

The sine of an angle is a ratio of the opposite side of a right triangle divided by the hypotenuse (Figure 14-16). Lines h were drawn perpendicular to the normal, so the two triangles are right triangles.

In Triangle OAP, the

$$\text{sine of angle I} = \frac{h}{OP}$$

In Triangle OA′P′, the

$$\text{sine of angle I}' = \frac{h}{OP'}$$

The line OP is called N.
The line OP′ is called N′.
Therefore:

$$\sin I = \frac{h}{N} \quad \text{and} \quad \sin I' = \frac{h}{N'}$$

By algebraic transposition:

$$h = N \sin I \quad \text{and} \quad h = N' \sin I'$$

Therefore Snell's law states:

$$N \sin I = N' \sin I'$$

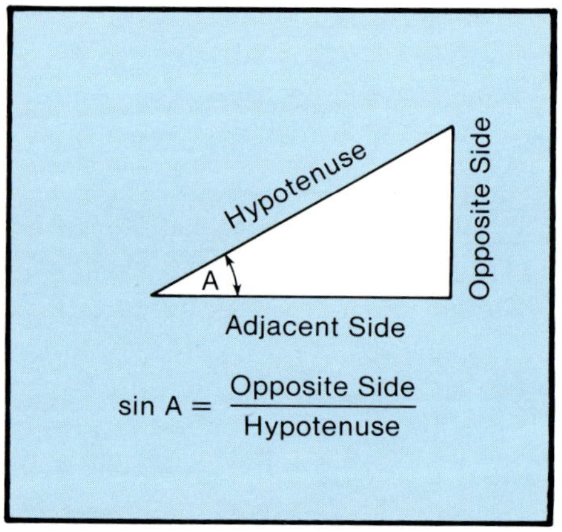

Figure 14-16. Trigonometric Function Of The Sine.

horizon to horizon is contained in a circular area. All around the circular area the surface of the water looks silvery, like a mirror. If there are any objects in the water beyond the circle's limits, they can be seen reflected in the surface of the water (it acts like a totally reflecting mirror). This happens because, beyond some critical angle, light is totally reflected.

Figure 14-17 shows the traces of several rays. The ray that is perpendicular to the surface of the water is not bent. It goes straight into the water. Several other rays are shown at various angles being refracted to the fish's eye. One ray is shown parallel to the surface of the water. It has an incident angle of 90° because incidence is measured to the normal. The angle within the water can be found by Snell's law; using an index of 1.000 for air and 1.333 for water:

$$N \sin I = N' \sin I'$$

$$1.000 \sin 90° = 1.333 \sin I'$$

The sine of 90° is 1.000. Therefore, $1 = 1.333 \sin I'$.

$$\sin I' = \frac{1}{1.333} = 0.75019$$

$$I' = 48.6°$$

The half angle of the cone formed with the apex at the fish's eye and the base of the circle on the water surface through which the outside world can be seen is 48.6° (Figure 14-18).

Beyond the 48.6° angle, all objects seen are reflections of things under the surface of the water.

The total cone angle is 97.2°.

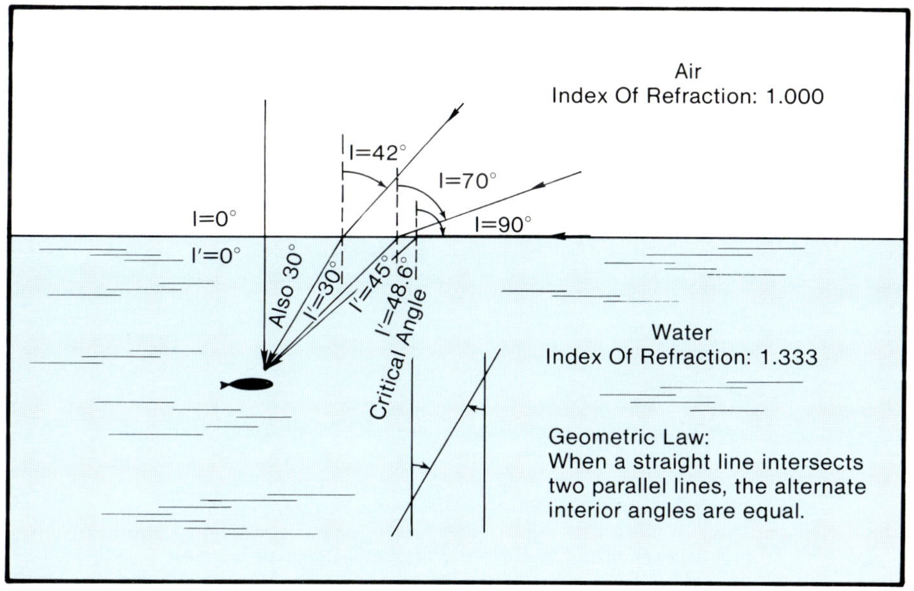

Figure 14-17. Refraction At A Water Surface And The Critical Angle.

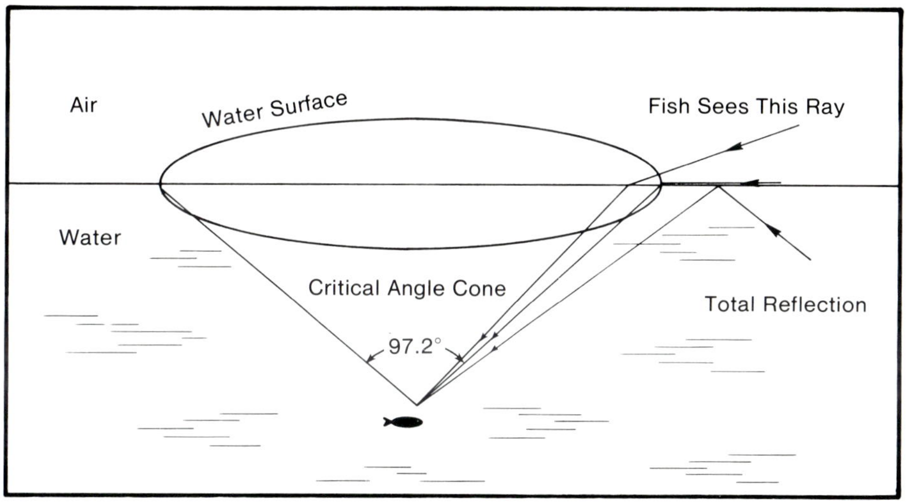

Figure 14-18. Fish-Eye View.

The Critical Angle in a Porro Prism

In Porro prisms the light is generally reflected within the prism. Most Porro prisms are made of borosilicate crown glass because it is a stable glass both physically and chemically. BSC-2 glass has an index of refraction of $n_D = 1.5230$.

$$N \sin I = N' \sin I'$$

$$1.000 \sin 90° = 1.5230 \sin I'$$

$$1 = 1.5230 \sin I'$$

$$\sin I' = \frac{1}{1.5230} = 0.6566$$

$$I' = 41°$$

In a pair of binoculars the axial ray is 45° to the reflective faces of the prism (Figure 14-19). Nearly all the incident rays are greater than the critical angle. In a wide-angle binocular, however, some rays are less than the critical angle and are transmitted through the reflecting surface. If the reflecting surface of the Porro prism is silvered, all rays are totally reflected. This subject will be discussed in more detail in a later chapter.

Using Snell's Law for Graphic Ray Tracing

In the schematic ray tracing discussed so far, the paths of rays (as they go through a lens) were determined primarily by two rules. A ray parallel to a lens axis passes through the focal point and a ray that goes through the optical center of the lens is not bent.

The major use of Snell's law is to trace the paths of light rays as they enter and exit optical elements such as lenses and prisms. The path of a ray can be calculated using trigonometry and the sines of the angles. There is also a geometric method

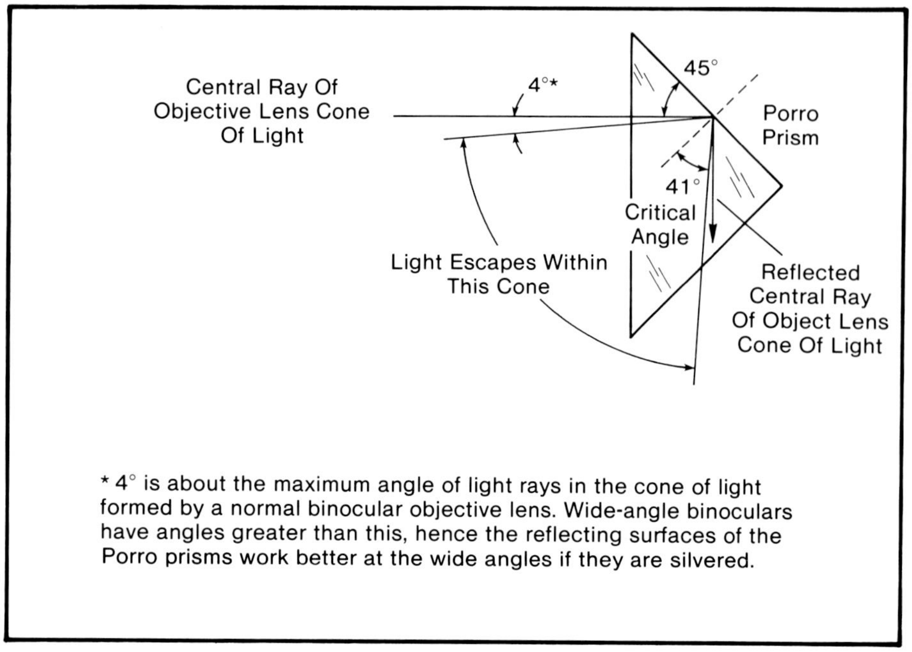

Figure 14-19. Critical Angle Of A Porro Prism.

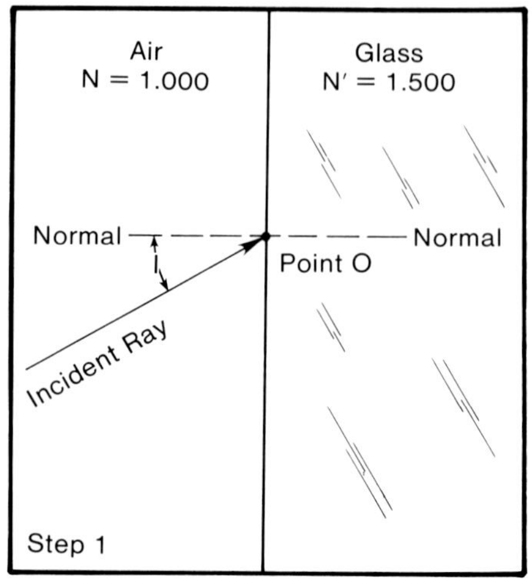

Figure 14-20. Graphic Ray Tracing: A Geometric Method Of Ray Tracing Using Snell's Law.

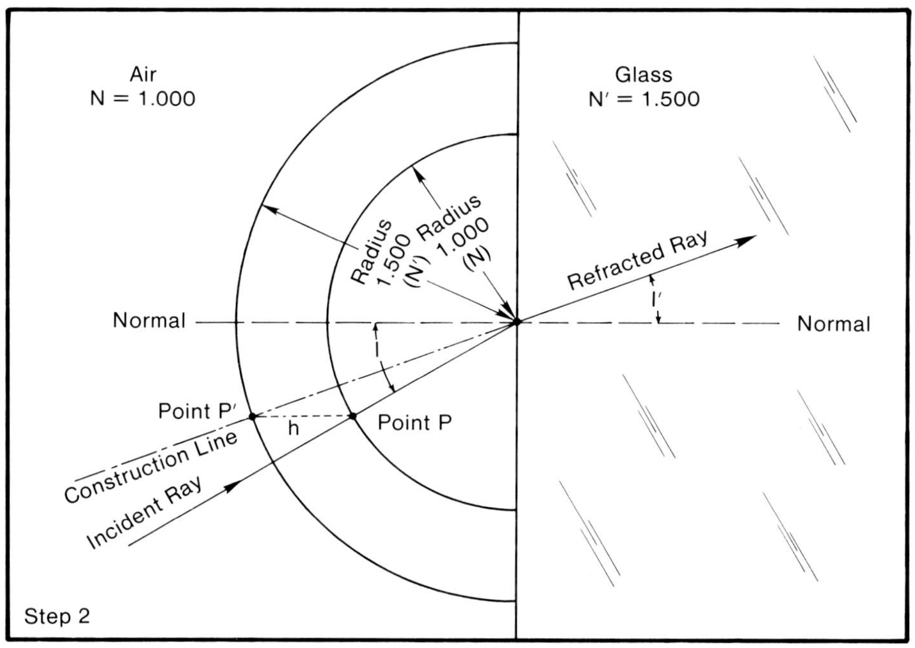

Figure 14-21. The Basics of Graphic Ray Tracing.

that supplements schematic ray tracing, which is called Graphic Ray Tracing.

In Figure 14-20, air is shown at the left, and glass with an index of refraction of 1.500 is at the right. Their indices of refraction are: N = 1.000 and N′ = 1.500. A ray enters the glass at Point O at an angle of I to the normal drawn through the point at which the ray enters the glass.

Using the Point O as a center (Figure 14-21), two semicircles are drawn with a compass on the air side of the glass surface. The smaller semicircle has a radius of 1.000, while the larger semicircle has a radius of 1.500. Any convenient units will serve; it is the 1.500 to 1.000 ratio that is important. These radii represent the indices of refraction of the air and the glass.

A short straight line, parallel to the normal, is constructed between the two semicircles and through the intersection of the ray and the smaller circle, Point P. Point P′ is where this line intersects the larger semicircle. A straight line is drawn through Point P′ and O. *This line, in the glass, is the path of the refracted ray*. The angle I′ is formed between the refracted ray path and the normal.

Problem

The finished drawing of the geometric version of Snell's law is given below. By drawing some additional lines and by referring back to the original development of Snell's law, prove that:

$$N \sin I = N' \sin I'$$

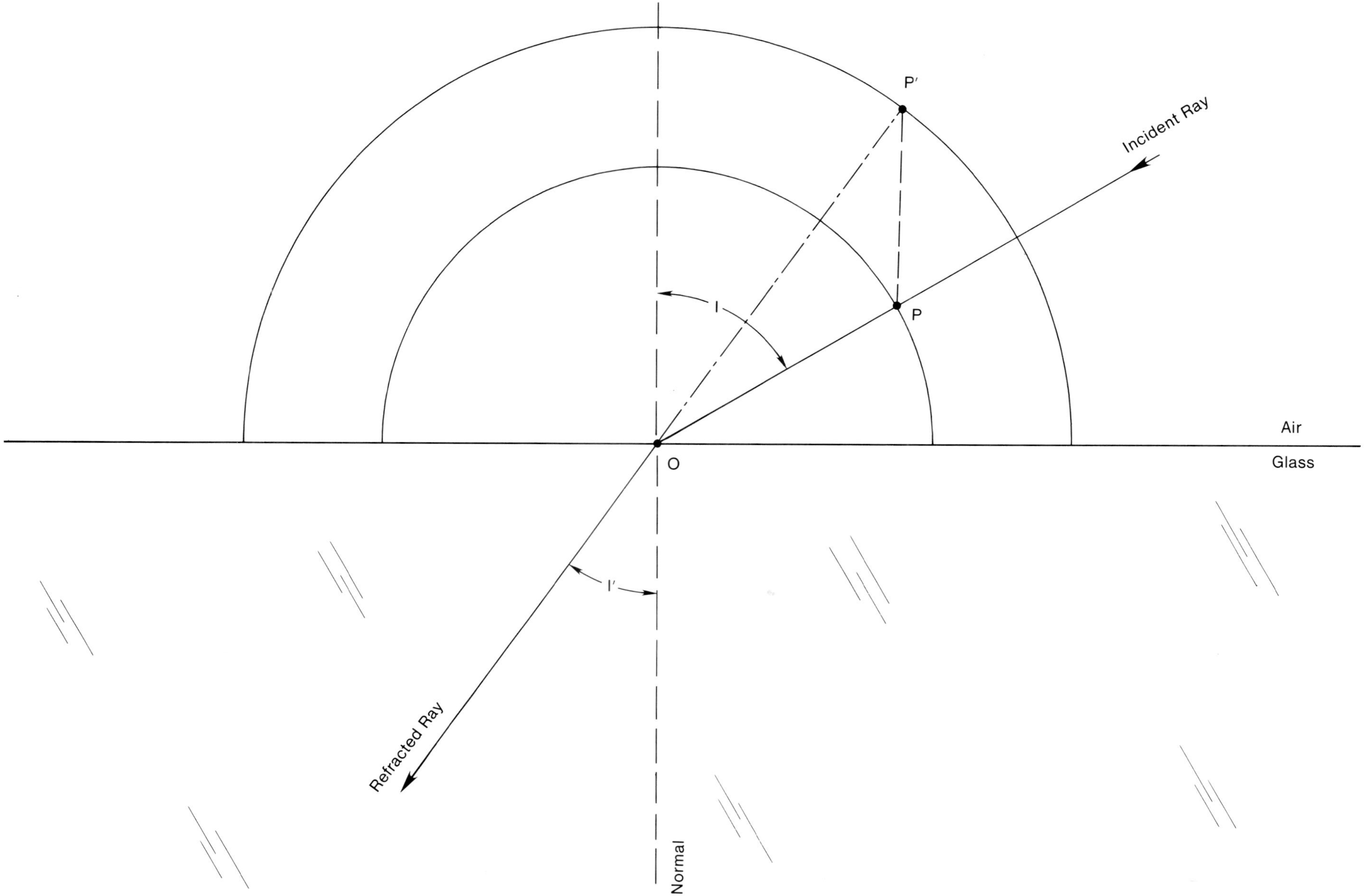

Problem
Chapter 14
Video Lesson 9

Chapter 15

VIDEO LESSON 10

Graphic Ray Tracing, Tracing a Compound Lens, Prisms vs. Mirrors, Some Uses of Prisms, Beam Splitters

Summary

The problem assigned at the end of the last chapter is reviewed, proving Snell's law using graphic ray trace geometry. The graphic ray trace method is shown for rays going from glass to air. The same procedure is used to determine the critical angle. The graphic method is expanded to trace the path of a ray through a single lens. A ray path through a compound lens is also illustrated. In right-angle prisms, there are two critical angles, one inside the glass and one outside the glass. Both plane mirrors and prisms are used to reflect beams of light and uses for each type of reflector are compared. Some gunsights with unity magnification are made with lenses. One such sight, made of prisms, is described. A stereo viewer, made from a prism block, is discussed. A beam splitter is defined, and a glass beam splitter, prism beam splitter and pellicle beam splitter are discussed.

Proving Snell's Law (Review of Assigned Problem)

Figure 15-1 shows the graphic

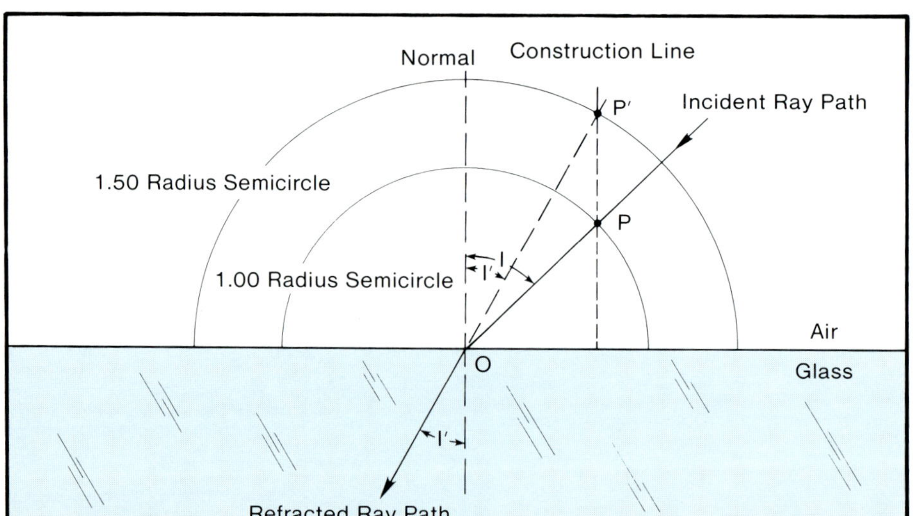

Figure 15-1. Geometric Proof Of Snell's Law (a).

method for determining the path of a refracted light ray as it passes from a less-dense to a denser optical medium.

The smaller semicircle represents the index of refraction of the less-dense medium, air, and has a radius of 1 inch. The larger semicircle represents the index of refraction of the denser medium, glass, which is given as 1.5. Its radius is 1½ inches. Angle I is the angle of incidence and it is measured to the normal drawn through Point O (the point where the ray enters the glass). A construction line, parallel to the normal, is drawn from the glass-air surface through Points P to P′ on the two semicircles.

I′ is the angle of the refracted ray, which is P′, and a straight line is drawn from this intersection through Point O. In air this line is dashed because it is a construction line. In the glass, it is a solid line because it is the refracted-ray path. This is an example of using the graphic ray trace method to find the angle of refraction.

The assigned problem was to use this drawing to prove Snell's law geometrically. In order to do this it was necessary to review the method by which Snell's law was developed. The problem presented Snell's law in the form of:

$$N \sin I = N' \sin I'$$

Right triangles are required because the sine is defined as the ratio of the side opposite the angle divided by the hypotenuse *in a right triangle.* In the original Snell's law development, two lines of equal length (designated h) were drawn. Draw these lines (labeled h) perpendicular to the normal from Points P and P′ (Figure 15-2). Since the construction line P to P′ and the normal are parallel, the h lines are equal in length.

The length of the line N, from Point O to the smaller semicircle represents the index of refraction of air, because its length is 1. The line N′ from Point O to the larger semicircle represents the index of refraction of the glass because its length is 1.5. It is the radius of the larger semicircle.

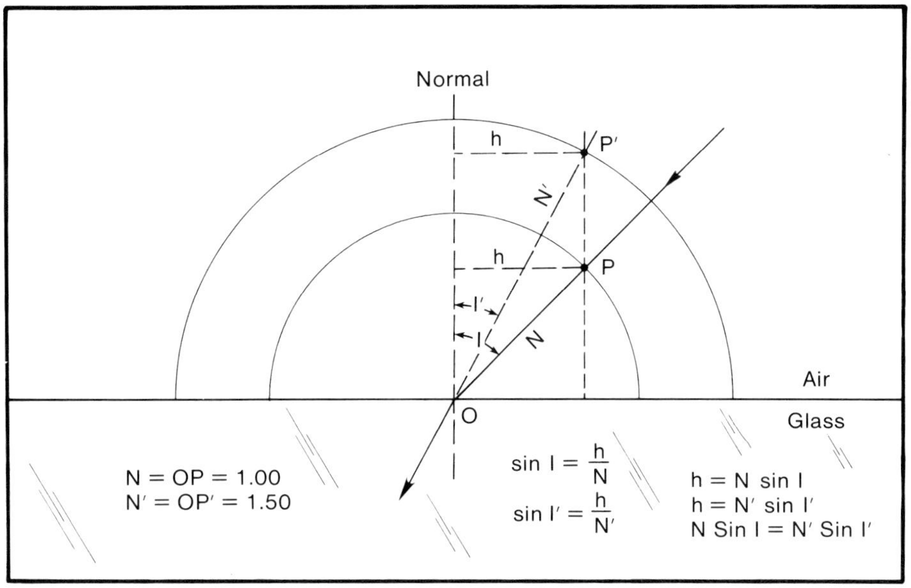

Figure 15-2. Geometric Proof Of Snell's Law (b).

The two needed triangles have been completed. On one, the angle is I, the hypotenuse is N and the opposite side is h. By the definition of a sine:

$$\sin I = \frac{h}{N}.$$

On the other triangle, the angle is I′, the hypotenuse is N′ and the opposite side is h. Therefore:

$$\sin I' = \frac{h}{N'}.$$

By cross-multiplying, the equation becomes:

$$N \sin I = h$$
$$N' \sin I' = h$$

According to the mathematical

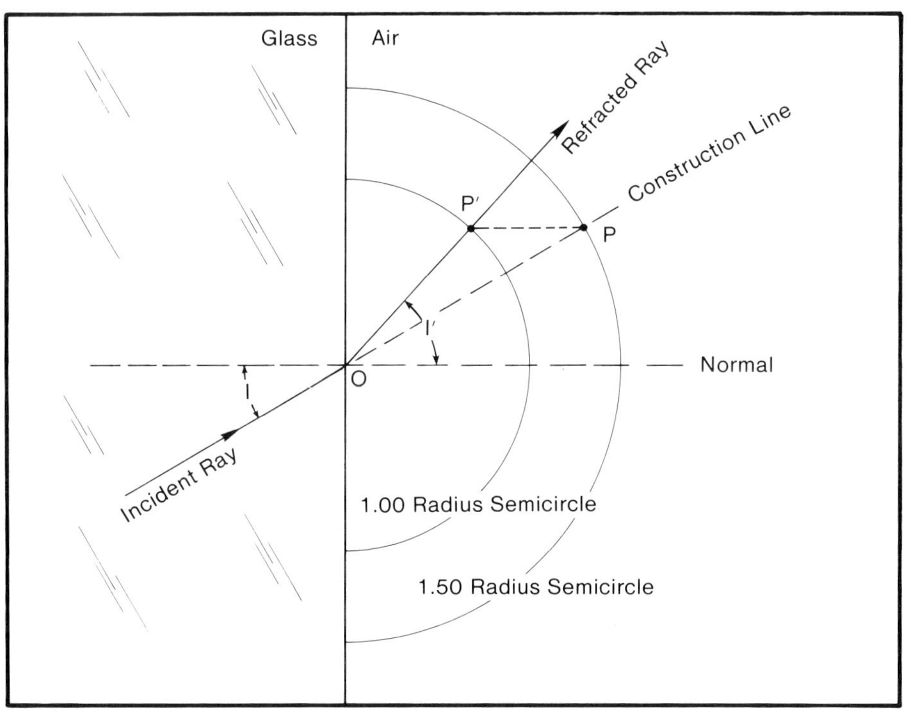

Figure 15-3. From Glass To Air.

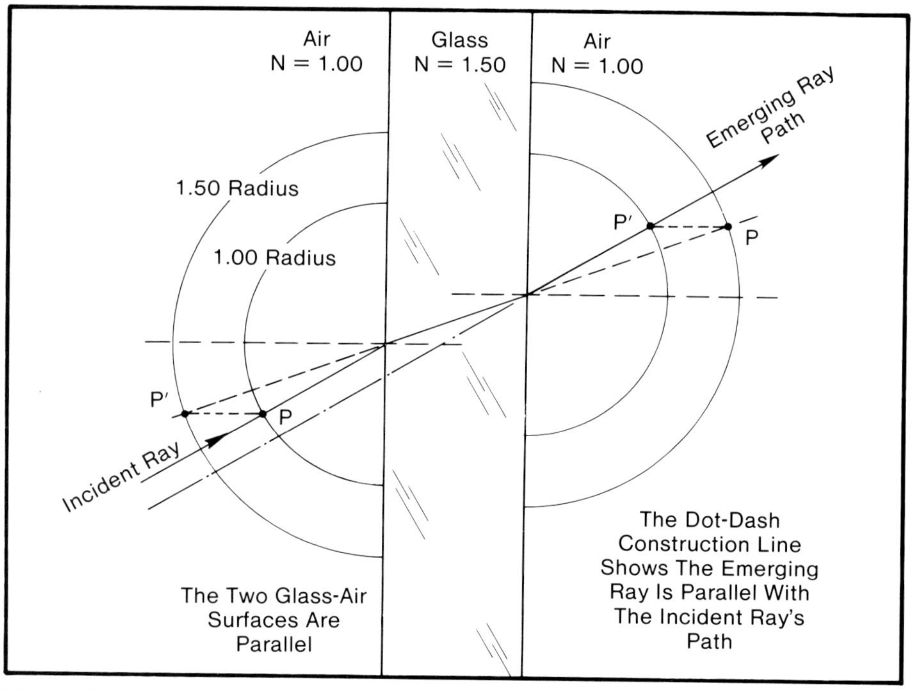

Figure 15-4. Graphic Ray Tracing Through A Window.

law that states "quantities which are equal to the same quantity are equal to each other":

$$N \sin I = N' \sin I'$$

which is Snell's Law. This is the geometric proof asked for in the problem.

From Glass to Air

In most instances, light rays enter glass and then leave it. Graphic tracing of a ray as it passes from a dense optical material to a less dense optical material is quite similar to the procedure described above.

Figure 15-3 shows a plane glass-air surface with the glass on the left and air on the right. The ray is shown striking the glass surface at Point O. The angle of incidence is I. Two semicircles are drawn with Point O as the center. The radius of the smaller semicircle is 1, while the radius of the larger one is 1.5. These represent the index of refraction of air and glass. The dashed construction line shows where the ray would go if it continued without being refracted. A construction line is drawn parallel to the normal from the intersection of the incident ray construction line and the large semicircle (P) to the smaller semicircle. A straight line is drawn from the intersection of the construction line and the small semicircle (P′) to the Point O. This is the ray path in air. The Angle I′ is the angle between the refracted ray and the normal.

This basic method can be used to trace rays through glass windows, prisms and lenses. Figures 15-4 and 15-5 illustrate a ray going through a window, and a ray passing through a right angle prism.

Finding the Critical Angle

Similar to the drawing above, Figure 15-6 shows the ray going from glass to air. It will be used to find the critical angle of glass with an index of refraction of 1.5.

At this time the construction line is drawn from the intersection of the small semicircle with the glass-air surface. It is parallel to the normal drawn through Point O. A ray is drawn from the intersection of the large semicircle and the construction line (P) to Point O. This is the incident ray. The refracted ray is drawn through the intersection of

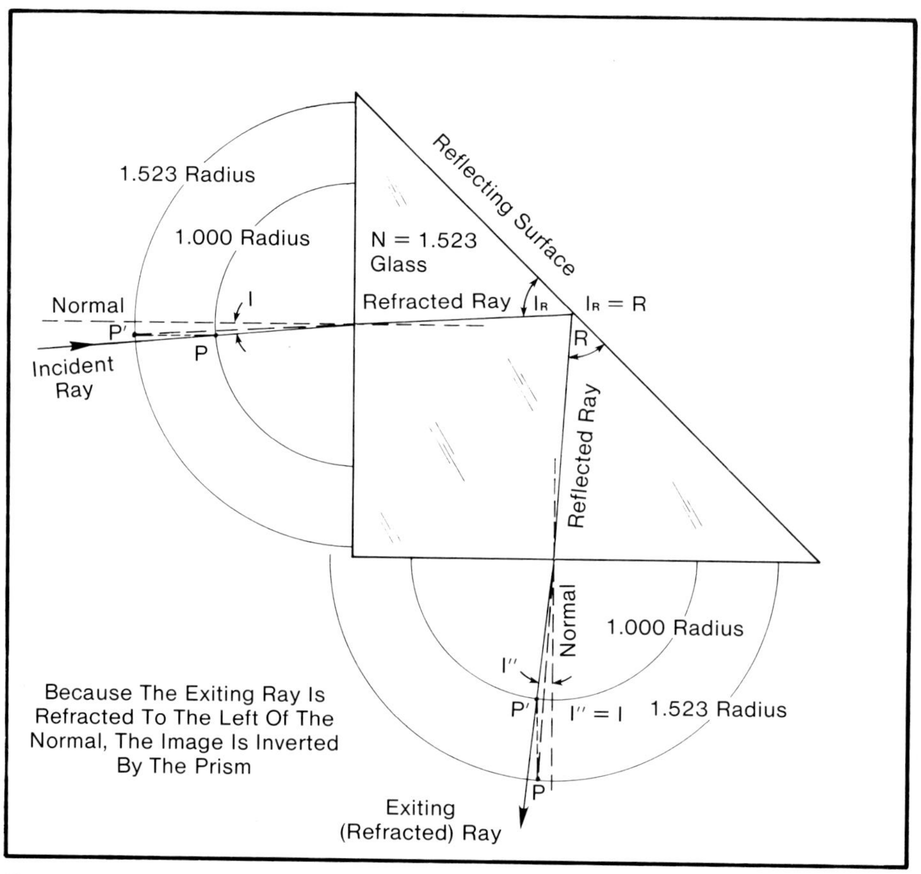

Figure 15-5. Graphic Ray Trace Through Right-Angle Prism.

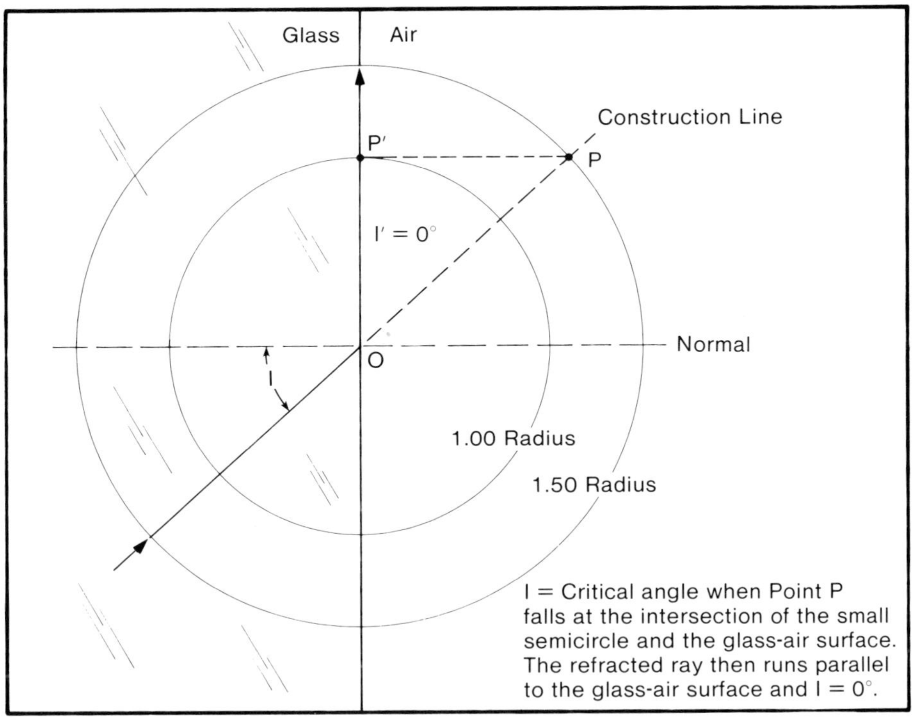

Figure 15-6. Finding The Critical Angle.

the construction line and the small semicircle (P′) to Point O. This ray coincides with the plane glass-air surface. Therefore, Angle I (the incident angle) is considered to be the critical angle and the refracted ray angle is 90°. This is the definition of the critical angle. On the drawing the critical angle measures about 42°. Using the equation:

$$\text{sine of critical angle} = \frac{1}{1.5} = .6667$$

$$\text{Critical angle} = 41.8°$$

The Path of a Reflected Ray

Rays whose incident angles are greater than the critical angle are totally reflected at the glass-air surface. The path of a reflected ray can be found by graphic ray tracing.

In Figure 15-7, the critical-angle trace is repeated and the semicircles have been completed to form full circles. A new incident ray is shown whose angle of incidence is greater than the critical angle and a construction line is continued to the large semicircle at P_2. A construction line is drawn from Point P_2 to the left, parallel to the normal. Because this construction line does not intersect the small semicircle, it is continued across the glass-air surface until it intersects the large semicircle at Point P_2''. A straight line is drawn from this intersection (Point P_2'') to Point O. The ray does not get into air because the angle is greater than a critical angle. The line to the left of Point O is shown as solid, because it is the path of the reflected ray.

It is evident from the drawing that Angle R is the angle of reflection since the law of reflection states that the angle of reflection equals the angle of incidence.

Graphic Tracing of a Ray Through a Positive Lens

Before actually tracing a ray through a lens it may be useful to review some of the geometry to be used.

A bi-convex lens has two convex spherical surfaces (Figure 15-8). In ray tracing, a two-dimensional model of the lens is used. Two arcs, that are sections of two circles, are used to represent the lens. The radius of curvature of each arc is the radius of curvature of the spherical surface of the lens. A cross-section of the lens at its axis is shown in the drawings. The lens is cut exactly in half, and the plane of the cut becomes the section of the lens used for graphic ray tracing. Since a lens is radially symmetric, this procedure is valid. Each arc, therefore, repre-

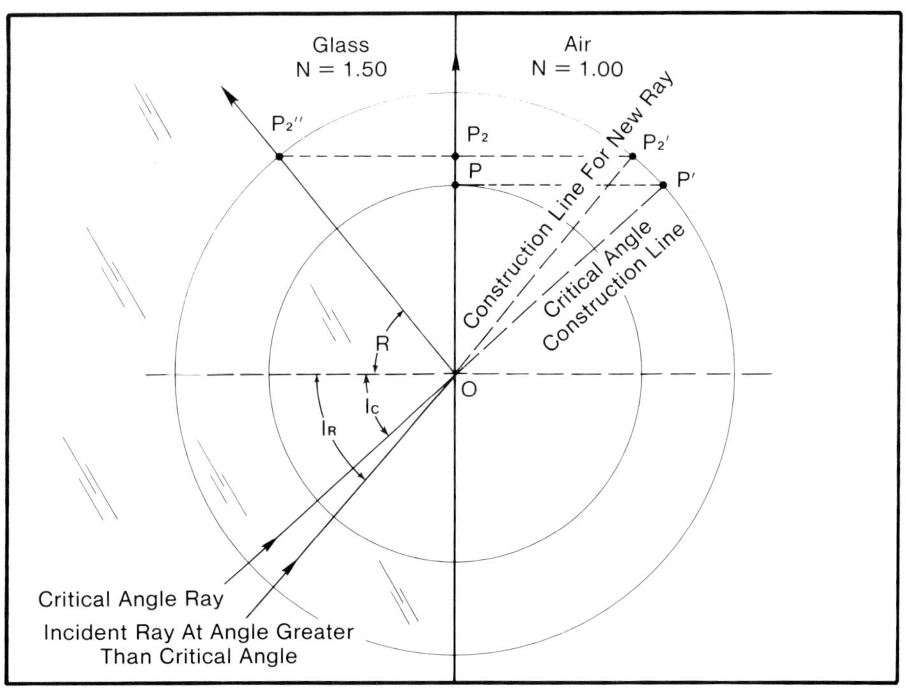

Figure 15-7. Graphic Ray Trace Of Total Reflection Within Glass.

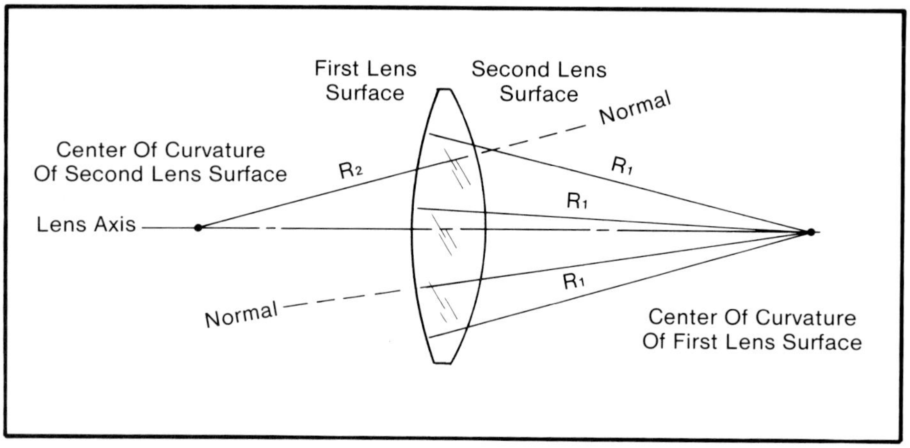

Figure 15-8. A Lens In Cross-Section For Graphic Ray Tracing.

sents the spherical surface of the lens. The drawing shows the positive lens including its radii. The center of curvature of each surface lies on the axis of the lens. A radius drawn through the arc representing the surface of the lens is normal to the surface at the point of intersection between the radius and the arc.

Each point on a circular arc surface can be considered a plane and is represented in Figure 15-9 by a line drawn perpendicular to the radius at the point of intersection. Since the radius is perpendicular to the plane, the radius is normal to the lens surface at that point.

In the previous examples of graphic ray tracing the glass-air surface was a plane. When performing such traces on a spherical surface, the point of incidence where the ray touches the surface can be considered a plane, with the radius of curvature at that point being the normal.

In Figure 15-10, the two semicircles have been drawn. The incident ray is shown going to the point of incidence (Point O). The construction line is drawn from the intersection of the ray and the smaller semicircle, parallel to the normal. When the ray passes from air to glass, the construction line is drawn through the air semicircle. The refracted ray is determined by drawing a straight line from the intersection of the construction line with the large semicircle and the point of intersection. In the glass, the refracted ray is shown at Angle I′ to the normal.

Figure 15-11 shows the radius of the second curve as drawn to Point O_2, the location of the refracted ray's intersection with the lens's second surface. The two semicircles are drawn with radii of 1 and 1.5 using O_2 as the center of curvature. A construction line extends the ray path to the right so that an intersection with the larger semicircle is achieved.

Since the ray is now passing from

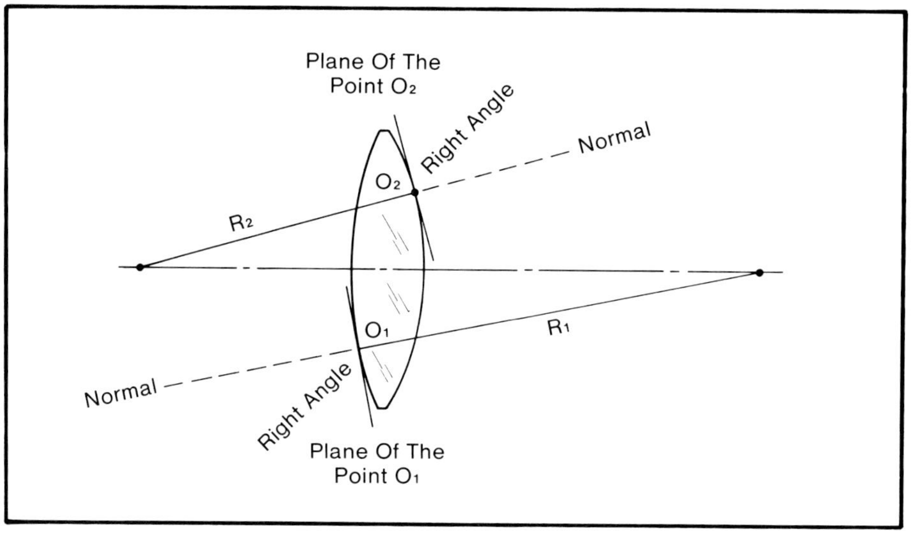

Figure 15-9. Each point on a lens surface represents a plane perpendicular to the radius at that point.

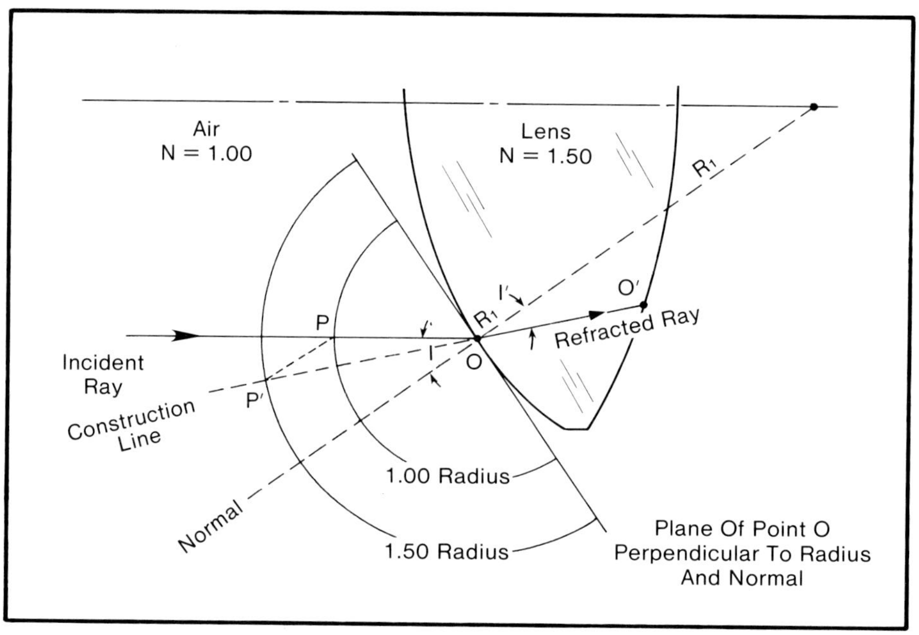

Figure 15-10. Graphic Ray Trace At The First Lens Surface.

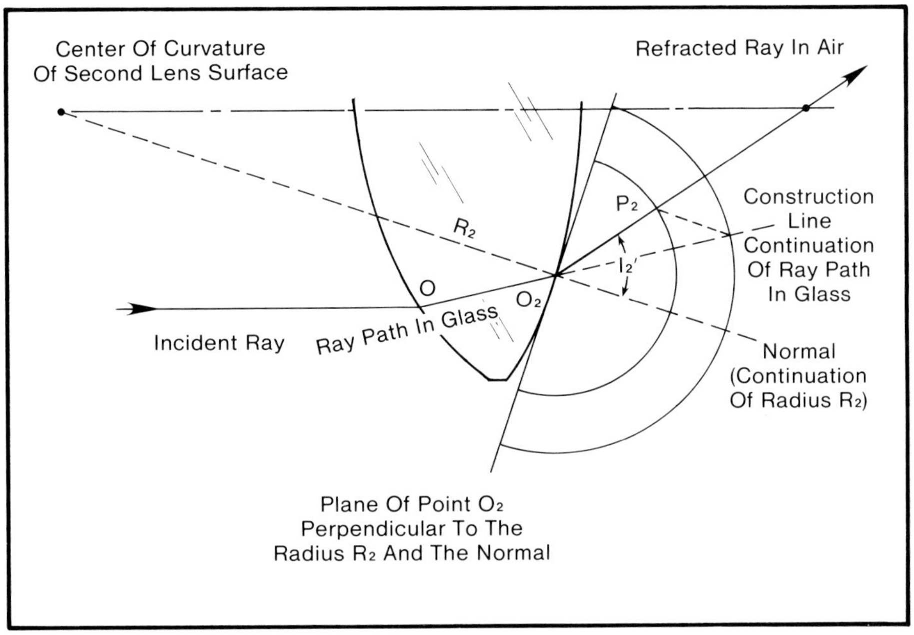

Figure 15-11. Graphic Ray Trace At Second Lens Surface.

glass to air, the construction line is drawn parallel to the normal at the intersection of the ray, and from the larger semicircle to the smaller semicircle. A straight line is drawn from this intersection through the point O_2. This line illustrates the path of the refracted ray as it goes from glass into air. Note that it is bent away from the normal in accordance with one of the laws of refraction given in an earlier chapter.

Figure 15-12 shows what can be found by this graphic ray trace of a lens. Because the incident ray is parallel to the axis, the exiting ray crosses the axis at the focal point. If both the incident ray and the exiting ray are extended, they will cross. Their crossing locates the lens's second principal plane. The second nodal point is located at the intersection of the principal plane and the axis. The distance between the principal plane and the focal point is the lens's focal length. By graphic ray tracing, we have found that a lens made of crown glass (N = 1.500) with a front surface whose radius of curvature is 5.75 inches, whose second radius of curvature is 8.75 inches, and that has a center thickness of 3.625 inches, will have a focal length of 5.0 inches.

Figure 15-13 shows the ray path as it goes through each lens and across each air space in a compound lens. It also shows the ray exiting the second surface of the rear lens. This ray helps form an image point in the optical image. Only the first surface shows the graphic trace of the ray.

Uses for Graphic Ray Tracing

Graphic ray tracing is not accurate enough for lens-design purposes. Such designs are done mathematically so that the positions of the rays can be determined to a hundredth of a millimetre (much closer than can be done by drawing).

However, there are useful purposes for graphic ray tracing of lenses. For example, it is often important in optical engineering to locate the principal planes. A ray can be drawn parallel to the axis entering the front surface of the lens and followed through until the ray intersects the axis at the focal plane. By extending the entering ray to the right, and the focal point ray to the left, the principal plane can be located (Figure 15-12).

If the angular coverage of a lens is known, the outer rays can be traced so that mount baffles can be designed to just clear the path of the outer rays and so that vignetting will not occur (Figure 15-14).

Prisms Versus Mirrors as Reflectors

Both right-angle prisms and first-surface mirrors with a plane surface can be used to reflect beams of light (Figure 15-15).

A mirror is generally less expensive to make because it requires only one polished surface, while a prism needs to have three polished surfaces (Figure 15-16). The prism's three polished faces must be exactly

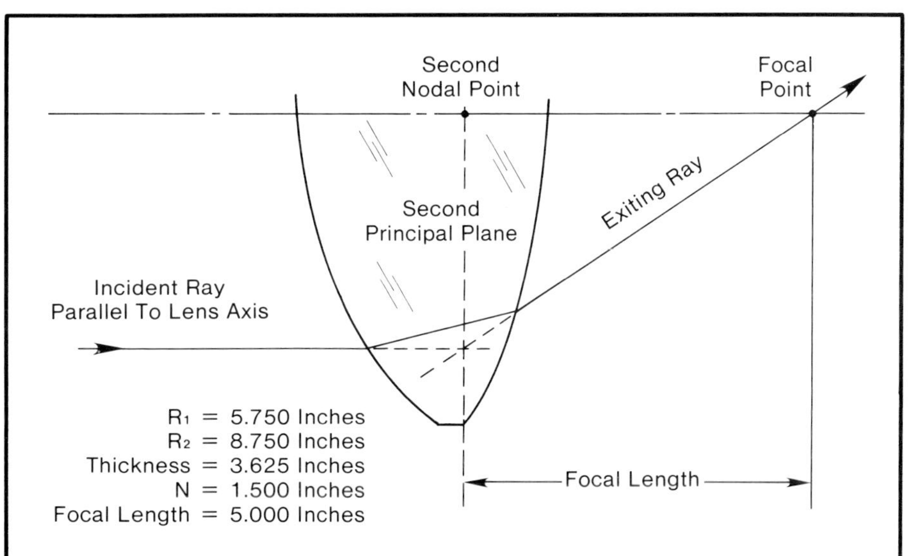

Figure 15-12. Scaled Values Of The Lens Designed By Graphic Ray Tracing.

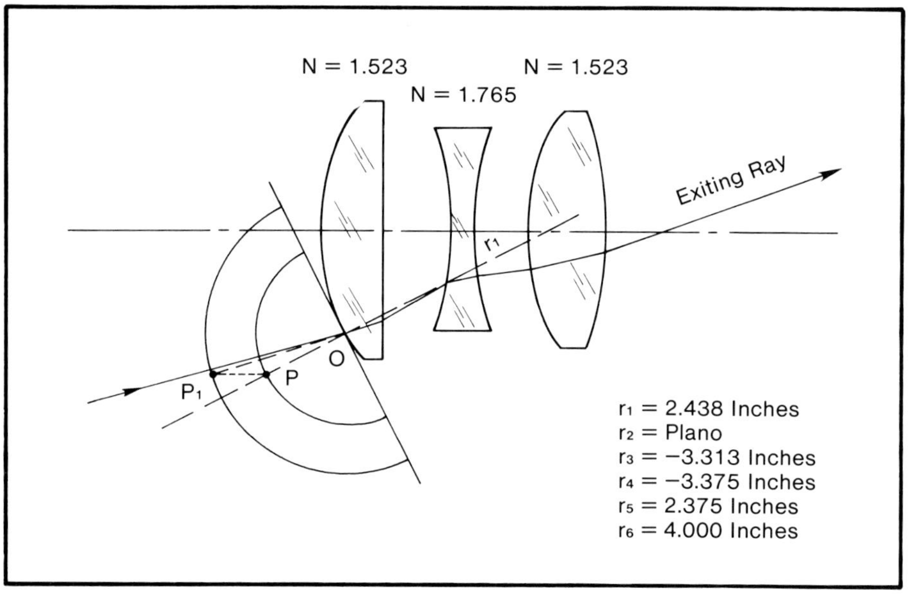

Figure 15-13. Ray Path Through A Compound Lens.

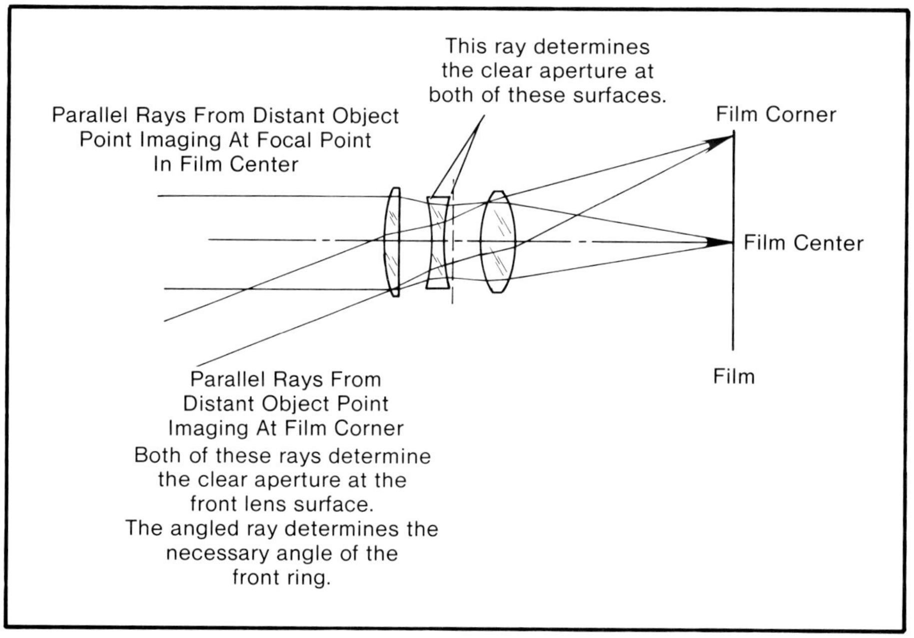

Figure 15-14. Graphic Ray Trace Showing Clear Apertures At Each Lens Surface.

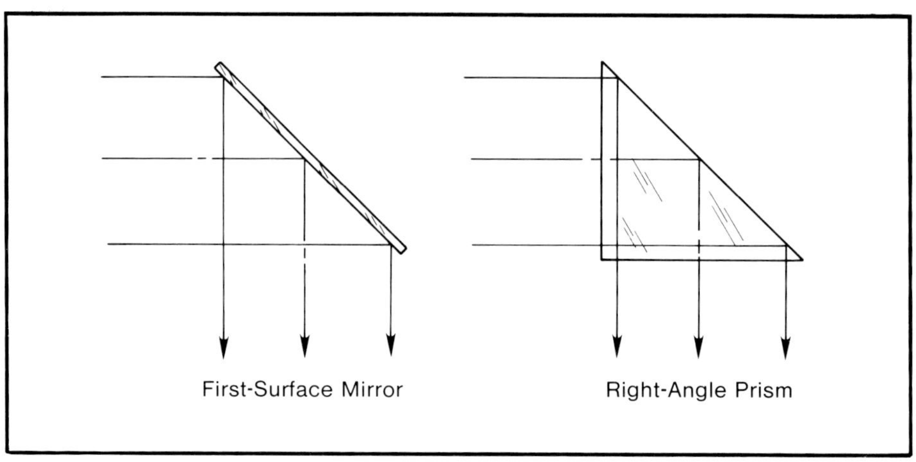

Figure 15-15. Comparison Of A Mirror And Prism.

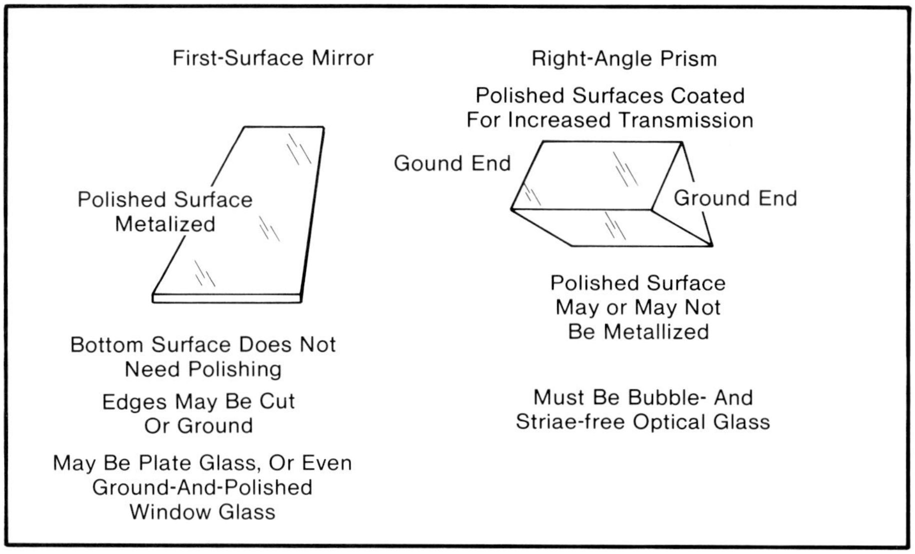

Figure 15-16. Comparison Of Prism And Mirror.

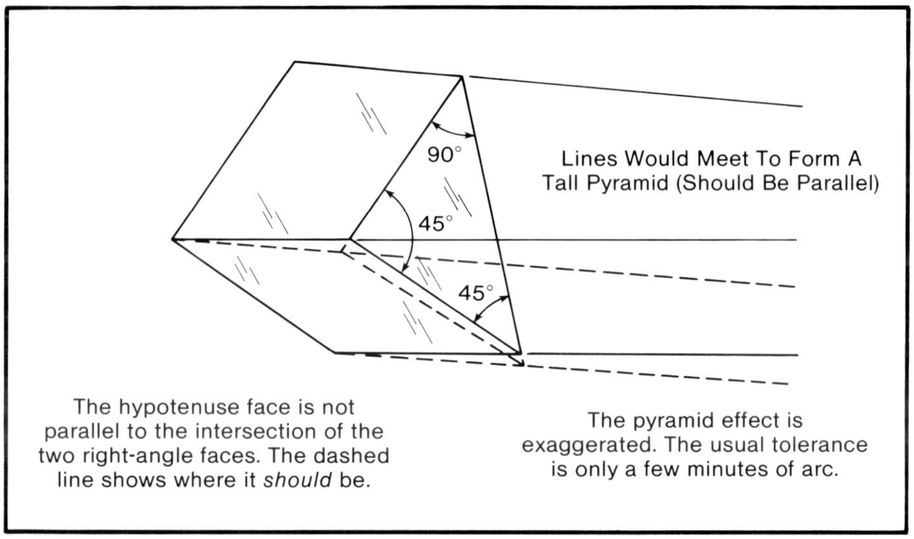

Figure 15-17. Prism Pyramiding.

placed at 45° and 90° to each other and the surfaces must not "pyramid" (Figure 15-17). Such accuracy is costly to manufacture.

If weight is a concern, mirrors can be made much lighter than prisms. However, a mirror must have a first-surface reflection coating (usually of aluminum) and is over-coated with a protective covering of a material such as silicon monoxide. If most of the internal rays are incident at less than the critical angle, a prism needs no reflective coating at all. If the angles of the rays are more than the critical angle, the reflecting surface of the prism may have to be silvered, then copper plated and enamelled for protection.

In most optical systems, the transmission faces of the prism must have anti-reflection coatings of magnesium fluoride, or perhaps even multi-layer anti-reflective coatings.

The glass used for making first-surface mirrors may be inexpensive glass because the light rays do not go through the glass. Prisms, however, must be made of high quality optical glass, free of bubbles and striae (distinct streaks of transparent material having a different refractive index), because the image-forming beam goes through the glass.

If a beam of light is formed of converging or diverging light rays, a mirror must be larger than a comparable prism (Figure 15-18). The refraction at the front surface of the prism causes the beam to have less spread than it does in air. The rays are refracted toward the normal which is usually parallel to the optical axis. Therefore, the prism can occupy less space than a mirror and serve the same purpose. An example of this is the reflector used in front of a camera lens to reverse the image left to right as is required in many copying duplicators.

Prisms are often used because they require less space.

Using the Critical Angle in a Prism

In the last lesson the critical angle of Porro prisms was discussed. It was found that when the prism was made of BSC-2 glass, with an index of refraction of 1.5230, some of the light rays incident to the reflective surface were at less than the critical angle and did not get totally reflected. This did not interfere with the functioning of a pair of binoculars.

In video lesson 10 a prism is shown and the loss of the image when the rays are less than the critical angle is demonstrated.

However, the fact that the light rays incident to the reflective surface at less than the critical angle do refract through the reflecting surface of a prism can be put to practical use. A silvered prism reflecting surface has no critical angle because the silver is in optical contact with the glass surface. All light rays, at whatever angle they strike the surface, are reflected.

Unity Power Gun Scopes

In many applications, it is useful to have a telescopic gunsight that has unity power; one that does not magnify. Rifles commonly have two sight pieces that are simply aligned with the eye and the target when sighting. However, such sights are subject to parallax problems. If the eye open-

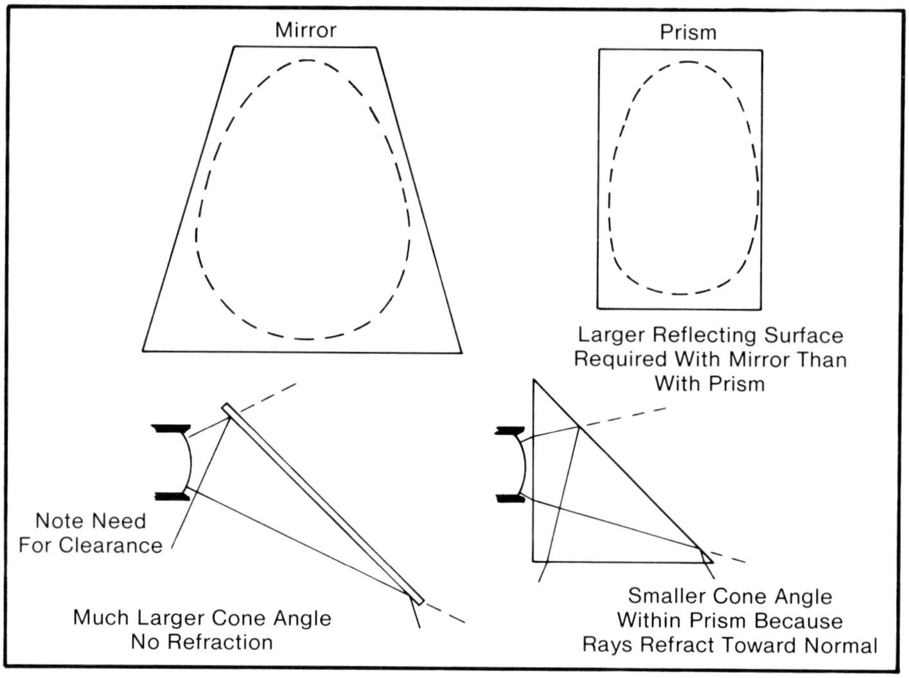

Figure 15-18. Mirror Reflecting Surface Compared To Prism Reflecting Surface.

ing is large enough for convenient sighting, the eye must be exactly aligned with the eye opening; misalignment causes sighting errors.

Unity power erecting telescopes are sometimes used. A reticle is placed in the scope in an image plane and is often at the field lens position which is in focus. Optically, this makes the reticle pattern appear to be on the distant object. Furthermore, the eye can move back and forth across the exit pupil without changing the position of the reticle on the image of the target. Parallax is eliminated.

Unity power gun scopes require many glass elements: an achromatic objective lens, a three-element eyepiece, a reticle, and a two or four element erector lens system. This sight can be replaced with a right-angle prism sight that uses rays beyond the critical angle.

The two prisms are put with their hypotenuse sides together and with a very slight space between the surfaces so that there will be neither physical nor optical contact (Figure 15-19). A block of glass is created. The bottom face of the prism closest to the eye is blackened by enamel or by placing some black plush material against it.

There is an area in the prism sight where the rays pass through the prisms. This is where the rays are less than the critical angle. The rays go through the prism block at an angle, but emerge parallel to their original paths.

At the critical angle, a black edge is seen because the black surface of the prism is seen by reflection. This edge can serve as a reticle line in the sight. The edge appears curved and shows color from chromatic aberration. The curve is straightened out if a low index liquid, such as water, is placed between the two prisms.

A Stereo Cube

Humans see objects in three dimensions because of stereo vision due to parallax. The viewpoint of the left eye differs slightly from the viewpoint of the right eye. The left and right eye images sent to the brain are, therefore, slightly different. The brain constructs a three dimensional visual image from the two slightly different stimuli.

Stereo photography uses this visual situation to provide photographs that can be seen in three dimensions. A stereo camera has two lenses placed about 2½ inches (65mm) apart, which is the average human interpupillary distance. Two photographs are taken simultaneously of the subject. A stereo viewer is used to present the photograph taken by the left lens to the left eye simultaneously with the one taken by the right lens to the right eye. The brain fuses the images into one, and a three dimensional image of the original subject is seen.

In the 1940's and 1950's, many 35mm stereo cameras, viewers and

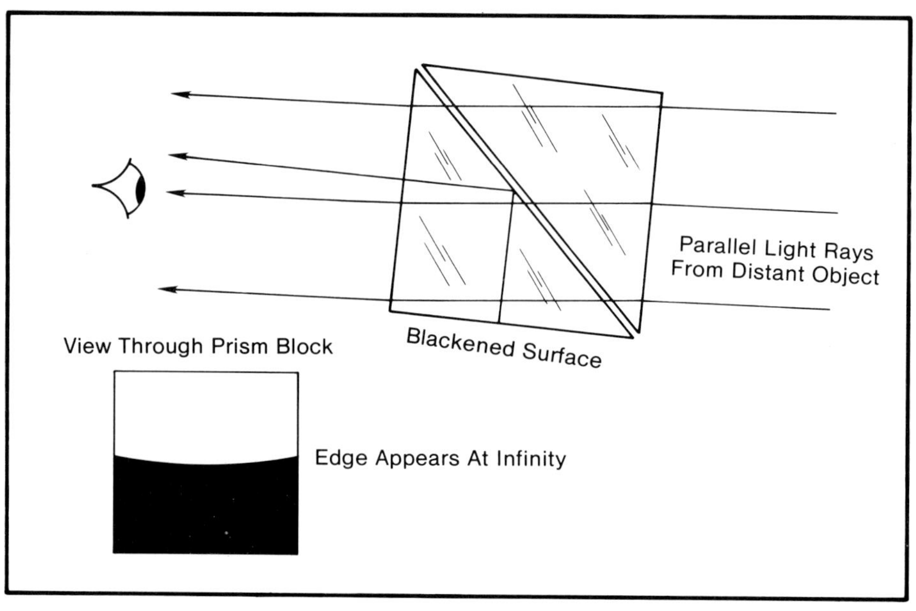

Figure 15-19. Prism Block Unity Sight.

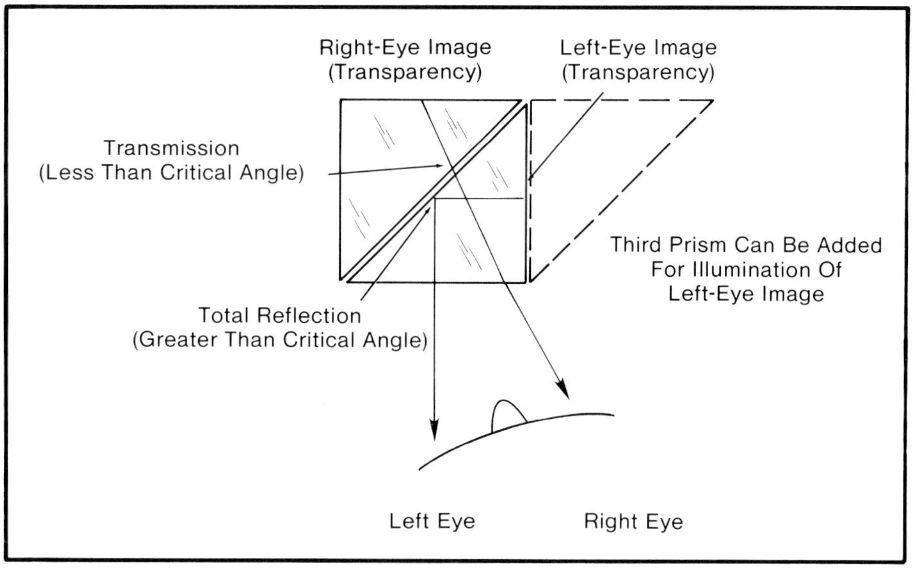

Figure 15-20. Prism Stereo Slide Viewer.

projectors were manufactured and sold to take and view stereo photographs. About the same time, children's stereo viewers became popular. These used stereo discs with a series of 16mm stereo photographic pairs mounted in them. Cartoon subjects, travelogues, fairy tales, and other subjects were, and still are, available today.

Another type of stereo photography viewer can be made with a pair of right angle prisms (Figure 15-20). The right eye photograph is cemented or laminated to the rear surface of a prism cube. The left eye photograph is reversed, left to right, and cemented or laminated to the right face of the cube. The prism is held so that the critical angle allows the right eye to see through the prism block while the left eye sees the reflection from the hypotenuse of the other photograph. If the photographs are properly aligned, the eyes and brain fuse the two pictures into a single stereo image.

Both images should receive the same illumination. This can be achieved by using a corner of diffuse white material, such as paper (one side behind and one to the right of the cube). Another method is to cement a third right angle prism to the left picture so that it reflects light from behind the prism block through the transparency. Any extended light object behind the cube will then illuminate both transparencies.

Beam Splitters

It has been seen that the hypotenuse of a prism becomes a reflector when the light rays are incident at more than the critical angle and it becomes a transmitter when the light rays are incident at less than the critical angle. It is also known that plane mirrors act as reflectors and change the direction of a light path under a variety of conditions.

In many optical devices it is necessary to take a beam of image-forming light and have it go in two directions simultaneously—usually at right angles to each other. A special type of mirror is used and is called a partially-reflecting mirror or beam splitter (Figure 15-21).

For total reflection, an opaque layer of metal (usually aluminum or silver) is applied to glass to form a mirror surface. Prior to World War II silver was applied to glass chemically for both first- and second-surface mirrors. At that time, beam splitters (partially-silvered mirrors) were made by reducing the chemical application time so that only a very thin coating was deposited. Today, chemical deposition is used only for second-surface mirrors.

Unfortunately, silver tarnishes and the silvered life of unprotected beam-splitter mirror surfaces is relatively short. Aluminum, on the other hand, is resistant to tarnish and can readily be applied in a vacuum by condensing vaporized aluminum on glass. Photocells can be used during evaporation to measure the amount of aluminum which is being deposited.

Whatever the type, a beam splitter is a mirror that reflects part of the incident light and transmits part of it (Figure 15-21). If a glass plate (window) is coated with a partially-reflecting metal surface and placed in an image-forming beam at 45°, one image can be seen by reflection

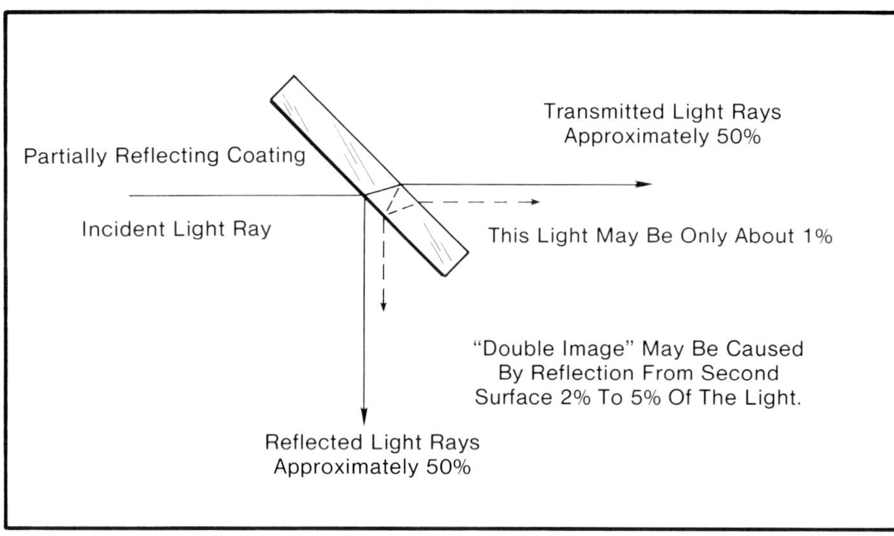

Figure 15-21. Glass-Plate Beam Splitter.

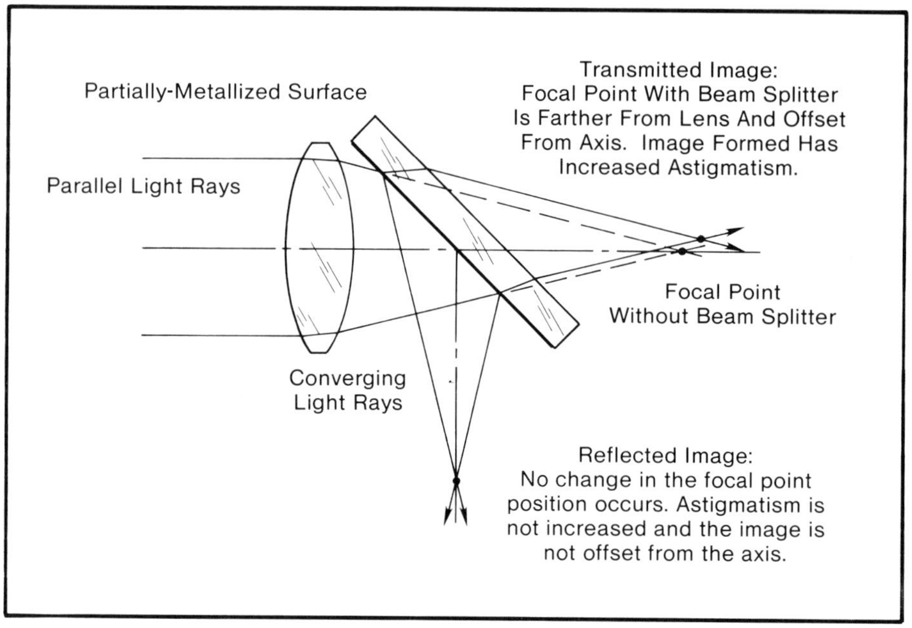

Figure 15-22. Beam Splitter In A Converging Beam Of Light Rays.

from the surface of the beam splitter and another by transmission through it. If the optical requirements are not critical, such glass-plate beam splitters work well.

Most beam splitters are used in optical instruments in conjunction with lenses. A glass-plate beam splitter can cause three types of difficulties when used in a converging beam of light (Figure 15-22). The thickness of the glass causes a change in the position of the image formed by the converging light in the transmitted beam. This moves it farther behind the lens. By going through the glass at an angle the aberrations of coma and astigmatism are introduced, reducing the sharpness of the image formed by the transmitted beam. The reflection from the second surface of the glass beam splitter causes a double image to be formed in the reflection image. The reflection image from the partially metallized surface is stronger, but the reflection image from the second surface can be about 10% as bright. This second image is offset from the primary reflected image by about $\sqrt{2}$ times the thickness of the glass plate.

The changed image size and position can be partially corrected by moving the lens and changing its focal length. However, if it is important that the transmitted image be exactly the same size as the reflected image, this correction cannot be used.

The intensity of the second reflected image can be reduced by coating the second surface of the beam splitter with a single- or multiple-layer coating. This reduces the comparative brightness from about 10% of the intensity of the primary reflected image to about 1%.

The coma (a lens aberration, resulting from different magnification in the various lens zones) can be corrected by placing another plate of glass in the beam at 45° to the axis of the beam, but at 90° to the beam splitter (Figure 15-23). This is called a corrector plate, and since it is not designed to form another beam, it is not metallized. If uncoated, it reflects about 8-10% of the light off of its two surfaces. However, once it is coated with magnesium fluoride, this loss is cut to about 1%. This plate changes the focal position still further and some astigmatism is still present.

Two prisms in the form of a prism block can also be used as a beam splitter. The hypotenuse of the first prism is partially coated with metal and cemented to the hypotenuse of the other (Figure 15-24). The hypotenuse does not have a critical angle any longer because the light is passing from glass to glass instead of from glass to air. The reflection is due to the partially-reflecting metal coating.

As can be seen in the drawing, the reflected light and the transmitted light both have the same path length in glass and leave it through a surface perpendicular to the axis of the existing beams. Therefore, both images are identical, and no astigmatism is formed. A slight amount of chromatic aberration and spherical aberration may result because of the dispersion of the glass forming

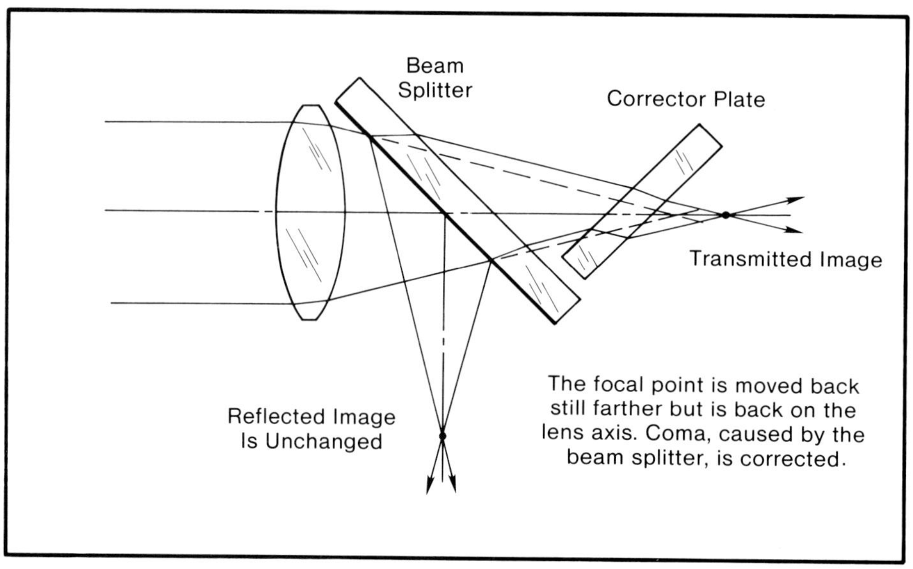

Figure 15-23. Beam Splitter In Converging Beam With Corrector Plate.

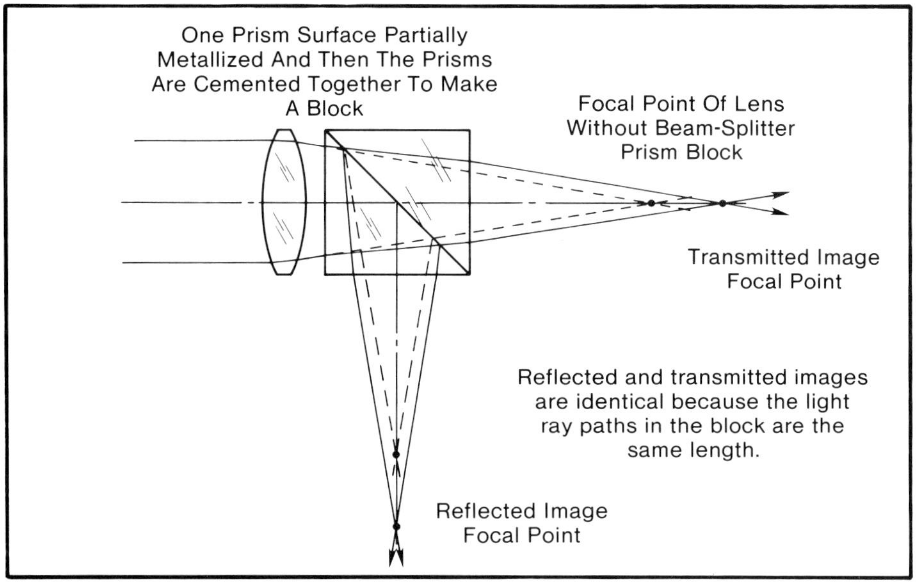

Figure 15-24. Beam-Splitter Prism Block.

the prism block.

Optical block beam splitters are used in microscopes when taking photomicrographs. The reflected image is used for aiming and focusing, while the transmitted image is used for forming the camera image.

A third type of beam splitter is called a pellicle. A pellicle is a *very* thin, (perhaps a thousandth of an inch thick) flat, partially-metallized membrane stretched over an optically-flat frame (Figure 15-25). The membrane is often made of collodion, which is guncotton dissolved in ether and alcohol. Guncotton (cellulose nitrate) is made by dissolving clean absorbent cotton in a mixture of nitric and sulfuric acids. A pellicle is made like this: A frame is cut from metal or glass and one side is ground optically flat. The frame is submerged in water and collodion is dropped on the surface of still water where it spreads into a very thin layer. The frame is lifted out of the water with the membrane clinging on top of the flat-frame surface. When the collodion dries, it is stretched as tight as a drumskin which insures its flatness. It is then vapor coated with a thin deposit of aluminum to make it a beam splitter.

Optically, the pellicle works better than either of the two other types of beam splitters because its thinness does not introduce any aberrations. However, it is extremely fragile, and must be handled very carefully. It is usually mounted in a protected position inside an optical device to minimize breakage. It must be used in a clean environment because it is extremely difficult to clean a pellicle without breaking it.

Prior to 1940 pellicles were often used in one-shot color cameras to form three optical images that were filtered with red, green and blue filters. These filters were used to expose three sheets of film to make separation negatives with one trip of the shutter (Figure 15-26). Color prints were made from the separation negatives by the three color carbro process or the dye transfer process. Sometimes the negatives were used to make halftones for printing press use. The high quality color photographs printed in fashion magazines in the 1930's were made with the use of pellicles.

Today, pellicles are used in automated "package printers" which print sets of school pictures with nine small, four medium, and one large-size picture in a set (Figure 15-27). Two pellicles are used to divide the light from a negative into three beams. One beam is focused through a single lens to make the large print. A second beam is focused through four lenses to make four smaller prints, and the third beam is imaged with nine lenses to make the nine smallest pictures. The beam splitters are completely enclosed to be dirt-free and to protect them from physical damage.

A single-lens-reflex 35mm camera, made in Japan for a short time, had a pellicle for a mirror so that the mirror would not have to swing out of the way when the photograph was

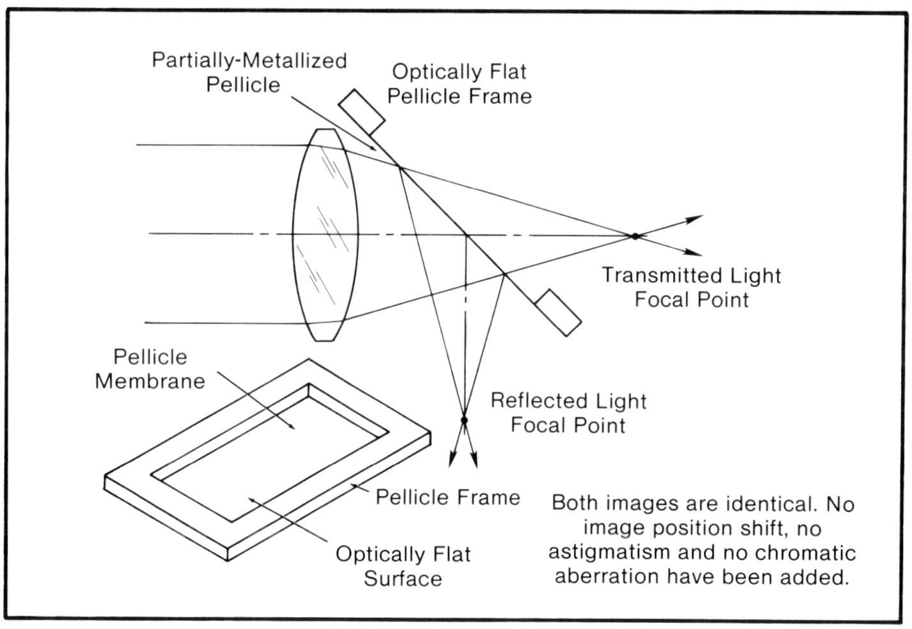

Figure 15-25. Pellicle Beam Splitter.

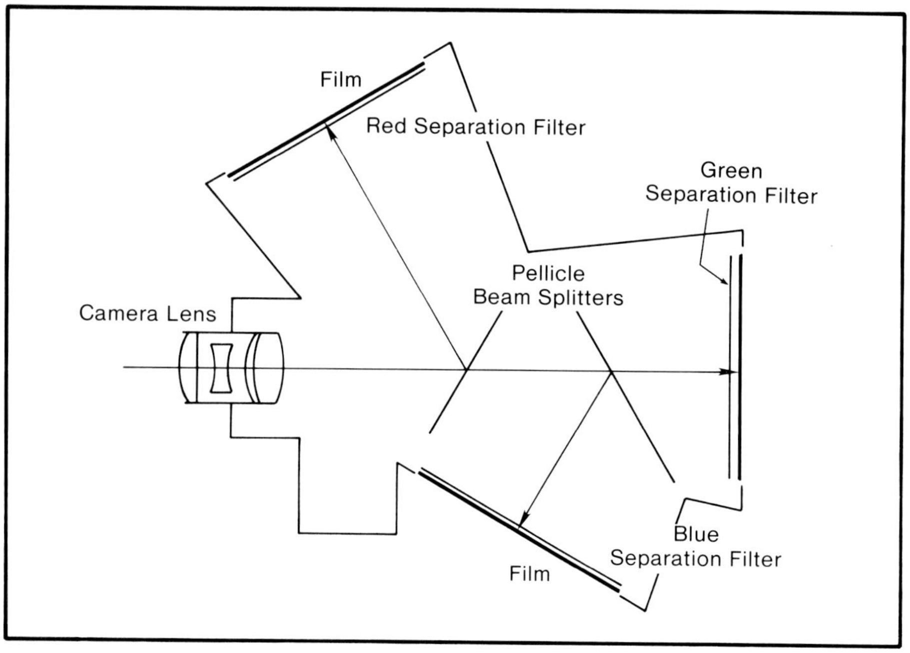

Figure 15-26. One-Shot Color Camera.

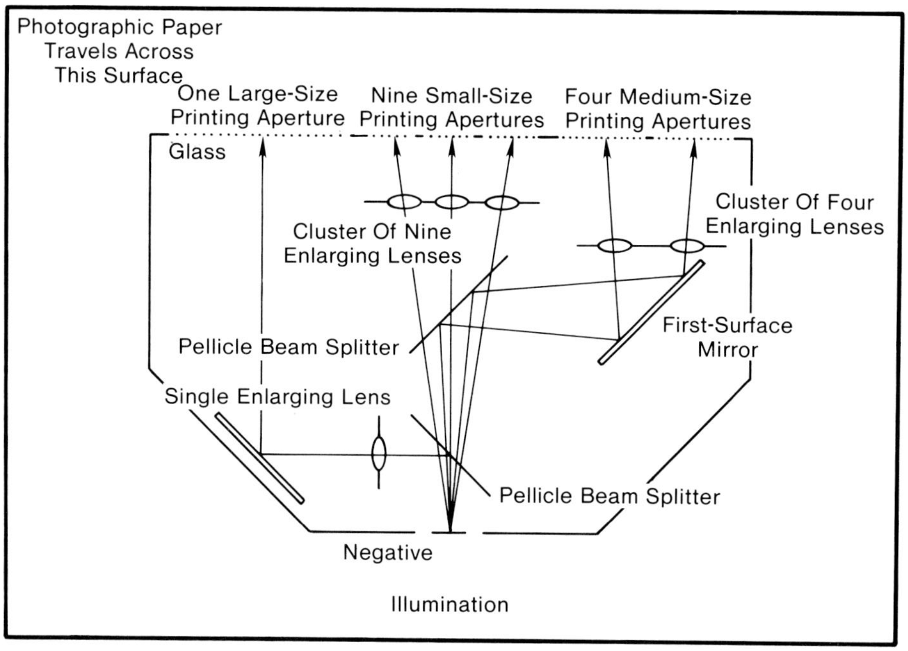

Figure 15-27. Optical Layout For Package Printer.

taken. Apparently the disadvantages outweighed the advantages since the camera is no longer made.

Problem One

A right angle prism is in air. The index of refraction of the glass is 1.500. Find the critical angle as viewed from outside the prism. Hint: This requires two graphic ray traces.

Problem Two

Find the critical angle as viewed from outside the prism when the index of refraction of the prism is 1.600.

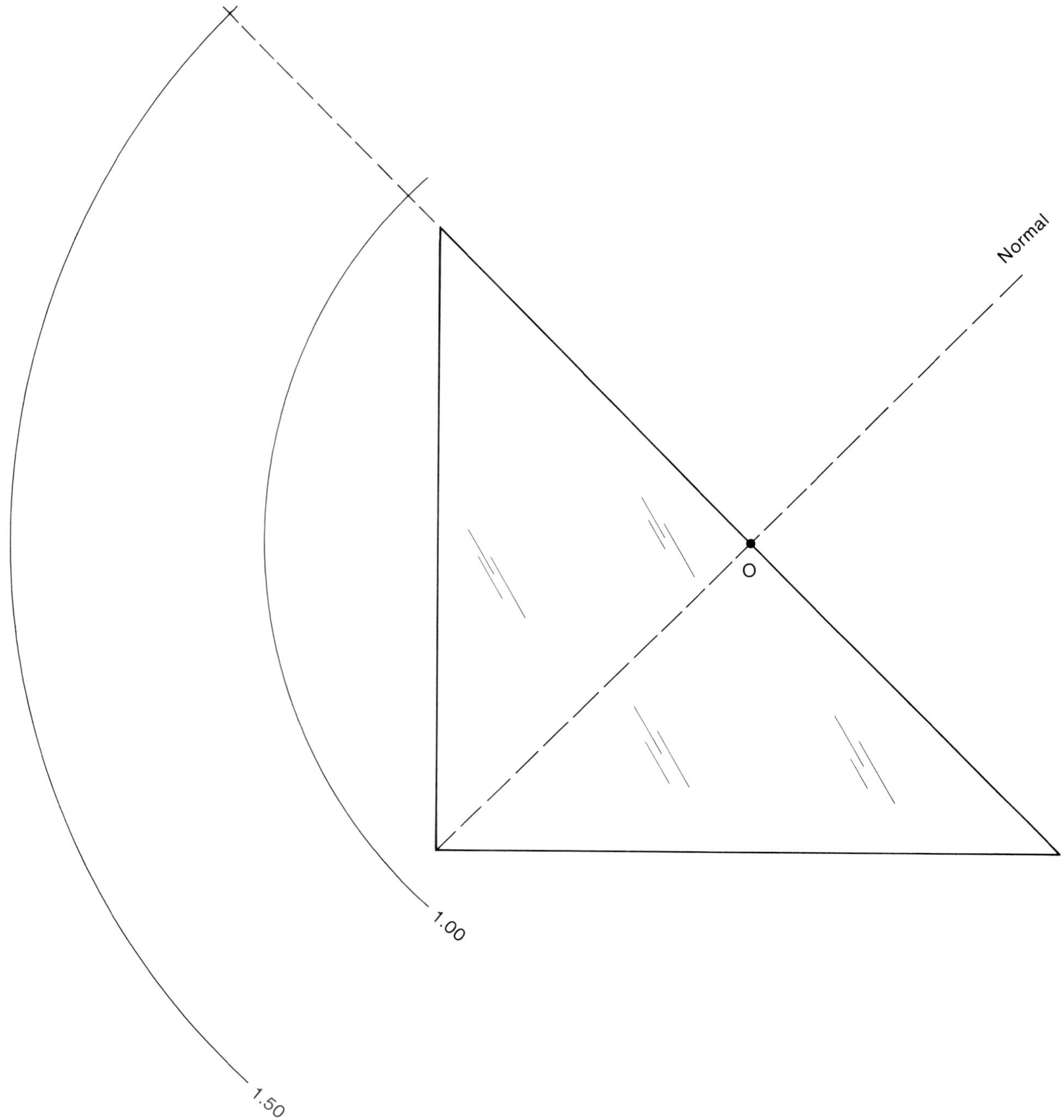

Problem 1
Chapter 15
Video Lesson 10

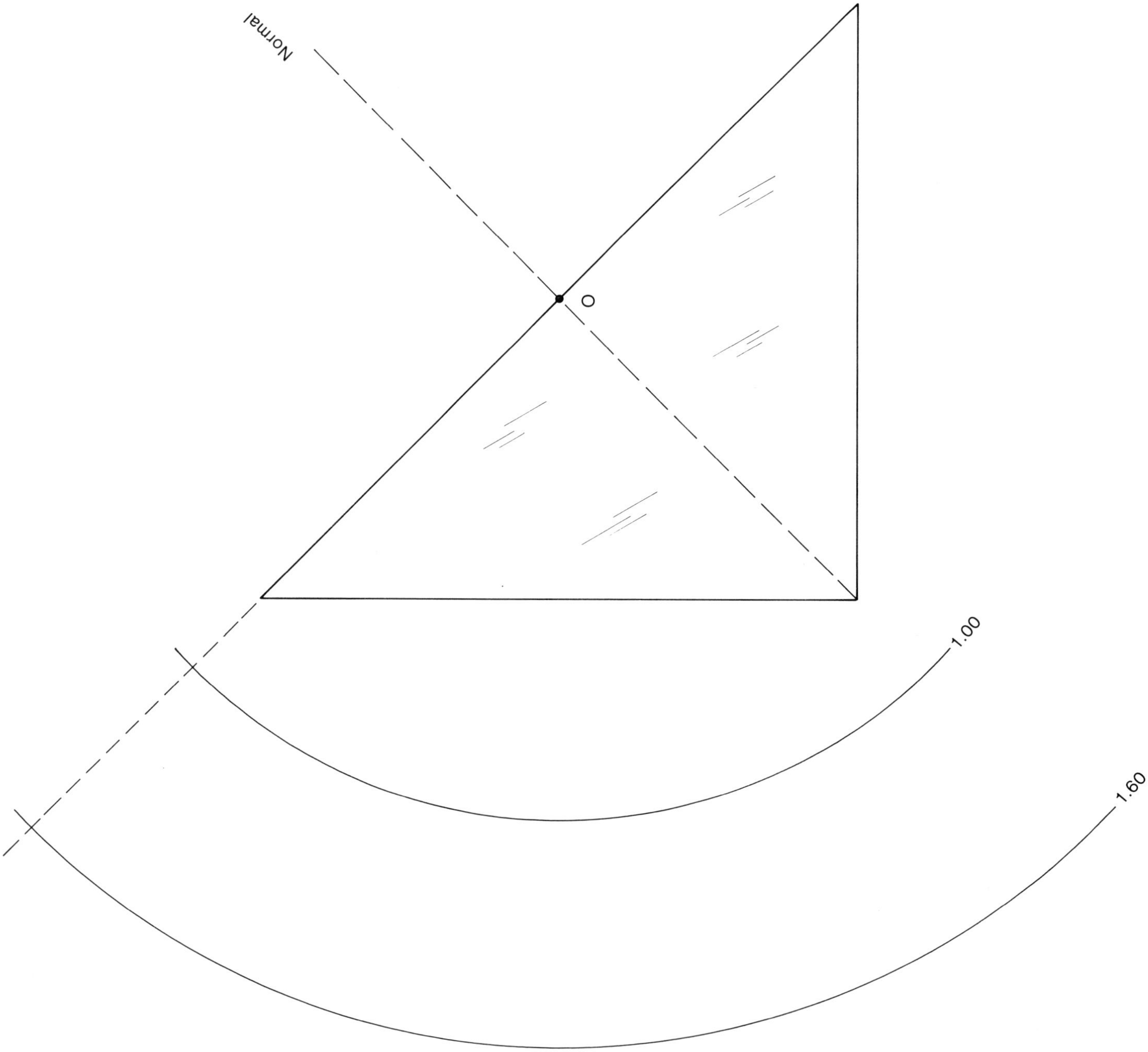

Problem 2
Chapter 15
Video Lesson 10

Chapter 16

VIDEO LESSON 11

Wedges, More Beam Splitters, Cylindrical Lenses and Anamorphic Systems

Summary

The two critical angles (inside and outside the glass) of right angle prisms are reviewed. When parallel light rays strike a glass-plate beam splitter, the transmitted beam of light is free of astigmatism. When a beam of light strikes a beam splitter made of a directionally-crazed styrene sheet at 45°, the light beam is divided into three beams. A wedge prism is a special type of prism having a relatively narrow included angle. A light beam entering a wedge prism at the minimum-deviation angle has minimum distortion and chromatic aberration. Two equal-angle wedge prisms with their bases placed at 180° and tilted, anamorphically change the image formed by a lens. Two such sets of prisms at right angles to each other form a variable-power supplementary optical system. Another anamorphic device can be made with a negative and positive cylindrical lens. A unit-power telescope can be made from two equal-power positive cylindrical lenses. This type of telescope reverses the image left to right without inverting it. When this telescope is rotated, the image rotates at twice the speed of the telescope's rotation.

Review of Assigned Problem

Figure 16-1 shows the first problem from the last chapter. The right angle prism is made of glass with an index of refraction of 1.5. You were asked to find the critical angle within the glass and the critical angle outside the glass by graphic ray tracing.

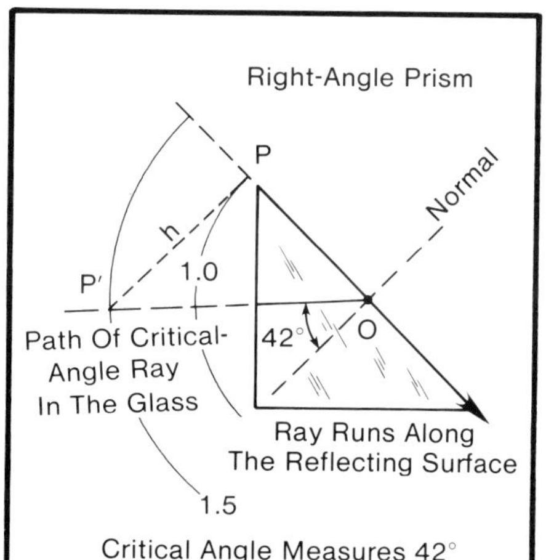

Figure 16-1. Finding The Critical Angle Inside A 90° Prism.

The problem is solved by the reversability principle. We know that the emerging ray runs parallel to the reflecting surface when the critical angle is reached, therefore we start with that ray and work backwards.

Using Point 0 as the center, two arcs are drawn with radii of 1.0 and 1.5. The 1.0 represents the index of refraction of air, and the 1.5 represents the index of refraction of the glass. Point 0 is where the ray emerges from the glass. A normal to the reflecting surface is drawn at Point O. Since the trace is being worked from air to glass, the h line is drawn starting at the 1.0 arc where a construction line extending from the reflecting surface intersects the arc and is drawn parallel to the normal at Point O. The h line's intersection with the 1.5 arc determines where the ray goes inside the glass. As

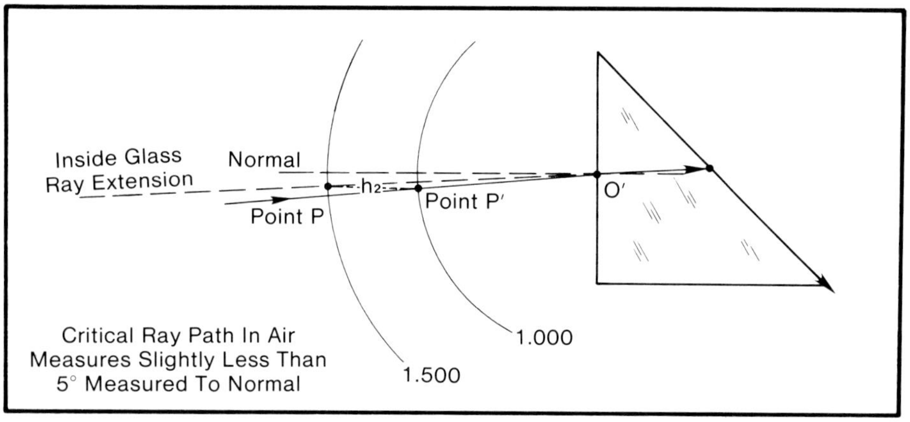

Figure 16-2. Finding The Critical Angle Outside The Glass In A 90° Prism.

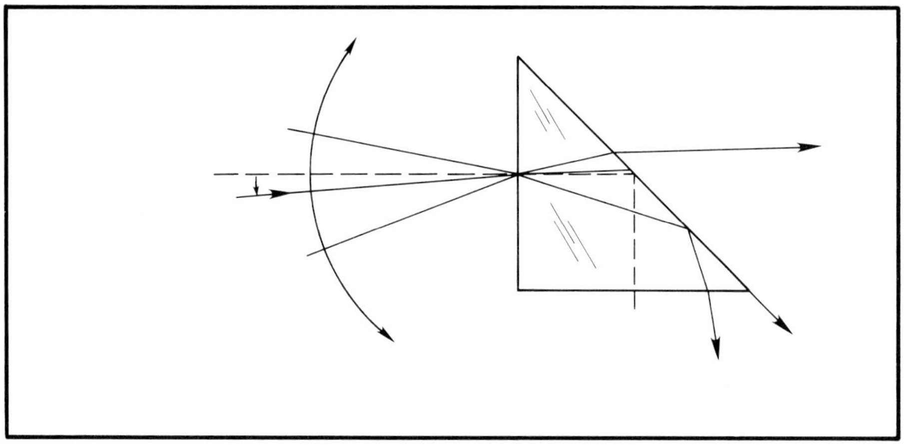
Figure 16-3. Rays Reflected Or Transmitted By The Reflecting Surface.

usual, the ray is drawn as a solid line in the glass and as a dashed line in air. In this drawing, the angle to the normal at Point O appears to be 42°, the critical angle in the glass.

The next step is to find the angle of the light ray in air. The reversed ray exits the glass at Point 0′. Using 0′ as the center, 1.0 and 1.5 arcs are drawn (Figure 16-2). The ray path is extended as a construction line (dotted) until it strikes the 1.5 arc because the ray is going from glass to air. This line intersects the 1.5 arc at Point P. Line h_2 is drawn parallel to a normal to the prism face from Point P to the 1.0 arc. This intersection is shown as Point P′. The actual ray path is then drawn from P′ to O. *Measured to the normal,* this angle is found to be slightly less than 5°. This is the critical angle in air, at Point O′, and rays entering the prism face at greater than this critical angle will not be totally reflected by the prism reflecting surface, but will be transmitted by the surface (Figure 16-3).

Snell's law can be used to verify the angles found by the graphic ray trace method. The critical-angle ray that goes parallel to the reflecting surface is 90° to the normal (I′). This ray is in air, with an index of 1.0. We need to find the refracted angle (angle I) in the glass. Snell's law sets it up this way:

$$N \sin I = N' \sin I'$$

$$1.5 \sin I = 1.0 \times \sin 90°$$

The sine of 90° is 1.0. Therefore:

$$1.5 \sin I = 1.0 \times 1.0 = 1.0$$

$$\sin I = \frac{1.0}{1.5} = .6667$$

Angle I = 41.81°

This is the critical angle in glass. Using graphic tracing we find the angle is about 42°. We have mathematically verified this angle.

The reflecting surface and the transmission surface form a 45° angle. Therefore, a normal to the prism face will intersect the reflecting surface at 45° and the refracted ray will strike the front face with an angle of incidence of 45°−41.81° = 3.19°.

Using Snell's law to find the angle in air, we get:

$$N \sin I = N' \sin I'$$

$$1.0 \sin I = 1.5 \sin 3.19°$$

$$\sin I = 1.5 \times .0556$$

$$\sin I = 0.0835$$

$$I = 4.79°$$

The trace showed an angle of about 5°, which we have verified.

When the Index of Refraction is 1.6

If the index of refraction is increased to 1.6 by using a flint glass or rare-earth glass, the critical angles change. The drawing shows that the graphic tracing will determine the critical angles of a right-angle prism made of glass with a 1.6 index of refraction. The critical angle inside the glass is found to be about 39° (Figure 16-4) while the angle outside the glass is approximately 10° (Figure 16-5).

Using Snell's law, the critical angles are found to be:

$$N \sin I = N' \sin I'$$

$$1.6 \sin I = 1 \times \sin 90° = 1$$

$$\sin I = \frac{1}{1.6} = .625$$

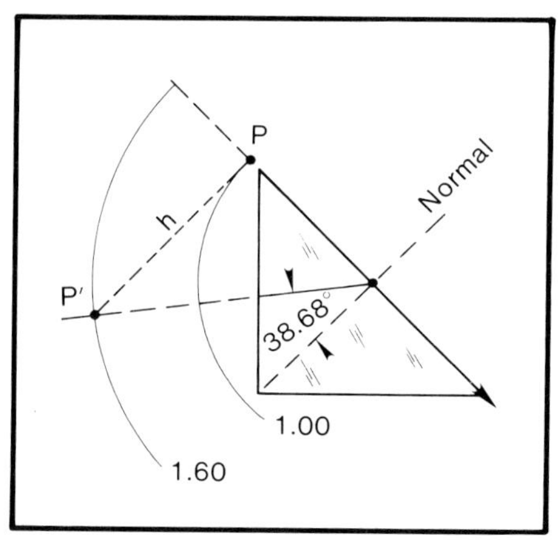

Figure 16-4. Critical Angle Inside The Glass Of A 90° Prism When N = 1.600.

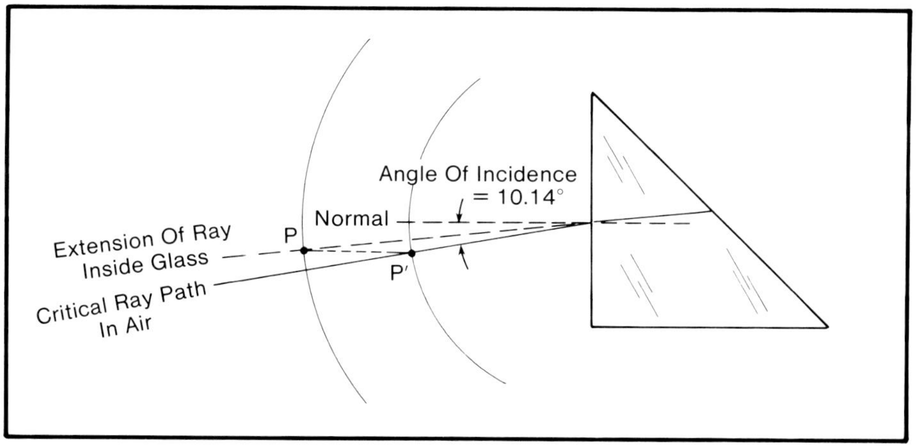

Figure 16-5. Finding The Critical Angle Outside The Glass In A 90° Prism With 1.600 Index Of Refraction.

$$I = 38.68°$$

$$45° - 38.68° = 6.32°$$

$$1.0 \sin I = 1.6 \sin 6.32°$$

$$\sin I = 1.6 \times .1101 = .1761$$

$$I = 10.14°$$

Thus, by increasing the index from 1.5 to 1.6 we have slightly more than doubled the critical angle outside the glass. This can be very useful in some optical instruments.

Glass-Plate Beam Splitters

In Chapter 15 it was noted that beam splitters made from glass plates with plane-parallel polished surfaces can cause the aberrations of astigmatism and coma. Plate-glass beam splitters do not cause astigmatism when the light rays of the beam are parallel.

Figure 16-6 shows a beam of parallel light rays being produced by a collimating lens. A point light source is placed at the focal point of a corrected lens. The light rays leaving the collimating lens are parallel. A corrected lens is a compound lens which has been designed to reduce the abberations to a tolerable level.

The light paths of two rays in the beam are shown going through a glass plate placed at 45° to the beam. Since the light rays are parallel, their incident angles to the front glass surface are identical. They are, therefore, refracted the same amount. The rays are parallel inside the glass and are refracted identically as they leave the glass. Because the two surfaces of the glass are parallel, the rays are parallel as they are refracted going into the air. They are the same distance apart, but have shifted slightly when compared with the axis. The dotted lines show the reflected rays, which are also parallel. As before, the reflected rays do not suffer from any aberrations.

The enlarged drawing (Figure 16-7) shows, however, that a plate-glass beam splitter causes chromatic aberration. Because blue light rays are bent more than red rays as they

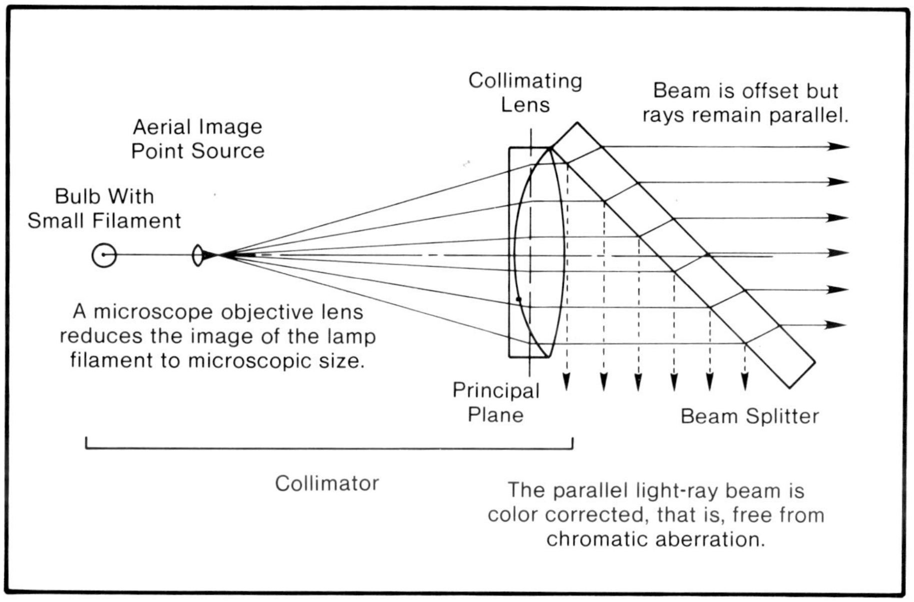

Figure 16-6. A Glass Plate Beam Splitter In A Beam Of Collimated Light Rays.

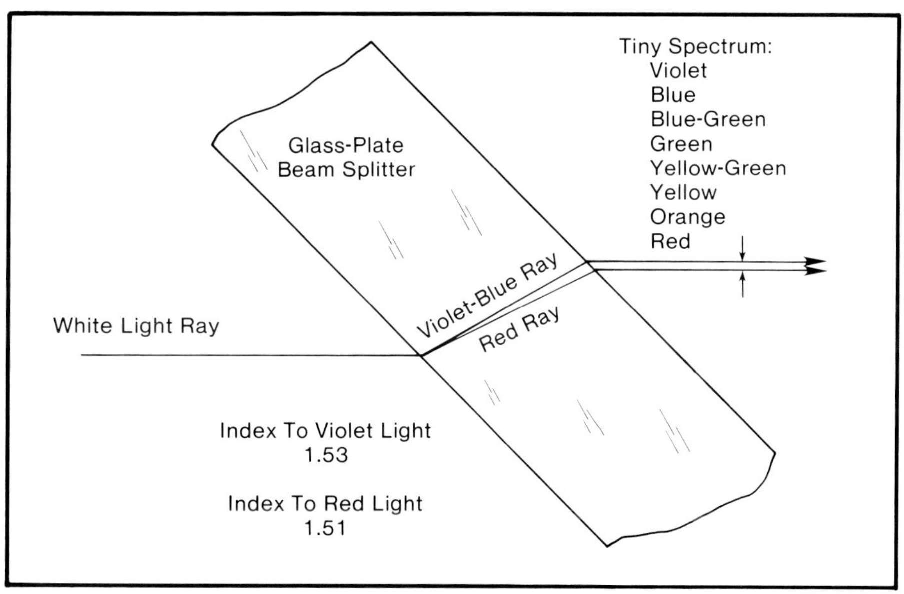

Figure 16-7. Chromatic Aberration In A Glass-Plate Beam Splitter.

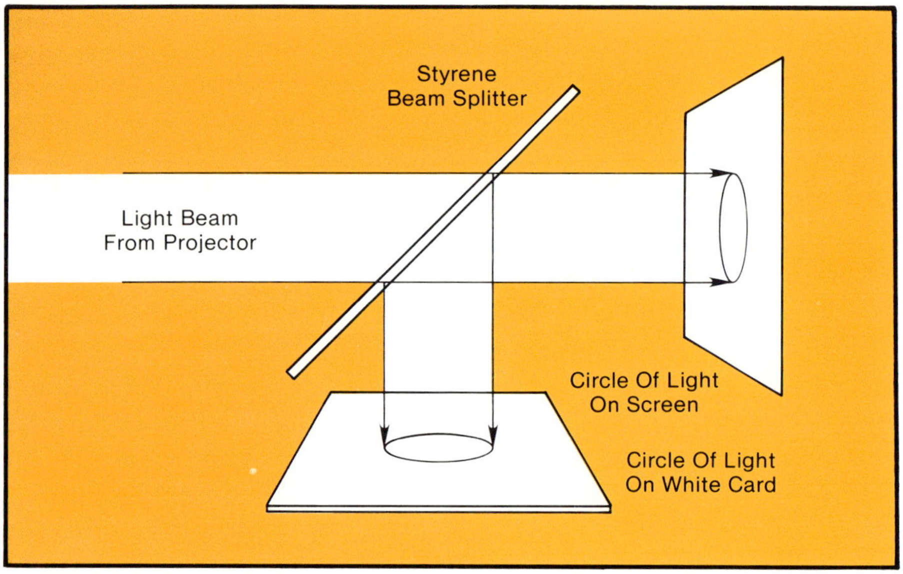

Figure 16-8. Perverse Beam Splitter — Seems To Reflect The Wrong Way.

enter the glass, they do not follow the same path in the glass as the red rays. Further, as they emerge into air, the red rays and blue rays are separated, but parallel. The ray of white light entering the beam splitter leaves as a spread of parallel light rays of color, forming tiny spectrum. The thinner the plate, the closer together the red and blue rays will be. Similarly, the thicker the plate, the greater their separation.

A Peculiar Beam Splitter

An interesting beam splitter is shown in the videotape lesson. Figure 16-8 shows a projector emitting a beam of light. A beam splitter is placed in the beam. A screen is placed in front of the projector showing that some light comes through the beam splitter. Another screen is placed at right angles to the first screen, and light shows on this screen too (Figure 16-9). However, the angle at which the beam splitter is placed should be reflecting light up instead of down. A third screen placed at the top, shows that light is being reflected in that direction too. Here is a beam splitter that forms three beams of light, not just two as conventional ones do. How is this possible?

Styrene is a brittle plastic that can be made transparent. If a thin sheet of styrene is pulled over the edge of a bench, it crazes into a series of parallel cracks nearly perpendicular to the surfaces of the sheet. The cracks run from surface to surface. When the crazed sheet is flattened it makes a three-way beam splitter that is useful for dividing light beams, but not for image forming beams. The cracks are not *exactly* parallel to each other and the polished surfaces cannot be made exactly flat after the crazing operation, so images passing through the material are distorted.

Figure 16-10 shows how the beam of light is transmitted in three different directions.

Wedge Prisms

Most of the prisms that have been discussed have been used to change the direction of light beams by reflection. They have been shown to invert and reverse beams of light formed by lenses.

Another type of prism is used only to bend the beam of light slightly. Such prisms transmit light and have no reflecting surface. The angle between the two transmitting surfaces, known as the *wedge angle,* or the *included angle,* is usually relatively small. These prisms are called wedge prisms or optical wedges (Figure 16-11).

Wedge prisms are made in a variety of powers. The power is determined by the included angle and the index of refraction of the glass. The power is measured in prism diopters.

Figure 16-12 shows a light ray entering a wedge prism. The ray is refracted at both surfaces and

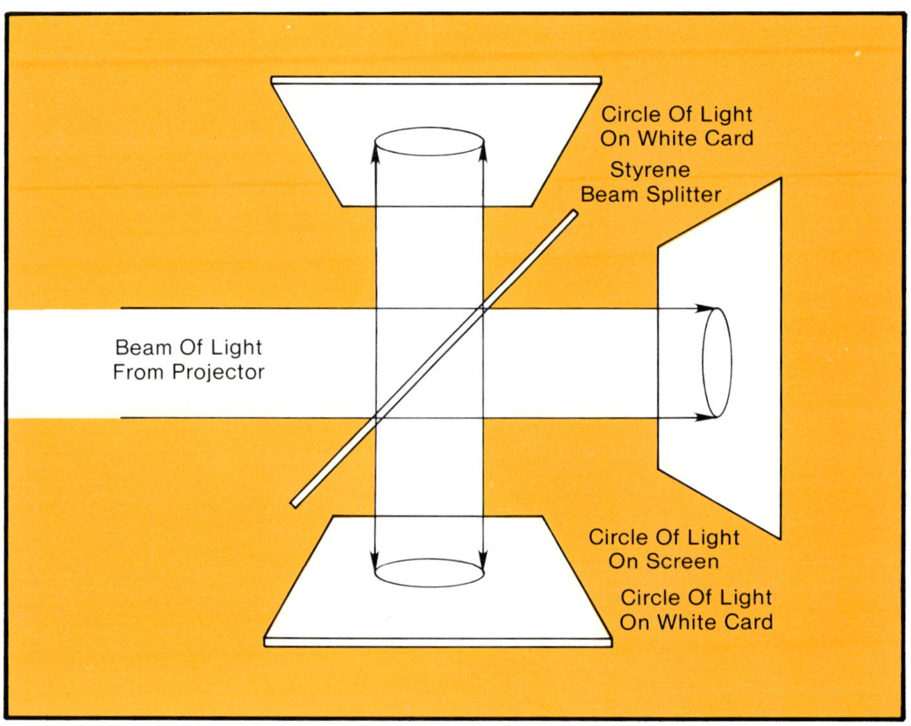

Figure 16-9. Three-way Beam Splitter.

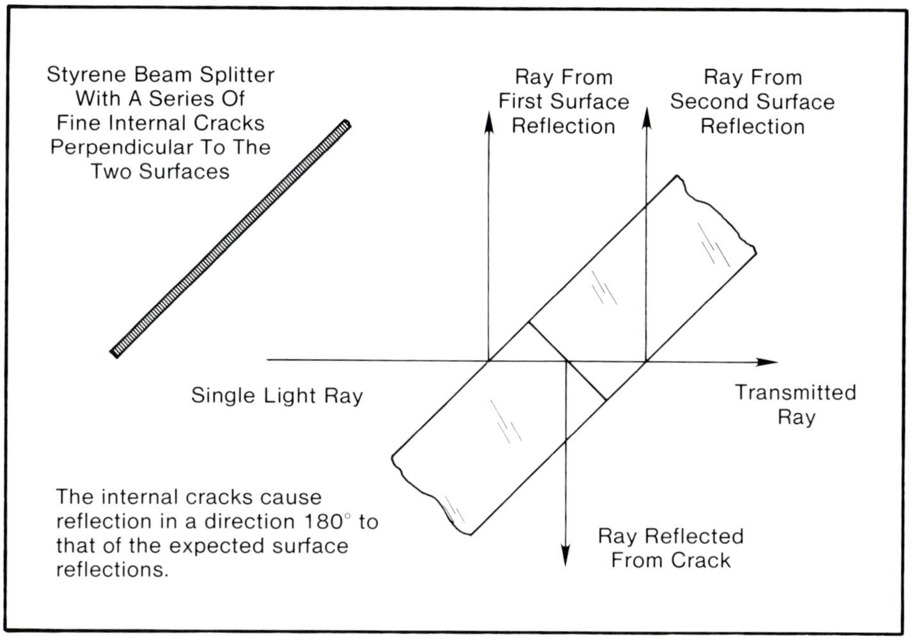

Figure 16-10. Cracked-Styrene Beam Splitter.

emerges in a direction different from which it entered. In the drawing, a construction line is shown continu-

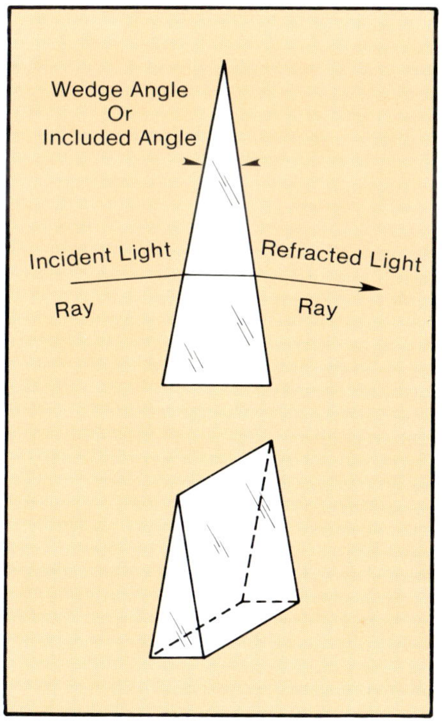

Figure 16-11. Wedge Prism.

ing the direction of the original ray. The angle between this construction line and the emerging ray is called the total angle of deviation. It is this angle that is used to measure the power of the wedge.

When a wedge bends a light ray so that it has deviated from its original path by 1 centimetre at a distance of 1 metre from the wedge, it is said to have a power of 1 prism diopter, or 1 Δ. The triangle is used as a symbol for prism diopters. The drawing shows the measurement of the angle of deviation in prism diopters.

Angle of Minimum Deviation

If a wedge prism is moved so that the incident angle is increased and decreased, the amount of deviation changes. If the ray is imaged on a screen, the image moves up and down as the wedge is tilted. The minimum deviation occurs when the path of the light ray within the wedge is perpendicular to the wedge-angle bisector. The power of a wedge is measured at the angle of minimum deviation.

Figure 16-13 illustrates this condition. The vertical dashed line bisects the wedge angle. The light ray passing through the prism is perpendicular to the bisector, as shown by the small square. In this position the angle of incidence of the ray entering the prism is equal to the angle of emergence of the ray leaving the prism.

$$\angle I = \angle E$$

A wedge prism produces the least chromatic aberration and distortion when the light goes through at the angle of minimum deviation. This is demonstrated in the tape session by projecting a beam of light on a screen.

Anamorphosis

Technically, anamorphosis is a directional change in magnification. It is usually used to compress an image made by a camera and to expand it back to its original proportions when projecting it (Figure 16-14). If the subject is a circle, and the image is compressed by anamorphic optics, it then becomes an ellipse.

Occasionally, an image is slightly changed anamorphically to make a fashion model look taller and slimmer, or to lengthen the image of a

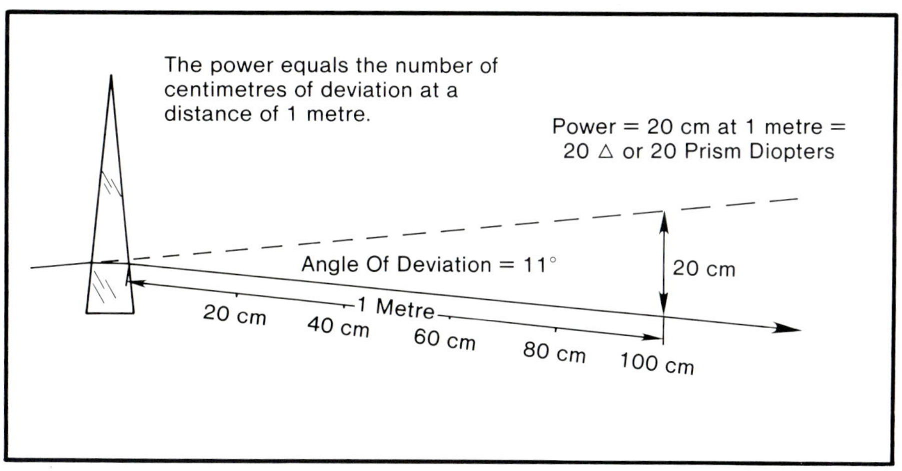

Figure 16-12. Power Of A Wedge Prism.

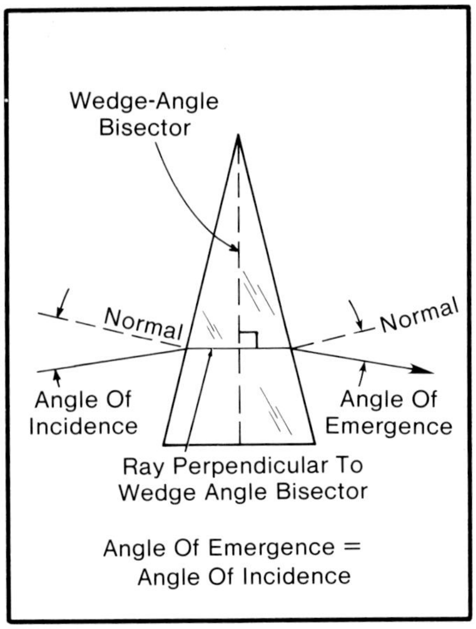

Figure 16-13. Minimum Deviation Condition In A Wedge Prism.

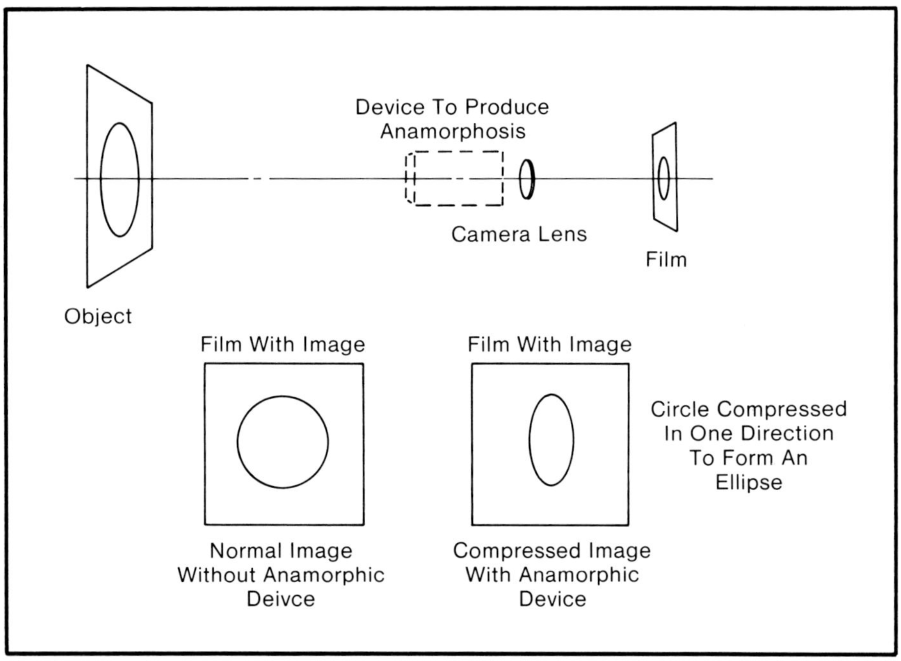

Figure 16-14. Anamorphosis.

new car to make it look more impressive.

Using Wedges to Cause Anamorphic Change

Figure 16-15 shows two wedges with their bases in opposite directions. The inner surfaces are parallel to each other and are almost in contact. The outer surfaces are parallel to each other, so the wedge pair is, in essence, a window. Light rays go through this wedge pair as through a window without the image being changed.

Figure 16-16 shows the wedges at an angle to each other, their tops touching, but their bases separated. In this position, the light goes through the wedges at angles considerably greater than the angle of minimum deviation. The distortion caused by each wedge is enhanced, therefore anamorphically changing the image so that it is compressed.

Since the wedge bases are oriented in opposite directions, the chromatic aberration is largely corrected and the images formed by the system are of reasonably good quality.

As demonstrated in the tape program, an anamorphic wedge pair must be used with an image forming lens. It is used as a supplementary optical device in front of a camera or a lens, as it does not have positive lens power to form a real image. The arrangement of the wedges for expanding the image, as is required for the projector, is shown in Figure 16-17.

Using Anamorphic Optics to Create a Zoom Lens

A variable-power supplementary lens can be made using two identical sets of wedges placed at right angles to each other (Figure 16-18). They are connected so that they pivot in unison. While one set is distorting

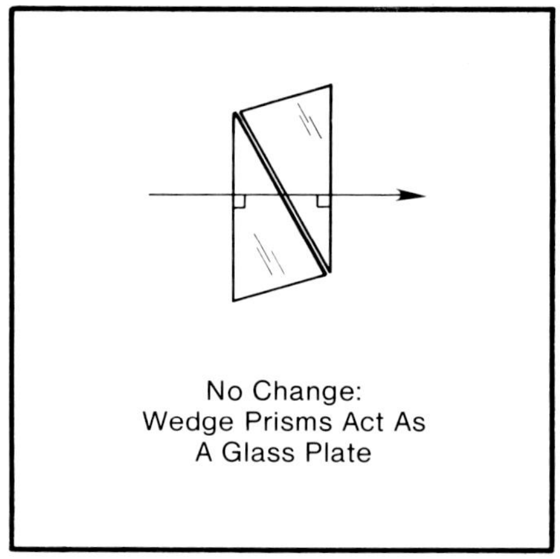

Figure 16-15. Anamorphic Wedge-Prism Afocal Devices (Part 1).

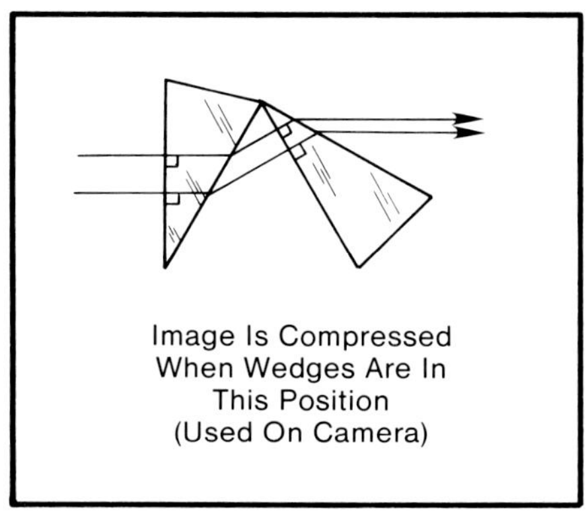

Figure 16-16. Anamorphic Wedge-Prism Afocal Devices (Part 2).

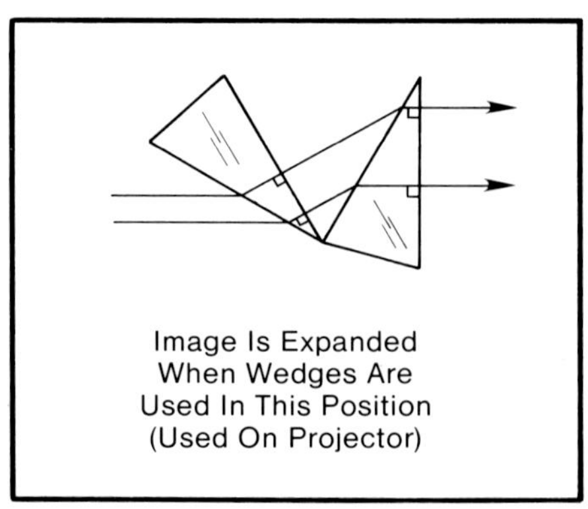

Figure 16-17. Anamorphic Wedge-Prism Afocal Devices (Part 3).

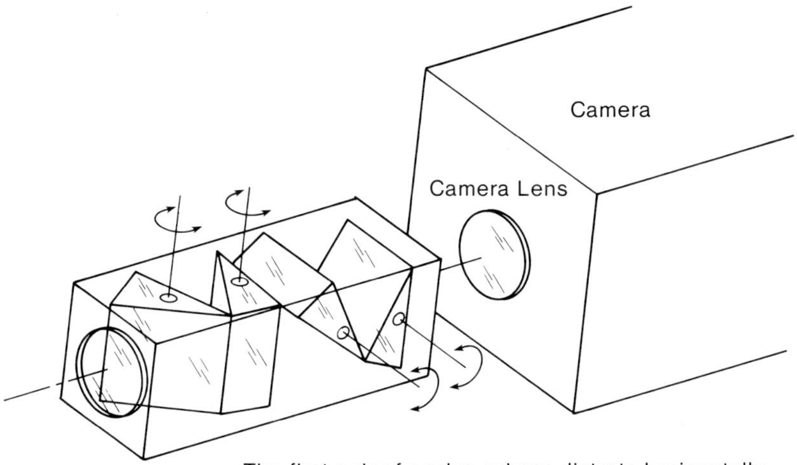

Figure 16-18. Wedge-Prism Zoom Afocal Attachment.

vertically, the other is distorting horizontally. As a result, the image size is changed as the wedges are rotated. This device acts like a zoom lens when placed in front of a camera or a projector lens. The image is distortion-free because it is expanded in two directions, at right angles to each other, at the same time. It is not a true zoom lens because the system's focal length changes slightly as the prisms are rotated, and the camera or projection lens must be slightly refocused.

When the wedges are together, the zoom device has zero power. When the wedges are angled, the image is enlarged. The image of the zoom lens, therefore, starts at a size created by the lens alone and is magnified. It is never smaller than the image made by the lens alone.

A Variable-Power Wedge Prism

A variable-power wedge is created when two wedges of identical power are cut into circles and mounted so that they can be rotated in opposite directions around an optical axis (Figure 16-19). When the base of one is at the apex of the other, the two outer surfaces are parallel, and a window—which in this instance can be called a wedge with zero deviation—is formed. As the wedges are rotated with respect to each other, the wedge angle of the pair increases. When the two bases are together, the combination wedge has twice the bending power of either wedge alone. When the two wedges are at 90°, the power is ½ the maximum. The power of the wedge is variable from zero, when the

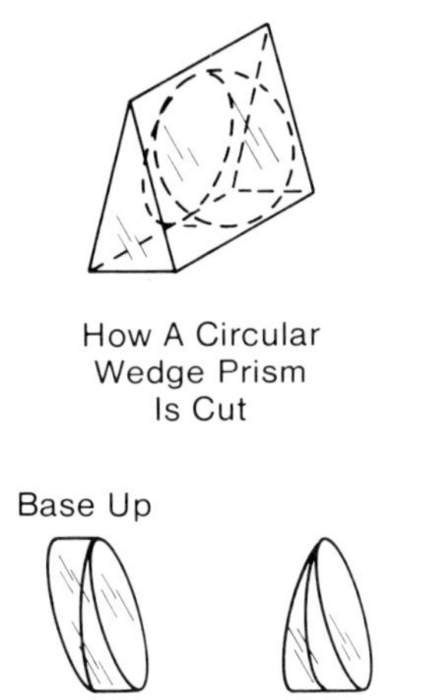

Figure 16-19. Variable-Power Wedge Prism.

wedge bases are 180° to each other, to twice the power of a single wedge when the bases are aligned.

One use for variable wedges is in rangefinders. In Figure 16-20, one sight path is straight through the eyelens and the finder lens. The other sight path is reflected from a partially-silvered elliptical beam splitter on the 45° window. This ellipse appears circular to the eye. The first-surface mirror is also at 45°. Both mirrors are fixed in position.

A filter covers one beam of light. When the wedge is in the position shown in the left drawing, with the bases at 180°, the path of light is parallel to the straight-through path and the distance setting is at infinity. In the second drawing, the bases are aligned. The path of the colored-image beam is now at the closest distance the rangefinder can read. The eye sees a colored double image in the circle in the center of the field until the two beams are set at the distance of the object. When this occurs both the colored and uncolored images coincide. The mount of the rotating wedge rangefinder is calibrated in feet or metres if the rangefinder is used to measure distances. The rotating wedge can be attached to a camera focusing mechanism, or to the aiming mechanism of an artillery gun as an aid in controlling the projectile trajectory.

Increasing the Stereo Effect with Wedges

The normal distance between the pupils in human eyes is about 2½ inches (65mm). The visual stereo effect at this spacing is considered "normal." If the interpupillary distance is increased, the stereo effect is exaggerated and the apparent distance between close and far objects is reduced. The pairs of Porro prisms in binoculars are used to accomplish this. This can also be done with a pair of wedges. The binoculars enlarge the images while increasing the interpupillary distance. Two wedges, properly ar-

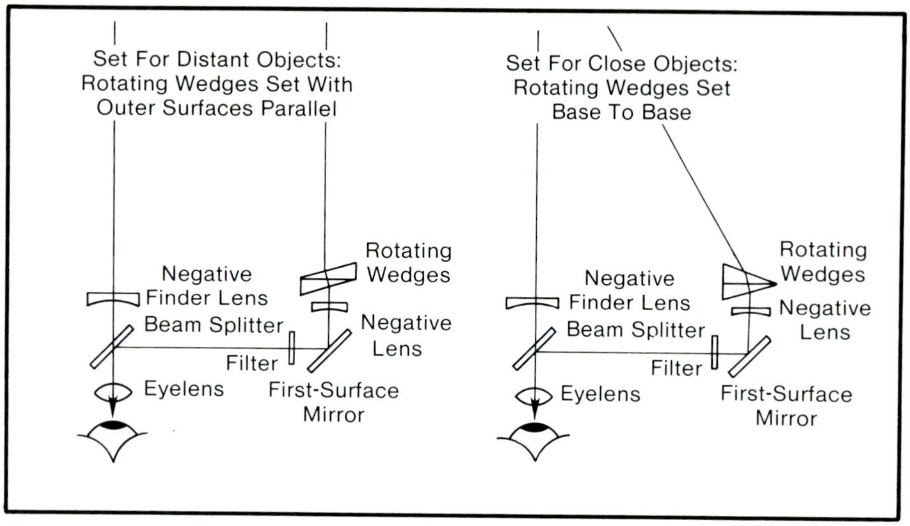

Figure 16-20. Rangefinder Using Rotating Wedges.

ranged (without additional optics) can increase the interpupillary distance while the magnification remains at 1X.

Figure 16-21 shows how this is accomplished. One wedge is placed in front of the eye and the other, with its base in the opposite direction, is moved away from the first wedge and to the side. This widens the effective separation between the eyes, so that the stereo effect is enhanced. Because the two wedges together are still, in effect, a window and because both are used at the minimum deflection angle, no distortion or chromatic aberration is seen.

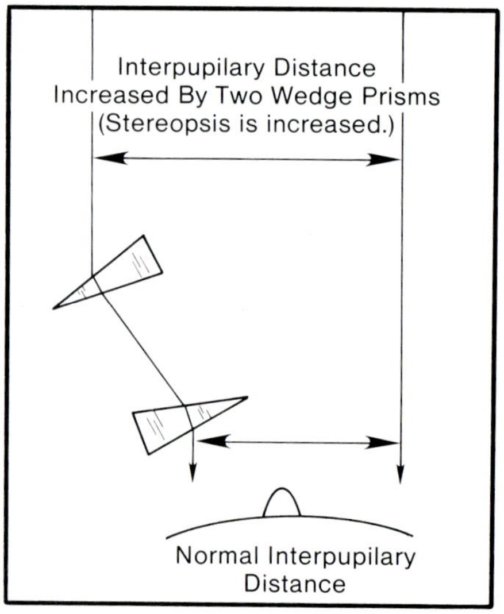

Figure 16-21. Increasing Stereo Effect Using Two Wedge Prisms.

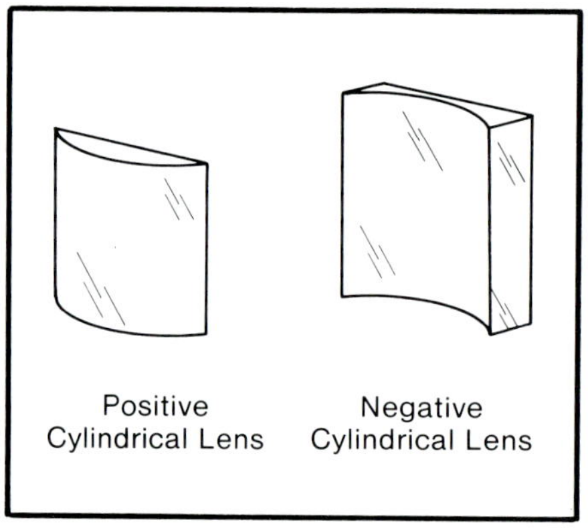

Figure 16-22. Cylindrical Lenses.

Anamorphic Optics with Cylindrical Lenses

When two wedges are used to stretch an image anamorphically, one image dimension remains constant while the other dimension expands. In the zoom supplementary lens, two pairs of wedges change the size of the image from one size to a larger, but never to a smaller than 1X size. Anamorphic optics can be made which compress the image in one direction, but this must be done with cylindrical lenses. A positive and negative cylindrical lens are shown in Figure 16-22.

In an earlier chapter, an afocal wide-angle attachment that behaves like a reverse Galilean telescope was described. Figure 16-23 shows the basic arrangement: a negative lens in front of a positive lens. The light rays leave the positive lens parallel, so that the system is afocal (without focal length). When looking through such a device, the eye's lens focuses the parallel light on the retina so that an image is seen. Since the negative lens is in front, the image seen is very small; that is, the device has a magnification of less than 1X.

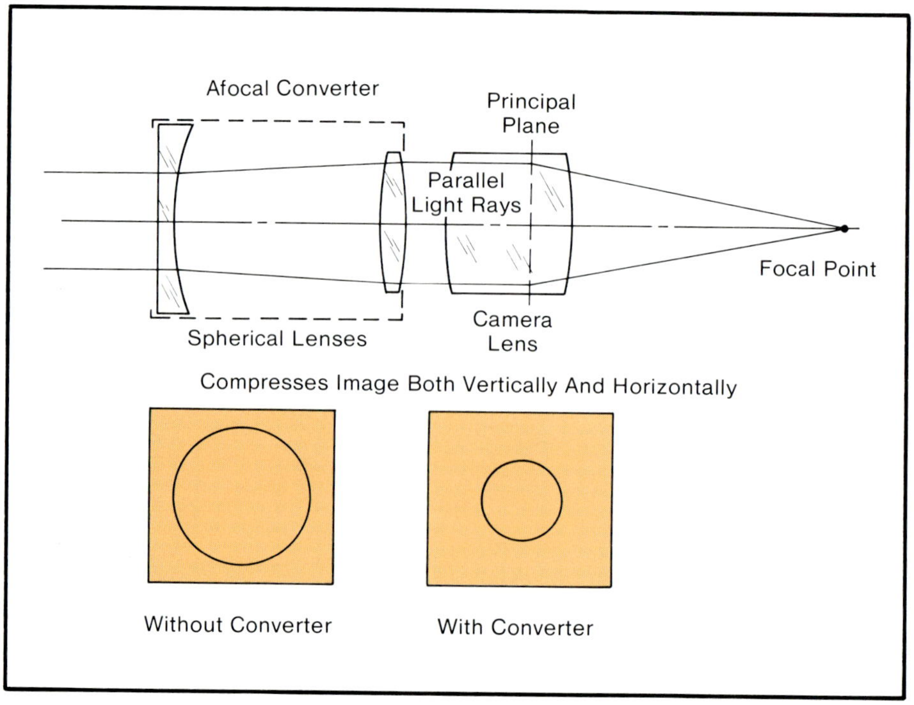

Figure 16-23. Normal Afocal Converter (Wide-Angle Usage).

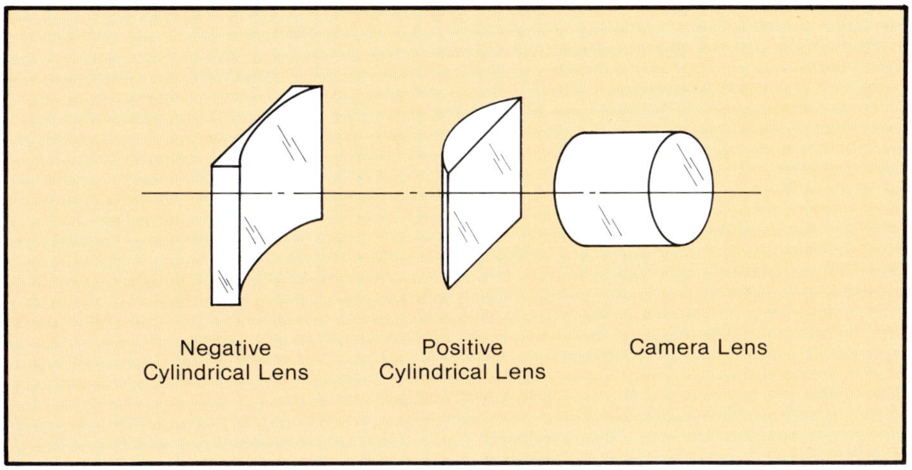

Figure 16-24. Cylindrical Afocal Converter.

If the afocal converter is placed in front of a camera lens, it shortens the focal length of the camera lens and converts it to a wide-angle lens. In both cases, the lenses are regular lenses with spherical surfaces.

The same type of device can be made using cylindrical lenses instead of spherical lenses. The lens surfaces, instead of being spherical, are ground and polished as segments of a cylinder–like the sides of a glass rod.

In Figure 16-24, the two cylindrical lenses of a cylindrical afocal converter are shown. They are drawn as square lenses because it is easier to visualize them than if they were drawn round. The negative lens is in front, just as in the wide-angle converter.

Figure 16-25 is a top view of the cylindrical converter and shows a beam of parallel light rays passing through the negative and positive lenses and emerging, still parallel. The beam is expanded.

The side view of the same converter shows the cylindrical lenses as windows because there is no surface curvature in that direction. The beam of light goes through without expansion or compression.

When this device is placed in front of a camera lens, the objects are imaged the same height as they would be by the camera lens alone. However, the width of the objects appears compressed. The degree of compression is 2X, so that twice as much subject is recorded in the image horizontally as would be recorded by the camera lens alone. The horizontal magnification is ½X.

If this device is placed in front of a motion picture camera and a movie is made and later projected without correction, the image on the screen appears as though everything is very thin. All the subjects have half the width they normally would have.

However, if a similar device is used over the projector lens in the same orientation, the width of the image is expanded. The width of the projected image is twice what it would be (if it is a 2X device) without the device, and the width of the subjects have been made normal. The height of the projected image remains unchanged.

Professional wide-screen motion pictures use these devices to acquire a wide screen format. The normal format size of a professional motion picture camera is 0.868 X 0.631 inches, which is an aspect ratio of 1.375:1. When this rate is doubled, the aspect ratio becomes 2.75:1, so that the projected image is 2.75 times as wide as it is high (Figure 16-26).

A Cylindrical Lens 1X Telescope

Figure 16-27 represents a 1X telescope. The focal lengths of the objective lens and eyelens are equal and the power is 1X. Parallel rays are brought to focus by the objective lens at the focal point, F, and the eyelens focuses on this point and emits the rays parallel. Because it has the same focal length, point F is midway between the two lenses.

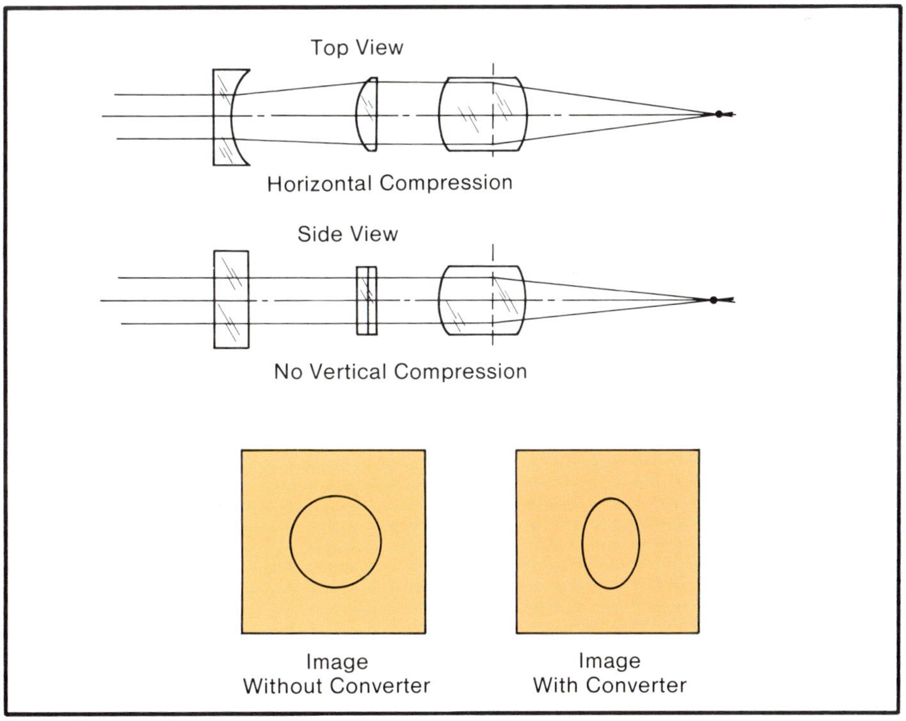

Figure 16-25. Cylindrical Afocal Converter.

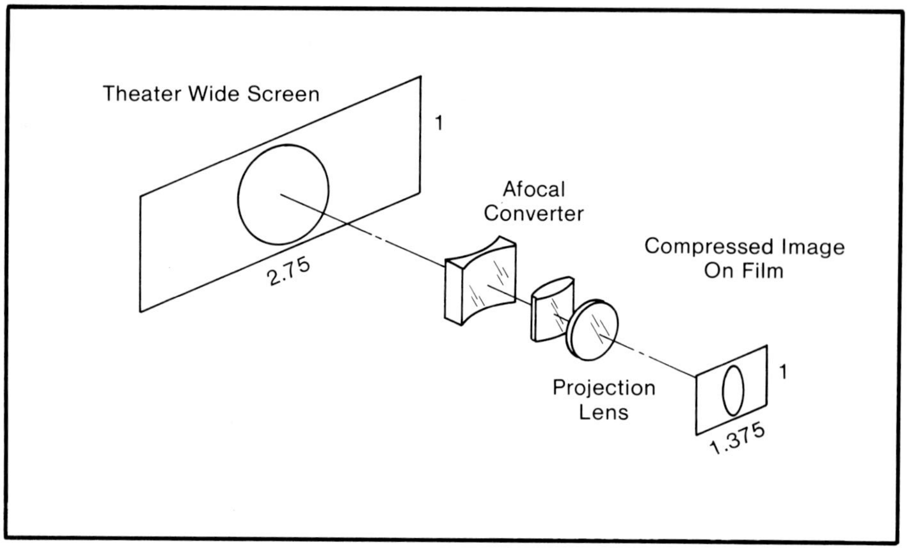

Figure 16-26. Wide-Screen Movies With Anamorphic Afocal Converters.

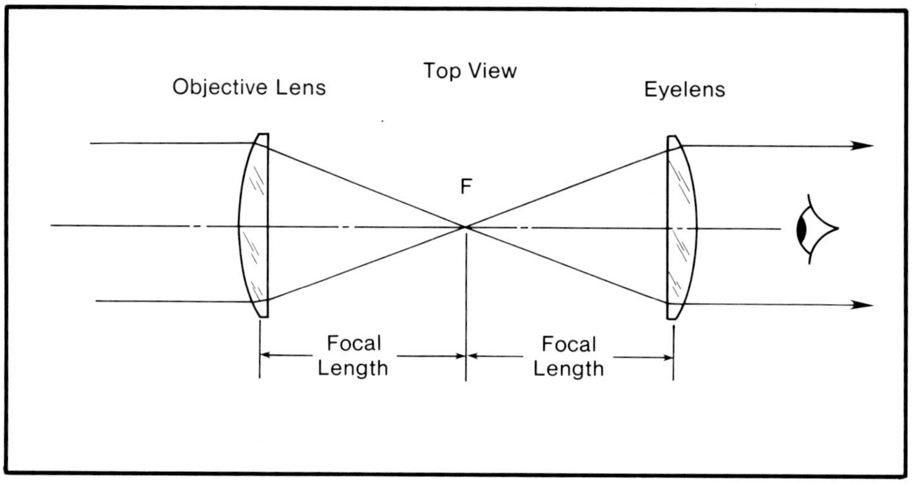

Figure 16-27. Unit-Power Telescope.

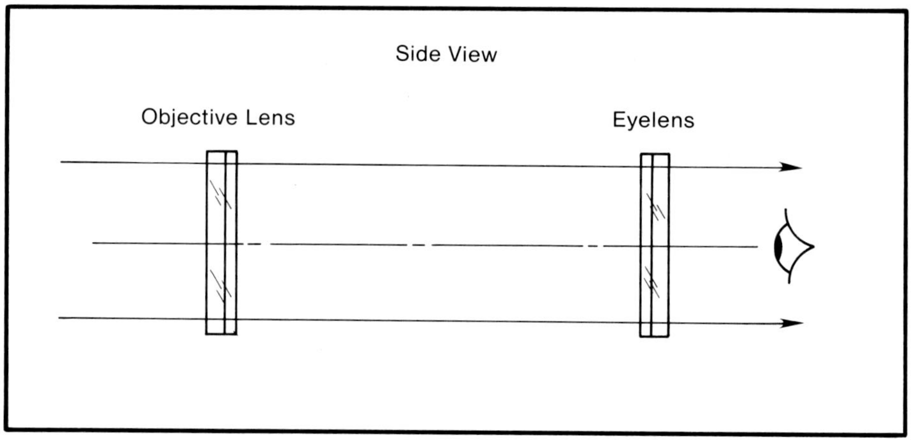

Figure 16-28. Unit-Power Telescope.

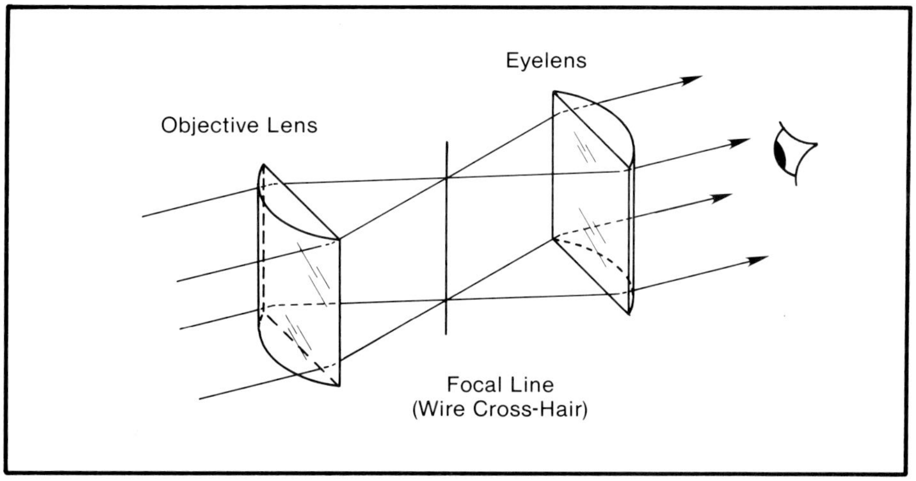

Figure 16-29. Unit-Power Cylindrical-Lens Telescope.

In Figure 16-29 it can be seen that the two lenses are actually cylindrical lenses, and that Figure 16-27 is the top view of a cylindrical lens telescope.

The side view of the telescope shows that the cylindrical lenses have no power in the vertical direction (Figure 16-28).

In the angular view (Figure 16-29) it can be seen that the parallel rays have been reversed crosswise, but not vertically or top to bottom. This unit telescope reverses the image left to right, but does not invert it top to bottom. If used as shown, the subject would be seen right side up, but backwards reading left to right. However, if it is turned 90°, the image will be seen upside down. In other words, as the telescope is rotated 90°, the image rotates 180°—the image rotates twice as fast as the telescope. In this way, the cylindrical lens unit telescope acts like a dove prism. However, it has more aberrations than a dove prism so it is rarely used to replace the dove prism in optical instruments. Figure 16-30 shows rays going through a dove prism with the upper incident ray emerging below and the lower incident ray emerging above.

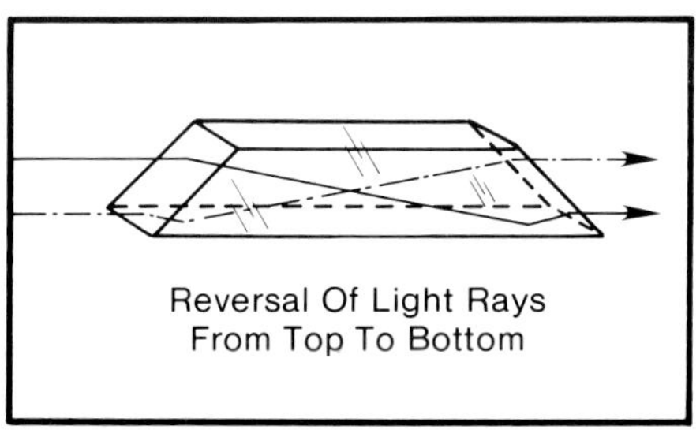

Figure 16-30. Dove Prism.

If a thin wire is placed in the line focal point (or more properly, focal line) of the two lenses, it acts like a reticle. Since it is in the focal line of both lenses, it will be seen sharply, and the image of the wire will be superimposed on the telescope image. Its apparent location will be at infinity.

A Unique Positive Lens

If two positive cylindrical lenses are placed close together so that their curves are 90° from each other, a lens equivalent to a spherical lens is formed (Figure 16-31). This lens has optical qualities similar to a simple spherical lens of the same diameter and focal length. The videotape lesson shows that this lens can be used as a projection lens. The image formed, while adequate for some purposes, actually suffers from the aberrations common to all positive single-element lenses.

Cylindrical-lens curves are more costly to grind and polish than spherical curves, so that a spherical lens costs less than a lens formed by two cylindrical curves at right angles to each other. While this is an interesting lens, there are no apparent practical uses for crossed cylindrical lenses.

Fresnel Cylindrical Lenses and Wedges

Spherical Fresnel lenses were discussed in a previous chapter. The spherical curve of a plano-convex lens is segmented and moved, segment by segment, to make a thin lens with many of the optical characteristics of a spherical lens. When large lenses with limited optical corrections are required, Fresnel lenses can be used. Fresnel lenses are used as spotlight lenses and as field lenses in rear-projection systems. A cylindrical lens can be made as a Fresnel lens. The cylindrical curve is moved over, segment by segment, to be close to the plano surface. The cross section of a spherical Fresnel lens cut through the center is similar to the cross section of a cylindrical Fresnel lens taken at right angles to the curve (Figure 16-32). While the spherical Fresnel lens segments are circular, the segments of a cylindrical Fresnel lens are straight lines—these segments are called *lenticules.*

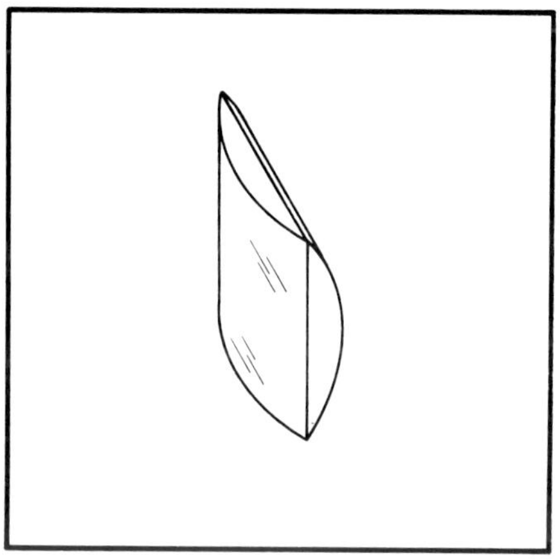

Figure 16-31. Crossed-Cylinder Positive Lens.

The same principle can be applied to make a Fresnel wedge prism. Each lenticule of a Fresnel wedge prism is identical. Figure 16-33 shows the wedge segments moved over to make a thin wedge.

There are only a few practical uses for Fresnel cylindrical lenses and wedges.

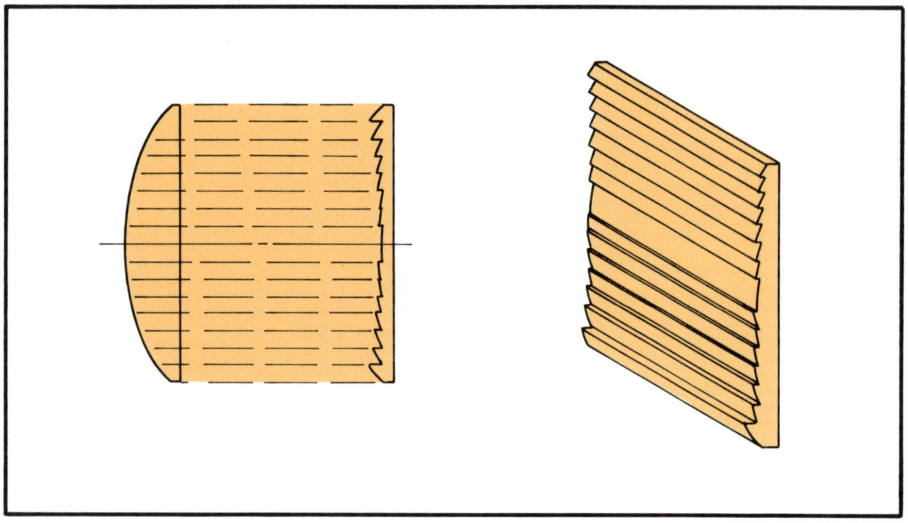

Figure 16-32. Fresnel Cylinder Lens.

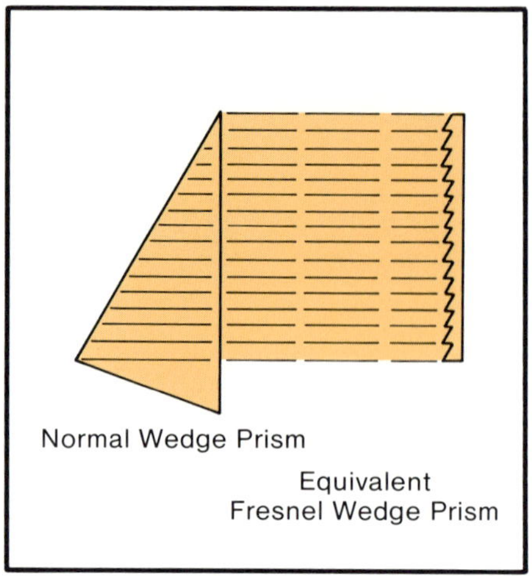

Figure 16-33. Fresnel Wedge Prism.

Problems

1. Using graphic ray tracing, establish the angle of minimum deviation with a wedge prism whose included angle is 30° and whose index of refraction is 1.5. As an alternative, you may calculate the angle using Snell's law.
2. Draw another ray with a slightly different angle, and show that the angle of deviation is greater than the angle of minimum deviation established in Problem 1. Use either graphic ray tracing or Snell's law.

NOTE: When using Snell's law, you will need to find the angle of minimum deviation once the angle of incidence of the entering ray and the angle of refraction of the exiting ray are known. An equation that gives this is:

∠Deviation = ∠Incidence (entrance ray) + ∠Refraction (exiting ray) − the wedge angle.

To find the angle of incidence of the ray inside the glass to the second surface of the wedge, use this equation:

Wedge Angle − ∠ of Refraction of Ray as it enters glass = ∠ of Incidence with the second wedge surface.

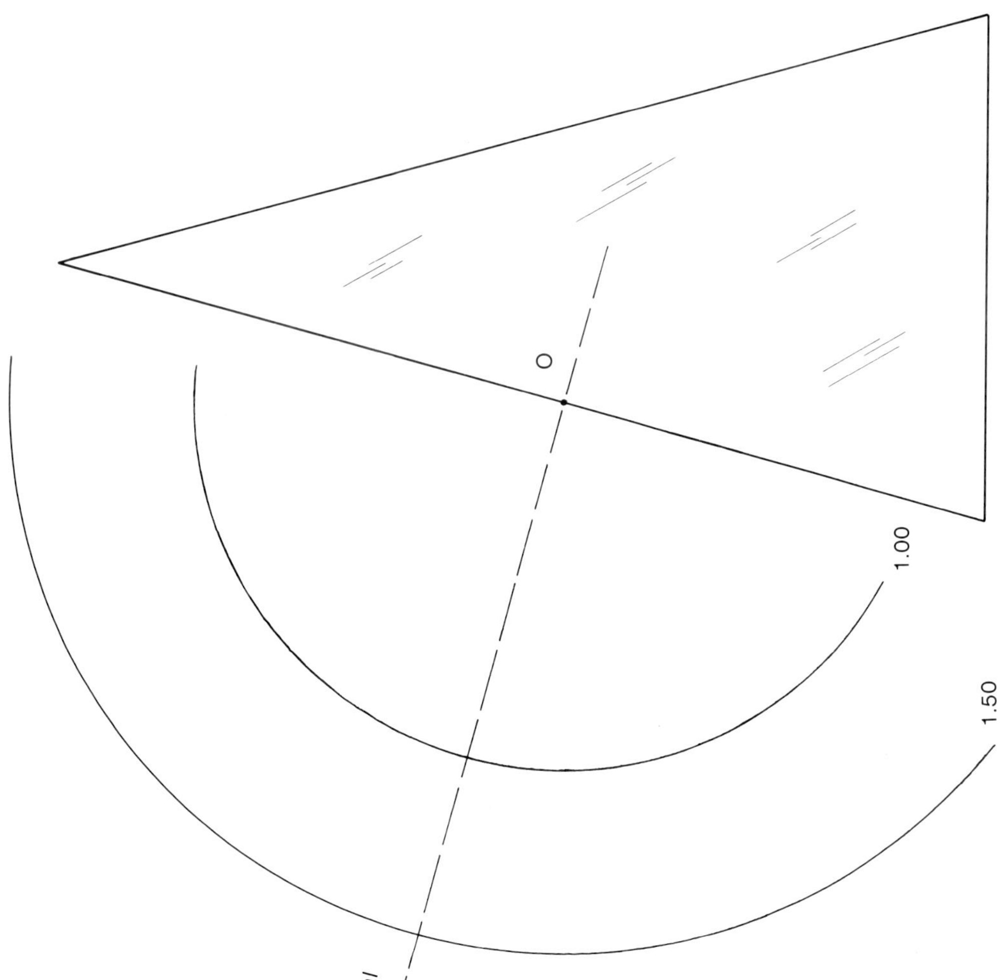

Problem 1
Chapter 16
Video Lesson 11

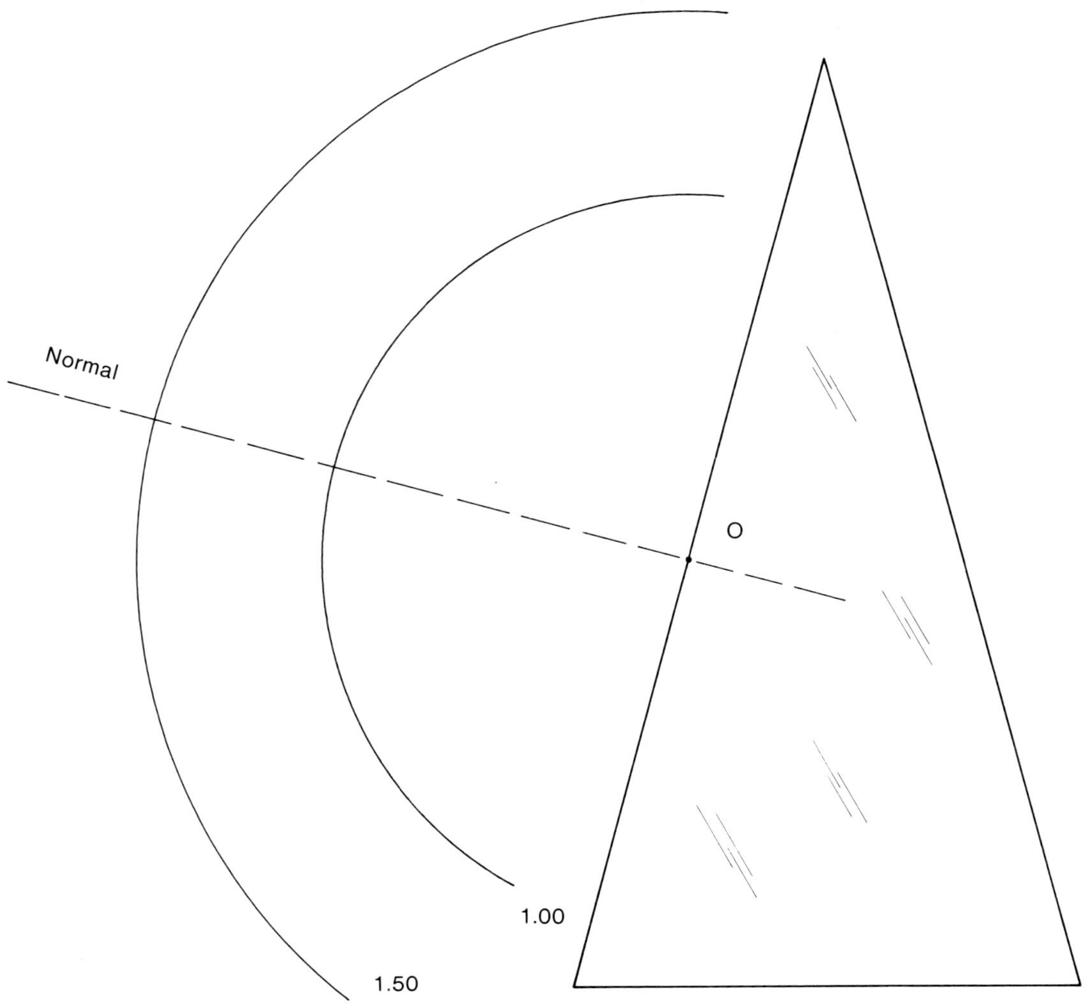

Problem 2
Chapter 16
Video Lesson 11

Chapter 17

VIDEO LESSON 12

Angles of Minimum Deviation in Wedge Prisms, Corrections for Chromatic and Spherical Aberration and Field Curvature

Summary

The ray path at the angle of minimum deviation of a wedge is ray traced and calculated using Snell's law. Tracing rays whose incident angles are slightly more and slightly less than that of the minimum deviation ray requires accuracy, but verifies the angle of minimum deviation. The problem of longitudinal chromatic aberration is studied and several solutions are described. Abbe's glass constant, based on the index of glass at certain wavelengths, is discussed. Fraunhofer's discovery of the solar spectrum lines, and their use by Abbe, are described. The Abbe numbers of various optical glasses are given. The glass-type numbers of optical glasses are based on their index to sodium light and their Abbe number (measure of dispersion). An unusual solution to the correction of chromatic aberration of a lens by means of a glass block is introduced. The effect of a glass block on spherical aberrations is discussed and several methods of correcting spherical aberration are described. A large negative lens, placed directly in front of the film plane, can be used to correct for the field curvature of a lens.

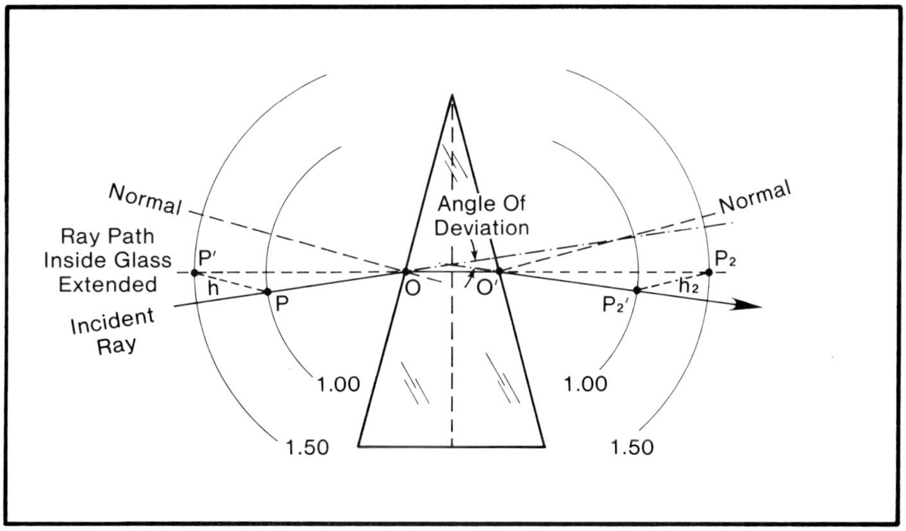

Figure 17-1. Finding The Angle Of Minimum Deviation Using Graphic Ray Tracing.

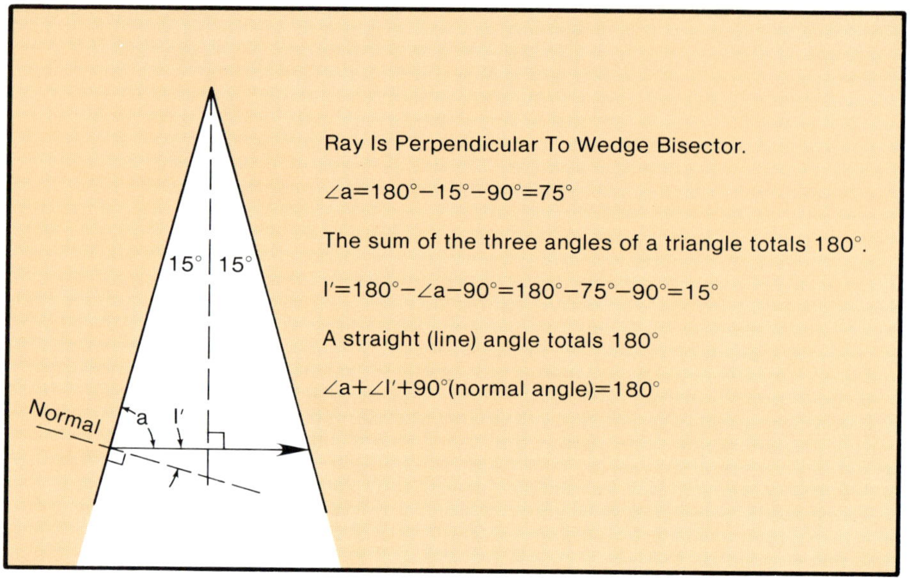

Figure 17-2. Finding I′ When Ray Is Perpendicular To The Wedge-Angle Bisector.

The second part of the problem is to verify that this is the angle of minimum deviation (Figure 17-3). The same procedure is used, however, the paths of the rays must be followed in sequence. A ray with an incident angle of 15° is drawn. Using the 1.0 and 1.5 arcs, the path inside the glass is found. Using two additional arcs, the exiting ray path is found and its angle of incidence is shown. By extension, the angle of deviation is measured to be just about 16°. This is slightly greater than the angle of minimum deviation.

Figure 17-4 shows a ray entering the wedge prism at 30° (slightly more than the angle of minimum deviation). The new angle of deviation is found by using the same method—and is found to be slightly less than 16° (also slightly greater than the angle of minimum deviation). By example, this is an indication that when the ray path in the prism is perpendicular to the wedge angle bisector, the angle of minimum deviation is found.

Problem Review: Angle of Minimum Deviation

Figure 17-1 shows the 30° wedge made of glass with an index of refraction of 1.5. The problem is to find the angle of minimum deviation. Using graphic ray tracing to solve the problems, the first step is to draw the ray path inside the glass. It is known that the path inside the glass is perpendicular to the bisector of the included wedge angle (shown as a dashed line). Since the wedge angle is 30°, the bisector is 15° from each wedge surface (Figure 17-2).

The entering ray strikes the left wedge surface at Point O (Figure 17-1). With O as a center, two arcs are drawn—one with a radius of 1.0 and the other with a radius of 1.5. They could be any convenient radii, so long as the 1.0 to 1.5 ratio is maintained. The normal is drawn to the wedge surface at Point O, and the ray path in the glass is extended to the left as a dashed construction line. Since the ray is being calculated as though it were going from the glass to air, the construction ray is drawn to the 1.5 arc. This intersection is shown as P′.

A construction line is drawn from P′ to the 1.0 arc parallel to the normal. This intersection is shown as P. A straight line is drawn from Point O through Point P. This is the ray path. We find the angle of incidence measured to the normal to be just under 23°.

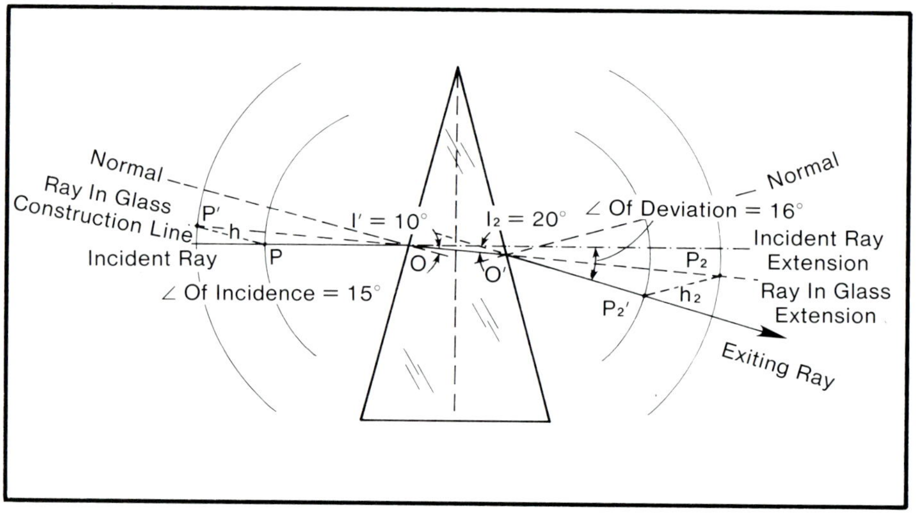

Figure 17-3. The angle of incidence is less than the angle of incidence when the angle of deviation is at minimum.

At the angle of minimum deviation the ray exiting from the prism leaves at the same angle to the second surface that the incident ray makes to the first surface. Therefore, the refracted ray can be drawn at just under 23° to the normal to the second surface, as well. The drawing also shows a graphic ray trace locating the exiting ray.

The incident and exiting rays are extended. The angle between the incident ray and the exiting ray is the angle of minimum deviation. It measures just under 16°.

Finding the angle of minimum deviation using Snell's law involves several steps. The first step is to find the angle of incidence when the angle of refraction inside the prism is 15°. Figure 17-5 shows how this is found.

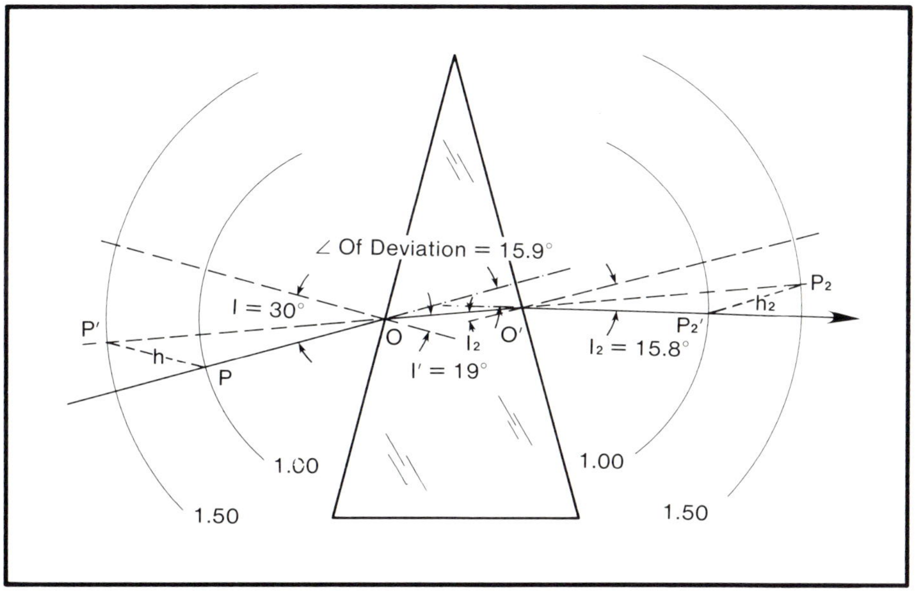

Figure 17-4. Finding The Angle Of Deviation When Angle Of Incidence Is Greater Than The Angle Of Incidence At Minimum Deviation Condition.

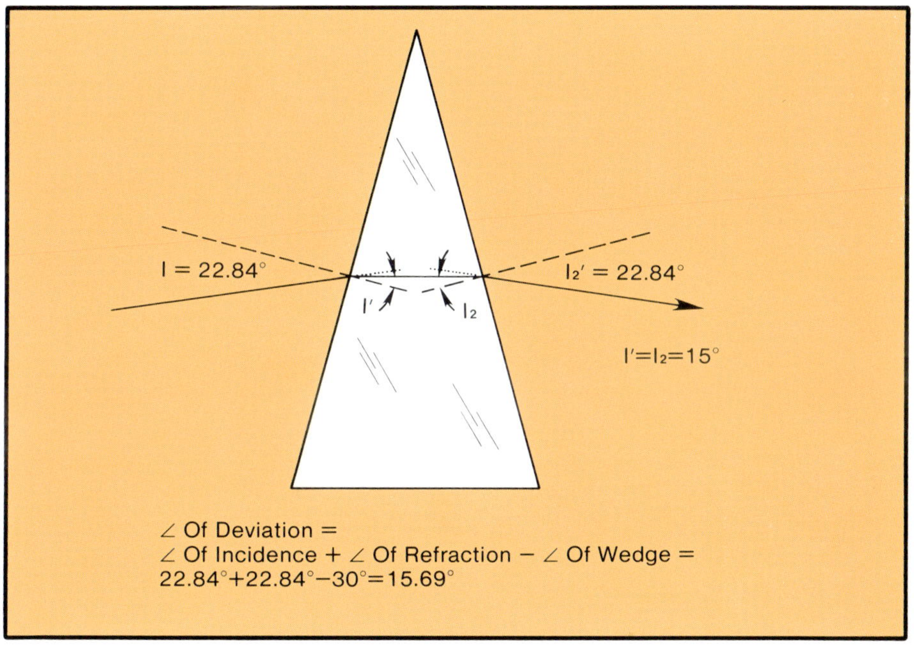

Figure 17-5. Geometry Used To Find Angles With Snell's Law For Minimum Deviation Condition.

$$N \sin I = N' \sin I'$$

$$1.0 \times \sin I = 1.5 \sin 15°$$

$$\sin I = 1.5 \times 0.2588 = 0.3882$$

$$I = 22.844°$$

By symmetry, the angle of refraction of the exiting ray is the same—22.84°. Using the equation given with the problem, the angle of minimum deviation is found.

∠deviation = ∠incidence (entering ray) + ∠refraction (exiting ray) − wedge angle.

∠deviation = 22.84° + 22.84° − 30°

∠deviation = 45.69° − 30° = 15.69

This is the angle of minimum deviation.

To find the angle of deviation when the incident angle of the entering ray is slightly less than 22.84°, we chose 15° as the angle of incidence (Figure 17-6).

$$N \sin I = N' \sin I'$$

$$1.0 \sin 15° = 1.5 \sin I'$$

$$\sin I' = \frac{\sin 15°}{1.5} = \frac{0.2588}{1.5} = 0.1725$$

$$I' = 9.94°$$

Geometrically, the ray in the glass has an angle of incidence of 20.06° with the second wedge surface.

$$1.5 \sin 20.06° = 1.0 \sin I'$$

$$\sin I_2' = 1.5 \times 0.341 = 0.5146$$

$$I_2' = 30.97°$$

Using the equation to find the angle of deviation:

∠deviation = ∠incidence + ∠refraction − 30° = 15° + 30.96° − 30° = 15.97°

This is slightly greater than the 15.68° of the angle of minimum deviation found earlier.

Next we chose an angle of incidence of 30°, greater than the 22.84° angle of incidence at the angle of minimum deviation.

$$N \sin I = N' \sin I'$$

$$1.0 \sin 30° = 1.5 \sin I'$$

$$\sin I' = \frac{\sin 30°}{1.5} = \frac{.5}{1.5} = 0.3333$$

$$I' = 19.47°$$

Using geometry we find that the ray in the glass has an angle of incidence I_2, of 30° − 19.47° = 10.53° with the second wedge surface.

$$N \sin I = N' \sin I'$$

$$1.5 \sin 10.31° = 1.0 \sin I_2'$$

$$\sin I_2' = 1.5 \times 0.1827 = 0.2741$$

$$I_2' = 15.91°$$

The ∠deviation = 30° + 15.91° − 30° = 45.91° − 30° = 15.91°. This is only slightly greater than the angle of minimum deviation of 15.69°.

It is understandable if you found no significant difference in your graphic ray traces. Changing the incident angle of the entering ray by even 5° changes the angle of deviation by only a fraction of a degree.

Correcting Chromatic Aberration in Lenses

White light is defined as a fairly even

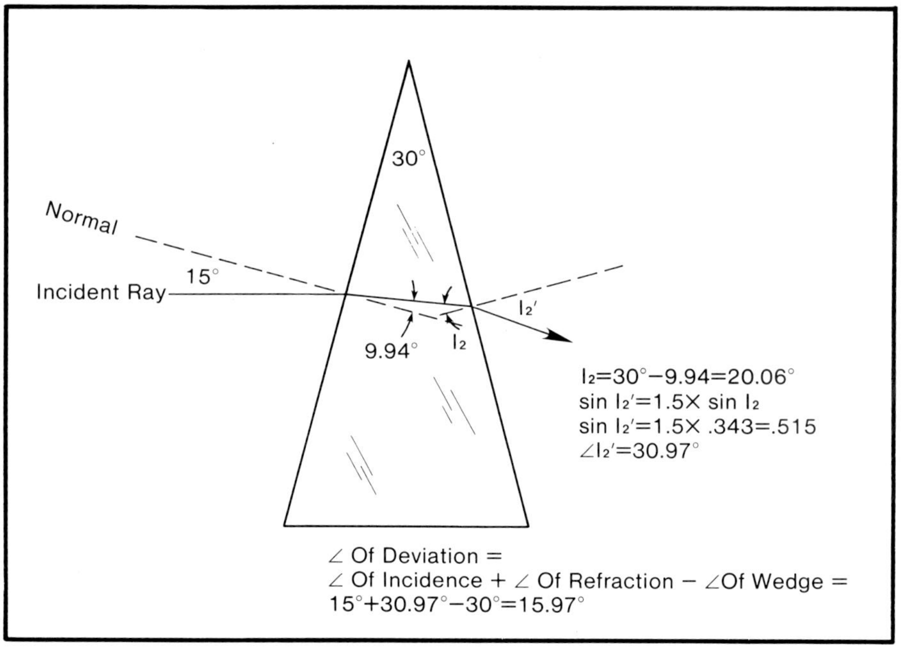

Figure 17-6. Geometry Used To Find Angles For Use With Snell's Law For An Angle Of Incidence Less Than The Angle Of Incidence At Minimum Deviation.

mix of all wavelengths of light. When a ray of white light is allowed to go through a prism, the wavelengths are separated into the colors of a rainbow.

The outer edge of a positive lens is shaped similar to a prism. If a ray of white light, parallel to the axis, passes through the outer edge of a positive lens, the blue rays are refracted more than the red rays. The focal length of the lens is shorter with blue light than with red light.

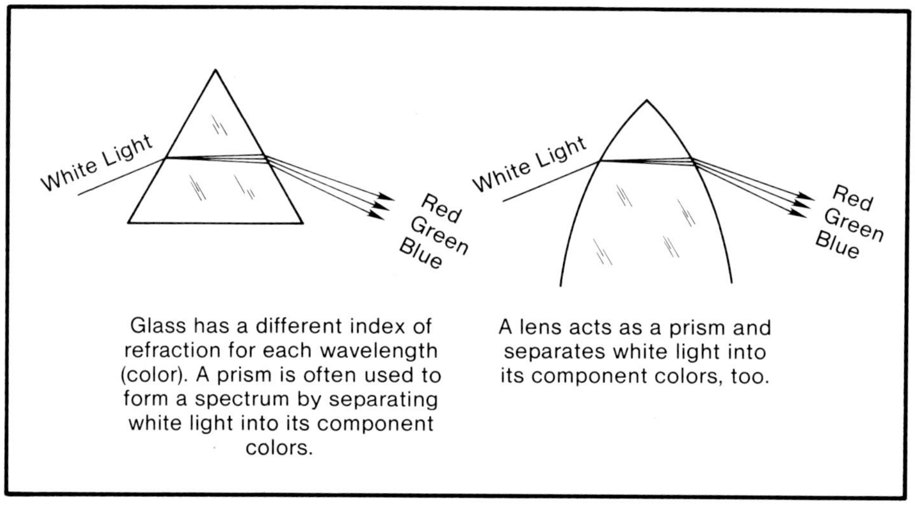

Figure 17-7. Dispersion.

In Figures 17-7 through 17-9 the degree of chromatic aberration is exaggerated to show the difference. In most of the schematic ray tracing with thin lenses, chromatic aberration, along with other aberrations, was ignored to show the basic principles. With real positive lens elements, the amount of chromatic aberration can be measured in tenths of a millimetre (Figure 17-8). This is enough to significantly reduce the resolving power of a lens. An uncorrected telescope objective lens with a long focal length, has enough chromatic aberration to seriously affect the image quality.

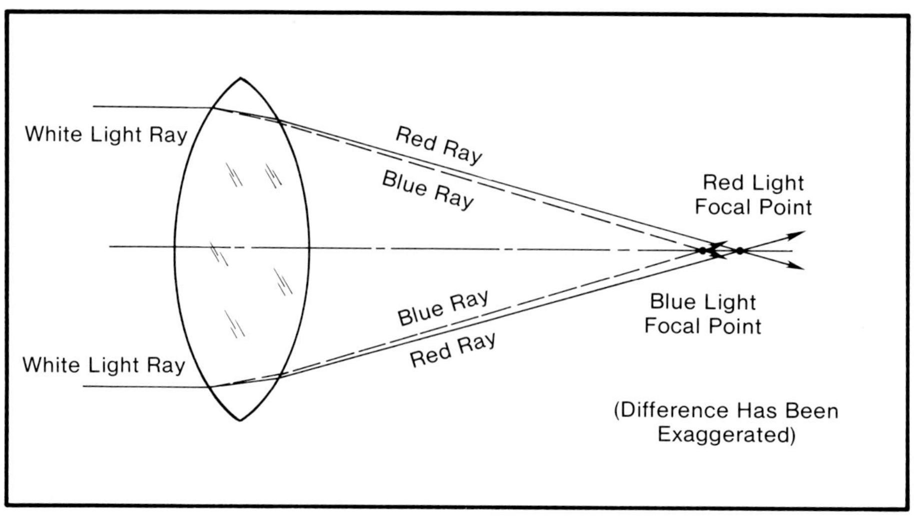

Figure 17-8. Longitudinal Chromatic Aberration.

A German scientist named Abbe, of the Jena glass works, was a leader in the early development of optical glasses. He developed the basic approach of "achromatizing," (correcting), lenses for chromatic aberration.

Abbe found that, in general, the higher the index of refraction of a glass, the greater its dispersion. Dispersion is the characteristic of glass that has to do with difference in index of refraction with different wavelengths. If a glass has very little difference in index between red and blue light, it has a low dispersion. If it has a great difference between red and blue light, it has a high dispersion. However, there are a few exceptions–glasses with low indices and high dispersion, and glasses with high indices and low dispersion.

A negative lens, because of its shape, bends the blue rays outward more than it bends the red rays

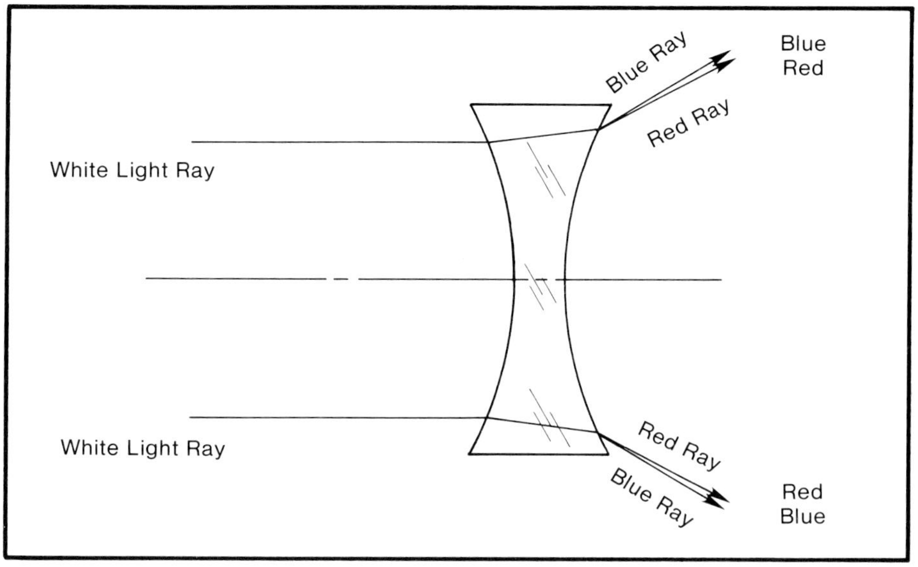

Figure 17-9. Dispersion By A Negative Lens.

(Figure 17-9).

If a positive lens has a surface with the same radius of curvature as a concave surface of a negative lens of the same power, and they are placed in contact (like a curved window) the positive power of the positive lens is cancelled by the negative power of the negative lens.

Abbe found that if you made a positive lens fairly strong (of a glass with relatively high index and low dispersion) and negative lens fairly weak (made of a glass with a lower index, but higher dispersion), the combination had the same focal length for red and blue light (Figure 17-10). It was corrected for longitudinal chromatic aberration. This discovery was very useful, but lenses of this type had to be made on a trial and error basis. He felt that there was a need to give a measure to the degree of dispersion in a glass.

At about the same time (1820's) a man named Fraunhofer was studying the solar spectrum using diffraction gratings made of fine silver wire. These silver wires were spaced evenly by winding them around finely-spaced screw threads. He discovered the absorption lines in the spectrum of sunlight. He was able to relate these lines to certain chemical elements. The dark lines in the Fraunhofer spectrum correspond to the wavelengths emitted when the chemical element is burned. Abbe used three of the wavelengths of spectral lines when setting up his measure of dispersion.

Color	Element	Line Name	Wave-length
Light blue	Hydrogen	F	4861 Å
Yellow	Sodium	D	5893 Å
Red	Hydrogen	C	6563 Å

He obtained light to make these measurements by burning hydrogen gas and salt (sodium chloride) as shown in Figure 17-11.

The Abbe constant, or V number, is an equation.

$$\text{V number} = \frac{n_D - 1}{n_F - n_C}$$

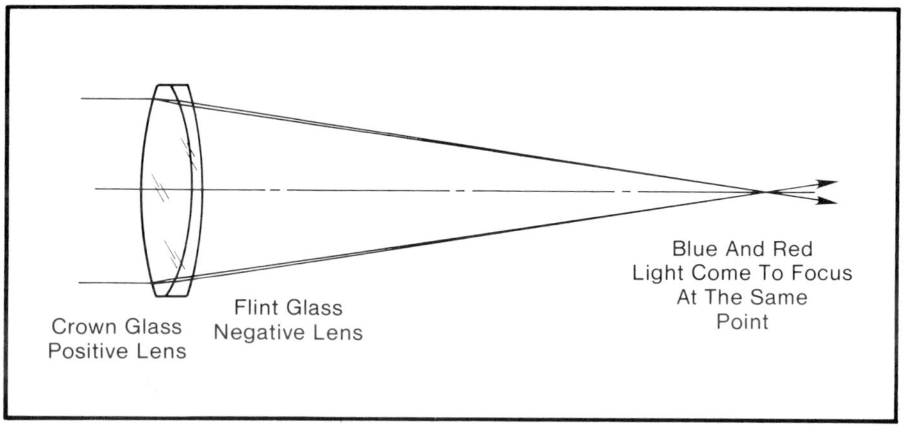

Figure 17-10. Color-Corrected Telescope Objective.

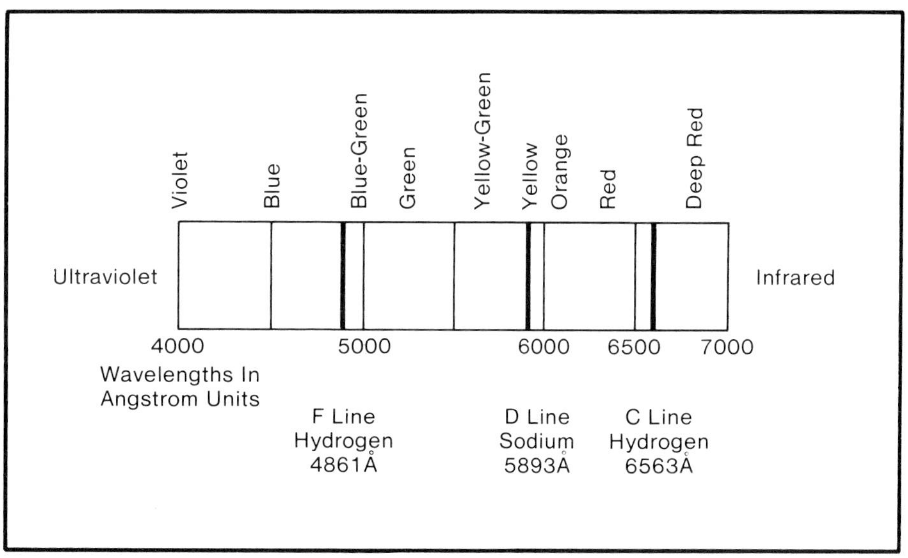

Figure 17-11. The Visible Spectrum Showing The Location Of The Three Abbe Lines.

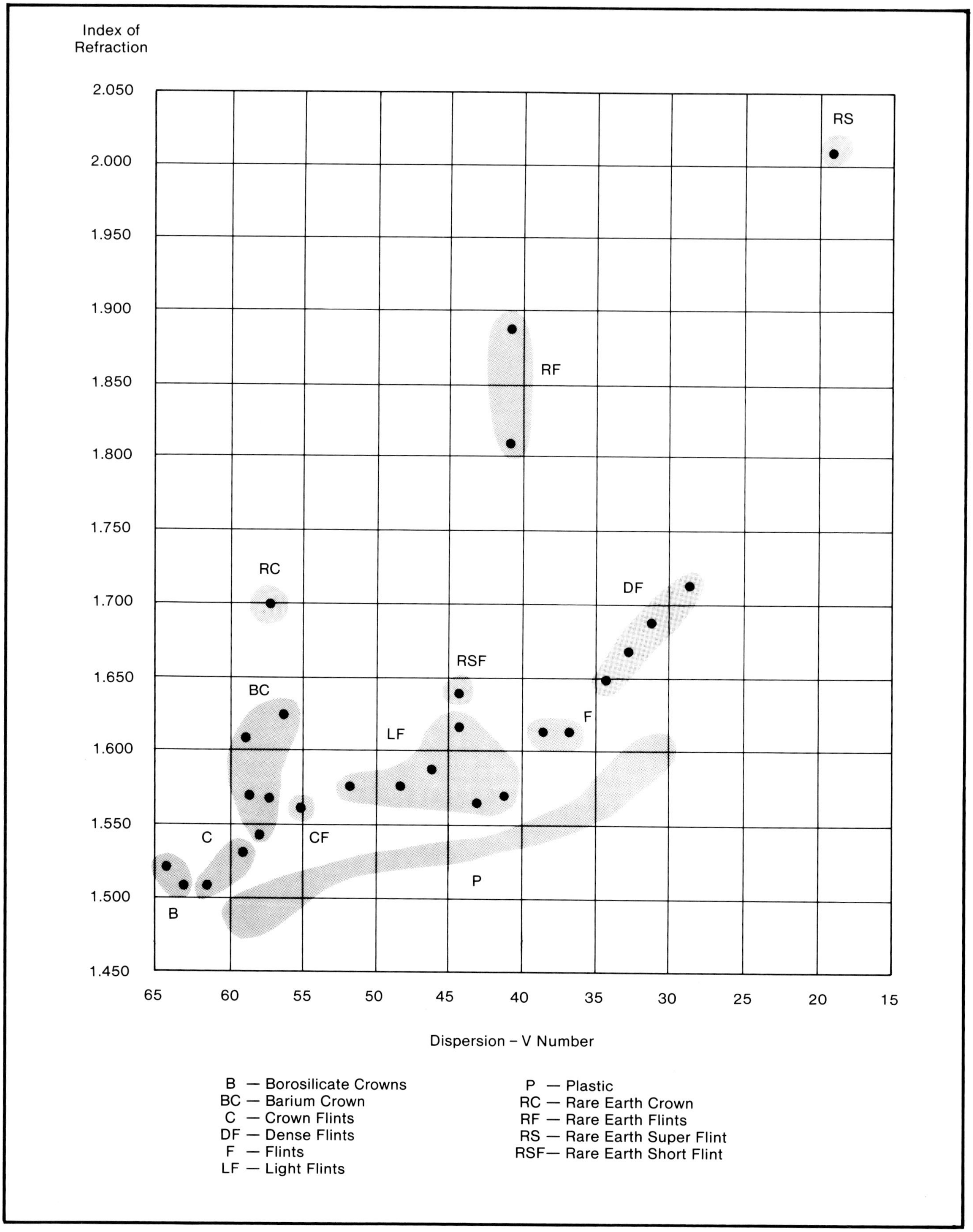

Figure 17-12. Optical Glasses, Index And Dispersion.

	n_F 4861 Å	n_D 5893 Å	n_C 6563 Å
A crown glass	1.5230	1.5170	1.5145
A dense flint glass	1.6637	1.6499	1.6444

The indices of two glasses are shown at each of the three wavelengths in the table above:

The Abbe dispersion numbers are calculated:

$$V\ (\text{crown}) = \frac{1.5170 - 1}{1.5230 - 1.5145}$$

$$V\ (\text{crown}) = \frac{.5170}{.0085} = 60.82$$

$$V\ (\text{flint}) = \frac{1.6499 - 1}{1.6637 - 1.6444}$$

$$V\ (\text{flint}) = \frac{.6499}{.0193} = 33.67$$

The flint glass (which has a high dispersion) has a low V number, while the crown glass (with a low dispersion) has a high V number.

In Figure 17-12, the graph shows the index (n_D) of a number of optical glasses plotted against their V numbers. Note that, in general, there is a curving line of optical glasses going up in index while the V number goes down. Note that there are some exceptions, such as the barium crown glasses, which have indices equivalent to many flint glasses, but dispersions equal to crown glasses.

Glass-Type Numbers

The optical industry often classifies glasses by a six digit glass type number. The first three digits are the first three digits after the decimal point of the n_D value of the glass. The last three digits are the value of the V number to one decimal place. In the example shown below, the glass type number for ordinary crown glass is shown.

$$n_D\ (\text{index}) = 1.5125$$

$$V\ (\text{Abbe number}) = 60.5$$

$$\underbrace{5\,1\,2}_{n_D}\ \underbrace{6\,0\,5}_{V}$$

The glass type numbers of several typical glasses are shown below:

Glass	**Glass Type Number**
Borosilicate crown	511635
Barium crown	541595
Extra Dense Barium crown	611588
Crown Flint	523551
Telescope Flint	528516
Light Flint	558455
Dense Flint	649338
Extra Dense Flint	720293

Producing a Spectrum

It is sometimes useful to actually produce a spectrum by passing light through a prism so that it is dispersed into the spectral colors.

If an unlimited choice of conditions existed, a 60° prism made of a high dispersion glass (low V number) would be ideal. The large angle tends to increase the spectral spread, and does not give total reflection. The high degree of dispersion also gives a wide separation of the colors.

However, right-angle prisms, usually made of a crown glass, can serve the purpose very well. A slide projector with a projection lens can serve as a light source (Figure 17-13). The incandescent bulb produces a continuous spectrum with all the colors in it.

It might be thought that a wide slit would produce a brighter spectrum. It does, but as shown in the video lesson, it produces a wide band of white in the spectrum with color fringes on the edges of the white band. A very narrow slit is required.

Such a slit can be made by using two pieces of aluminum foil with sharply cut edges. They are located in a slide mount with a very narrow space between them. An additional slit placed on the lens, parallel to the slit in the slide, will result in a spectrum with even purer colors.

This slide is placed in the projector with the slit positioned horizontally. The projector lens is focused to place an image of the slit on the screen (which is placed near the projector) and then is aimed upward at about a 45° angle. The prism is held in the light beam emitted by the

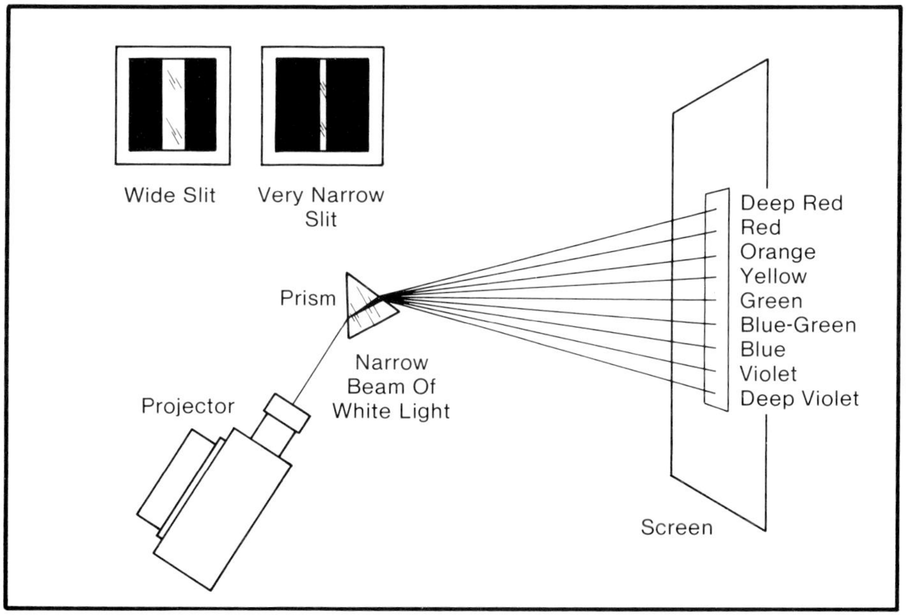

Figure 17-13. Using a projector, slit, and prism to produce a spectrum.

projector and rotated until the image of the slit is seen on the screen. The image appears as a long narrow spectrum and all the spectral colors (violet, blue, blue-green, green, yellow-green, orange, and red) are distinguishable.

Another Method of Achromatizing

The usual method of color correcting is to match a weak negative lens made of flint glass with a positive lens made of crown glass (i.e., achromatizing a lens) as shown in Figure 17-10. There is, however, another way that might have some applications.

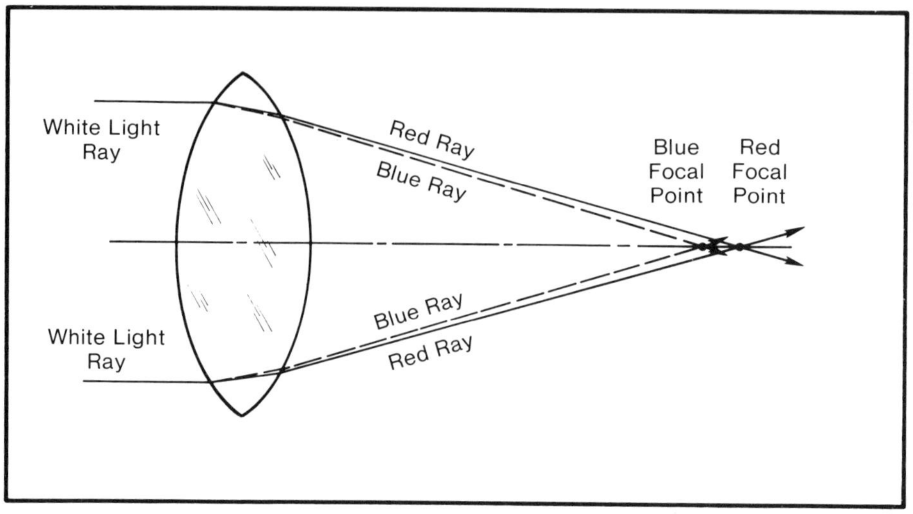

Figure 17-14. Chromatic Aberration Of An Uncorrected Positive Lens.

Figure 17-14 shows a simple bi-convex lens imaging a ray of light parallel to the axis. The red light (Hydrogen C light) and blue light (Hydrogen F light) rays are shown. The blue light is refracted more by the lens than the red light because the glass has a higher index of refraction to the blue light.

In Figure 17-15, a block of glass is placed in the light path between the lens and the focal point. In an earlier chapter, when a glass-plate beam splitter was placed in the beam of light, it was stated that astigmatism resulted and that the thicker the glass, the more astigmatism. This only occurs when the glass plate is at an angle of about 45° to the axis. When the two surfaces of the glass block are perpendicular to the lens axis, no astigmatism is created.

In the previous context it was also stated that the rays emerge from a glass plate parallel to the incident ray when the surfaces of the glass plate are parallel.

In Figure 17-15, you can see that this refraction and parallel emergence from the glass block have brought the focus of the red ray and the blue ray together, therefore correcting the chromatic aberration of the uncorrected lens. The power of the lens, the dispersion of the glass from which the lens is made, and the thickness of the glass block must be matched for an exact correction. Even if these elements are not matched, some correction will always take place.

A word of warning: If a glass block is placed behind a lens that is already color corrected, the chro-

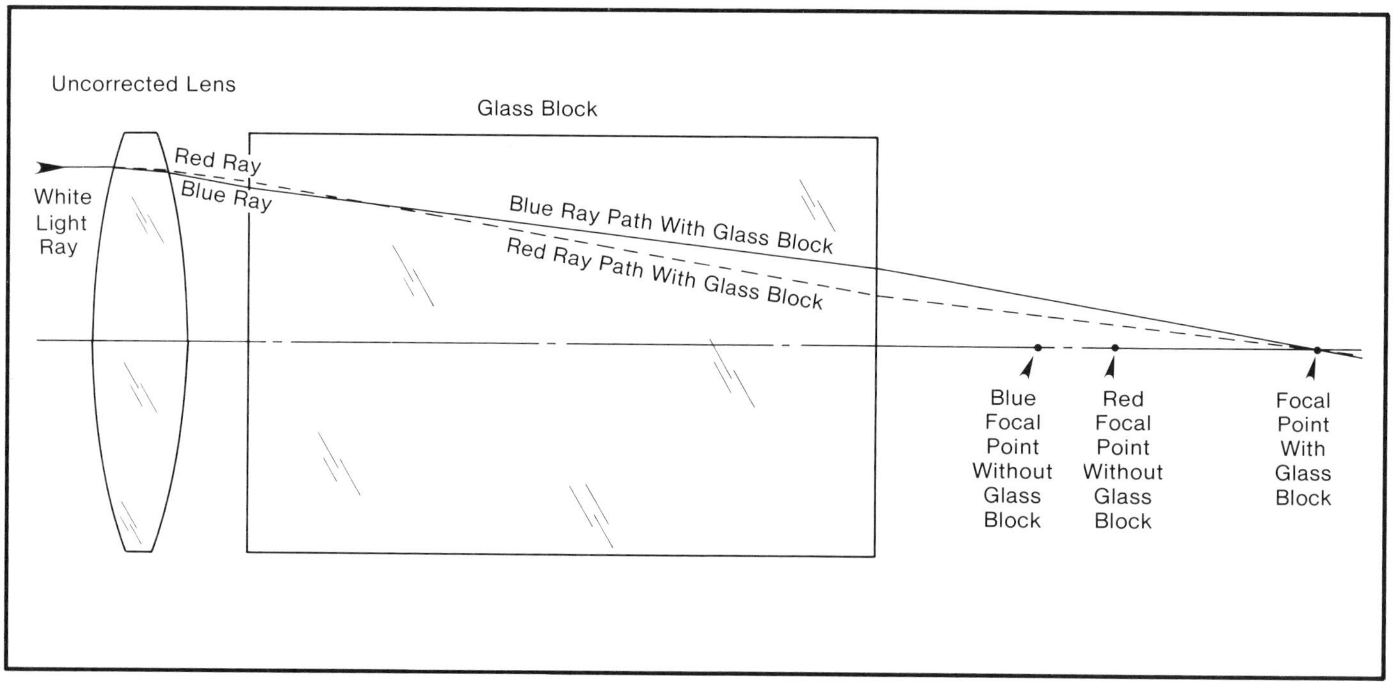

Figure 17-15. Glass Block Correction Of Longitudinal Chromatic Aberration.

matic aberration will be over corrected—that is, the blue ray will focus farther from the lens than the red ray. This is something to watch out for when designing optical systems.

Optical Systems with Glass Blocks

One might question whether glass blocks are actually used in optical systems. Right-angle prisms are commonly used in many types of optical instruments. An optical beam going through a prism is essentially going through a glass block.

In Figure 17-16, a right-angle prism is shown behind a lens. Several image forming light rays are shown.

A dashed area representing an equivalent glass block is shown behind the prism with dashed ray paths going through it. The rays go through the prism in the same manner they would go through a glass without a reflecting surface.

One example of an optical instrument with prisms is a binocular. The light from each objective lens goes through a pair of Porro prisms before being imaged at the field lens. The eyelens is a magnifier which enlarges any chromatic aberration along with the image. In a well designed binocular, the objective lenses must be under-corrected for longitudinal chromatic aberration because of the correction influence of the Porro prisms.

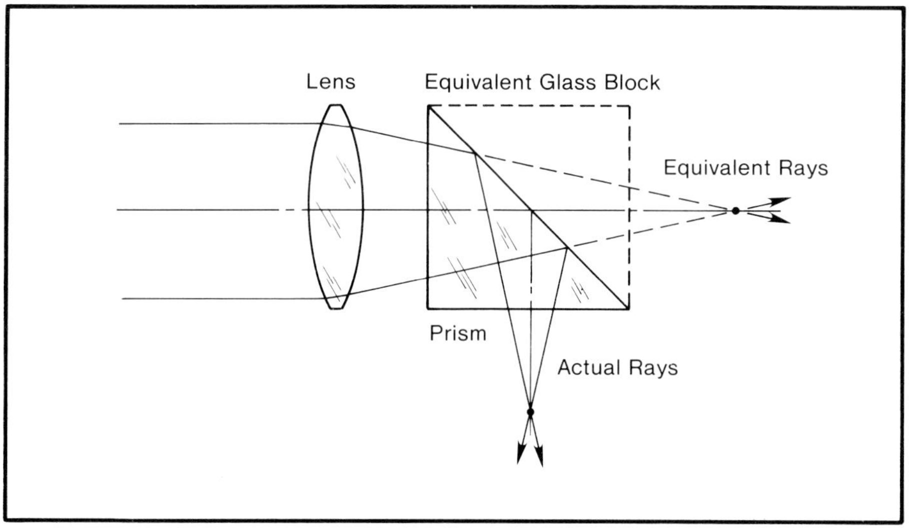

Figure 17-16. A prism is equivalent to a glass block.

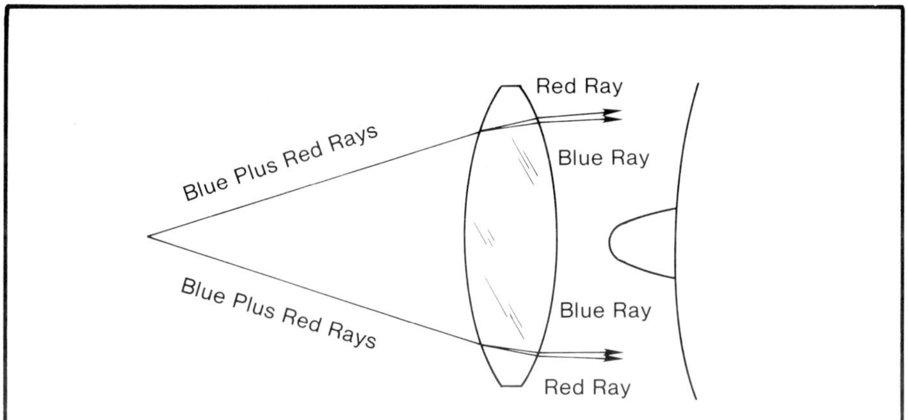

Blue rays are bent more than the red rays. The blue rays reaching the eyes are parallel, in this example, and appear to be coming from a distant object. The red rays, bent less, are diverging and are interpreted as coming from a close object. Thus a red object appears closer than a blue object.

Figure 17-17. A magnifier with chromatic aberration makes red objects appear to be closer than blue objects.

Stereo by Chromatic Aberration

If a simple magnifying lens is large enough to be used with both eyes, the chromatic aberration in the lens will cause a stereo effect. On a road map, for example, the red road lines appear to be elevated above the plane of the map. Blue and black road lines appear to remain on the paper.

Figure 17-17 shows how the chromatic aberration of the lens causes the rays of light from a red dot to focus closer to the lens than a blue dot. This causes an angular difference in the rays that the eyes see as a difference in distances from the lens, therefore giving a three-dimensional effect.

Slides can be designed to make use of this effect. A red rose against a blue sky or against a green foliage background will appear to be closer than the background when viewed with a two-eyed magnifying lens. However, a blue object against a red background gives the opposite effect; the blue object appears to be "behind" the background.

Chromatic Aberration of the Eye

The lens in the human eye is not achromatized. For this reason, red objects appear closer than they actually are and blue objects appear farther than they actually are.

If a red (Wratten No. 25), a green (Wratten No. 58) and a blue (Wratten No. 47B) filter are masked and placed over an illuminator and viewed in a darkened room from a distance, the red filter will appear definitely closer than the green, and

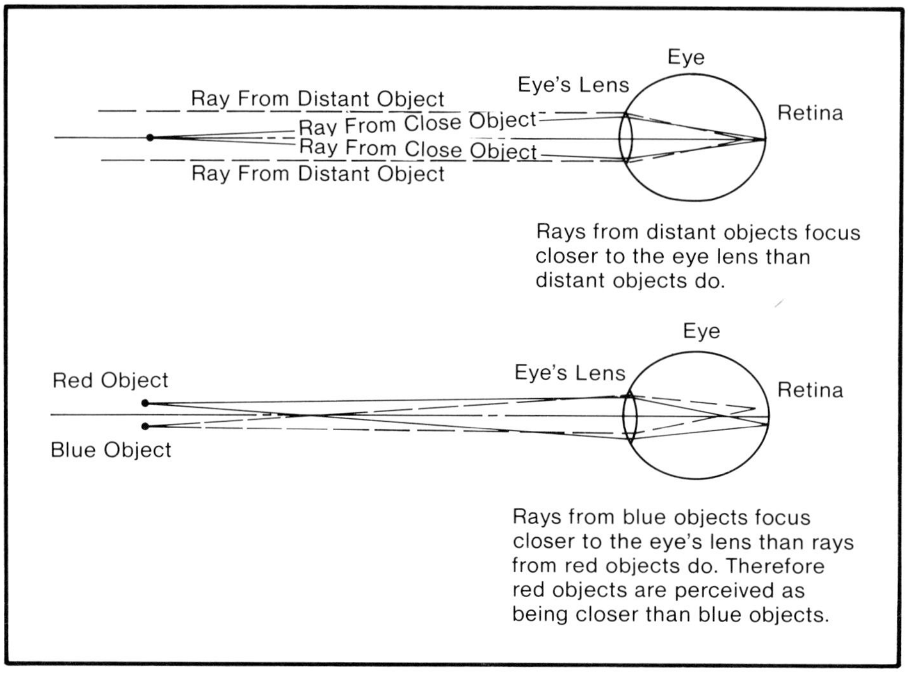

Figure 17-18. How The Eye Lens's Chromatic Aberration Makes Red Objects Appear Closer Than Blue Objects.

the blue will appear farther away than the green. The spacial difference will be very noticeable.

This is because the eyes' lenses actually bring the blue rays to focus closer to the lenses in the eyeballs than the red rays (Figure 17-18). When a close object is viewed, the rays from the object focus farther back. (In a camera, the lens to focal plane distance is also greater when a close object is focused than when a distant object is focused.) Thus, the red rays take a path similar to the rays from a close object. In order to bring the red rays in focus on the retina, the lens must thicken. The eye interprets this as being "close."

Artists and interior decorators call reds and oranges "advancing colors," while they call blues "receding colors," because the warm colors look closer than the cool colors.

Spherical Aberration

In most of the thin-lens examples given earlier it was assumed that rays parallel to the axis, going through all portions of a lens, come to a focus on the axis at the same point. It has just been seen that this is not true for light rays of different colors. It is also not true because of an aberration called spherical aberration.

In Figure 17-19, two rays are shown going through a single positive lens. The ray going through the central portion of the lens actually comes to a focus farther from the lens than the ray going through the outer portion of the lens. When a film is placed in the system there will be no place where the image of the subject point forming the parallel rays can be imaged as a point. This aberration is caused by the spherical surfaces of the lens.

With the object at infinity, and if a plano-convex lens (Figure 17-19a) is placed so that the curved side faces the object and the plano side faces the film (image plane), the spherical aberration becomes least. If the lens is reversed (Figure 17-19b) so that the plano side faces the object, the aberration is greatest. If a double convex lens (Figure 17-19c) is used with the same power (focal length), the aberration is intermediate. As a general rule, if the bending power at each of the spherical surfaces is the same, the spherical aberration is least.

As the outer rays going through a lens are removed (the lens is stopped down by putting a diaphragm in the system), the amount of spherical aberration is reduced (Figure 17-20).

Spherical aberration is not always a disadvantage. For example, portraits of people are often better if

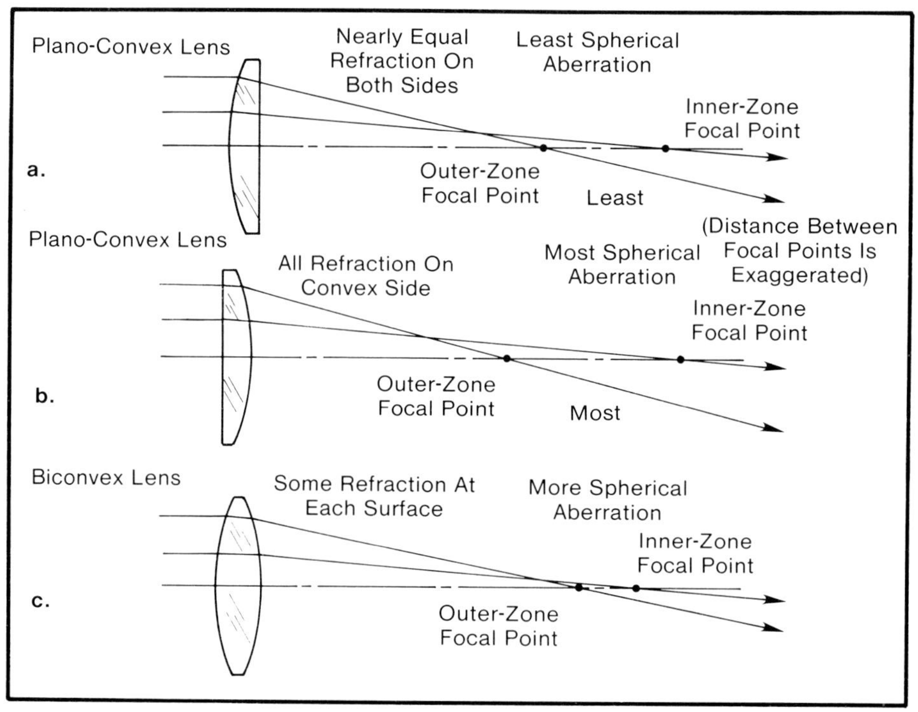

Figure 17-19. Spherical Aberration.

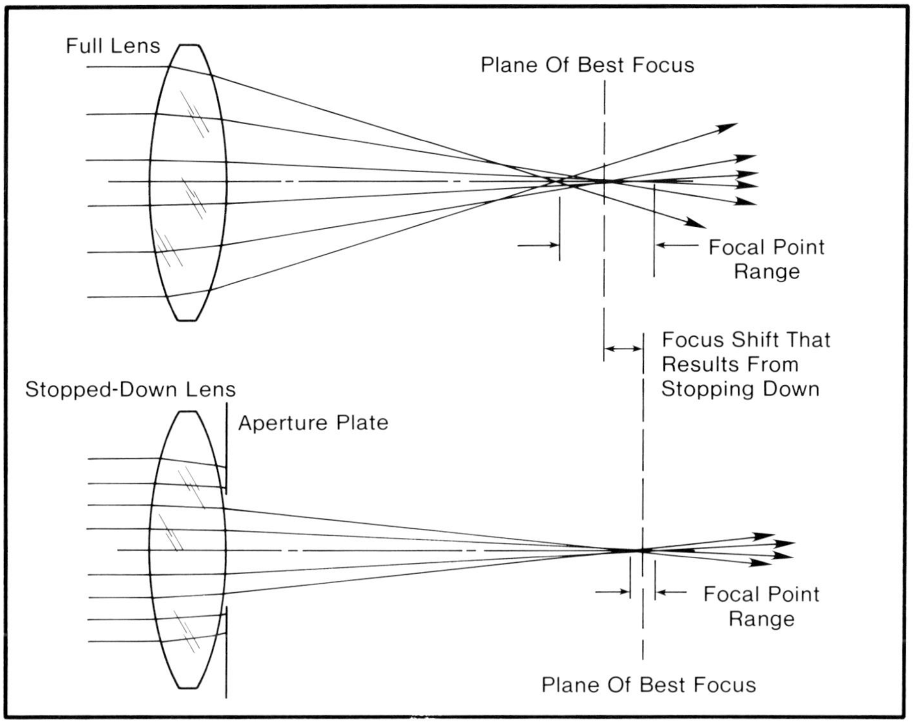

Figure 17-20. Image Improvement Resulting From Using A Small Aperture.

there is a degree of image softness that reduces the imaging of wrinkles and other skin defects. Portrait lenses are designed to have spherical aberration to provide this image softness. Because the amount of spherical aberration changes with the size of the diaphragm opening, the amount of softness is variable. An $f/4.5$ portrait lens used wide open produces a very soft image. Stopped down to $f/16$, the image is almost as sharp as that created by a fully-corrected lens.

In the portrait lens shown in Figure 17-21, several points may be considered. The lens is made of two elements cemented together. It is an achromatic lens corrected for chromatic aberration. The positive element is crown glass and the negative element is flint glass. It is designed so that the F and C hydrogen line wavelengths come to focus at the same distance from the lens.

The lens is meniscus shaped, and is placed behind the diaphragm. Such a lens shape produces a relatively flat field (see next section). A flat field is a requirement for camera lenses so that the sharpest possible image will be obtained all across the film. In the case of a portrait lens, about the same degree of softness will be recorded across the film.

In a rear-meniscus lens, such as the portrait lens, the principal plane is between the lens and the film. The back focus of the lens is greater than the equivalent focal length. The lens must be a considerable distance from the film.

Early box cameras used this type of achromatic doublet lens stopped down to about $f/16$. Such box cameras were necessarily quite long compared to the film size (Figure 17-22).

In order to reduce the camera size, later box cameras were made with front-meniscus lenses. In these, the principal plane is in front of the lens, resulting in a relatively short distance between the lens and the film. Unfortunately, used this way, considerable field curvature results. To compensate for this the roll film was held in a cylindrical curve by the camera which approximated the spherical image field produced by the lens.

When designing high aperture lenses for modern cameras, spherical aberration is corrected by the use of multiple elements. However, nearly all lenses exhibit a small degree of spherical aberration when the lens is wide open. This lessens when the lens is stopped down and is virtually gone by the time the lens is stopped down two or three stops. For example, most $f/2$ lenses are free from spherical aberration when stopped down to $f/5.6$.

In an earlier section of this chapter, the chromatic aberration of an uncorrected lens was improved by placing a glass block between the lens and the image. The problem for this chapter investigates what hap-

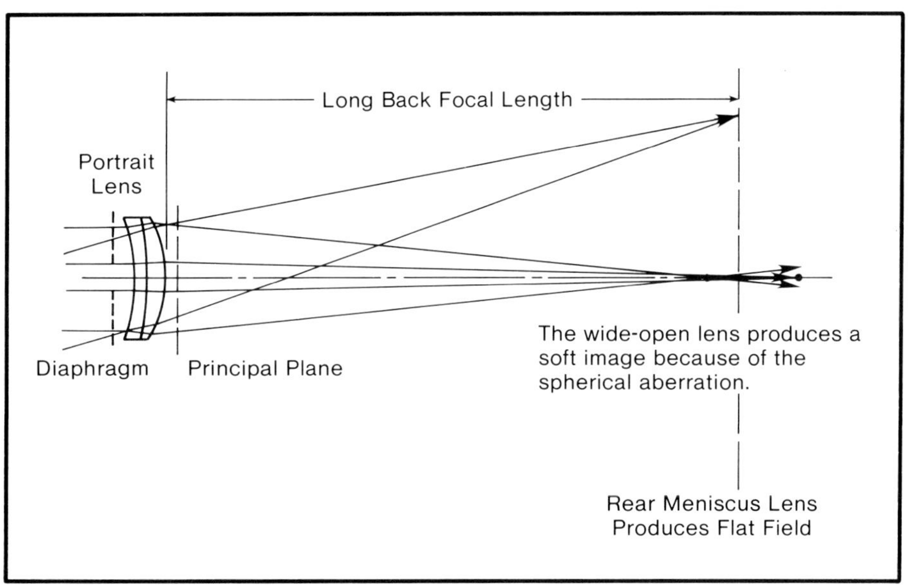

Figure 17-21. Achromatic Rear-Meniscus Portrait Lens.

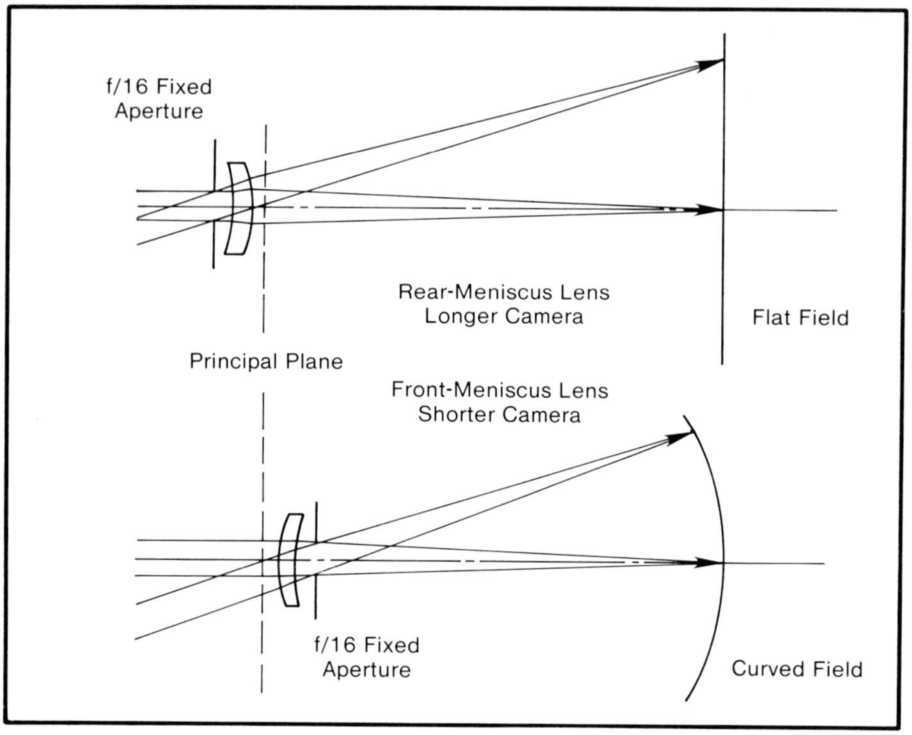

Figure 17-22. Comparison Of Rear- And Front-Meniscus Camera Lenses.

pens to spherical aberration when a glass block is placed behind the lens.

Field Curvature

In the previous section, two methods for correcting field curvature were discussed. One method used a meniscus-shaped lens as a rear-meniscus lens. The second used a curved film plane to match the curvature of field of the lens.

Field curvature is more of a problem in camera lenses than in optical instruments such as telescopes. Telescopes cover very small field angles so, within the small angle, the curvature is very slight. A telescope objective lens often has about equal-strength curves on each side to minimize spherical aberration—the rule being that when both surfaces provide the same amount of bending, spherical aberration is minimized. Stopping down reduces the amount of light, and it is desirable to have the maximum amount of light. So it is better with telescopes to allow field curvature to remain.

In cameras, the field is relatively large. Field curvature reduces the sharpness of the image in the field if the focus is on axis. To get sharp pictures over the entire field, the image field must be fairly flat.

The maximum aperture of meniscus lenses that produce acceptably sharp images (not diffused images like portrait lenses) is from *f*/16 to *f*/11, depending on several factors. What can be done when a high aperture lens is designed to correct for all the other aberrations and still has field curvature (Figure 17-23)? Would a glass block placed between the lens and the film flatten the field?

As the lens "looks" back at the film through a glass block, the film appears to be closer to the lens than it really is. Actually, the glass block moves the focal plane farther from the lens than it would be without the glass block, and it does this fairly evenly across the field. However, it makes the field *look* 1/3 closer to the lens than it really is. What is needed to correct for field curvature is a glass block of variable thickness that makes the edges of the film appear to be closer to the lens than the center.

Such a glass block is a large concave-plano negative lens as large as the film. The glass is thick at the edges so it makes the edges look closer to the lens. The center of the correcting lens is very thin—it does not appear to bring the film forward very much. Therefore, as the camera lens sees the film through the correcting lens, the film appears curved like the interior of a bowl. A properly proportioned correcting lens corrects the field curvature of the lens.

Field-flattening lenses are used in projection lenses for motion pictures. Such projection lenses may be as fast as *f*/1.2. They have a negative lens in a spring-loaded mount that presses against the front of the film gate in the projector. As the lens is focused, the main part of the lens moves back and forth, but the field flattener stays in the same position, just in front of the film (Figure 17-24).

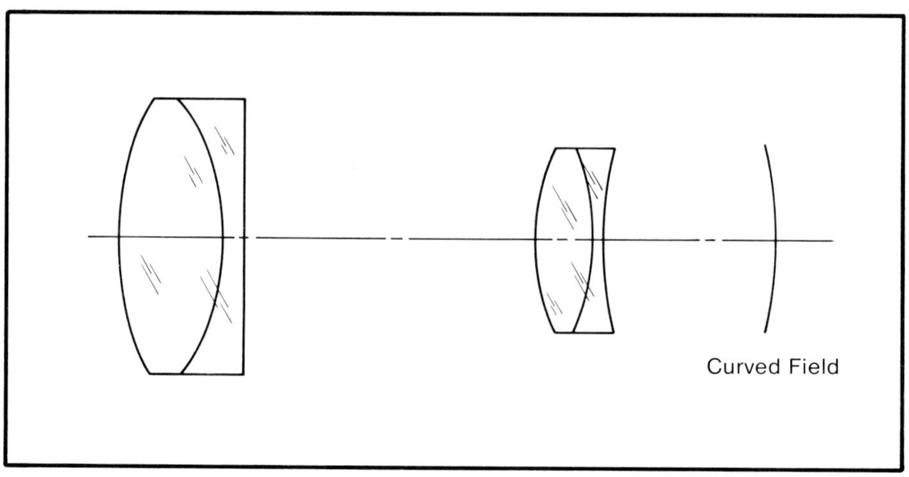

Figure 17-23. High-Aperture Motion Picture Projection Lens.

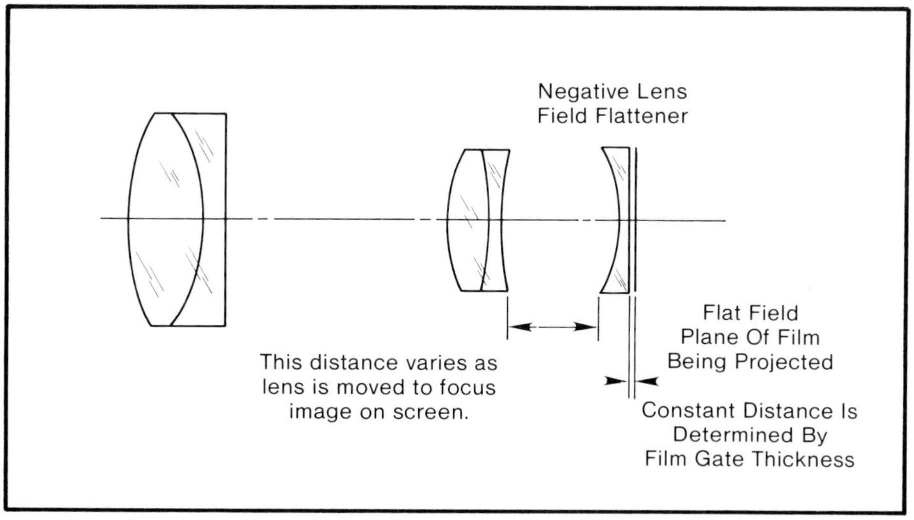

Figure 17-24. High-Aperture Motion Picture Projection Lens With Field Flattener.

Problem

In the drawing, two light rays parallel to the axis are incident to a 2-inch thick equibiconvex lens. One is 2 inches from the axis and the other is 4 inches from the axis. The 2-inch ray would normally fall 7 inches from the principal plane, while the 4-inch ray would fall 6.75 inches from the principal plane due to spherical aberration. These paths are shown by the dashed lines. A glass block 5-inches thick is placed behind the lens. Using graphic ray tracing, Snell's law, or both, locate the new paths of the two rays. State whether a glass block can be used to correct for spherical aberration.

A drawing of this problem is provided half-size on the following page. However, for increased accuracy, you may want to work this problem out on a larger sheet of paper, with the dimensions of the lens and the glass block to scale.

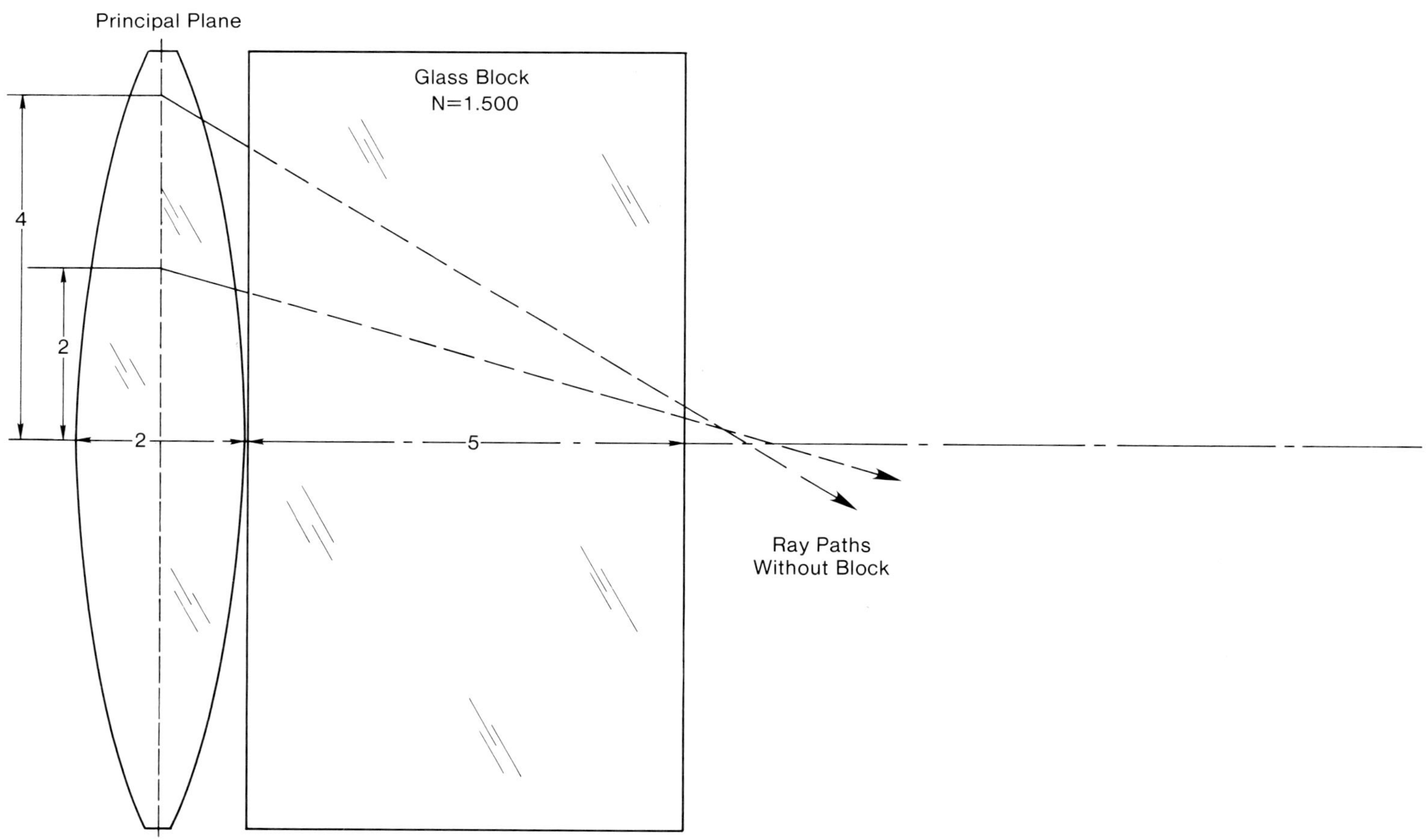

Problem
Chapter 17
Video Lesson 12

Chapter 18

VIDEO LESSON 13

Spherical Aberration, Distortion, Image Illuminance, Beam-Splitter Rangefinders, Loupes, and Gunsights

Summary

A glass block is found to reduce spherical aberration. Lenses having one molded aspheric curve have less spherical aberration than ones with two spherical curves. A mirror objective has less spherical aberration when the curve is changed from a spherical to a parabolic curve. Positive distortion gives "pincushion" shapes; negative distortion gives "barrel" shapes. The *f*-number determines the central area image luminance. The actual image luminance is found using an equation. Three geometric factors result in image illuminance fall-off by a cosine fourth factor. Undersized lenses or mounts can cause additional image illuminance fall-off by vignetting. A single reflection from a plane mirror reverses the image. Virtual image location behind a mirror is found by schematic ray tracing. This principle is used in rangefinders. Several beam splitting devices (a rangefinder, loupe and gunsight) are described.

Spherical Aberration Correction (Graphic Ray Tracing)

The problem presented in the last lesson requires accurate drawing. Using a large piece of paper and making a 2X drawing helps to improve the accuracy. The paper size needed is about 14 x 24 inches. This discussion is based on the figures given in the problem.

The two rays are refracted at the principal plane of the lens. The ray entering 2 inches from the axis is drawn to the axis 7 inches from the principal plane. The ray entering 4 inches from the axis is drawn to the axis 6.75 inches from the principal plane. This establishes the direction of the rays without the glass block, and between the principal plane and the front surface of the glass block after it has been inserted into the light beam behind the lens. In the drawing given with the problem, the surface of the block is 1 inch from the principal plane (Figure 18-1).

In Figures 18-2 and 18-3, the 1.0 and 1.5 arcs are drawn from the points formed by the intersection of the two rays and the front surface of the glass. Because the rays are going from air to glass, the h lines are drawn from the intersection of the ray with the 1.0 arc to the 1.5 arc. The intersection of the h lines with the 1.5 arc establishes the directions

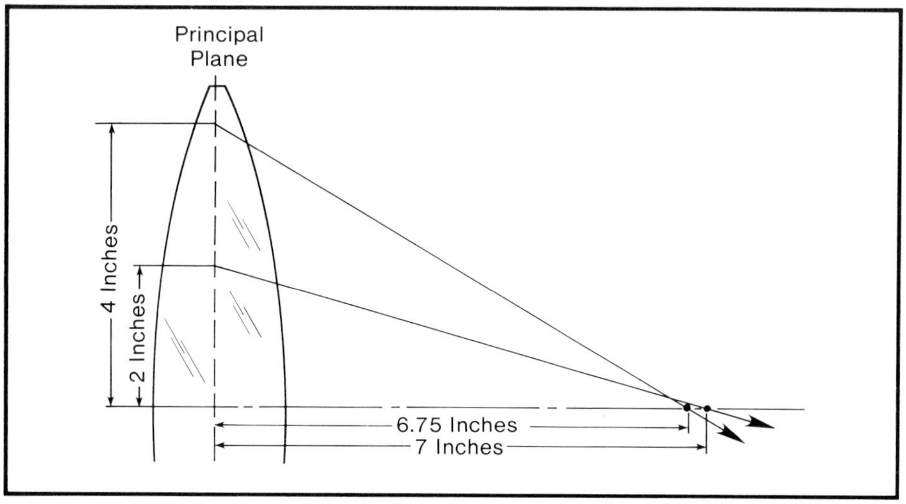

Figure 18-1. Lens Without Glass Block Showing Spherical Aberration.

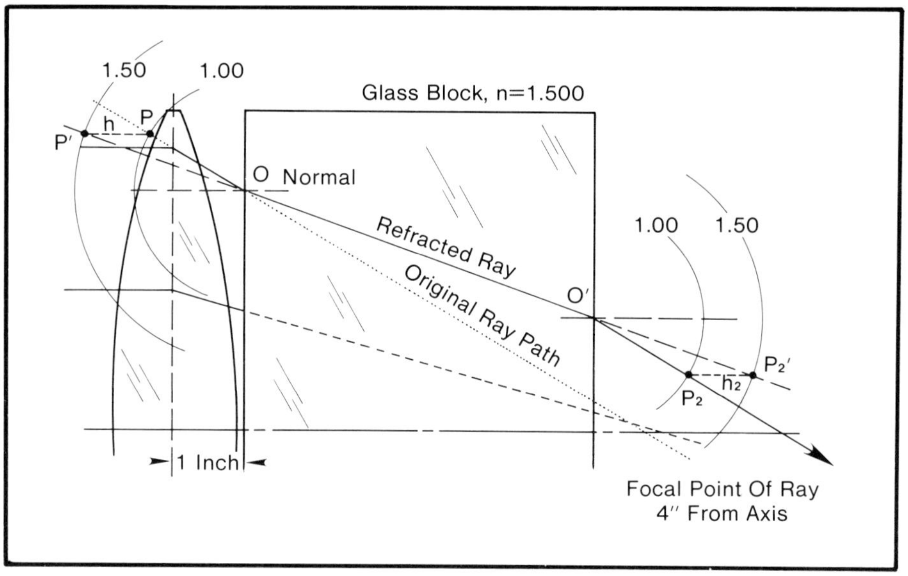

Figure 18-2. Graphic Ray Trace With Glass Block In Place. Ray Trace Of Ray 4 Inches From Axis.

of the ray paths in the glass. They are drawn from these intersections through the O points where the rays enter the glass-block surface. Both rays are refracted towards their normals and are drawn through the points labeled P′.

The two refracted rays intersect the second surface of the block at the points labeled O′. The refracted rays emerge parallel to the original ray paths because the two surfaces of the glass block are parallel. The final paths are simply drawn from the O′ points parallel to the original ray paths until they intersect the axis.

If the work was done very carefully, the ray which originated 2 inches from the axis should intersect the axis 8.74 inches from the principal plane. The ray which started 4 inches from the axis should intersect the axis 8.70 inches from the principal plane. The difference is 0.04 inches. The glass block *has* improved the spherical aberration, but has not corrected it completely. It has reduced the difference between the two focal points from 0.25 inch to less than 0.04 inch.

Solving the Problem Using Snell's Law

Snell's law uses the sines of the angles to establish ray paths. It is necessary in this problem to use basic trigonometry to establish the angles accurately.

The 2-inch ray reaches the axis 7 inches from the principal plane. The ray, the principal plane, and the axis form a triangle (Figure 18-4). The angle that the ray makes with the axis, angle a, can be found by using the tangent.

$$\tan a = \frac{2}{7} = 0.285714$$

$$\angle a = 15.945°$$

Notice that three decimal places are used in order to increase the accuracy of the calculations. The angle of incidence at the front surface of the glass block is this same angle because the normal is parallel

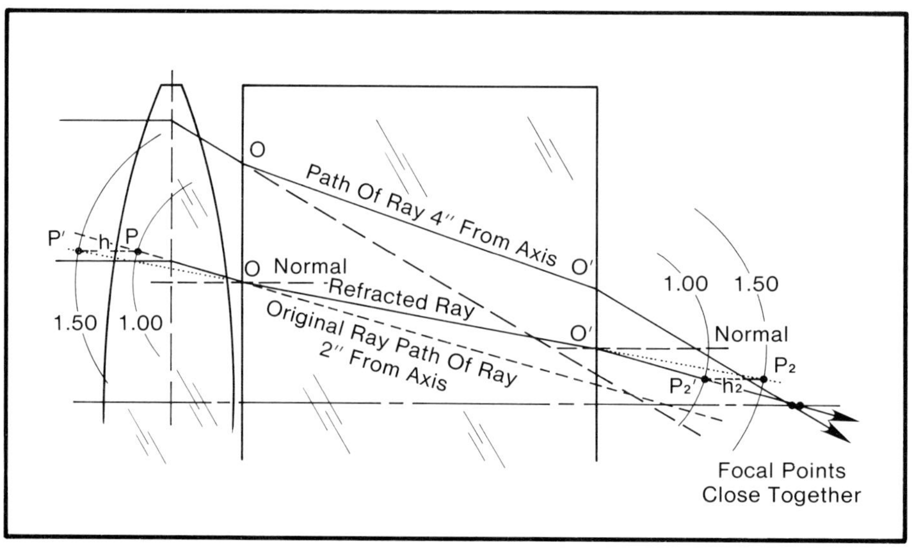

Figure 18-3. Graphic Ray Trace Of Ray 2" From Axis.

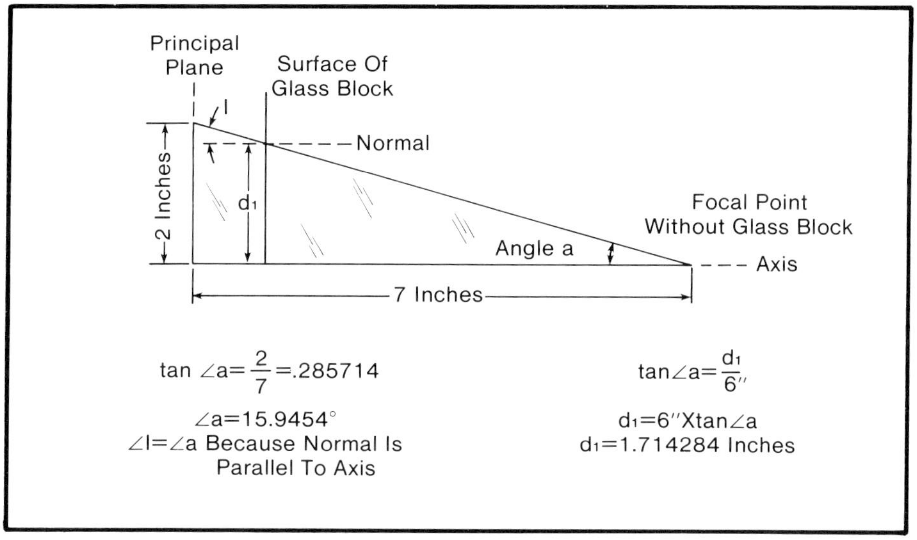

Figure 18-4. Finding Angle Of Incidence Of Ray 2 Inches From Axis.

to the axis. Snell's law is used to find the angle of the refracted ray inside the glass.

$$n \sin I = n' \sin I'$$

$$1.0 \sin 15.945° = 1.5 \sin I'$$

$$\sin I' = \frac{\sin 15.945°}{1.5} = \frac{0.285714}{1.5}$$

$$= 0.1831426$$

$$I' = 10.56286°$$

It is necessary to find the distance, d_1 between O and the axis. A triangle exists whose base is 6″ (from the front surface of the block to the focal point of this ray) and whose angle is the original 15.945°.

$$\tan a = \frac{d_1}{6''}$$

$$d_1 = 6'' \times \tan 15.945 = 6'' \times 0.285714 \text{ inches}$$

$$d_1 = 1.714284 \text{ inches}$$

If we assume that the ray is entirely within the glass, and if the angle of refraction (10.56286°) and the distance (d_1) are known, the focal point of the ray can be found. This is needed to find the distance d_2.

$$\tan 10.56286° = \frac{1.714284}{y}$$

$$0.186474\, y = 1.714284$$

$$y = \frac{1.714284}{0.186474} = 9.19315 \text{ inches}$$

The small triangle to the right of the block is used (Figure 18-5) to find the distance d_2. The distance from the second surface of the block to the intersection of the ray (as if it were in glass all the way) is:

$$9.19315 - 5 = 4.19315 \text{ inches}$$

$$\tan 10.56286° = \frac{d_2}{4.19315}$$

$$d_2 = 4.19315 \times 0.186474$$
$$= 0.781897 \text{ inches}$$

The angle of refraction at surface d is known to be 15.945°. The final ray path is found (Figure 18-6).

$$\tan 15.945° = \frac{0.781897}{y}$$

$$y = \frac{0.781897}{0.285714} = 2.737 \text{ inches}$$

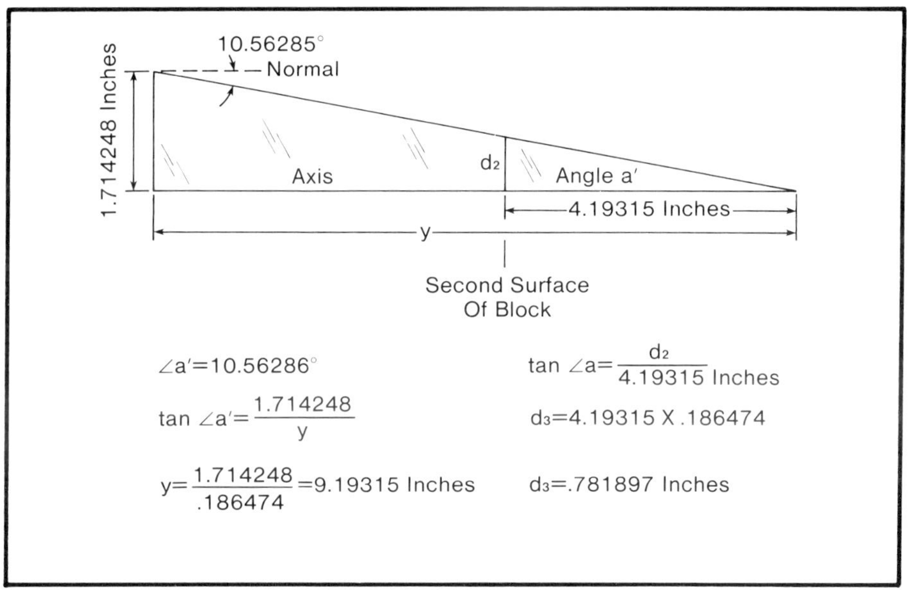

Figure 18-5. Using Triangles And Tangents To Find d_3.

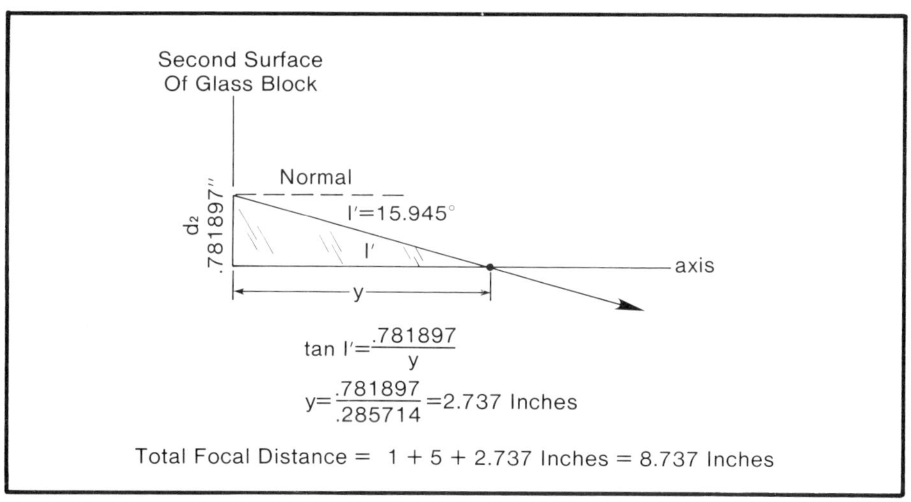

Figure 18-6. Finding Focal Point Distance From Glass Block.

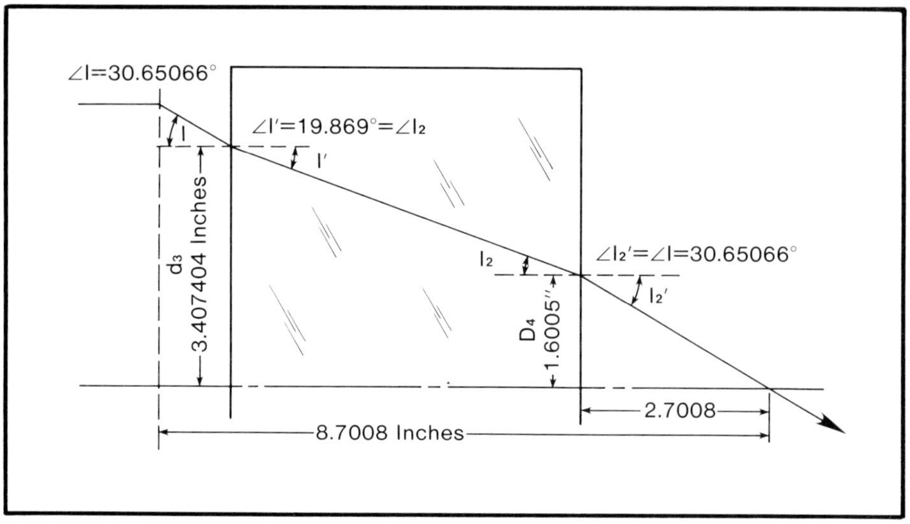

Figure 18-7. Dimensioned Path Of Ray 4 Inches From The Axis.

The distance between the principal plane and the intersection of the ray with the axis is 1 inch + 5 inches + 2.737 inches = 8.737 inches.

Using the same procedure, trace the path of the ray that starts 4 inches from the lens axis (Figure 18-7). The calculations are shown below:

$$\tan a = \frac{4}{6.75} = 0.5925925$$

$$\angle a = 30.65066°$$

$$n \sin I = n' \sin I'$$

$$1.0 \sin 30.65066 = 1.5 \sin I'$$

$$\sin I' = \frac{0.509802}{1.5} = 0.339868$$

$$I' = 19.869°$$

To find distance d_3:

$$\tan 30.65055° = \frac{d_3}{5.75}$$

$$0.59259 \times 5.75 = d_3$$

$$d_3 = 3.407404$$

To find the focal point in glass:

$$\tan 19.869° = \frac{3.407404}{y}$$

$$0.361383 = \frac{3.407404}{y}$$

$$y = \frac{3.407404}{0.361383} = 9.42879$$

To find d_4 using a small triangle:

$$9.42879 - 5 = 4.42879 \text{ inches}$$

$$\tan 19.869° = \frac{d_4}{4.42879}$$

$$d_4 = 4.42879 \times 0.36138$$
$$= 1.6005 \text{ inches}$$

The final angle of refraction is known to be 30.65066°:

$$\tan 30.65066° = \frac{1.6005}{y}$$

$$y = \frac{1.6005}{.5925925} = 2.7008 \text{ inches}$$

The total distance from the principal plane to this focal point is:

1 inch + 5 inches + 2.7008 inches = 8.7008 inches

Rounded off, this is 8.701 inches

The 2-inch ray's focal length is 8.737 inches. The 4-inch ray's focal length is 8.701 inches. The difference is 0.036 inches, which is close to the 0.04 inch difference found by graphic ray tracing.

Other Methods of Correcting Spherical Aberration

As mentioned in the previous lesson, spherical aberration can be used to advantage when an unsharp or "soft" image is desired, as with a portrait lens. Stopping down the lens is one method of correcting the spherical aberration. If the rays going through the outer zone of the lens (which focus at a shorter distance than the inner-zone rays) are removed by the diaphragm, the inner-zone rays exhibit much less spherical aberration by themselves.

A lens designer uses degrees of freedom when designing a lens. Degrees of freedom are the radii of the curvature of each lens element, the type of glass in each element, the thickness of each element and the spacing between the elements. By manipulating these factors, the designer minimizes the various aberrations, such as spherical aberration. As a practical matter, the designer may leave a small amount of spherical aberration at the widest aperture

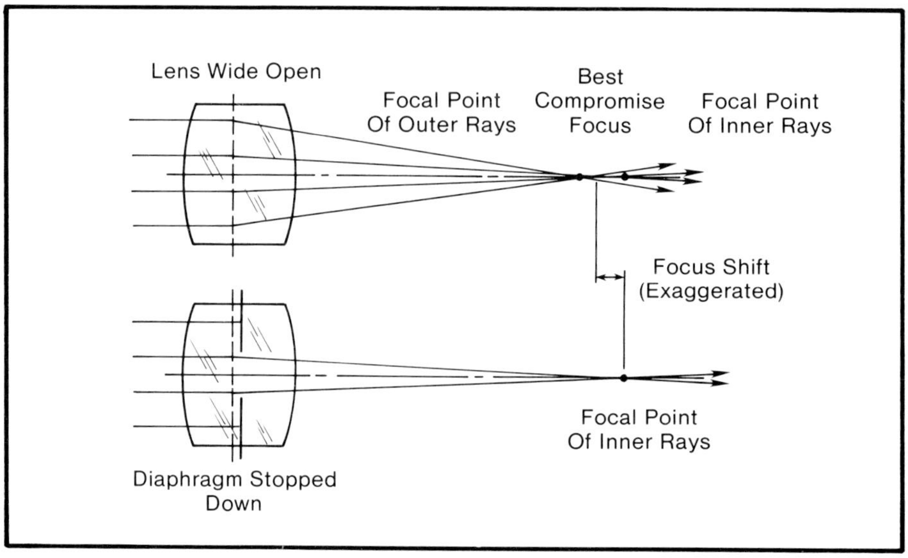

Figure 18-8. Focus Shift In A Corrected Lens.

(Figure 18-8). When the lens is stopped down this disappears.

However, this can result in what is known as "focus shift." Many camera lenses are focused on a ground glass. If the lens is focused at its largest aperture, the rays from the outer zone make the best focus position appear closer to the lens. We have seen that this is because there is more power in the outer zone. The photographer who is focusing with the aperture wide open is using these outer-zone rays. When the lens is stopped down to take the photograph, the best plane of focus of the lens shifts farther back, so that the photograph is not focused accurately. Some enlarger lenses also contain a focus shift due to spherical aberration. If a photographer becomes aware that a lens has focus shift, he or she learns to compensate for it by making a final check of the focus on the ground glass or easel with the lens stopped down to the aperture needed to make the photograph or to make the enlargement.

In some instances, a simple lens must be corrected for spherical aberration. A condenser lens system provides the best illumination if it focuses an image of the lamp filament on the rear surface of the projection lens. Condenser lenses are very strong lenses, with relatively short radii of curvature for their diameters, which are used to distribute the light evenly over the object. They are low *f*-number single lenses. As we have seen, such strong lenses have great amounts of spherical aberration.

It is not critical (for image sharpness reasons) that the condenser lens systems form images that are free of spherical aberration. However, in such a lens, it is possible that the axial rays may come to focus as much as an inch from the focal point of the outer zone rays. This can result in unevenness of illumination across the projected image.

Fortunately, it is possible to correct a condenser lens for spherical aberration. This is done by changing the spherical curve to an aspheric curve (Figure 18-9).

The outer zone of a positive lens element has a shorter focal length than the inner zone. It has more power and to correct for this, one of the spherical curves can be changed so that it has less power at the edges than it does in the center. The radius of curvature is shortened in the central area and lengthened in the outer zone.

In a projection condenser lens, an aspheric curve is visibly different from a spherical curve because the change is great. It is extremely difficult to grind and polish such aspheric surfaces; hence, very costly. It is economical, however, to spend the time and care on grinding and polishing a stainless steel molding tool that can be used to mold the aspheric surface of many condenser lenses. Fire-polished rod glass is used with polish on the glass and mold so the molded lens ends up with a polished surface. The other surface of aspheric condenser lenses is conven-

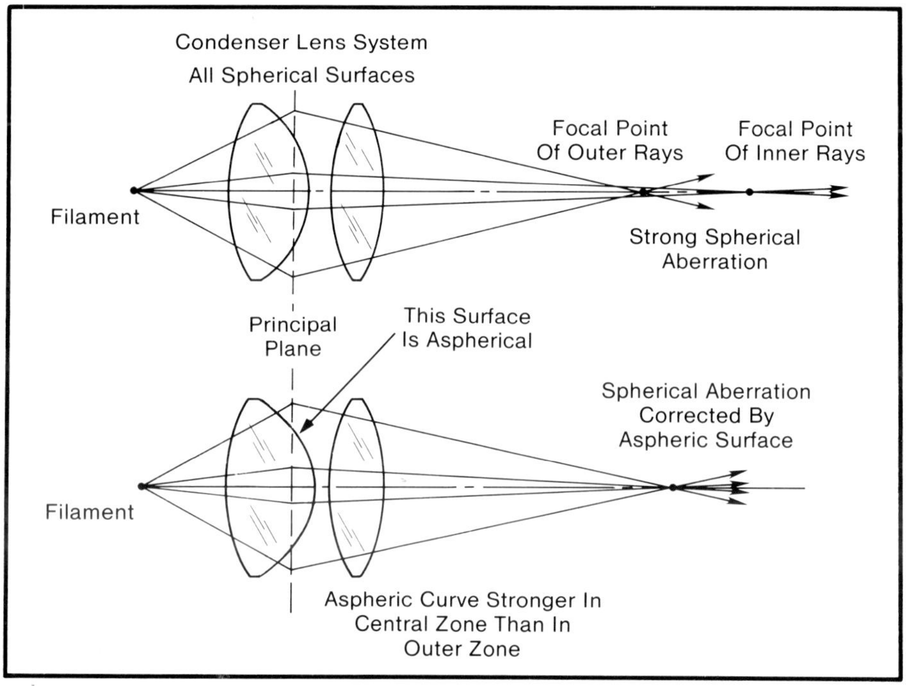

Figure 18-9. Correction Of Spherical Aberration Using An Aspheric Curve.

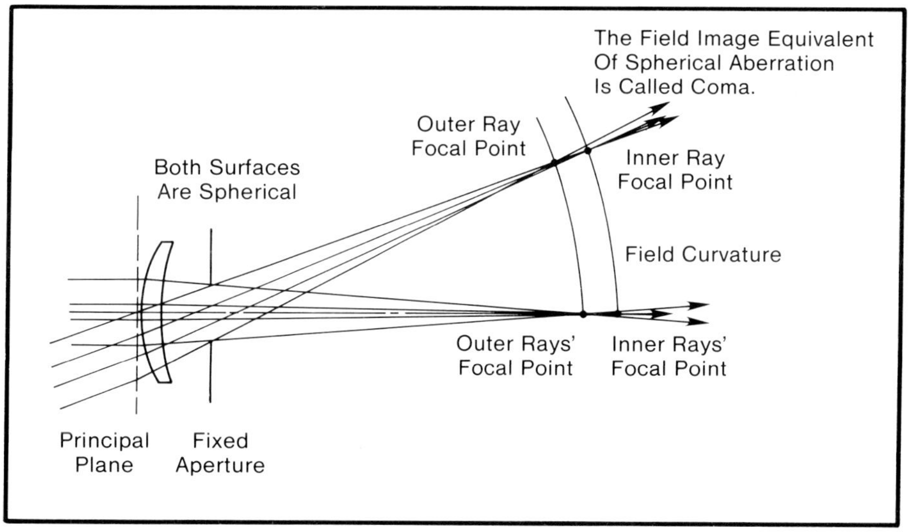

Figure 18-10. Field Curvature And Coma In A Front Meniscus Camera Lens.

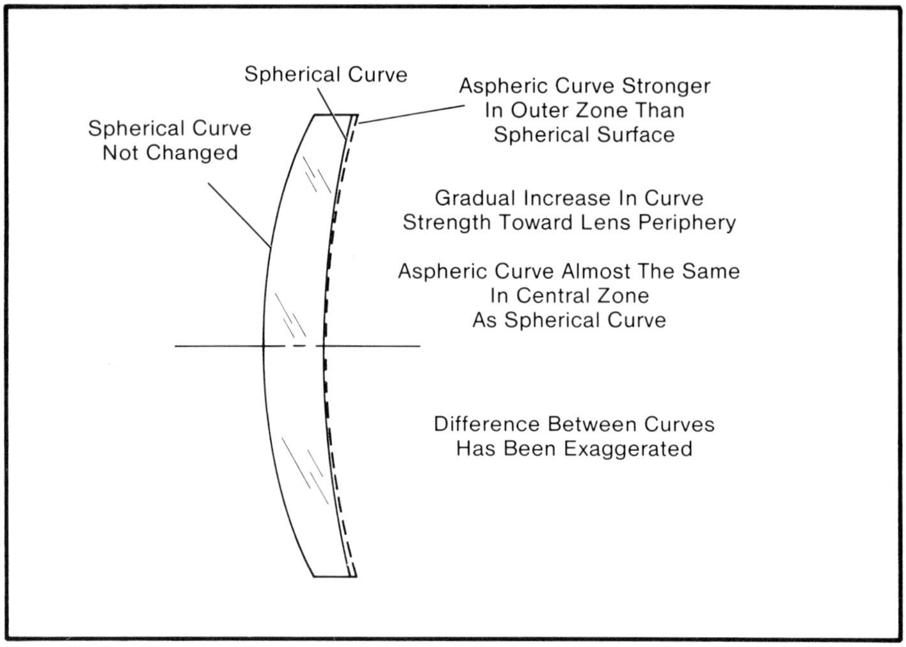

Figure 18-11. Changing The Meniscus Lens's Concave Surface To An Aspheric Curve.

tionally ground and polished with a spherical curve; one aspheric surface is all that is needed in a condenser lens system. The aspheric lens is often used in conjunction with a conventional spherically surfaced condenser lens.

Front-meniscus lenses are commonly used as camera lenses. Spherical aberration and field curvature are two aberrations that are inherent in such lenses. If one of the two curves of a meniscus lens is made into an aspheric surface, both of these aberrations can be reduced.

In Figure 18-10, the rays show the field curvature and spherical aberration caused by the lens' greater strength at the edges as compared to the central area. Both curves are spherical. In Figure 18-11, the concave curve is shown to have been changed in the outer zone so that it has a shorter radius. By reducing power in the outer zone of the lens, the focal length is made longer in this zone. Figure 18-12 shows how the weakened outer zone corrects both the spherical aberration and the field curvature.

Acrylic plastic can be molded to make satisfactory camera lenses for many uses. The front-meniscus lens discussed above can be made with a mold that is spherical on one surface and aspherical on the other surface, and such a lens shows less of these two aberrations than a lens molded with two spherical surfaces.

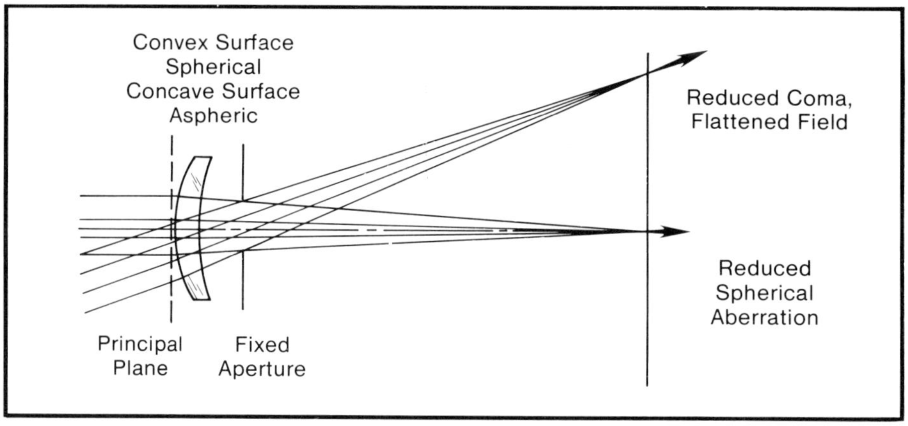

Figure 18-12. Aspheric Meniscus Lens Image Characteristics.

Aspheric Mirrors

Mirrors are used in reflecting telescopes and in catadioptric camera lenses. A spherical concave mirror introduces severe spherical aberration when imaging parallel light rays from distant subjects (Figure 18-13).

Parallel rays should image at a point that is half the radius of curvature. The law of reflection (the angle of incidence equals the angle of reflection) shows rays striking the outer zone of the mirror cross the axis closer to the mirror than rays

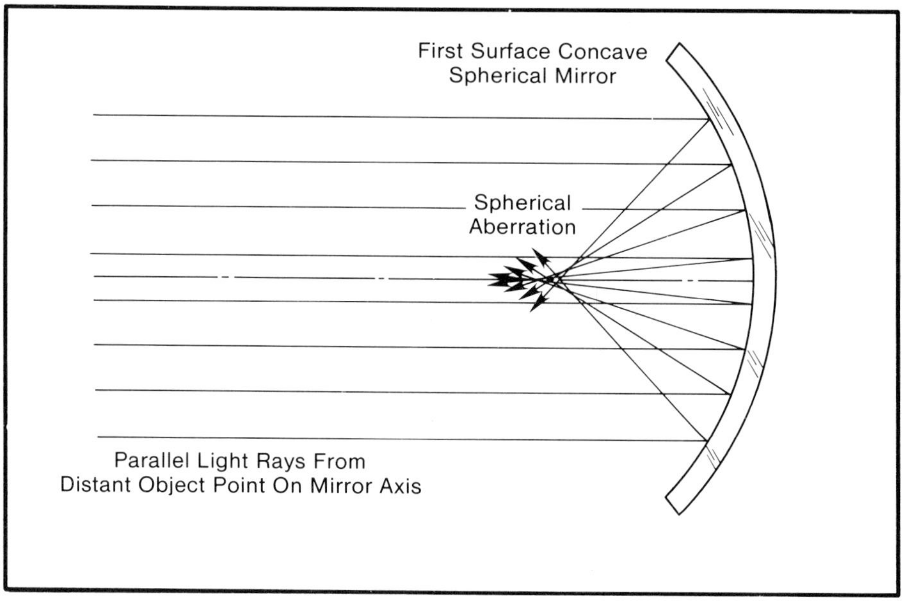

Figure 18-13. Spherical Aberration In Image Formed By Spherical Concave Mirror.

near the axis. The spherical mirror is stronger (has a shorter focal length) in the outer zone than in the inner zone.

A parabolic curve gives a concave mirror equal power all over (Figure 18-14). This is one form of aspheric curve—actually it is one of the conic curves.

Telescope mirrors have long focal lengths. The difference between the spherical curve and the appropriate parabolic curve is slight. In practice, telescope mirrors are ground and polished to a spherical surface and the spherical surface is changed to a parabola by hand polishing the central area to strengthen the curve. In a mirror with a focal length of 25 feet, the amount of glass removed to reach the required parabolic shape is slightly more than a hundredth of a millimetre (less than a thousandth of an inch).

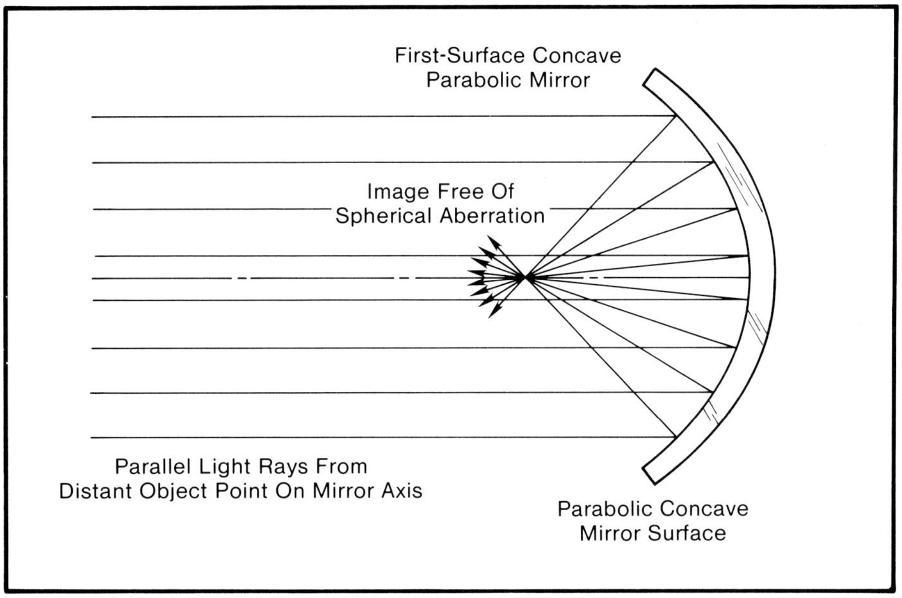

Figure 18-14. Image Formed By Concave Parabolic Mirror Surface Free From Spherical Aberration.

Distortion

Spherical aberration, chromatic aberration, coma, astigmatism and field curvature cause a loss in image sharpness. Distortion is the one aberration that does not cause a loss in sharpness. Distortion is a change in image shape. In most of the discussion on imaging, we have assumed that magnification remains constant across the imaged field. This is true in most cases, but there are some exceptions.

If the magnification gradually increases from the axis to the corners of the field, a square imaged by the lens is imaged as a pincushion. All straight lines perpendicular to a radial line in the field, and not on axis, will be imaged as curved lines, concave outward. Because the magnification increases as the distance from the axis increases, the distortion is called positive distortion (Figure 18-15).

If, on the other hand, the magnification decreases as the distance from the axis increases, the distortion is called negative distortion. A rectangle imaged by a lens having negative distortion resembles a barrel; therefore, the common name for negative distortion is barrel distortion. Tangential lines in the field are imaged as curved lines concave toward the axis.

When a plano-convex lens is used as a magnifier, it exhibits pincushion distortion. When the plano side is toward the object, the distortion is less than if the convex side is toward the object. When a biconvex lens is used, the distortion is reduced considerably. If an aspheric condenser is used as a magnifier, the distortion is barrel distortion. This can be expected because there is greater magnification in the central area as a result of the shorter radius of curvature there.

If a single lens is used as a camera lens, the amount of distortion depends upon the position of the diaphragm, or lens stop. If the diaphragm opening is in front of the lens, the distortion is negative (barrel). If the lens opening is behind the lens, the distortion is positive (pincushion). This is shown in Figure 18-16.

This leads to the obvious conclusion that if a lens system has two similar lenses, one placed in front of the diaphragm and one placed behind the diaphragm, the distortion can be reduced to zero (Figure 18-17). This is especially true for lenses used at close distances, such as lenses used in graphic arts reproduction

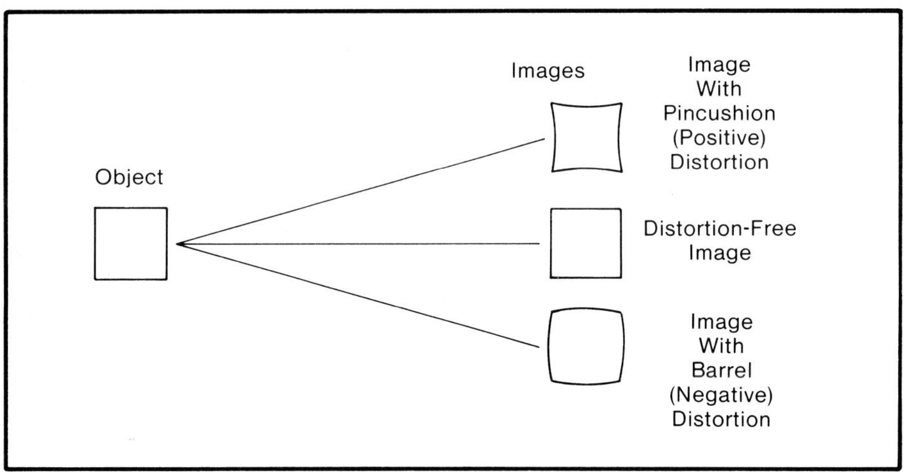

Figure 18-15. Pincushion And Barrel Distortion.

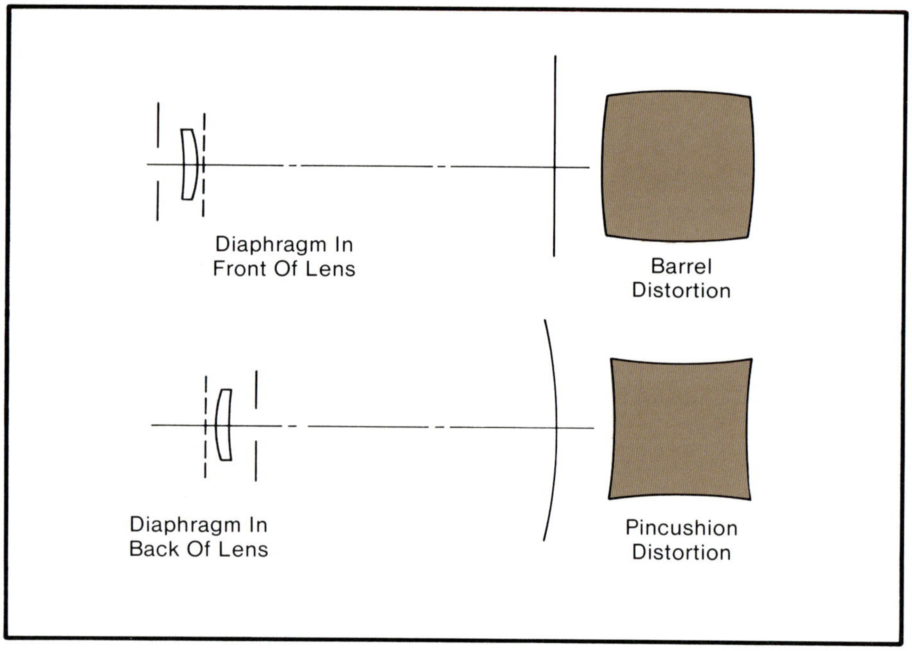

Figure 18-16. Lens Aperture Position Determines Type Of Distortion.

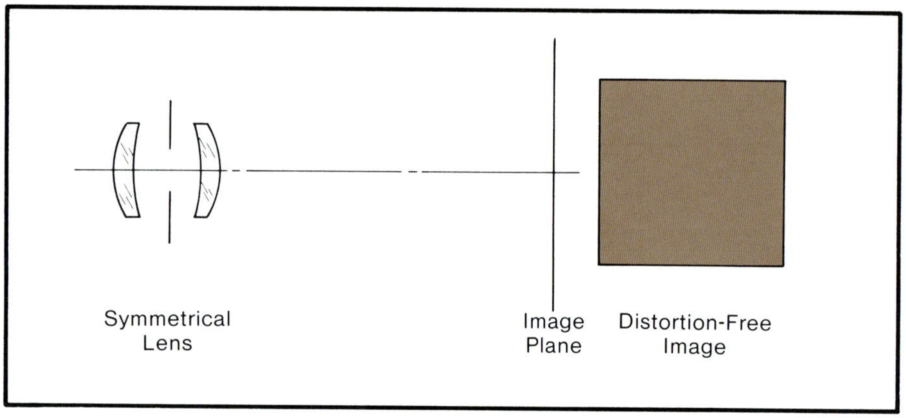

Figure 18-17. Symmetrical Lens — Distortion-Free.

and copy making (Figure 18-18a).

Aerial mapping photography (photogrammetry) requires almost completely distortion-free images. Aerial camera lenses for mapping must also be wide-angle lenses. Wide-angle lenses are typically hard to make and difficult to correct for distortion. The Zeiss Metrogon lens shows one solution to the twin problems of wide-angle coverage and minimizing the distortion. Note the nearly-perfect symmetry in the drawing. The Metrogon lens has a fixed aperture of *f*/8 (Figure 18-18b).

Extreme wide-angle lenses with field angles near 180° are allowed to exhibit barrel distortion. Because of the distortion, such lenses are called fish-eye lenses.

Zoom lenses are very difficult to correct for distortion at their extreme focal lengths. Generally, they exhibit slight pincushion distortion at their long focal lengths and barrel distortion at their short focal lengths.

Telephoto lenses usually have a positive lens in front of a diaphragm and a negative lens in back of the diaphragm. This condition is not symmetrical. As a result, telephoto lenses usually exhibit slight pincushion distortion. Since the field angle is relatively small, however, this is not very noticeable.

Typical normal-focal-length moderate-aperture photographic objectives are designed as triplets such as Tessar-type lenses, which are moderately but not exactly symmetrical. Their degree of distortion (positive or negative, depending on the design) is of the order of .2 millimetre distortion in the corner of an image formed by a 100 mm lens.

Large aperture photographic objectives of the double Gauss design are much more symmetrical and exhibit even less distortion (Figure 18-18c).

Image Illuminance

f-numbers were defined earlier as the focal length of a lens divided by the lens diameter.

$$\frac{\text{focal length}}{\text{lens diameter}} = f\text{-number}$$

No matter what size the lens, if the *f*-number is the same, the image illuminance will be the same.

This seems to go against intuition. If there is a large lens and a small lens, intuition says that since the large lens obviously lets more light through, its image *must* be brighter. For example, if one lens has a 3-inch diameter, and another lens has a diameter of 0.3 inch, reason dictates that the 3-inch lens must create a brighter image. It *is* true that the total amount of light accepted by the larger lens is greater because the amount of light accepted is proportional to the area of the lens.

0.3-inch diameter lens

$$\text{area} = \pi r^2$$
$$\text{area} = 3.1416 \times .15^2$$
$$\text{area} = 3.1416 \times .0225$$
$$\text{area} = 0.0707 \text{ sq. inches}$$

3-inch diameter lens

$$\text{area} = \pi r^2$$
$$\text{area} = 3.1416 \times 1.5^2$$
$$\text{area} = 3.1416 \times 2.25^2$$
$$\text{area} = 7.07 \text{ sq. inches}$$

The 3-inch lens has 100X the area of the small lens. It lets 100X as much light pass through to form the image. However, this does not mean that the image is 100X as bright.

If the focal length of the 0.3-inch diameter lens is 0.6 inches while the focal length of the 3-inch diameter lens is 24 inches, what is the *f*-number of each?

$$f\text{-number} = \frac{\text{focal length}}{\text{diameter}}$$

$$f\text{-number} = \frac{0.6}{0.3} = f/2$$

$$f\text{-number} = \frac{24}{3} = f/8$$

The image illuminance is related to the *f*-number, not the diameter alone. Which lens provides greater image illuminance? How much greater illuminance?

In Chapter 11 the *f*-number series was given:

f/1.0	*f*/11
f/1.4	*f*/16
f/2	*f*/22
f/2.8	*f*/32
f/4	*f*/45
f/5.6	*f*/64
f/8	*f*/90

The smaller the *f*-number, the higher the image illuminance. Each successive number in the series has one half the illuminance, going down the series, or 2X the illuminance, going up the series. Each number in the series is the previous number multiplied by $\sqrt{2}$. Every other number is two times the smaller number, but one-fourth times the illuminance.

The small lens, which has an aperture of *f*/2, creates a brighter image than the large lens which has an aperture of *f*/8. How much brighter?

f/5.6 provides 2 times the illuminance that *f*/8 does.

f/4 provides 4 times the illuminance that *f*/8 does.

f/2.8 provides 8 times the illuminance that *f*/8 does.

f/2 provides 16 times the illuminance that *f*/8 does.

A film placed in the image plane of the large lens would require 16 times as much time to yield the same exposure as the same film placed behind the small lens.

Actual Illuminance Values

f-numbers deal with focal lengths and aperture sizes that control relative values of light. It is sometimes useful to be able to deal with real or absolute values.

When the sun has been up for about two hours (on a clear day), and for the period of time until two hours before sunset, the sunlight falling on a vertical illuminated surface has an illuminance of about 7200 foot candles (Figure 18-19).

The average reflectance of outdoor objects is about 0.13 (13% reflection). The luminance of an average subject is found by the following equation:

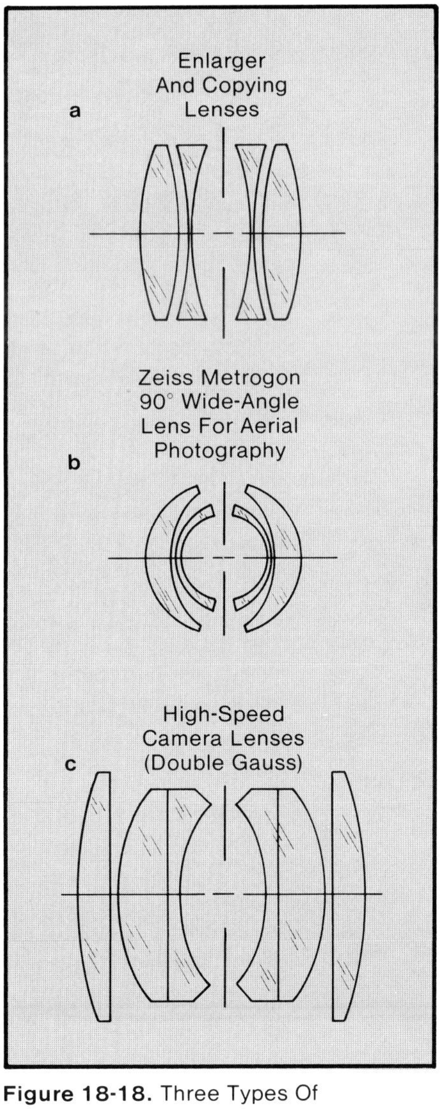

Figure 18-18. Three Types Of Symmetrical Lenses.

$$\text{Luminance (candles per sq. ft.)} = \frac{\text{Illuminance (f.c.) x Reflectance}}{\pi}$$

$$\pi = 3.1416$$

$$\text{Luminance (candles per sq. ft.)} = \frac{7200 \text{ x } .13}{\pi} = 298 \text{ candles per sq. ft.}$$

Hence, the average luminance of the scene is 298 candles per square foot.

The following equation is used to find the illuminance on the film plane of a camera in foot candles:

$$\text{Illuminance} = \frac{\pi LT}{4N^2}$$

L = The subject luminance in candles per sq. ft.

T = The transmittance of the lens

N = the *f*-number

Suppose the *f*/2 lens is imaging an average subject and the transmittance of the lens is .98 (98%):

$$\text{Illuminance} = \frac{\pi \text{ x } 298 \text{ x } .98}{4 \text{ x } 2 \text{ x } 2}$$

$$\text{Illuminance} = \frac{917.5}{16} =$$

57.3 foot candles at the film plane.

Comparatively, the *f*/8 lens illuminance is found:

$$\text{Illuminance} = \frac{\pi \text{ x } 298 \text{ x } .98}{4 \text{ x } 8^2}$$

$$\text{Illuminance} = \frac{917.5}{256} =$$

3.6 foot candles at the film plane.

3.6 x 16 = 57.3

The ratio of 16 to 1 found earlier by *f*-numbers is verified.

The exposure on a film is usually given in lux-seconds (metre-candle-seconds). One lux (metre-candle) is 1/10.76 foot candle. To convert foot candles to lux, multiply by 10.76:

57.3 ft. candles x 10.76 = 616.5 lux

3.6 ft. candles x 10.76 = 38.7 lux

If a film is exposed to each image for 1/25 second, the amount of

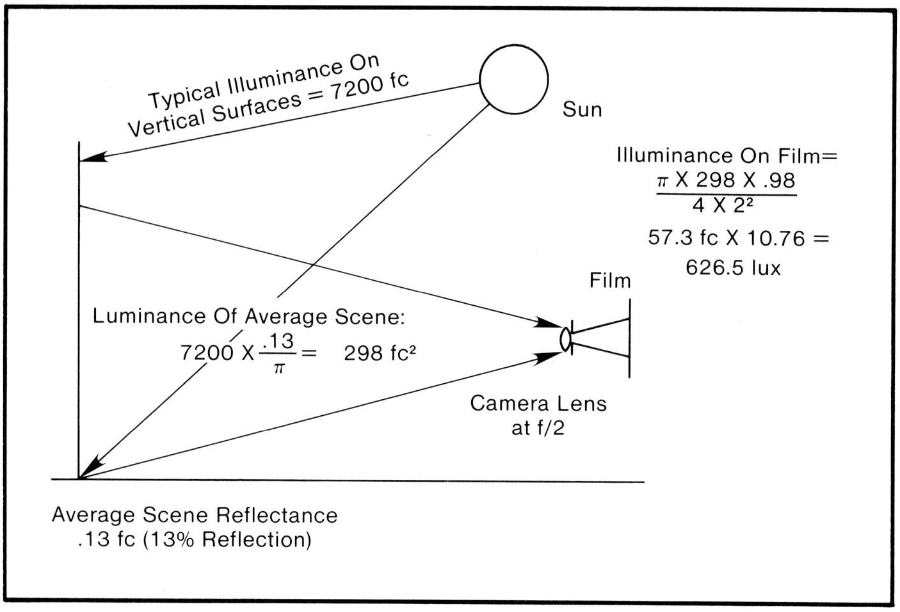

Figure 18-19. Typical Light Values In Outdoor Photography.

exposure the film will receive is:

616.5 lux x 1/25 sec. =
24.7 lux seconds

3.6 lux x 1/25 sec. =
1.5 lux seconds

1.5 lux seconds x 16 =
24.7 lux seconds

Measuring Illuminance and Luminance

In practical optics it is sometimes useful to know the luminance or illuminance in a given lighting situation. Special meters are made for measuring illuminance, but unless many measurements are needed, it is not necessary to obtain such a meter.

Photographers use Kodak neutral test cards, which have a reflectance of .18 (18% reflection), with a reflection-type exposure meter. If the gray card is placed where either illuminance or luminance is to be measured, and the meter calculator dial is set at ⅛ second with an ISO 125 film speed, the *f*-number found for the exposure, used with the following table, gives the illuminance and luminance. (The values between the marked *f*-numbers are one-third stop increments. Most exposure meters are so calculated.)

f-number	Illuminance (ft. candles)	Luminance (candles per sq. ft.)
1.4	3.52	.20
	4.43	.25
	5.58	.32
2.0	7.03	.40
	8.85	.51
	11.20	.64
2.8	14.10	.81
	17.70	1.02
	22.30	1.28
4.0	28.10	1.61
	35.40	2.03
	44.60	2.56
5.6	56.30	3.22
	70.90	4.06
	89.30	5.12
8	112.00	6.45
	142.00	8.12
	179.00	10.20
11	225.00	12.90
	283.00	16.20
	358.00	20.50
16	450.00	25.80
	567.00	32.50
	715.00	40.90

The gray card does not have to be used to measure luminance (the light reflected from a surface). If the exposure meter is pointed at any surface and the *f*-number is found for ⅛ second with an ISO 125 film, the luminance is as indicated in the table. With other than the 18% gray card, however, the illuminance cannot be found from the table.

Exposure Correction for Close Distances

The above section applies directly when the subject distance is from about 8 times the focal length up to infinity. When the lens is focused on objects closer than 8 times the focal length, corrections must be made.

The *f*-number is calibrated with the lens focused at infinity because it is the focal length that is used for calculating the *f*-number. When focusing at close distances, the lens opening stays the same size, but the distance between the lens opening and the film plane increases. The effective *f*-number then becomes:

$$\text{Effective } f\text{-number} = f_e = \frac{\text{diameter of lens opening}}{\text{image distance}}$$

$$f_e = \frac{d}{v}$$

To use this equation you must know the lens opening. Another equation for finding f_e without the use of d is:

$$f_e = \frac{v \times f_n}{f}$$

f_n = the *f*-number
v = image distance
f = focal length

In some situations the magnification may be easier to obtain than the image distance. The equation using magnification to find the effective aperture is:

$$f_e = f_n (m + 1)$$

For example, if an image is half the size of the original, the magnification is .5X. If the marked *f*-number is *f*/8, what is the effective *f*-number?

$$f_e = 8\,(.5 + 1)$$

$$f_e = 8 \times 1.5 = f/12$$

Although the diaphragm is set at *f*/8, the exposure is calculated as though the lens opening is *f*/12.

Another way to calculate the exposure of close subjects is to find the

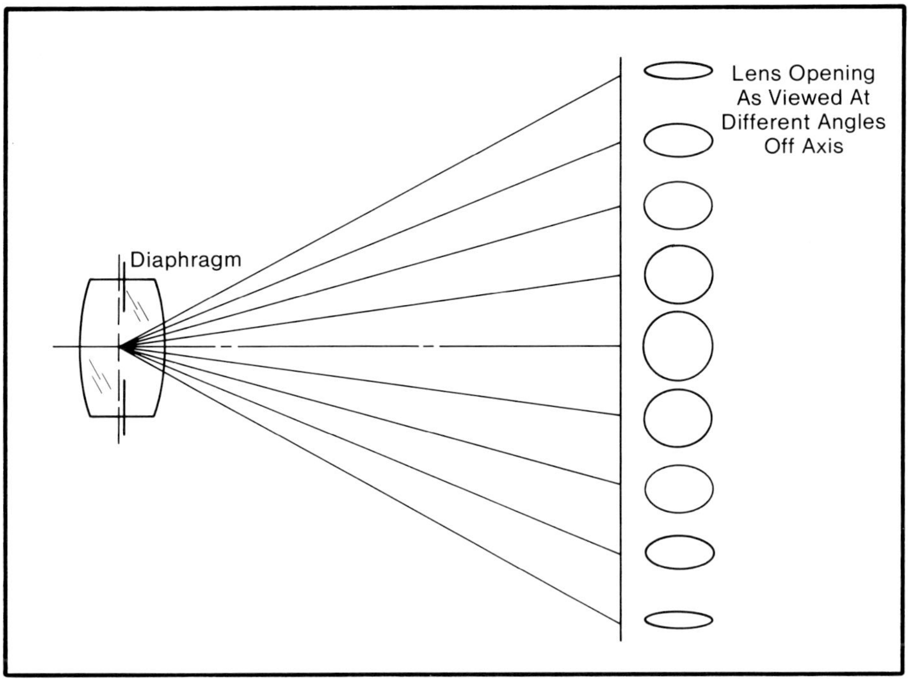

Figure 18-20. First-Cosine Factor: Shape Of Lens Opening At Different Field Angles.

exposure factor. The product of the factor and the normally-calculated exposure time yields the correct exposure time. The effective aperture need not be calculated with this method:

Exposure factor $= (m + 1)^2$

Example: The image is the same size as the object. The magnification is, therefore, 1.0X. The exposure factor is:

Exposure factor $= (1+1)^2 = 2^2 = 4X$

The calculated exposure based on the luminance of the subject must be multiplied by 4 to obtain the corrected exposure.

Cosine Fourth Law

The above sections dealing with the illuminance of the image and the exposures based on that illuminance assume that the image has the same illuminance all over. This is true enough for most practical purposes, but it is not *absolutely* true. The axial region of an image has the greatest illuminance, and the illuminance falls off the farther you go from the axis (it is least in the film corner).

In black-and-white photography, there is a compensating factor. The negatives are somewhat less dense along the edges than they are in the center due to this illuminance fall-off. However, when the negative is placed in the enlarger, the enlarging lens also has a fall-off, which has a compensating effect. With color slides and transparencies, there is not this correcting factor and the fall-off can often be seen as darkened edges on the transparencies.

Why does this fall-off occur, and how much does the illuminance diminish?

If you open the shutter (with the back off the camera) and look along the lens axis, the lens opening appears round. If you move your eye out into the field, the lens opening appears elliptical (Figure 18-20). One dimension remains the same, but the other dimension gets smaller. The area of the lens opening as seen off axis is less. In the corner of the field the narrow dimension of the ellipse may be very narrow. At 90°, the narrow dimension of the ellipse would become zero—so there would be no light (actually, with most lenses, the lens mount cuts off the image long before a 90° angle is reached).

It has been found that the area of the ellipse decreases as the cosine of the angle to the axis, at which the ellipse is viewed, increases. The cosine of an angle in a right triangle is the adjacent side divided by the hypotenuse (Figure 18-22). At 0°, that is on axis, the hypotenuse and adjacent side are equal in length. Therefore, at 0° the cosine is 1. As the angle increases, the adjacent side stays the same length, while the hypotenuse increases in length. Therefore, the cosine decreases as the angle increases. When the angle is 90°, the hypotenuse is infinite in

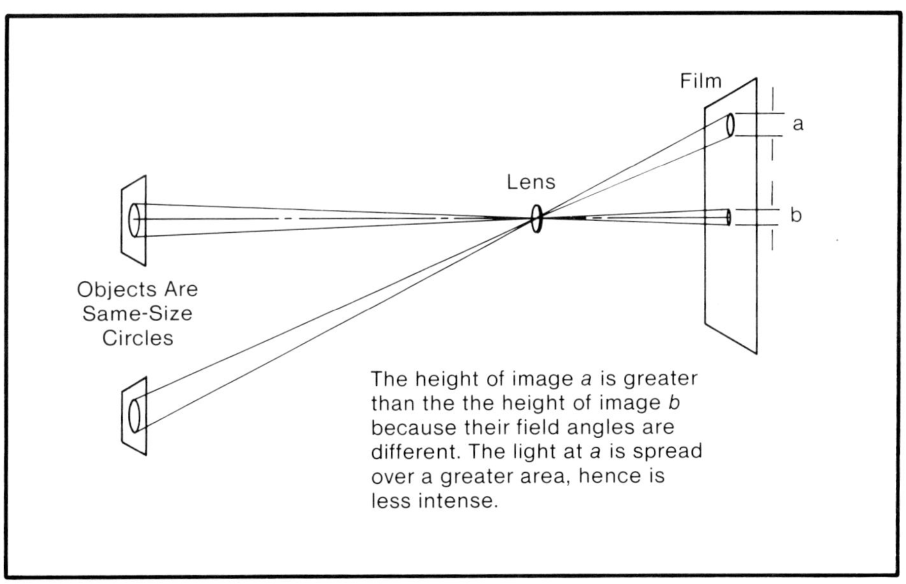

Figure 18-21. Second-Cosine Factor: Circle Image Length Increases With Increase In Field Angle.

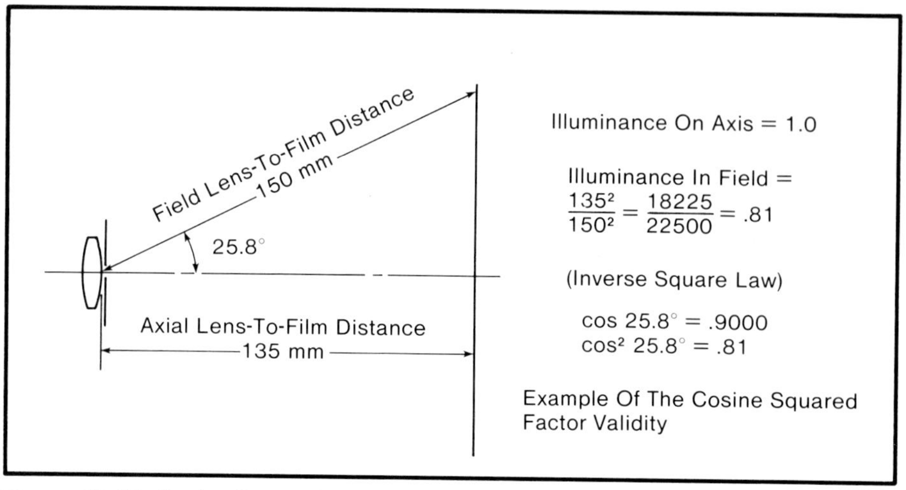

Figure 18-22. Cosine Square Factor Resulting From Lens-To-Film Distance.

length, so the value of the cosine is 0.

As the off-axis angle increases, the area of the ellipses decreases, and this accounts for one of the four cosine factors. This agrees with the area of the ellipse as viewed.

On axis, the image (film) is at 90° to the beam of light. The image of a round beam is round. At a field angle of 10°, the same size beam of light strikes the film at an angle of 10°. The same beam of light forms an ellipse in the image plane which has the same width as the circle on axis, but is longer.

Since the same amount of light must cover a larger area, its illuminance on each unit of area is decreased (Figure 18-21). It has been found that the illuminance decreases with the cosine of the angle. From these two causes, the illuminance decreases by a cosine times a cosine, or a cosine squared ($\cos^2$).

On axis, the beam travels a given distance. It travels a longer distance from the lens to the corner of the film. In an earlier lesson it was determined that the illuminance decreases by the square of the distance.

In this case, the adjacent side stays a constant length (the distance to the focal plane along the axis). The hypotenuse is the distance from the lens to the film. It turns out that the decrease in illuminance is equal to the square of the cosine of the angle (Figure 18-22).

The first two factors reduced the illuminance by a $\cos^2$ and this third factor also reduces the illuminance by $\cos^2$. The total loss is, therefore, $\cos^2 \times \cos^2 = \cos^4$.

Simplified, the illuminance decreases by a $\cos^4$ factor of the field angle.

Most lenses have positive elements in front of and behind the diaphragm. Optically, this reduces the illumination even more than by the $\cos^4$ factor. In Figure 18-23, the $\cos^4$ function is plotted every 10° from 0° to 70° in two directions from the axis (0°). One line shows the actual illuminance of a typical lens at *f*/1.4. It stops at 23° on each side because that is the typical field angle of a normal lens for 35 mm cameras. The dotted line shows the illuminance across the field of the same lens but set at *f*/5.6 (Figure 18-24).

The dashed line shows the illuminance of an extremely wide-angle lens like the Metrogon aerial lens discussed above in the section on distortion. You can see that the illuminance is less than the $\cos^4$ function out to 45° (which is the half angle of the 90° field angle it covers). This lens is nearly always used with a graduated-density filter which absorbs light in the central zone to compensate for more than the $\cos^4$ losses of the lens. Without this filter, the central portion of the film would receive much more light (exposure) than the corners.

When lenses have strong negative power in the front elements, the illuminance loss is much less than the $\cos^4$ normal loss. The reverse-telephoto wide-angle lenses used on 35 mm cameras are typical of this situation. If you view the diaphragm through the rear element of one of these lenses, the ellipse does not become narrow as rapidly as with a normal lens. A 21 mm lens covers a

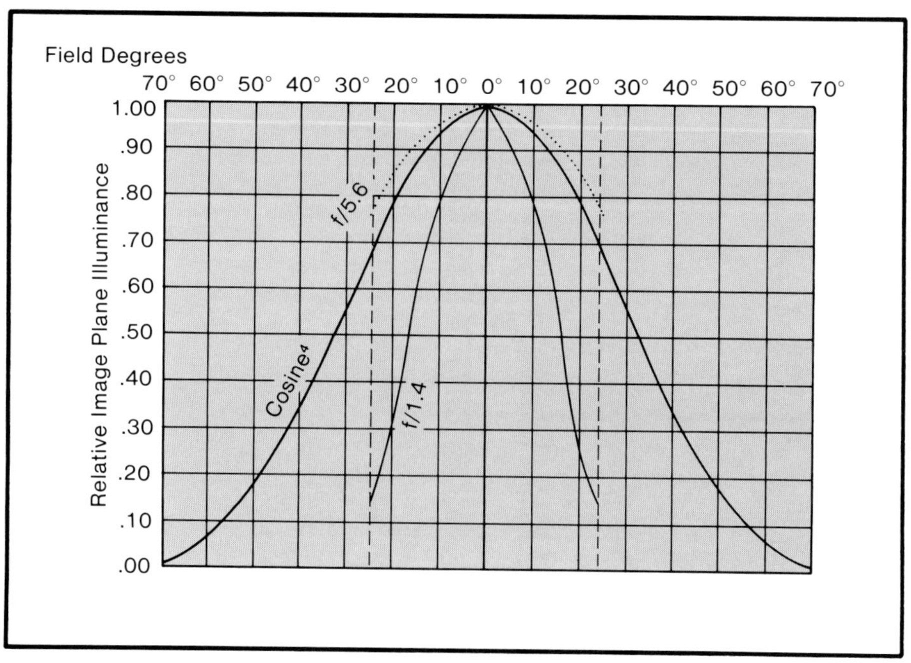

Figure 18-23. Image Plane Illuminance for 50 mm f/1.4 Lens For 35 mm Camera.

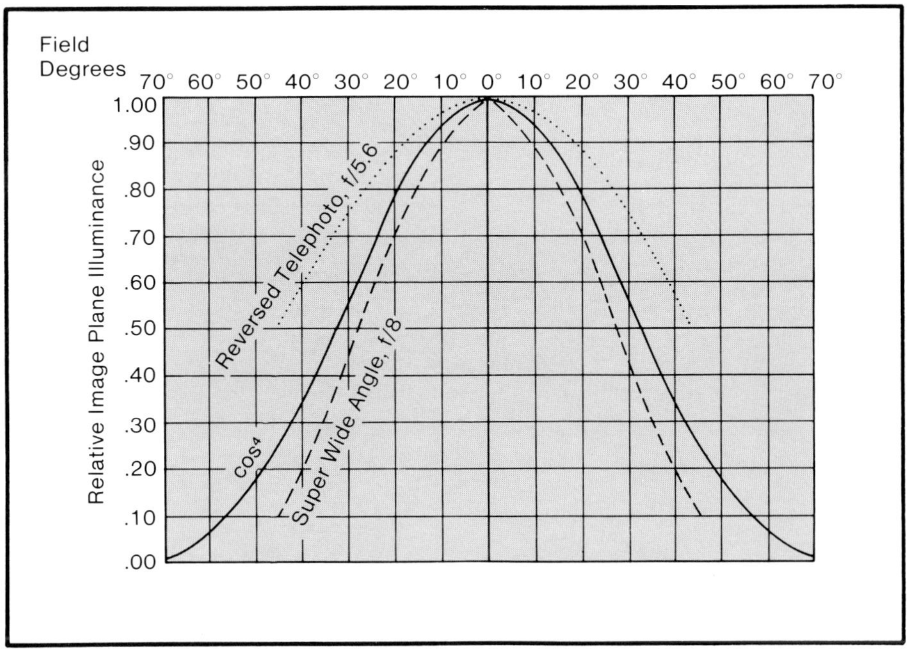

Figure 18-24. Image Plane Illuminance.

90° total field angle on a 35 mm frame just like the Metrogon lens covers on its larger format. However, no special filter is required with its use. The dotted curve in Figure 18-24 shows a typical illuminance curve of such a lens, and indicates that its illuminance does not diminish as fast as the $\cos^4$ loss.

In the extreme, a fish-eye lens is a super wide-angle reverse-telephoto lens that covers a total field of 180° with a half angle of 90°. The $\cos^4$ law says that there is zero illuminance. However, because of the extreme negative power in front of the diaphragm, the beam is bent to 90° from the axis. The angles behind the lens are far less—but it is the total angle that determines the lens coverage.

For most photographic applications, lenses that do not lose more than 50% of the central illuminance (*f*/stop) at the corners of the image frame are satisfactory for color photography at most of their apertures.

Vignetting

The diameters of the glass elements of lenses and their lens mounts are kept small to keep the manufacturing cost down, and to reduce the weight and bulk. There are two factors that favor increasing the diameters of the lenses for a given focal length. One is, of course, to maximize relative aperture, or *f*-number. The other is to minimize a fall-off in image illuminance in the field due to vignetting.

Vignetting is a form of shadowing by the lens mount that narrows the beam of light forming the image. It can happen at both the front and the rear of the lens.

Figure 18-25 shows a simple two-element lens and the beam of light illuminating the corner of the image. The diameters of the lens mounts are just large enough to allow the full beam to illuminate the corner of the image. If you placed your eye in the corner and looked at the lens, you would see the aperture of the lens as an ellipse. The cosine^4 law is working.

In Figure 18-26, the effective opening of the mount has been reduced. Part of the beam of light has been "shadowed" by the mount so that the illuminance of the corner is reduced. If you look at the aperture of this lens, the shape of the aperture is no longer elliptical—it has changed to a pointed oval shape as shown.

These drawings show how the original ellipse, vignetted by the edge of the lens mount, is changed in shape and reduced in area. This reduction in area reduces the illuminance of the image.

The front mount can also cause vignetting. When this happens, it is the opposite edge of the ellipse that is cut off.

When the vignetting is slight, it can be eliminated by stopping the lens down. Figure 18-27 shows how the vignetted ellipse changes when the aperture is reduced. On the other hand, if the vignetting is severe, the corner may receive no light at all when the diaphragm is stopped down to a very small lens opening

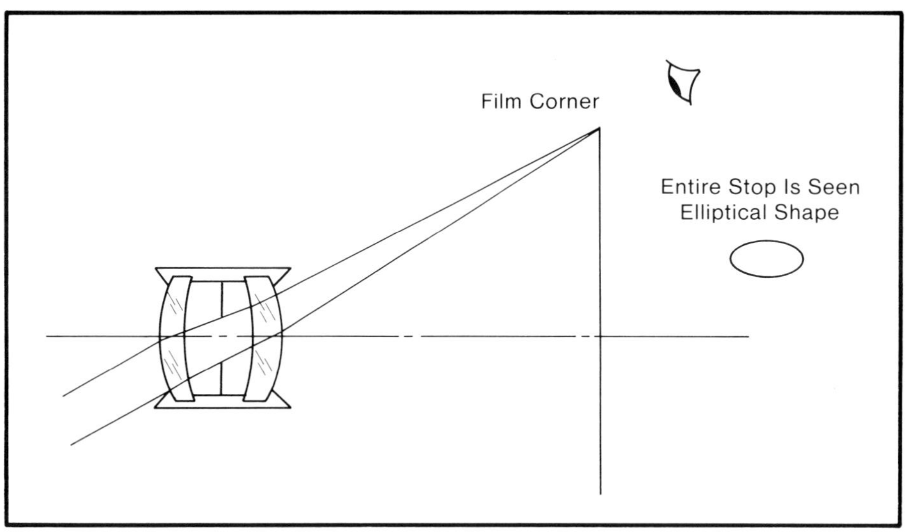

Figure 18-25. Lens With Well Designed Mount: No Vignetting.

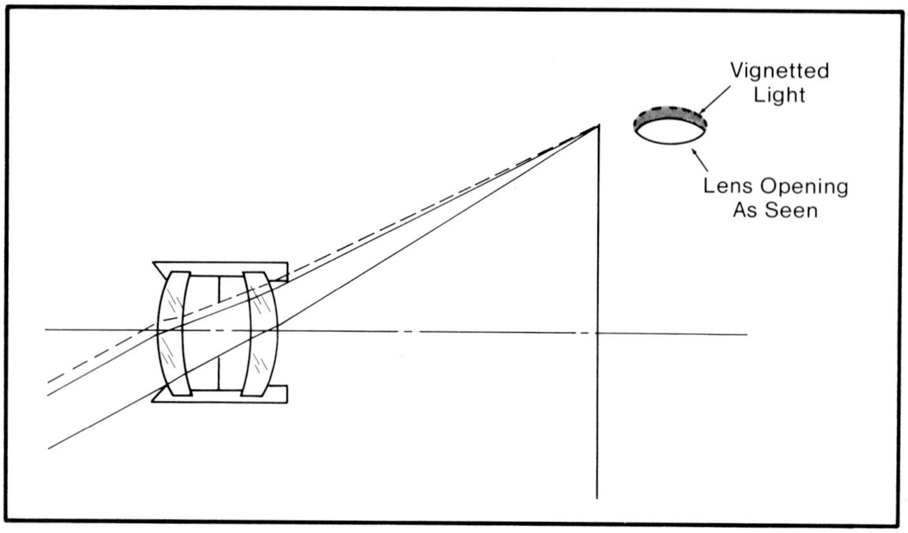

Figure 18-26. Lens With Back Mount That Vignettes.

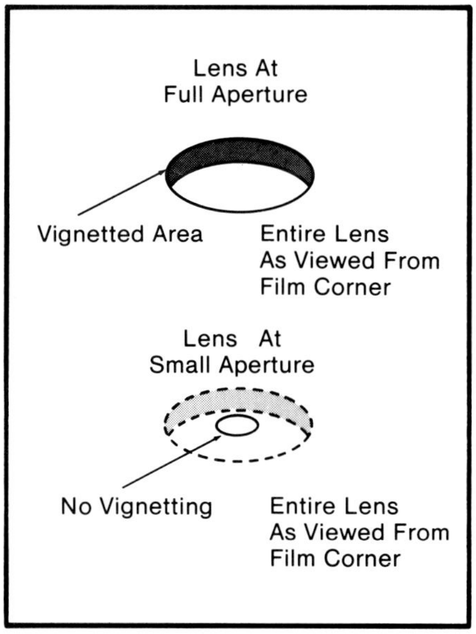

Figure 18-27. Slight Vignetting.

(Figure 18-28).

A photographic lens may be relatively free of vignetting, but a photographer can cause vignetting by adding an accessory such as a lens hood. A lens hood is designed to reduce image flare when there are bright areas such as the sky, or worse yet, the sun, just outside the image field. However, if the hood is too long, or too small in diameter, it can cause vignetting of the image. The image may be viewed with the lens wide open and the vignetting may not be too obvious. However, when the photograph is taken with a smaller aperture (larger *f*-number), the corners of the image may receive no light at all. Vignetting must be especially watched when view cameras are used and the camera adjustments, such as a rising front, are employed.

Lens Fields and Coverage

The image of a scene formed by a lens is projected on the image plane as a circular area of light. Vignetting reduces the illumination on the film to zero at the outer limits of the image field. The area within this circle is called the circle of illumination. In most cases, the definition (image sharpness) in the outer part of the circle of illumination is poor. The area over which the definition is acceptable is called the circle of acceptable definition (Figure 18-29).

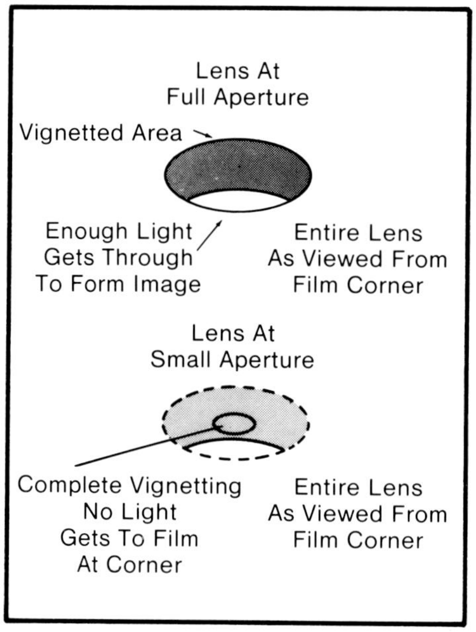

Figure 18-28. Large Amount Of Vignetting.

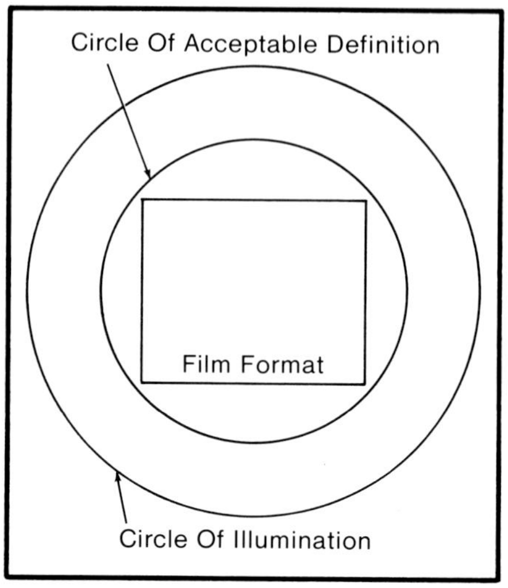

Figure 18-29. Entire Lens Field At Full Aperture.

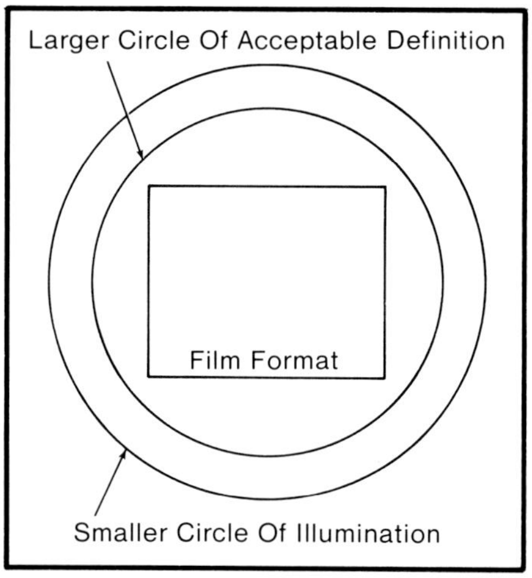

Figure 18-30. Entire Lens Field With Lens Stopped Down.

As the lens is stopped down, the circle of acceptable definition increases in diameter, while the diameter of the circle of illumination decreases (Figure 18-30). In most cases, the lens is designed so that the film format fits just inside the circle of acceptable definition. For view camera lenses, where the lens is tilted or moved from its normal position, the circle of acceptable definition must be quite a bit larger than the film format.

Field Angle and Covering Power

Photographers refer to these same factors from another viewpoint. With a normal focal length lens, where the focal length equals the film diagonal of about 6 inches angle is 53° (a half angle of 26.6°). The lenses are made to cover 55° with a little safety factor (Figure 18-31).

A 4 x 5-inch film size camera has a film diagonal of about 6 inches (152.4 mm). A lens must have a circle of acceptable definition of over 6 inches for use with this camera. Some lenses cover a 4 x 5-inch film with the aperture wide open, and a 5 x 7-inch film with the aperture stopped down (Figure 18-32). This means that the lens has a circle of acceptable definition of just over 6 inches with the aperture wide open, and that the circle of acceptable definition increases to a diameter of 8.6 inches when the aperture is stopped down.

Simple lenses have a relatively large circle of illumination and a relatively small circle of acceptable definition. An achromatic telescope objective may have a circle of illumination that covers a 90° field angle or more, but a circle of acceptable definition of less than 5°. Camera lenses, which are highly corrected, have circles of illumination only slightly larger than their circles of acceptable definition.

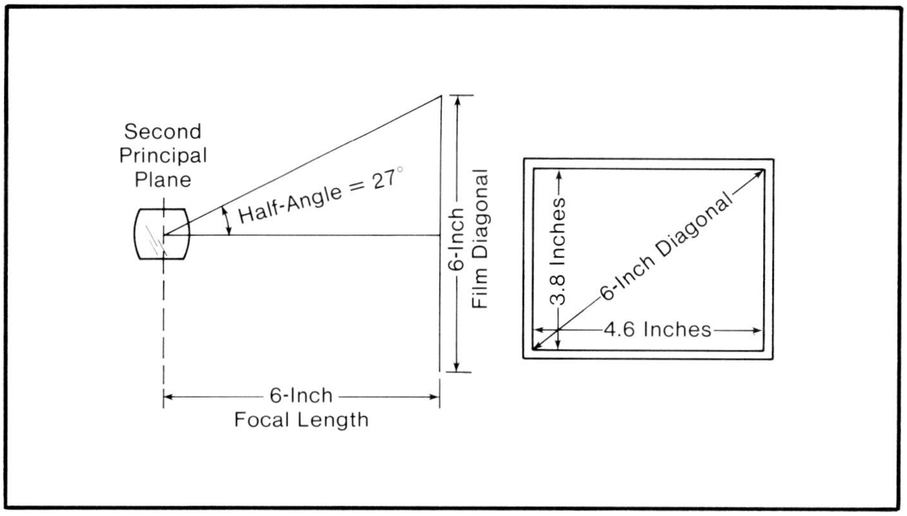

Figure 18-31. Normal Focal-Length To Film-Diagonal Relationship.

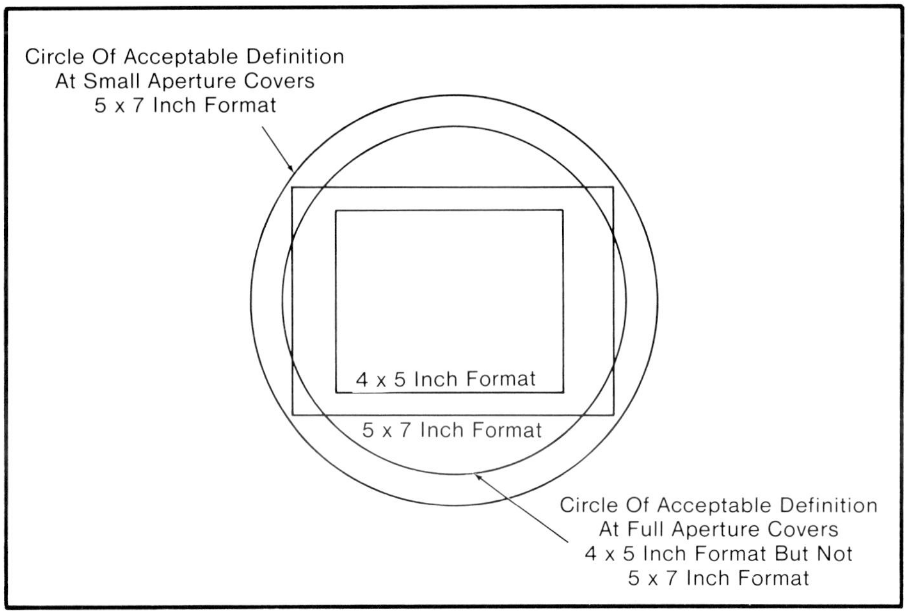

Figure 18-32. Lens Fields Of Lens That "Covers" 4 x 5 Inch Film At Full Aperture And 5 x 7 Inch Film Stopped Down.

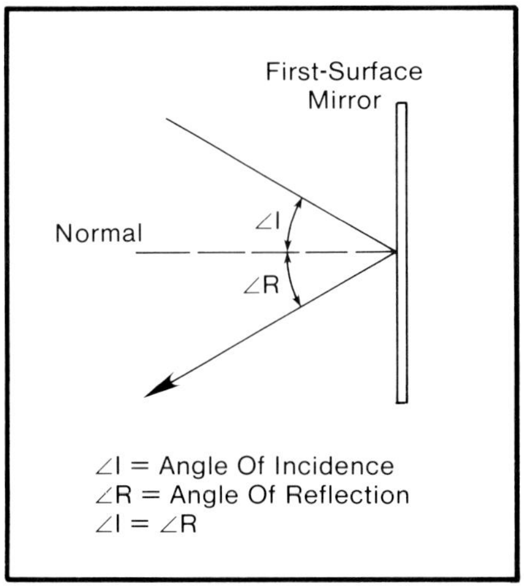

Figure 18-33. Law Of Reflection.

Images With Plane Mirrors

Plane mirrors are often used to reflect images for various optical reasons. The principles of reflection are based on one basic rule: the angle of reflection equals the angle of incidence. Figure 18-33 shows a ray of light reflected from a plane mirror surface. The angle of incidence and reflection are measured to the normal.

However, with complex subjects, determining how the image will be oriented after reflection is not always easy. The two drawings in Figure 18-34 show the letter F being reflected. Which of the two drawings is correct?

You may have said that the left drawing is correct because you remember that a single reflection in a mirror reverses the image left to right. If you look at your own image in the mirror, you see your hair parted on the opposite side. You know that the part is on the left side of your head, but in the mirror image it is on the right side.

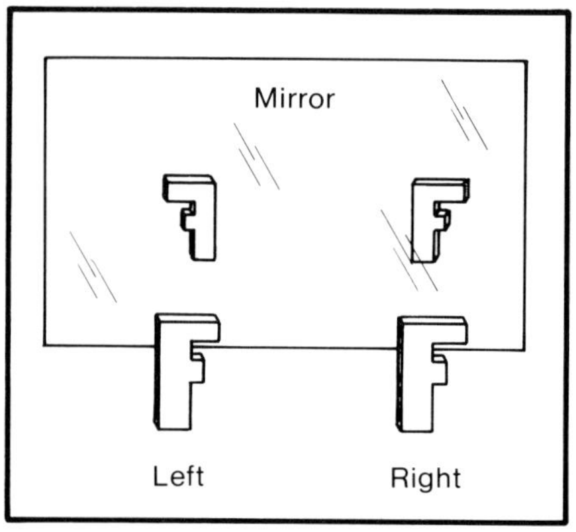

Figure 18-34. Which Reflection Is Correct?

However, the right-hand drawing is correct. What has fooled you is that you are looking at the back side of the letter F (Figure 18-35). From the front (the side toward the mirror) the letter F in the right drawing is reversed, left to right. Therefore, in the mirror it will be reversed and will look correct.

Schematic Ray Tracing With a Plane Mirror

In Figure 18-36, a plane mirror, an object, and an observer are shown (the observer is indicated by the eye). Where will the observer see the image of the object? Trial and error shows the rays can be found that follow the law of reflection—but there is an easier way.

Draw a normal to the mirror surface so that it intersects the object. Carry it on through the mirror. At a distance behind the mirror equal to the distance between the object and the mirror, locate the image of the object. Draw a ray from the observer to the image. This ray strikes the mirror at Point P. Now continue the ray from P to the object. Draw a normal at Point P, and it will be found that the angle of reflection equals the angle of incidence.

If the observer moves to another position, the image stays at the same location, but the ray paths change; angle I still equals angle R. Note that wherever the observer is, the image appears to be the same distance behind the mirror as the object is in front of it.

Two Rangefinders

An object is located at an unknown distance outside a window (Figure 18-37). The distance is fairly close to the window. If another object inside the room is placed so that its reflection seems to be the same distance from the window as the first object, its distance from the window can be measured and is the same as the outdoor object's distance from the window. In the drawing, the outdoor object is found to be 12 feet from the window.

A lighted lamp makes a good indoor object to use because its reflection is easily seen in the window.

A beam splitter and a measuring tape can be used to make another type of rangefinder. Figure 18-38 shows the arrangement. As you look through the beam splitter, both the tape and objects beyond the beam splitter are seen. The object is seen

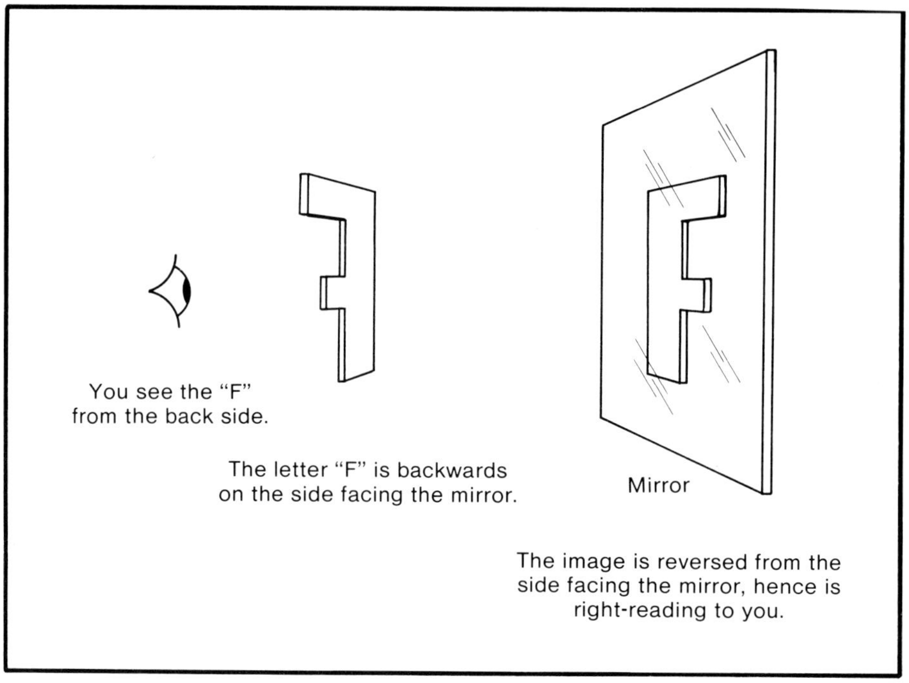

Figure 18-35. Explanation of "F" Reflection.

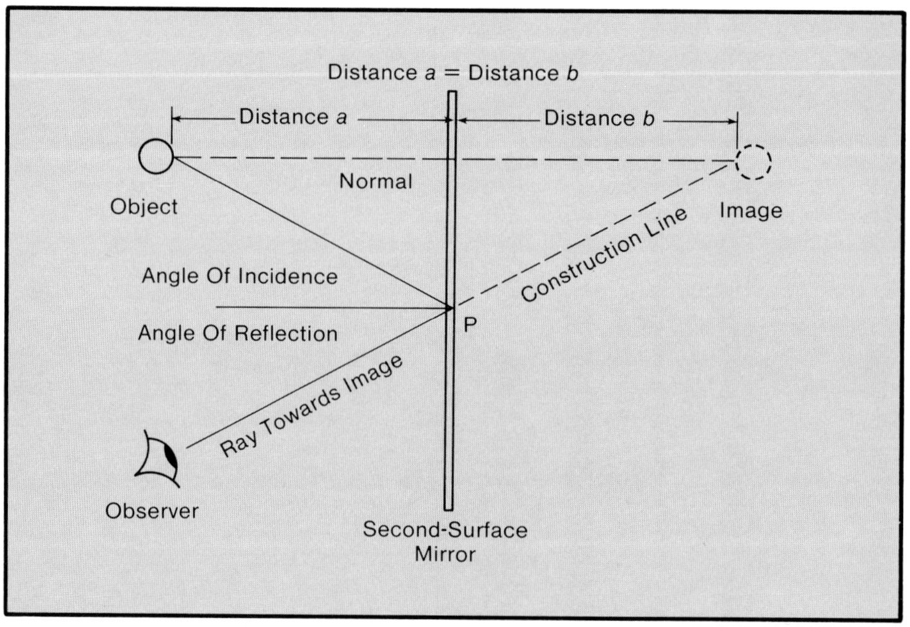

Figure 18-36. Schematic Ray Trace Of Mirror Reflection.

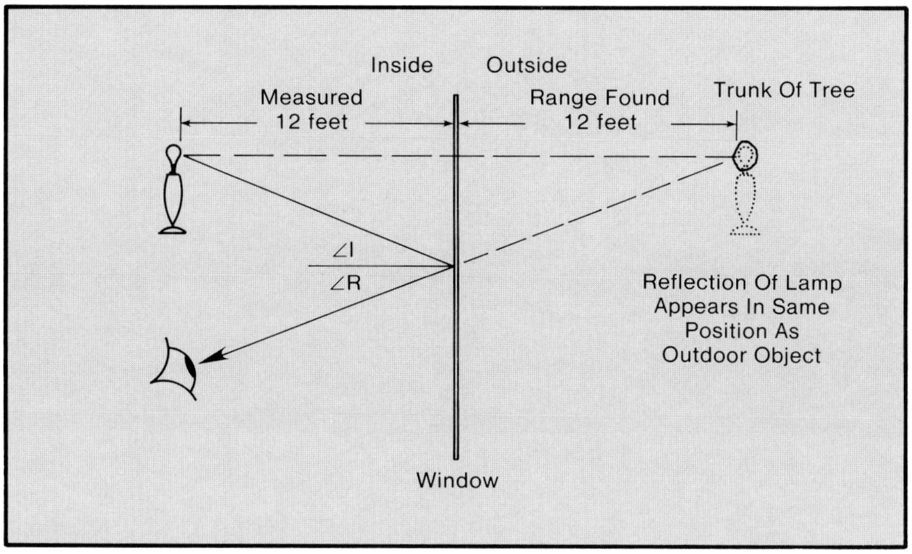

Figure 18-37. Using A Lamp As A Rangefinder.

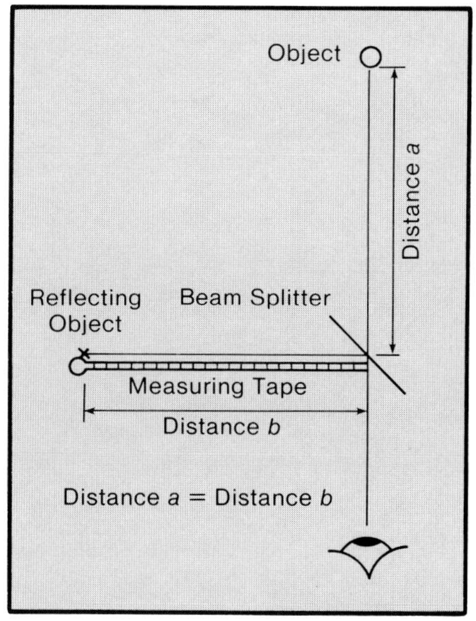

Figure 18-38. A Beam-Splitter and Tape-Measure Rangefinder.

superimposed on the image of the tape. The distance of the object behind the beam splitter is read directly from the tape.

A Beam Splitter Measuring Loupe

In an earlier lesson, a measuring loupe was described. It consisted of a magnifying lens and a reticle that contained calibrated linear and angular scales. In use, the reticle end of the loupe is placed on a surface, and the desired measurements are made under the magnification provided by the lens.

If the object to be measured is in a cavity, or a depression, the regular magnifying loupe cannot be used. However, a beam splitter arrangement can be used satisfactorily.

In Figure 18-39, the location of the magnifying lens, beam splitter and calibrated reticle are shown. The beam splitter is elliptical so that its 45° projection appears circular. The reticle is placed in the focal plane of the lens. The loupe is placed at a distance equal to its focal length from the recessed object to be measured. The image of the object is thus superimposed over the image of the reticle seen through the magnifying lens. Using this device an angular or linear measurement can easily be made of an otherwise inaccessible object.

Reflex Gunsights

When aiming various types of guns it is often useful to have the image of a reticle located at infinity, where it is superimposed over the target object. A terrestrial telescope with a reticle placed at the focal point of the objective and eyelenses is one way of accomplishing this. The reticle pattern is often etched on the surface of a field lens, however, there are other ways of accomplishing the same purpose.

In one device, a concave mirror and beam splitter are used to image a reticle. Figure 18-40 shows the arrangement. Note that the mirror is a second-surface mirror, which is unusual.

In an earlier discussion of image-forming concave mirrors, it was shown that a concave spherical first-surface mirror suffers from spherical aberration. A common way of correcting the spherical aberration is to make the mirror surface parabolic

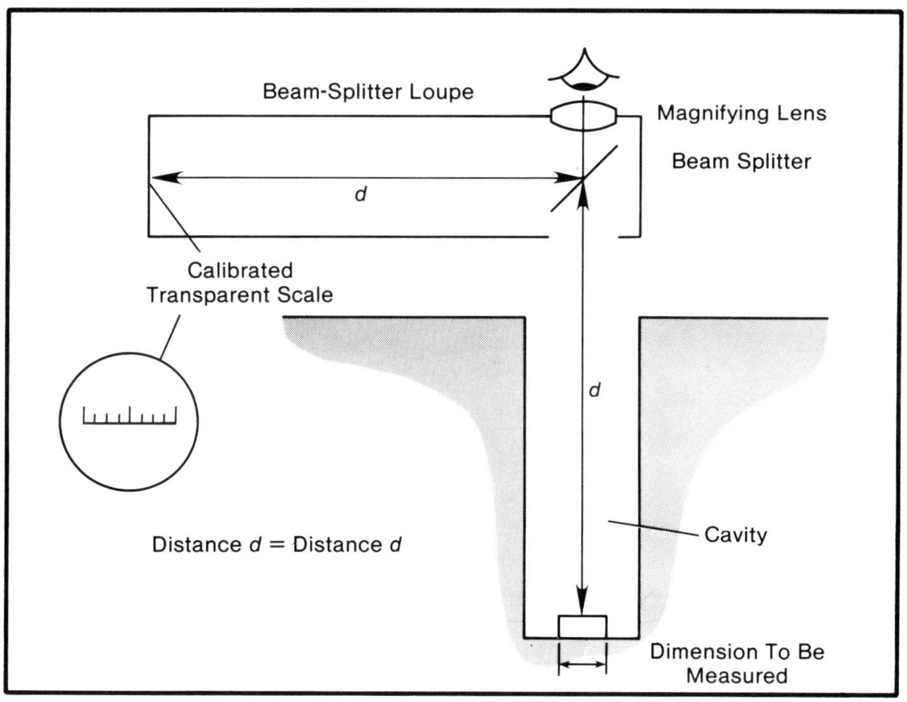

Figure 18-39. Cavity Loupe With Beam Splitter.

instead of spherical.

For some purposes, two spherical surfaces, as shown, correct the spherical aberration. The front surface has a curvature related to the second surface in such a way that the degree of spherical aberration is reduced. This is similar to using a glass block to correct the spherical aberration of a lens.

The reticle is illuminated. The light goes through the beam splitter which makes the diverging light rays to the mirror parallel. The reticle, of course, is at the focal point of the mirror. Some of the parallel light rays are reflected by the beam splitter to the eye. Both appear at infinity.

The eye sees both the object and the reticle through the beam splitter. The light rays from both are parallel, so the eye can focus both simultaneously. Because the light rays are parallel, there is no parallax, and the eye can move back and forth without changing the position of the reticle pattern on the object.

A similar gunsight uses a lens, a beam splitter and a reticle (Figure 18-41). The illuminated reticle is located at the focal point of the lens. Parallel light rays are emitted by the lens and are reflected by the beam splitter to the eye. The object is seen through the beam splitter with the reticle image superimposed over it. This gunsight is also parallax free.

The lens required is a relatively simple one, but such lenses usually exhibit curvature of field. The reticle pattern is, therefore, etched or scribed on a curved surface to match the field curvature. For correction of spherical aberration of the objective lens, a positive meniscus lens may be used for the reticle, and the reticle pattern scribed on the convex surface. This correction is similar to the solution for the concave mirror described above.

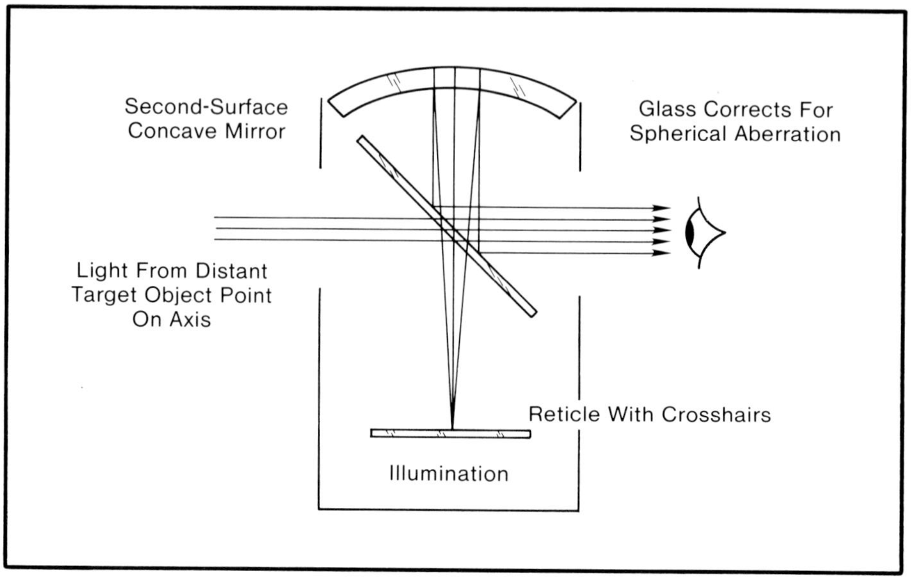

Figure 18-40. Concave-Mirror Gunsight With Unit Power.

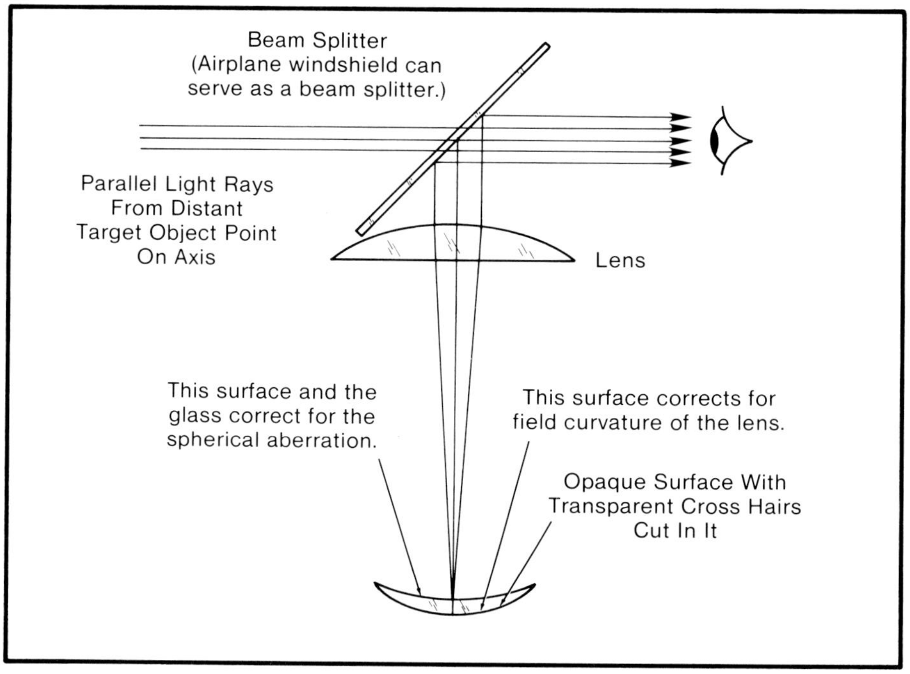

Figure 18-41. Projected Cross-Hair Unit-Power Gunsight.

Military aircraft use this type of gunsight. The windshield can be sloped at 45° and serve as the beam splitter. The pilot sees the reticle as he looks through the windshield. The pattern does not move in relation to a distant object even if the viewer moves his head. The sight is parallax free.

Problems

1. Here are two lenses with the following characteristics:

Characteristic	Lens A	Lens B
Diameter	.3 inches	3 inches
Focal Length	.6 inches	24 inches
Relative Image Illuminance	16	1
Aperture	*f*/2	*f*/8

   Lens B has 10X the diameter of Lens A. Hence, its area is 100X that of Lens A. It allows 100X as much light to enter, yet the image of Lens A is 16X as bright. This is a 1600X discrepancy. Explain using drawings and math.

2. You are watching a weather forecaster on television. The forecaster is behind a transparent map. He easily marks the weather symbols on the maps. The map appears right-reading both to the forecaster behind the map and to you looking at your television screen. There is an optical trick involved. Explain how this is accomplished.

Clue: The forecaster is left handed. Does he look left handed to you?

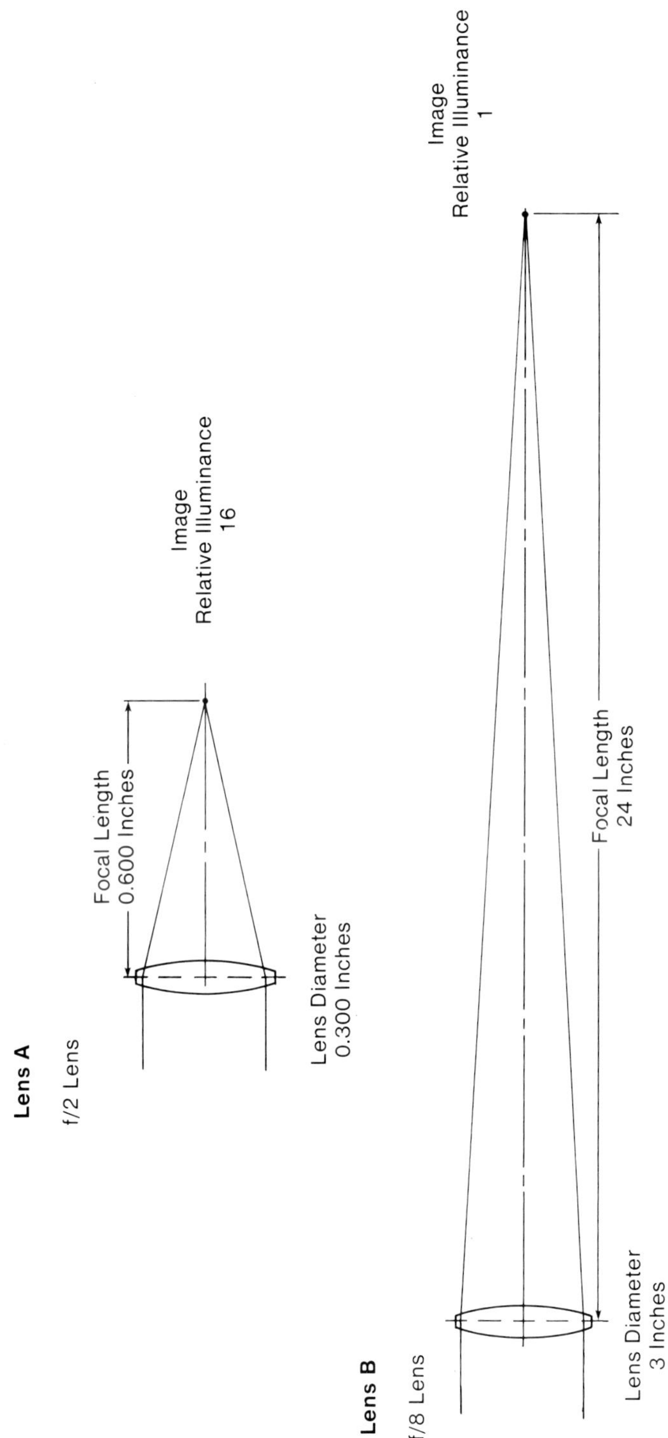

Problem 1
Chapter 18
Video Lesson 13

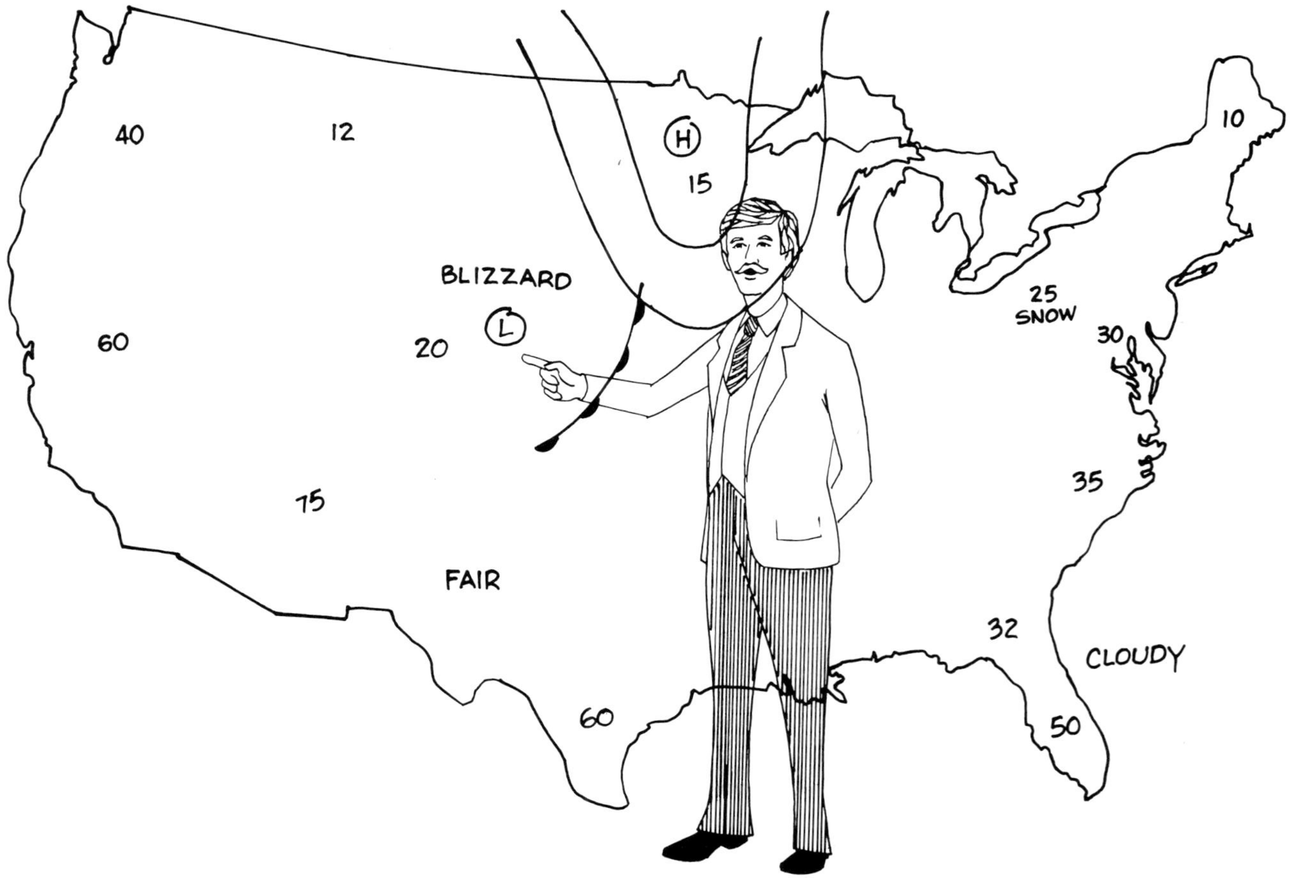

Problem 2
Chapter 18
Video Lesson 13

Chapter 19

VIDEO LESSON 14

Image Illuminance and Reflected Image Orientation

Summary

Two factors explain how the image of a small lens can be much brighter than the image of a large lens. A single mirror reflection reverses a camera image. Two adjoining mirrors at 90° to each other reflect a right-reading image. A light funnel is only partially helpful in increasing the amount of light at the funnel's apex. A device is described that permits studying the rotation of an image reflected from one, two or three mirrors. A transparent reference cube helps find image rotation when an image is reflected from a single mirror. Cascading cubes aid in the analysis of reflections from two mirrors. A transparent reference hemisphere permits the study of image orientation at many reflection angles.

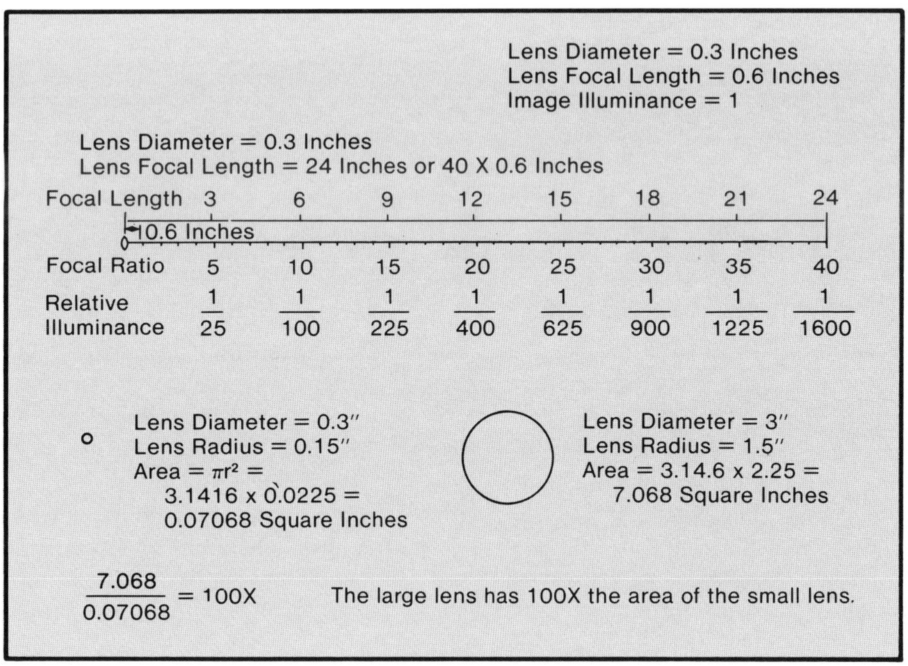

Figure 19-1. Comparison of the image illuminance of two lenses.

f-numbers and Illuminance

In the first problem, two lenses had the following characteristics:

Lens	Diameter	Relative Area	Focal Length	Relative Illuminance
Small	0.3″	1	0.6″	16
Large	3.0″	100	24.0″	1

The relative areas of the two lenses were found (in the last lesson) to be 100 to 1. This is a valid indication that the large lens allows 100 times as much light to enter as small lens. The *f*-number of a small lens is $\frac{.6}{.3} = 2$, while the *f*-number of the large lens is $\frac{24}{3} = 8$. By the difference in *f*-numbers, the image of the small lens is 16 times as bright as the image of the large lens. This is an apparent discrepancy of $100 \times 16 = 1600$ times.

The answer lies in the distance from the lens to the image—which is the focal length. The lens acts like a source of light so that the inverse square law applies, just as it does in calculating the cosine-fourth law.

As Figure 19-1 indicates, at a distance of 3 inches from a lens, the illuminance is 1/25 as bright or it is 1/5 as bright as it is at 0.6 inches. At 6 inches, it is 1/100 the relative illuminance or is 1/10 as bright as it is at 0.6 inches. Twenty-four inches is 40 times 0.6 inches. So, at 24 inches, the light is $1/40^2$ or 1/1600 as bright if the two lenses have the same *f*-number. However, because the area of the small lens is only 1/100 as large as the area of the large lens, the image of the large lens is 100/1600 or 1/16 as bright as the image of the small lens.

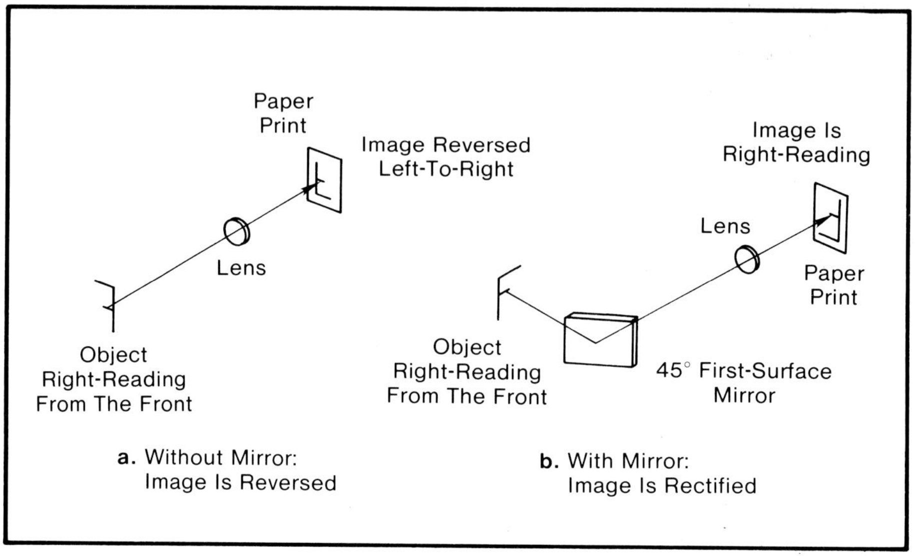

Figure 19-3. Image Rectification In A Booth Camera.

TV Trickery

In the second problem you were to explain how the TV weather reporter could be writing on a transparent map that read the right way for him, and yet viewed on the opposite side by the TV camera (and you, looking at your TV screen) it is also read the right way.

The answer should be clear. A single reflection in a mirror reverses the image from left to right (Figure 19-2). As the drawing shows, the TV camera is aimed at 90° to the map and there is a 45° mirror (it could also be a right-angle prism) in front of the camera lens. Viewed from the camera side, the map is backwards. When the image is reversed by the mirror it becomes right-reading.

The mirror also reverses the reporter so that while he writes with his left hand, it appears to you that he is writing with his right hand. If you look carefully, his hair is parted on the wrong side, his coat buttons the wrong way, and the breast pocket of his suit coat is on the wrong side.

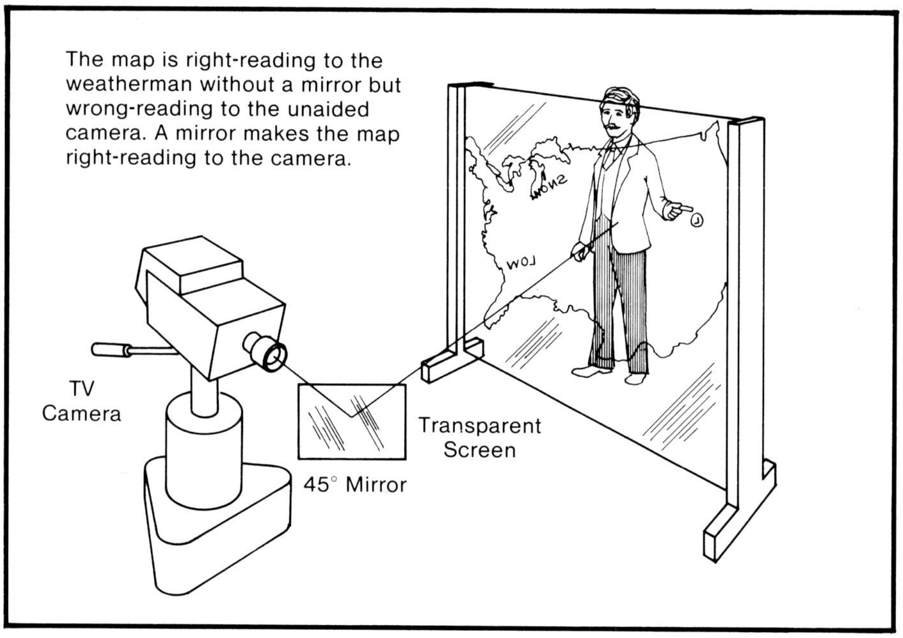

Figure 19-2. TV Magic.

Photo Booth Images

A photo booth is a small portrait studio in which a coin initiates the operation of lights, camera and processing. A sitter enters the booth, primps in the mirror placed on the wall opposite the seat, the lights come on, the sitter poses, and click, a picture is taken. Usually several poses are made in a short period of time. Soon out of a slit in the wall comes a strip of small portraits.

The portraits are made on direct-positive photographic paper. The process is a chemical reversal process that produces positive images on the paper. No film is involved.

Since the photo is imaged directly on the photographic paper, the portraits are reversed left to right (Figure 19-3a). Because people are used to seeing their own images reversed in a mirror, such portraits look natural to the sitter but look odd to other people who are used to looking at the sitter "unreversed."

Just as with the TV weather map, these images can be rectified with a single reflection in a mirror (Figure

19-3b). While such an arrangement makes the pictures right-reading, they do not look quite natural to the sitter. Most photo booths produce reversed portraits, which seem to satisfy most users of the photo service provided.

A Right-reading Mirror

As stated above, most people like the way they look in a mirror because that's the way they see themselves every day. Men use their mirrors while shaving and combing their hair. When parting it on the left side, it looks to them as though they have parted it on the right. Women apply makeup, do their hair, etc., looking at their reversed images in their mirrors, too.

Figure 19-4. Right-Reading Mirror.

A direct-vision rectifying mirror which allows you to see yourself as others see you is simple to make. It consists of two mirrors butted together and fastened at exactly 90° to each other (Figure 19-4). While a single mirror reflection reverses the image from left to right, the second mirror "rectifies" the image and people see themselves in the same way others see them. Such mirrors are used in optical instruments but are not practical for everyday use because of the space they require.

Roof Prisms

Right-angle prisms, as discussed previously, are used to bend beams of light 90°. As with a single mirror reflection, a right-angle (Porro) prism reverses the image left to right, or top to bottom, depending on prism orientation.

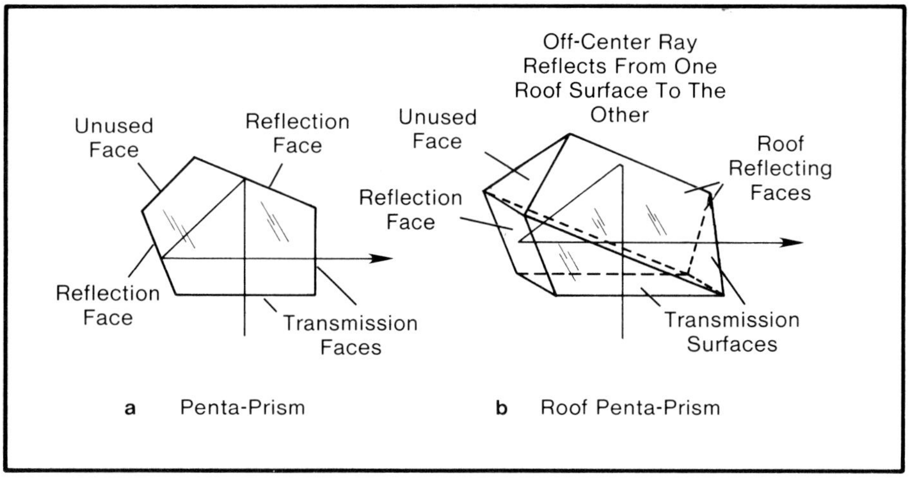

Figure 19-5. Penta- and Roof Penta-Prisms.

If the reflecting face of a right-angle prism is changed to two surfaces at 90° to each other and 45° to the transmission faces, the image light is bent 90° without reversal. Such prisms are called roof prisms because the two surfaces come together like the roof of a house.

A penta (five-sided) prism is made to provide two reflections, as shown in Figure 19-5a. If one of the reflecting surfaces is converted to a "roof" with its surfaces at 90°, the prism is called a "roof penta prism" (Figure 19-5b). The prism in the viewfinder of 35mm single-lens reflex cameras is usually a roof penta prism.

The image on the ground glass of a reflex camera is right-side up, but is reversed left-to-right. The roof penta prism folds the path by 90° so that by looking in the eyepiece (parallel to the objective axis) the image is seen right-side up and is correctly oriented left-to-right (Figure 19-6).

The light path through the prism

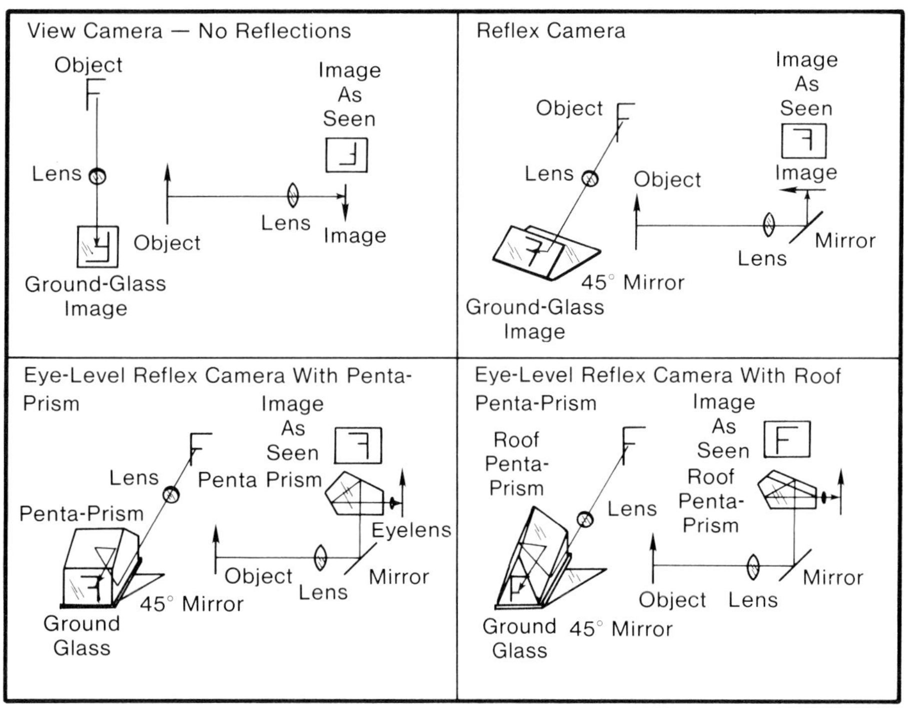

Figure 19-6. Camera Ground-Glass Imaging.

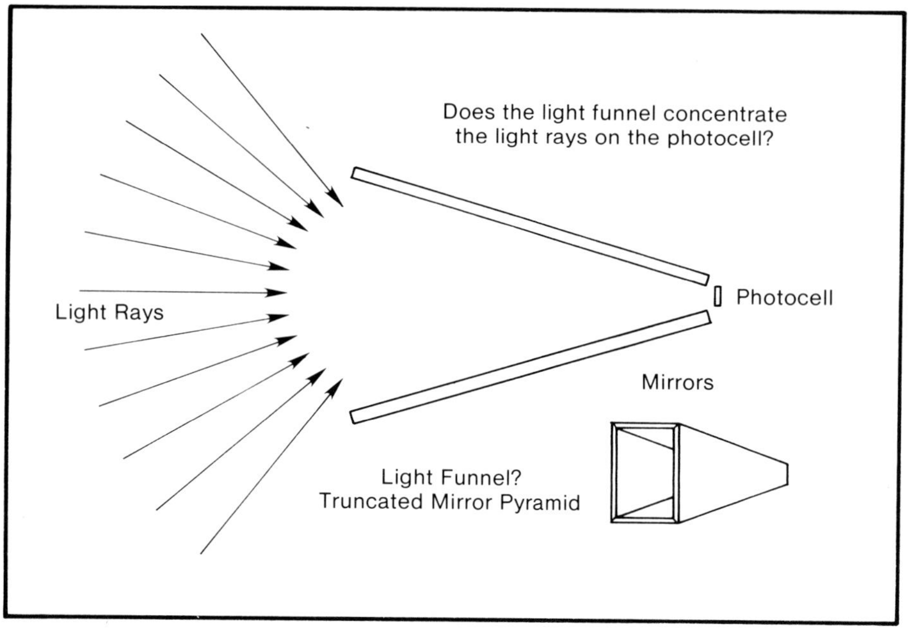

Figure 19-7. Light Funnel?

is equal to the focal length of the eyelens. The eyelens acts as a simple magnifier of the ground-glass image.

Most such finders have a very finely ridged fresnel lens next to the ground glass to act as a field lens so that the image is seen brightly at the corners of the ground glass.

Light Funnels

It might be thought that four plane mirrors, made in a funnel shape, could be used to concentrate light on a small area.

One such use might be in a camera to concentrate light on a small cadmium sulfide or silicon blue photocell to increase its illuminance.

Figure 19-7 shows such a funnel with the small area cell at the truncated apex. You might think that the light from the scene in front of the funnel bounces back and forth between the mirrors and eventually reaches the photocell. This does not happen. Actually, each light ray reflects back and forth between the mirrors until it reverses direction and is reflected back toward the scene!

The side view of the light funnel (Figure 19-8) shows the paths of two rays of light. It shows how one ray reverses direction and is reflected back in the general direction from which it came, while the other ray (which strikes the mirror near the apex) reaches the photocell.

Only if the sides of the funnel are parallel will the light rays continue to reflect back and forth. However, the rays will not concentrate the light at the light cell, so this device is impractical.

Most of the light rays that strike the mirrors are reflected back toward the scene. However, a few that strike the mirrors near the cell are reflected onto the cell and increase its response. If you look into the funnel toward the cell, you see the cell itself and reflections of the cell in the mirrors around the cell. It makes the cell appear as a multifaceted ball. You see the area of the cell as being increased somewhat. By the reversability principle, it is this larger area that accepts a larger amount of light which falls on the cell and increases its response.

Therefore, the light funnel can be made quite small. It only needs to be large enough to include the mirror areas that reflect light to the cell.

Multi-mirror reflections

Many optical instruments make use of reflections from plane-polished surfaces. Sometimes the reflections are produced with mirrors and sometimes with prisms.

In general, mirrors are less costly to make and weigh less. When multiple reflections are required, the angle accuracy of prisms requires less precise mounting. In some instances, prisms require less space than mirrors. In many cases, either mirrors or prisms can be used.

In discussing what happens to image orientation with single and multiple reflections, we shall use

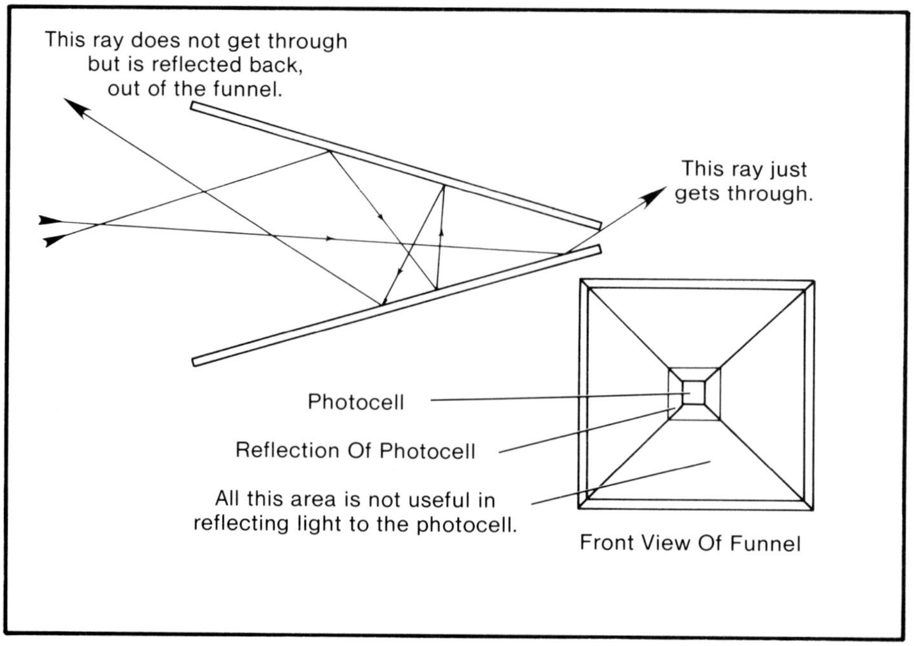

Figure 19-8. Ray Trace Of A Light Funnel Showing Useful Area.

mirrors as examples, but it should be kept in mind that in many instances prisms can be used instead.

It is fairly easy to determine the placement and angles of mirrors to get light beams to go where they are needed, but it is more difficult to get the image orientation that is required.

Two angles must be considered when orienting a mirror to the axis of a light beam. The first is the angular rotation about the vertical axis, and the second is angular rotation about the horizontal axis. Figure 19-9a shows a vertical mirror. Its surface is parallel to the vertical axis. Therefore, its vertical angular rotation is 0°. It is perpendicular to the axis of the light beam. Figure 19-9b shows the mirror rotated 45° about its vertical axis. The light beam is reflected 90°. The mirror's rotation about the horizontal axis is 0°.

Figure 19-9c shows the mirror rotated 45° about the horizontal axis, while the rotation about the vertical axis is 0°.

Note that the limit of rotation about either axis is 90° in either direction—a total of 180°. Since the angle of reflection equals the angle of incidence, the light beam is always reflected twice the angle of rotation of the mirror. If either of the two rotations is 0°, the other is easy to figure. However, if both angles are at angles other than 0°, the final angle of the mirror surface to the beam is difficult to calculate.

The orientation of the image, whether it is reversed or right reading, whether it is upright, upside down, or at an angle, is also difficult to determine.

When there are two or more mirrors, this becomes even more difficult. One of the methods used is the trial and error (empirical) method.

A Three-mirror Empirical Device

The videotape lesson shows a three-mirror device with which the image orientation can be found for reflections with one, two or three mirrors with limited rotations. The mirrors pivot and swing, providing the rotation. The device is designed so that you can use one, two or three mirrors.

The transparency of the letter F provides the object. A positive lens is placed in front of the letter F so that an image can be projected through the optical system. With the first mirror set at 45°, and the other mirror swung out of the way, the image is reflected 90°. By rotating the mirrors, the image can be projected up, down, to the left, to the right, and at all points in between.

The second mirror can be swung into place so that it intercepts the beam reflected by the first mirror. If this mirror is set at 45° on the second axis, the beam can again be reflected at 90° to its original direction just as with one mirror, but now it is being reflected twice. The orientation of the image at the same positions can be seen. The third mirror can be swung into place to intercept the beam, and using three reflections, the image orientation can be determined.

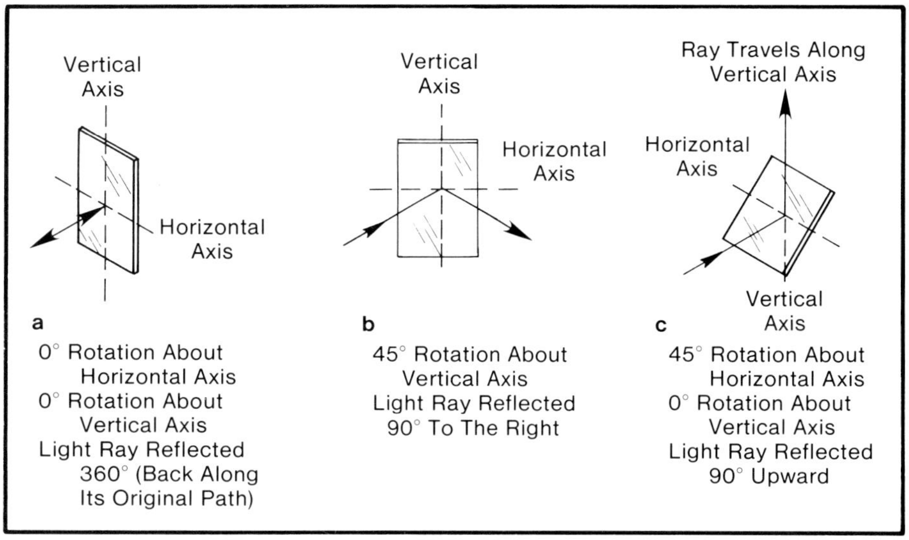

Figure 19-9. Mirror Axes.

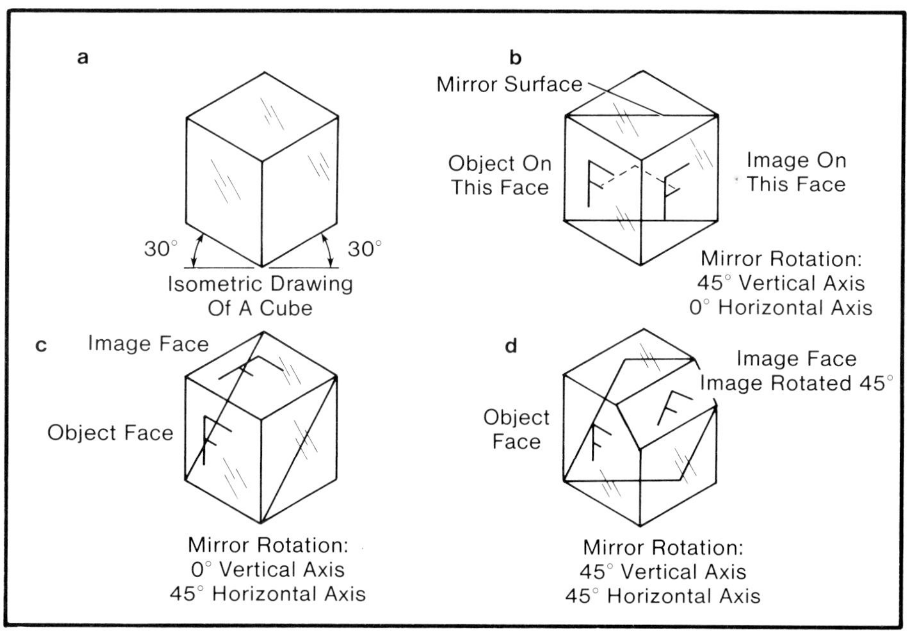

Figure 19-10. Reference Cubes

When a desired orientation is found, a bread-board model can be made using the same mirror placement with an assurance that the same image orientation will result. Many of the same mirror rotation and image orientation results can be determined with drawings. Two devices make this easier.

The Transparent Reference Cube

An isometric drawing of a cube is made (Figure 19-10a). The left front face is used to show the orientation of the object. Again, the letter F provides an easy object to use because it is not symmetrical, vertically or horizontally.

A mirror is drawn inside the cube (Figure 19-10b). It is vertical, so it is rotated 0° about its horizontal axis. It is rotated 45° about its vertical axis so that the image is visible on the right front face of the cube. The optical axis is shown by the dashed line. Seen on the front face, the image is correct reading. Remember, that looking at the image from the mirror position, the letter F appears reversed left to right. The image has been reversed left to right by the reflection.

Figure 19-10c shows the mirror rotated 45° about the horizontal axis and 0° about the vertical axis. The

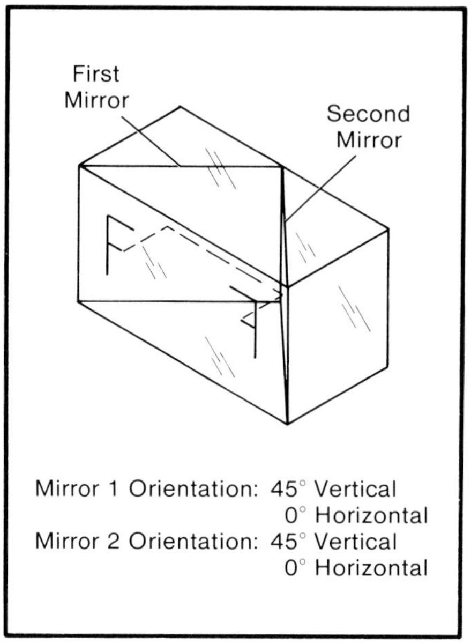

Figure 19-12. Horizontal Cascading With Two 45° Mirrors.

drawing shows the orientation of the image.

Figure 19-10d shows the upper-right edge of the cube cut off at a 45° angle. In order for the mirror to reflect the image up to the new face caused by cutting the cube, the mirror is rotated 45° on both axes. The image is reversed, and is rotated 45° clockwise.

The Cascaded Cube

To find the image orientation when a second mirror is added, an additional transparent cube can be added to the first. Adding cubes in this manner is called cascading (Figure 19-11).

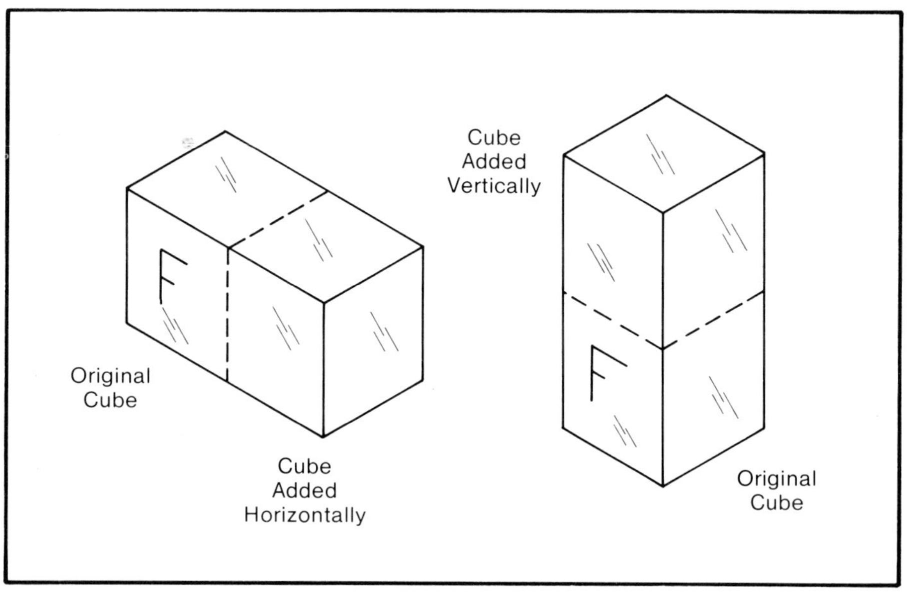

Figure 19-11. Cascading Cubes.

Figure 19-12 shows a second cube added to the right front face of the first cube. The first mirror is rotated 45° about its vertical axis and 0° on its horizontal axis. The second mirror is rotated 45° about its vertical axis and 0° about its horizontal axis. The image is seen on the front face of the second cube.

In Figure 19-13, the cubes are cascaded vertically. Both mirrors are rotated (45° on their horizontal axes) and the image is now seen on the front face of the second cube.

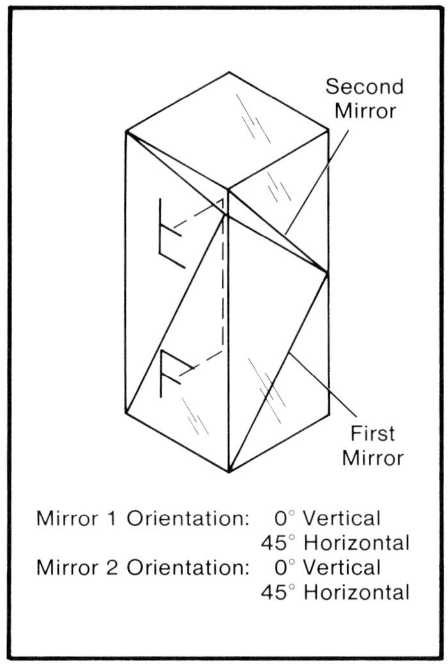

Figure 19-13. Vertical Cascading With Two 45° Mirrors.

Following the Orientation

The orientation of the image is not always readily apparent. Lines representing rays within the cubes can be used to determine the orientation. In the drawing of the single cube (Figure 19-14), a thin solid line represents a ray from the upper left of the object. A dashed line represents a ray from the upper right, while an alternating long-short dashed line represents a ray from the bottom of the F. By following the

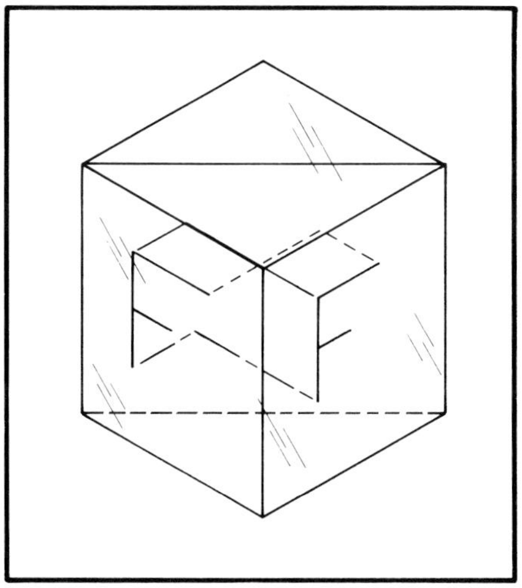

Figure 19-14. Rays That Establish The Image Orientation.

path of each ray the orientation of the image can be traced.

When drawing the ray lines it is necessary to visualize where each one hits the mirror. Both of the upper lines strike the mirror at the top, while the bottom line hits the mirror at the bottom. Because the mirror is not rotated about its horizontal axis, they stay the same distance from the top and bottom of the cube. The F is not inverted top to bottom. The left upper line hits the mirror near the front face, and therefore is reflected to the near edge of the right face. The upper right ray hits the mirror farther back, so it strikes the right face near the back edge. Therefore, the vertical line of the F is to the left and the open part of the F is to the right. This establishes the orientation of the image.

The several small drawings (Figure 19-15) show the reflections in two cascaded cubes with the mirrors arranged differently—but all at 45°.

Since all the faces of the cubes are 90° to each other (unless one is cut as in Figure 19-10d) using the cubes has certain limitations. An alternative method, that allows the use of mirrors at angles other than 45°, is the transparent hemisphere, which will be discussed later.

"F" Objects

The transparent cubes need not be drawings. Real cubes can be made of transparent plastic sheets cemented together. Only five sides of the cubes are used—the bottom is left open for placing over the mirror (Figure 19-16). A mirror can be mounted on a small base as shown in the drawing. The mirror is mounted so that it can rotate about a horizontal wire axis place in a yoke. The base is rotated for vertical axis rotation.

A letter F can be cut from black paper and waxed so that it will stick on the face of a cube and yet may easily be removed. When the mirror is positioned, and the cube set over it, you simply look in the proper face and see the orientation of the letter by reflection.

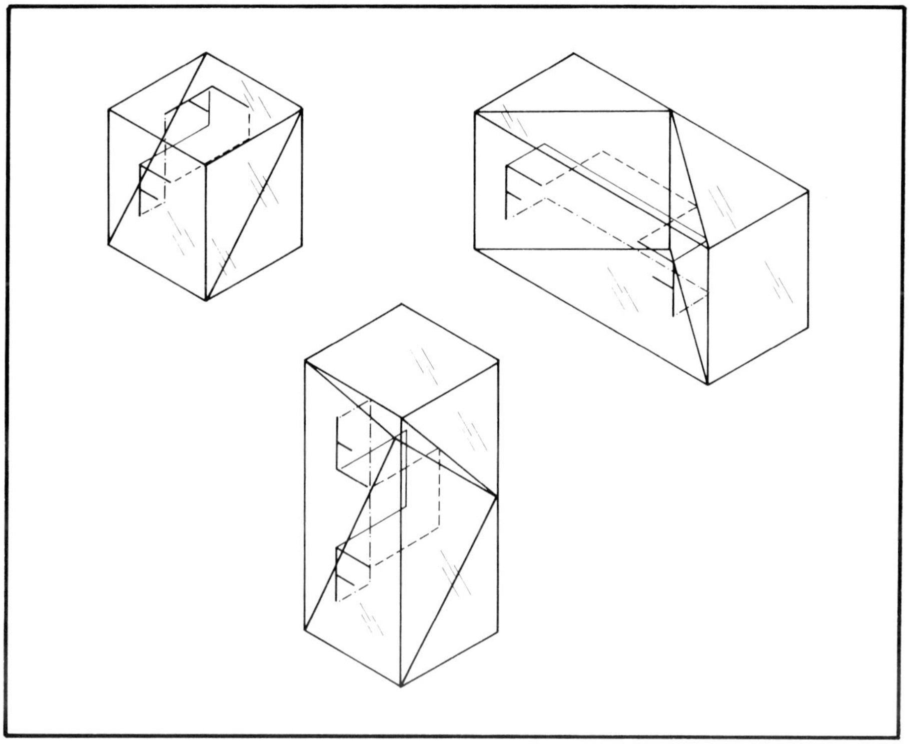

Figure 19-15. Tracing Rays To Find Image Orientation.

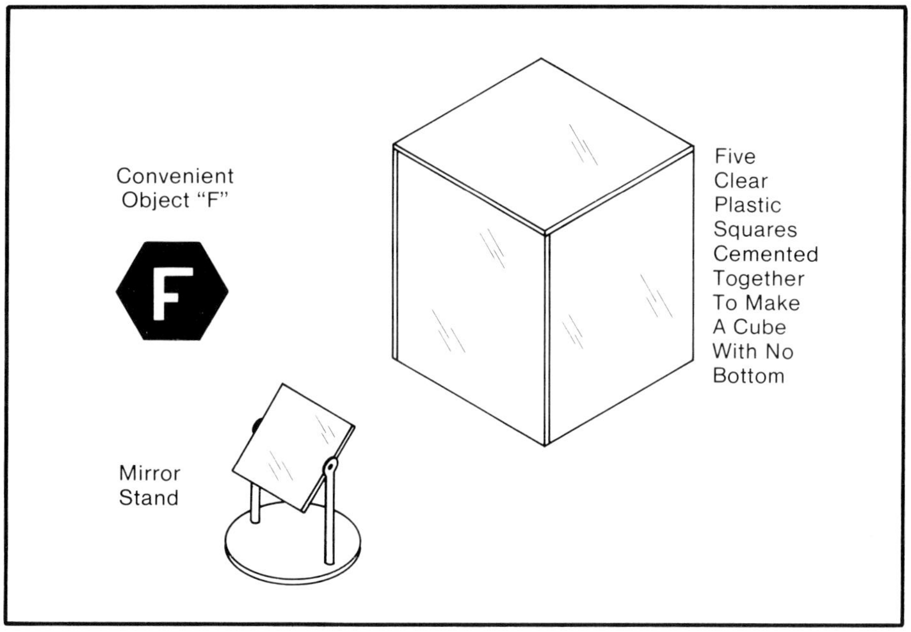

Figure 19-16. Plastic Reference Cube With Mirror Stand And Object.

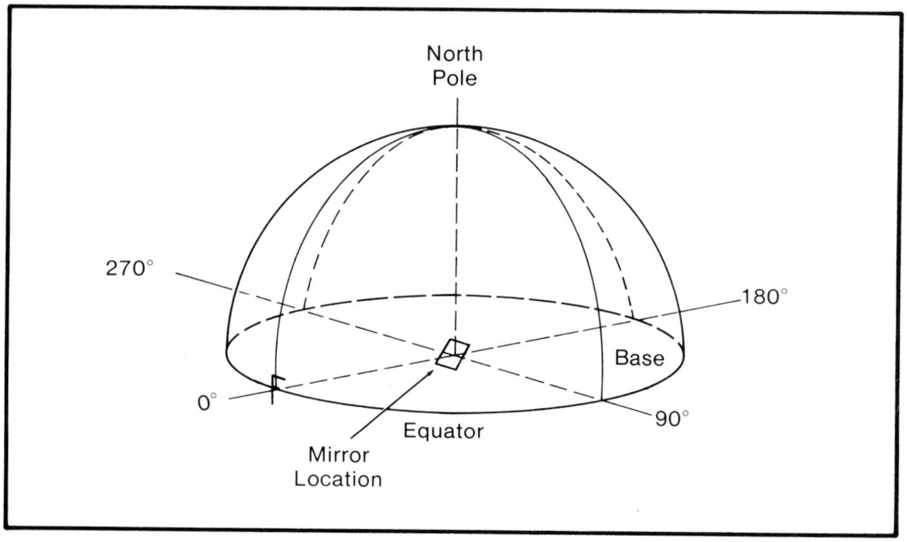

Figure 19-17. Reference Transparent Hemisphere (Front View).

The Transparent Hemisphere

A transparent hemisphere can be used to replace the cube as a device for finding the orientation of the image as reflected by a mirror. It is useful when the mirror angles are other than 45°.

Like the cube, the transparent hemisphere can be used in drawing form or as a real test device. Glass or transparent plastic hemispheres can sometimes be found as vases, fishbowls, hanging planters, etc.

The hemisphere is marked with two great circles at 90° to each other (Figures 19-17 and 19-18). These intersect at the "North Pole." The base plane is marked with two lines at 90° to each other that intersect at the center and that join the great circles at the equator (the periphery of the base). The object (letter F) is placed on the front left great circle arc just above the equator as shown.

When a vertical mirror (0° horizontal rotation) is placed in the center of the base and rotated about its vertical axis in small increments, a series of image positions can be found. These are shown in Figure 19-19. The F images seen on the front surface of the hemisphere are shown as solid images. Those on the back surface are shown as dotted images.

If the mirror is left fixed at 0°-vertical and is rotated in increments about its horizontal axis, the series of F images are seen on the 0° to 180° great circle (Figure 19-20). As they go on the opposite side, they are shown as dotted.

The image does not rotate as it travels on this great circle. At the North Pole it is horizontal. The upright of the F is parallel to the great circle over which the image has been moving, and the base is toward the front. As it moves down the back of the hemisphere, the top line of the F leads. At the back equator, the reflection is a grazing one with the mirror nearly horizontal on the base, and the F is upside down.

If the mirror is rotated 45° about its vertical axis, and the rotation about the horizontal axis is zero, the image is on the equator at the 90° great circle (Figure 19-21). If the mirror is now rotated about the horizontal axis to make the image move up to the 90° great circle, the image rotates. The amount of rotation is equal to the number of degrees displacement from the base. If the image is halfway up, it is up 45° from the equator, and the angle of image rotation is 45°. At 90° displacement, the image has rotated 90° and is in the same position as it would have been if it had traveled up the front (0°) of the great circle.

The equation for the amount of rotation is this:

Angle of Rotation = Angle of Rise

This is a simplified form of a more general equation:

$$\text{Angle of Rotation} = \frac{\text{Angle of Rise}}{90^\circ} \times \text{Angle of Rotation at } 90^\circ$$

In this case, the angle of rotation at 90° is 90°. The two 90° angles cancel each other. This equation works fairly well for angles up to 90°, but it is not exactly true. However, it works satisfactorily for a first approximation of image rotation.

For angles near 180°, it fails completely. Fortunately, most reflections used in optical instruments are 90°

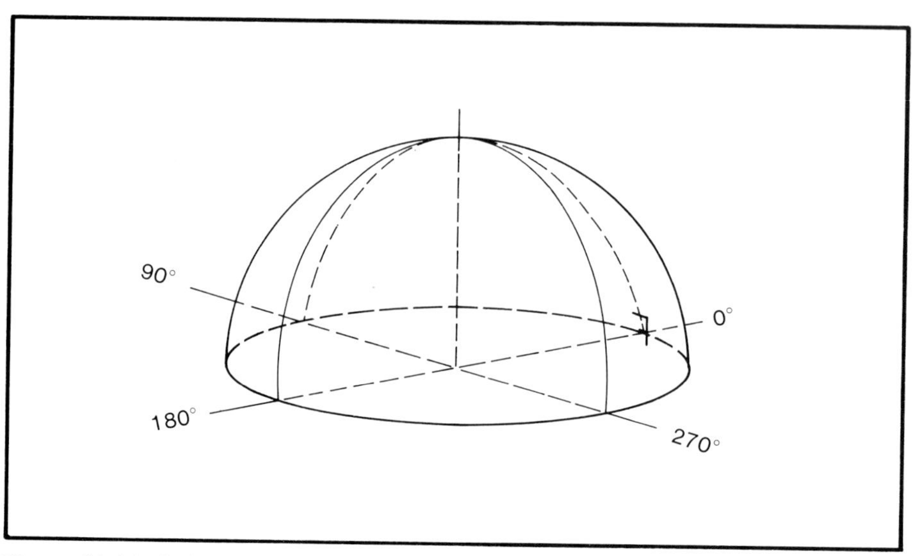

Figure 19-18. Reference Transparent Hemisphere (Left-Rear View).

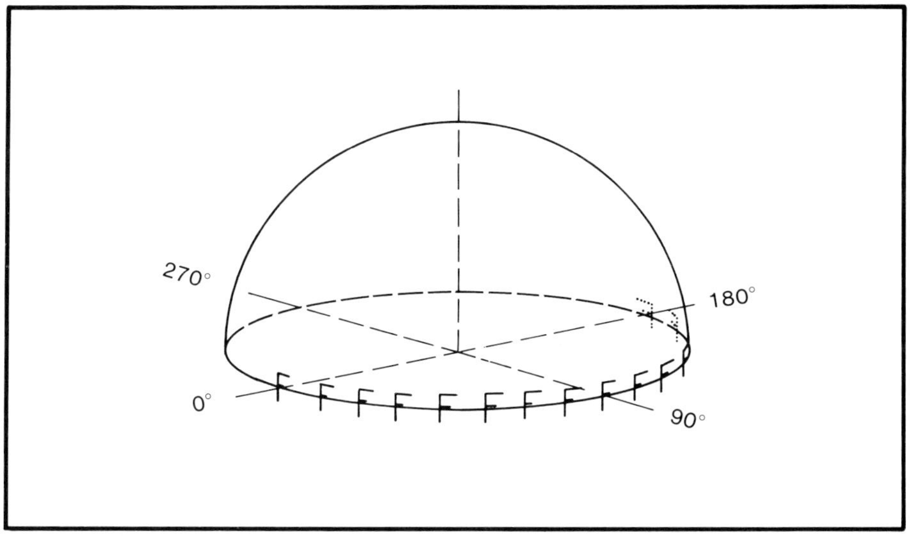

Figure 19-19. Image Orientation Around The Equator.

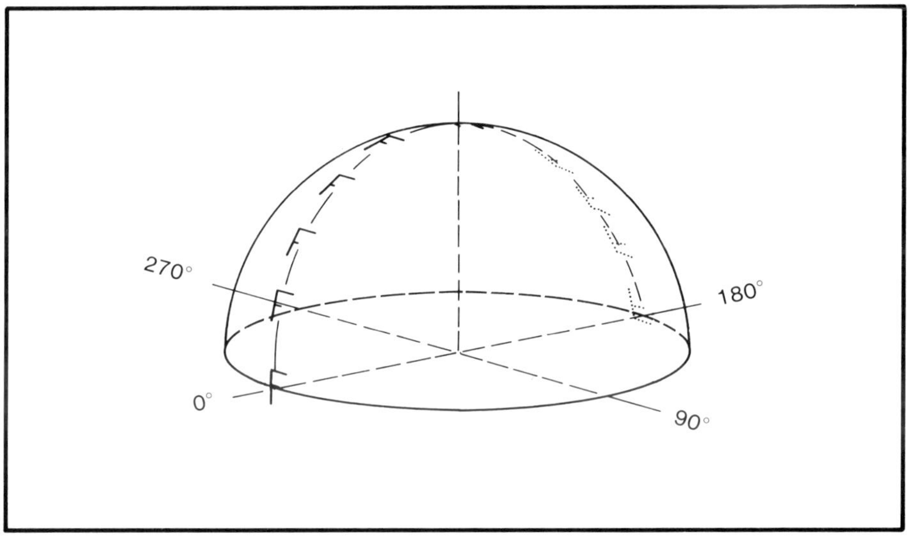

Figure 19-20. Image Orientation On The 0° – 180° Great Circle.

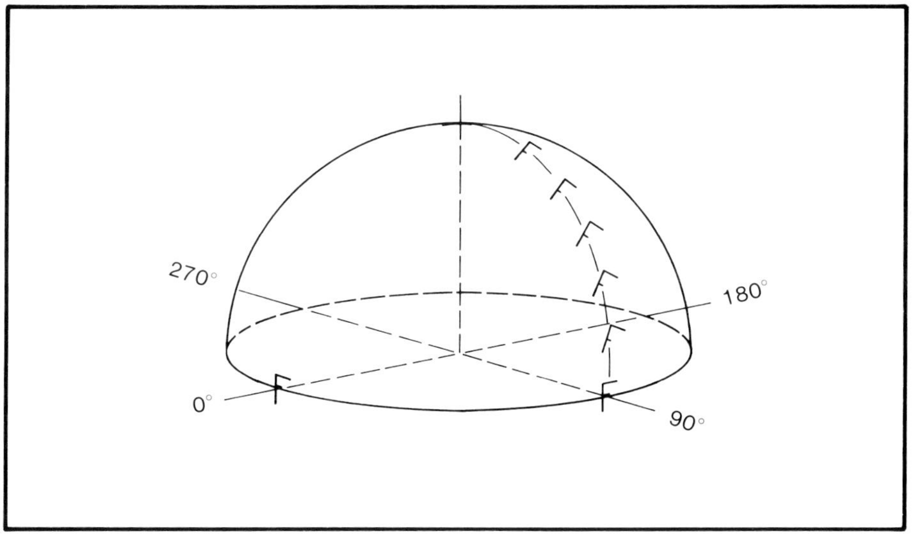

Figure 19-21. Image Orientation On The 90° – 270° Great Circle.

reflections where the image rotation is straightforward.

Problems

1. In the drawings, the angle of reflection at the base is 165°. The polar great circle at 165° is shown (drawing a) as are the image positions at the equator and at the north pole. Compare what you know must happen to the image rotation with the results you get using the equation and discuss how the equation must fail.
2. Drawing b shows the rear of the hemisphere with the 165° polar great circle. Draw F's in the spaces with the orientation you think must occur.

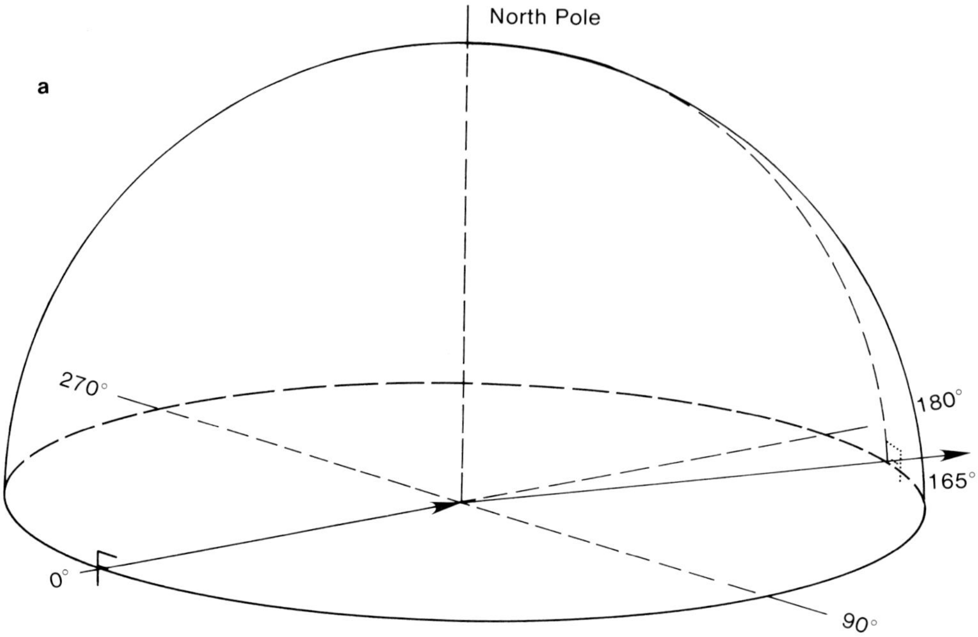

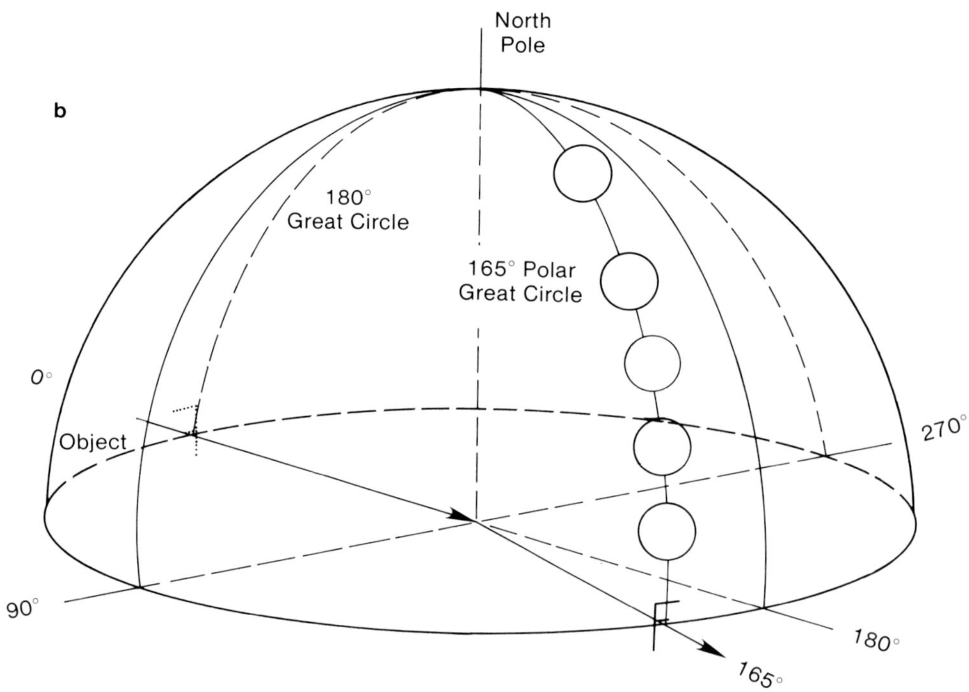

Problem
Chapter 19
Video Lesson 14

Chapter 20

VIDEO LESSON 15

Reference Hemispheres, Paper Path Folding, and Signal Mirrors

Summary

If a great circle band (rail) is pivoted at the 0° and 180° equator points, it can be used to find the orientation of reflected images. The degree of rotation at 180° is twice the rotation at 90°, where the degree of rotation is equal to the elevation. At other points on the rail, the degree of rotation is twice the degree of rotation at 90° times the number of degrees along the rail divided by 90°. Ray paths in drawings sometimes are difficult to visualize. A very useful way to find the position and size of mirrors is to make a paper model of the light beam and fold the paper where the mirrors are located. Signal mirrors for rescue purposes are provided with a method for aiming the beam of reflected sunlight.

More on the Reference Transparent Hemisphere

Reflections from mirrors, prisms and beamsplitters are used in many optical instruments. Every modern eye-level single-lens reflex camera has a movable mirror in the camera and a roof penta prism in the finder (Figure 20-1a). Every submarine has a periscope with complicated mirror and/or prism reflections to permit seeing out of the submarine while it is submerged just below the surface of the water (Figure 20-1b). The long telephoto lenses used to make images of the earth from satellites in space are likely to be folded lenses in which mirror reflections are used to physically shorten the optical path so that it will fit in the satellite (Figure 20-1c). Microfilmers use strip mirrors to record both sides of a

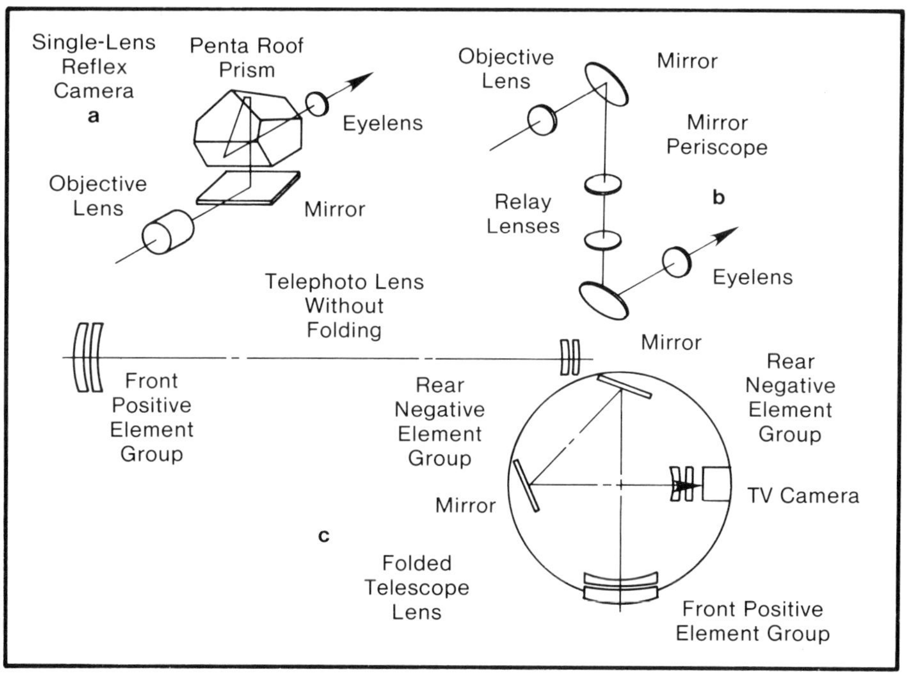

Figure 20-1. Reflections in Optical Devices.

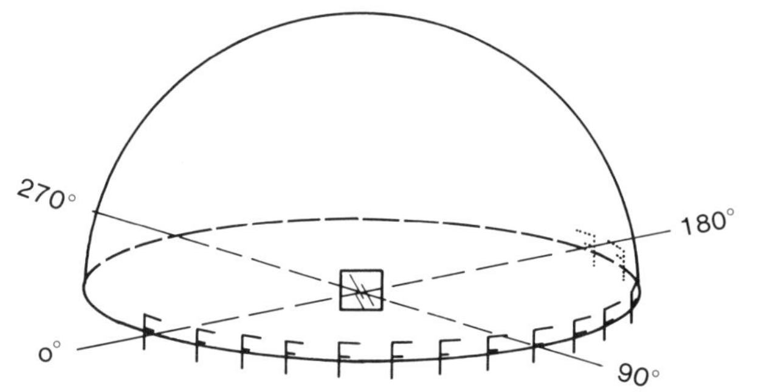

Figure 20-2. Mirror Rotation: 0° About Horizontal Axis; Various Angles About Vertical Axis.

Therefore, the orientation of a reflected image is important. The transparent reference hemisphere is one way of working out what the orientation will be before a breadboard model of a new design is made.

Rotation Around the Vertical Axis

If the mirror in the hemisphere is vertical it means that it has 0° rotation around its horizontal axis. If its rotation around the vertical axis is also 0°, it faces the object letter F directly, and the reflection comes back to the object letter F. If it rotates around the vertical axis in steps, the

document at the same time. Contour projectors, used to measure the profiles of complicated parts, use several large mirrors with extremely flat surfaces to permit imaging from a convenient horizontal stage and viewing from an equally convenient vertical rear projection screen. "Package" printers use beam splitters, mirrors and multiple enlarging lenses to image one negative in three sizes simultaneously. "Package" printers are used to simultaneously image 14 prints in several sizes from a single negative for school-package portraits. Many instant-picture cameras today use mirrors in the image beam to rectify the image and/or to make the camera more compact and, in some, to permit viewing.

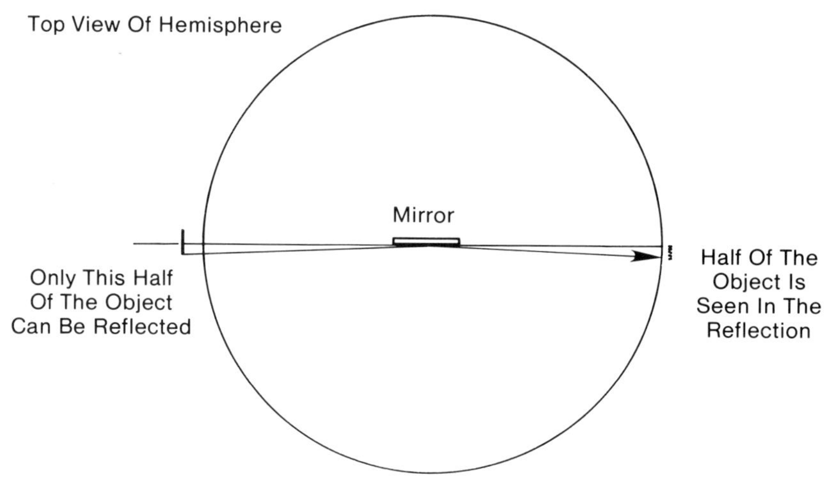

Figure 20-3. Grazing Incidence With Vertical Mirror.

Reflection Paths on the Hemisphere

1. **Great circle:** any circle that divides a sphere into two equal portions. The radius of any great circle is equal to the radius of curvature of the sphere. A sphere cut at any great circle is divided in half.
2. **Polar great circle:** any great circle that goes through the pole. On the globe, these are longitudinal lines.
3. **Equatorial great circle:** the great circle equivalent to the earth's equator. The plane of this great circle is perpendicular to the polar axis, and divides the sphere into the north and south hemispheres.
4. **Rails:** all great circles that intersect the 0° and 180° positions on the equatorial plane. A rail is defined by the angle its plane makes with the equatorial plane (hemisphere base). This is measured on the 90° polar great circle.
5. **Tropic circles:** any circle whose plane is parallel to the equatorial plane, but which is not a great circle on a globe. The Arctic Circle, Antarctic Circle, Tropic of Cancer and Tropic of Capricorn are tropic circles. The radius of any tropic circle is less than the radius of a great circle.
6. **Non-circular excursions:** most excursions are not circular. Some may be nearly circular for a short way and then veer from the circular path.

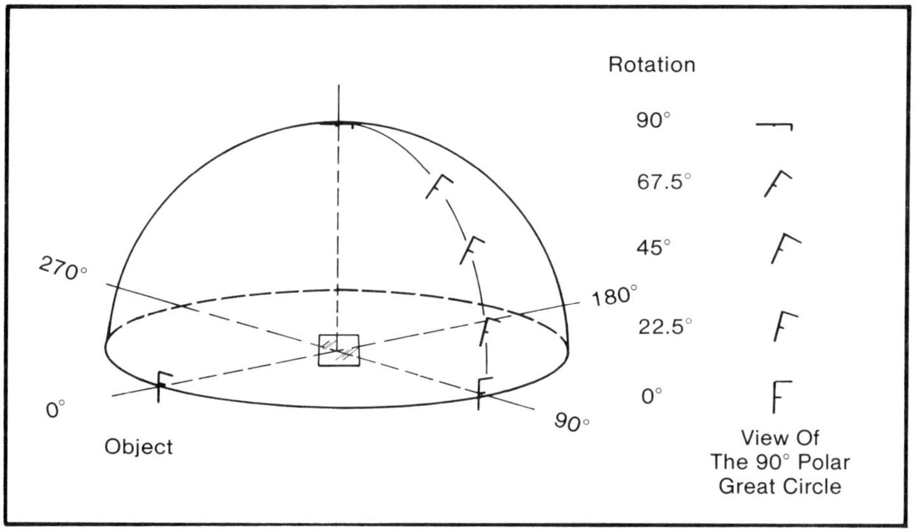

Figure 20-4. Mirror Orientation: 45° About The Vertical Axis To Start; Rotated About The 0°—180° Axis.

letter F is imaged about the equator with vertical F's (Figure 20-2). The image does not rotate around the optical axis. When the mirror is at 45°, the reflection is at 90° and the letter is upright and right reading as viewed from the outside. When the mirror is at 90°, the reflection is a grazing one, and the letter F is still upright and right reading from the outside.

This grazing position is a theoretical one, not a real one. If the mirror surface is vertical and on the 0°-180° line, light rays from only half of the F can be reflected (Figure 20-3). This position might be called a limiting position. Imaging takes place up to the 90° rotation position of the mirror. As the 90° position is reached only partial imaging can occur.

When the mirror is at 45°, the image is at 90° on the equator, and is vertical and right reading from the outside (Figure 20-4). When the mirror is rotated in steps around a horizontal axis (0°-180° line), the image moves up the 90° great circle, towards the North Pole. The image rotates the same number of degrees by which the mirror is rotated around the horizontal axis. At the pole, the image has rotated 90° so that it lies flat on the polc, parallel to the plane of the 0°-180° polar great circle.

Again, if the mirror is vertical and at 0° rotation around the vertical axis, the image comes back directly to the object. If the mirror is then rotated around its horizontal axis, (90°-270° line), the image climbs from 0° to the North Pole great circle (Figure 20-5). It does not rotate. At the pole it is the same image as the image arrived at by going up the 90° great circle, and the mirror is in the same position. However, as the mirror is rotated past 45°, the image travels down the great circle to the 180° position. Again, the image is only a partial image and, in this case, the image is upside down (the partial image is caused by the mirror being in the

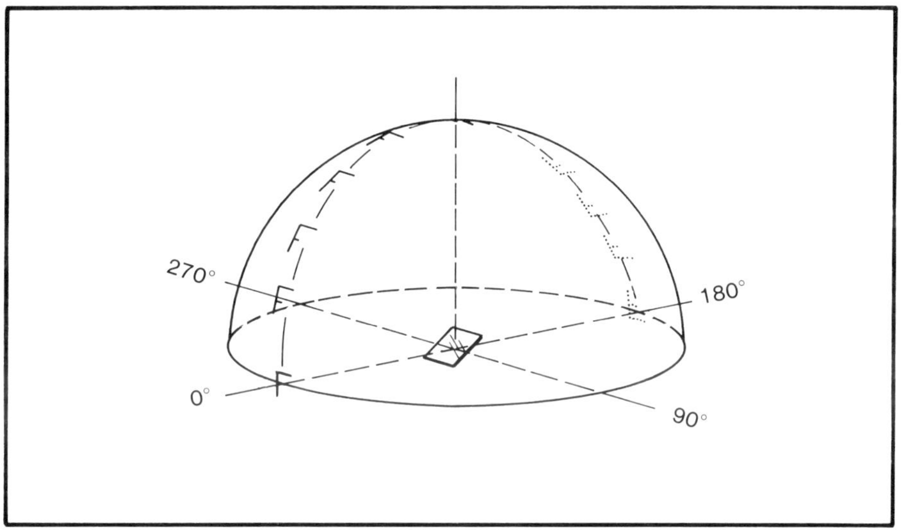

Figure 20-5. Mirror Orientation: 0° About Vertical Axis; Rotated About The Horizontal (90°—270°) Axis.

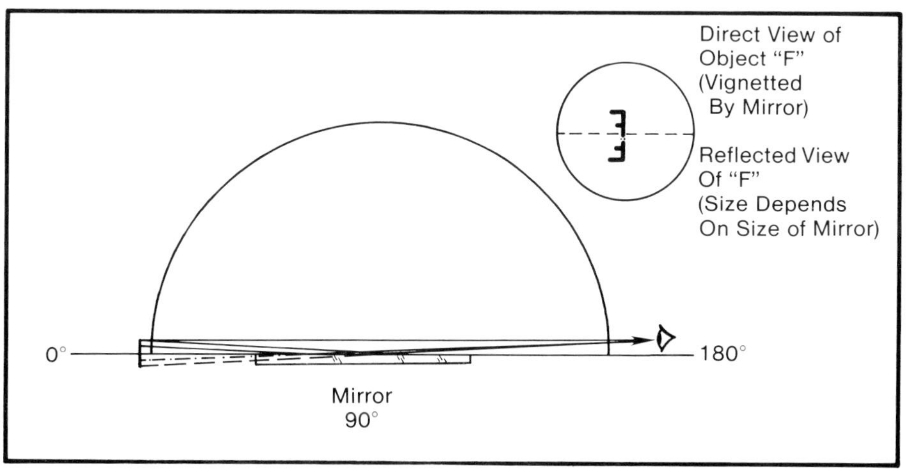

Figure 20-6. Mirror Orientation: 0° Rotation About Vertical Axis; 90° Rotation About Horizontal Axis (Lying Flat On The Hemisphere's Base).

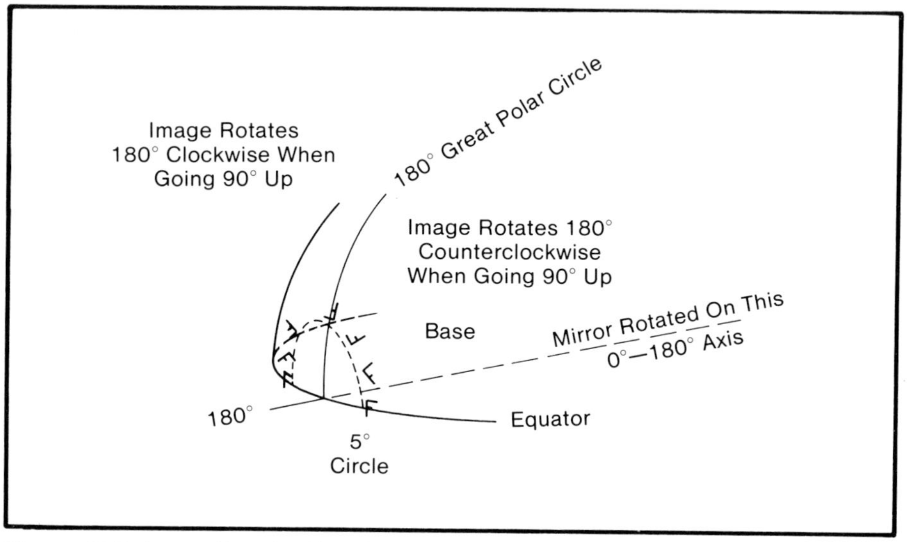

Figure 20-7. Image Rotation Near The 180° Axis.

grazing position, lying flat on the hemisphere's base) as shown in Figure 20-6.

The image can be placed at the 180° position only by grazing (Figure 20-7). However, if the mirror (in grazing position) is rotated on an axis equivalent to the 0°-180° line, the partial image rotates twice as fast as the mirror. The image at 180° can be oriented in any position depending on the grazing orientation of the mirror.

If the mirror is lying flat on the hemisphere's base and is rotated by 2° on the 90°-270° horizontal axis, the image rises 4° (Figure 20-6). If the mirror is rotated on the 0°-180° axis, the image is found to rotate rapidly. It must rotate 180° when going from the 180° great circle to the equator. It rotates twice as fast as its change in position.

If the mirror is again placed vertically and faces the object, the image returns to the object (Figure 20-8). If the mirror is rotated 2° around the 90°-270° horizontal axis, the image moves up the 0° great circle, but does not rotate. If the mirror is rotated about the 0°-180° axis, the image moves in a small circle around the 0° position, but does not rotate at all. It is always vertical.

Thus, we find that (1) the image does not rotate around the optical axis near the 0° position, and (2) that the image rotates at the same speed as the mirror speed when it is on the 90°-270° great circle, and (3) that it rotates twice as fast as the mirror when at the 180° position.

In the positions between 0° and 90°, therefore, the image must rotate around the optical axis at less than the speed of the mirror when going from the equator to the pole. Between 90° and 180°, the rotation is between one and two times the mirror's speed. This is the reason it is difficult to relate the degree of rotation to the movement of the mirror when using polar great circles.

Solution to the Chapter 19 Problem

In the problem for Chapter 19, the image moves from the equator to the North Pole along the 165° polar great circle. At the equator, the image is vertical. At the pole the image is aligned with the 0°-180° great circle. The image must rotate 165° clockwise, when passing from

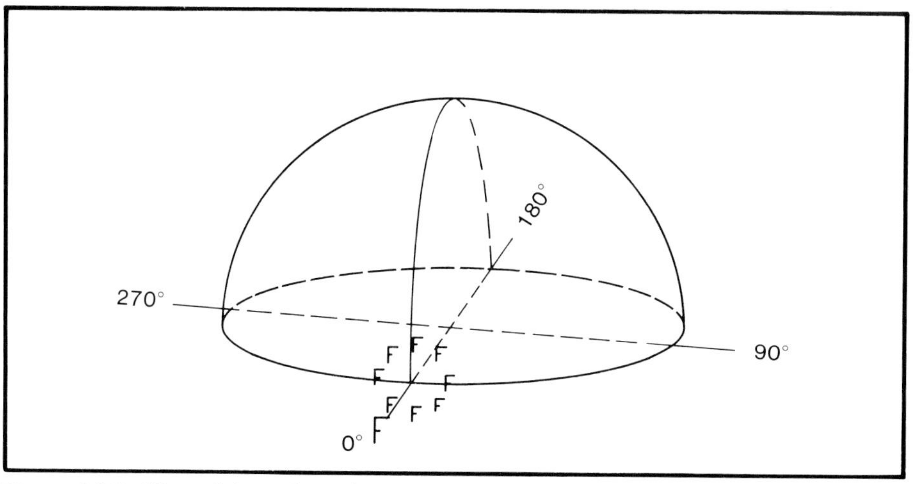

Figure 20-8. Mirror Orientation: 2° About The Horizontal Axis (270°—90°); Rotated About The 0°—180° Axis.

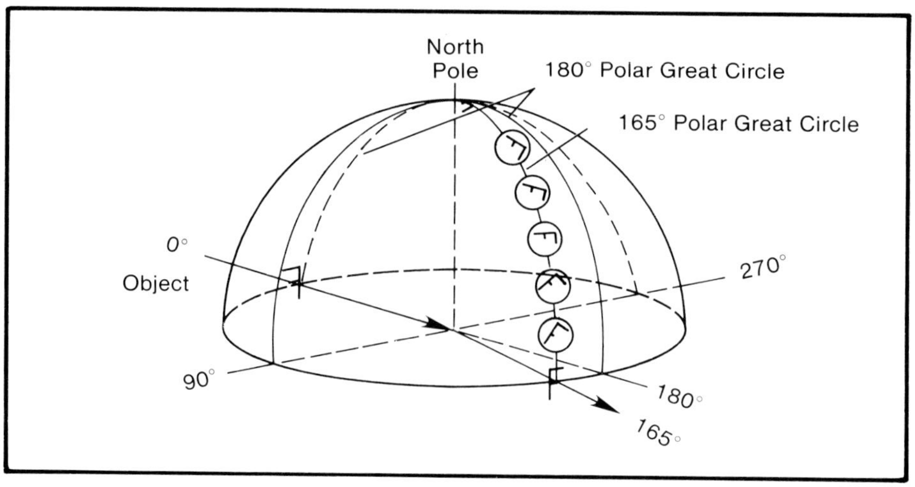

Figure 20-9. Answer to Chapter 19 Problem.

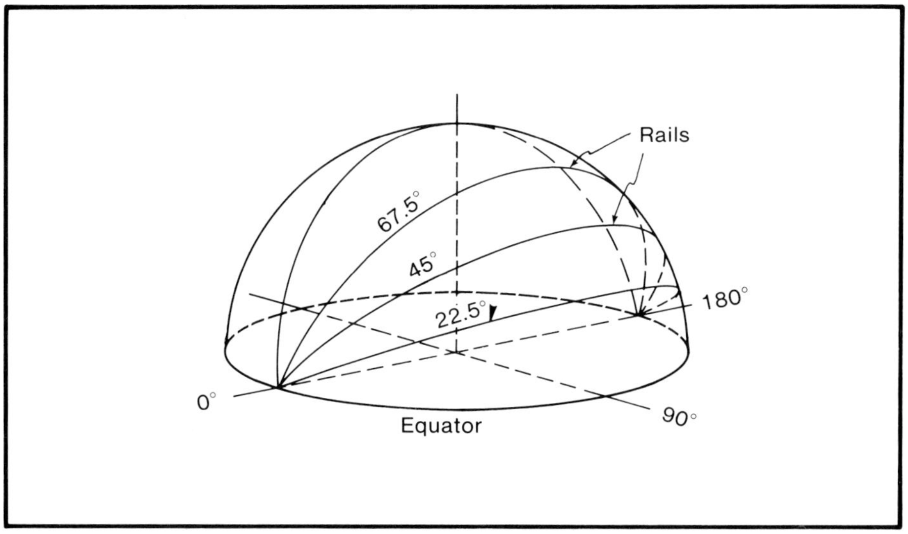

Figure 20-10. Great Circles That Intersect The 0° And 180° Equator Points.

the equator to the pole. Figure 20-9 shows the approximate rotation of the image at various elevations along the 165° great circle.

Other Great Circles

There are other great circles that help predict the degree of rotation of the image at any position on the reference hemisphere. These go through the 0° and 180° points on the equator. Figure 20-10 shows some of these great circles, which can be called rails. The equatorial and polar rails that go through these reference points, in which there is no rotation, have already been discussed.

The 45° rail can be used to show how you can find the amount of rotation of the image. Halfway around this rail is the 90° great circle position. We have already determined that halfway up (45°) this rail, the angle of rotation is 45°. At this point, the 45° rail intersects the 90° polar great circle. If we draw the 45° polar great circle, it will intersect the 45° to 18° great circle. At this position, the angle of rotation will be half of what it is at the 90° polar great circle, or ½ x 45° = 22½° (Figure 20-11).

The general rule is:

$$\text{Angle of rotation} = \text{amount of rotation at } 90^\circ \times \frac{\text{No. degrees along rail}}{90^\circ}$$

Figures 20-12 and 20-13 show how these locations are found. The Chapter 19 problem can be used as an example. All of the points are on the 165° polar great circle. The positions of the images shown are on the 0°, 15°, 30°, 45°, 60°, 75°, and 90°-180° great circles. Figure 20-9 shows these locations. The following table shows the calculations.

Figure 20-13 shows the image rotation at the 22½°, 45°, and 67½° (0° to 180°) great circles. It also shows the 22½°, 45°, 67½°, 90°, 112.5° and 135° polar great circles. For calculating the image rotation on the left side of the hemisphere (180° to 360°) use the minus angles shown in the top view in Figure 20-12, and as stated earlier, the rotation from the equator up the polar great circle is counterclockwise. Calculate other angles by subtracting from 360°.

0-180 great circle	Image orientation on 90° circle	Calculation	Amount of Rotation
0° (equator)	0°	$0^\circ \times \frac{165}{90} =$	0°
15°	15°	$15^\circ \times \frac{165}{90} =$	27.5°
30°	30°	$30^\circ \times \frac{165}{90} =$	55°
45°	45°	$45^\circ \times \frac{165}{90} =$	82.5°
60°	60°	$60^\circ \times \frac{165}{90} =$	110°
75°	75°	$75^\circ \times \frac{165}{90} =$	137.5°
90°	90°	$90^\circ \times \frac{165}{90} =$	165°

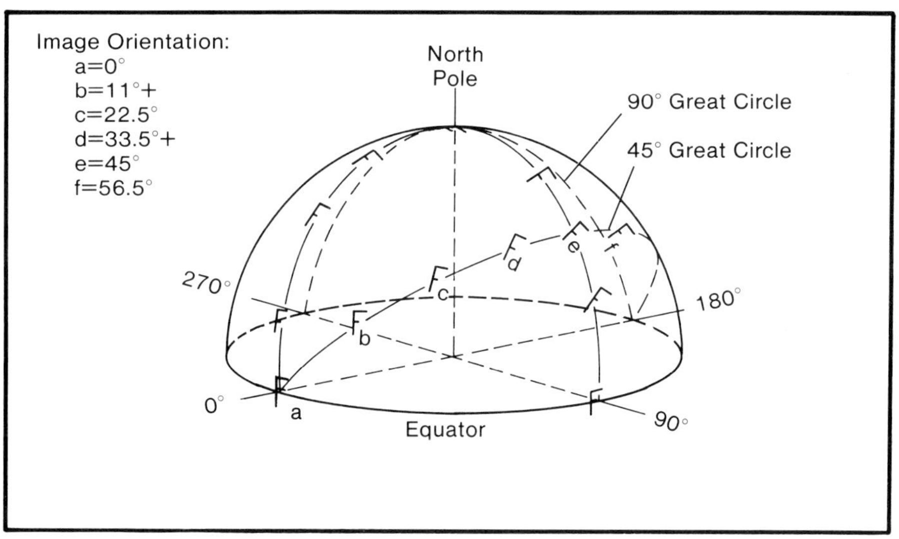

Figure 20-11. Image Orientation On The 45°—0° to 180° Great Circle.

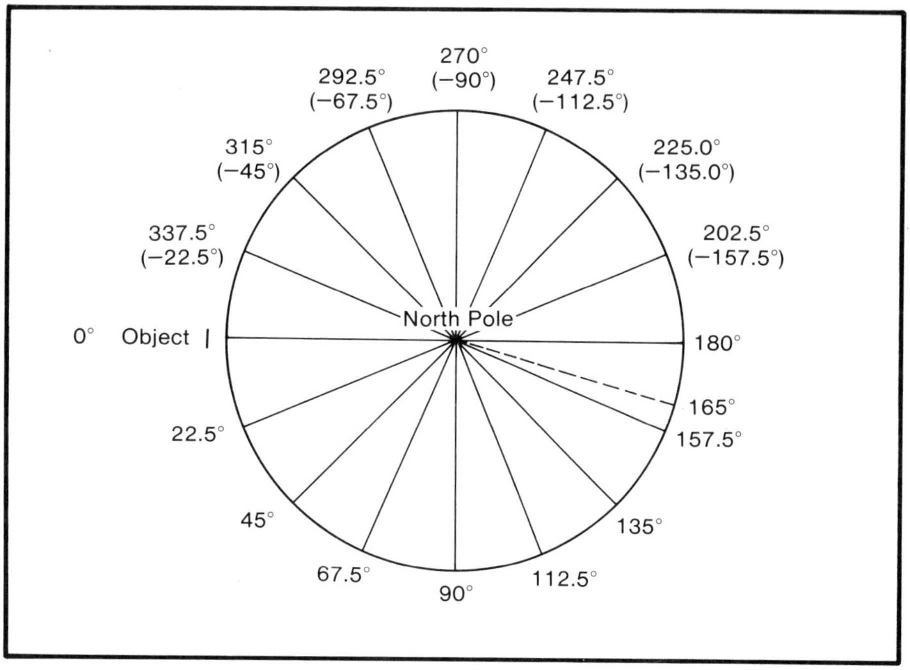

Figure 20-12. Top View Of Hemisphere Showing Polar Great Circles.

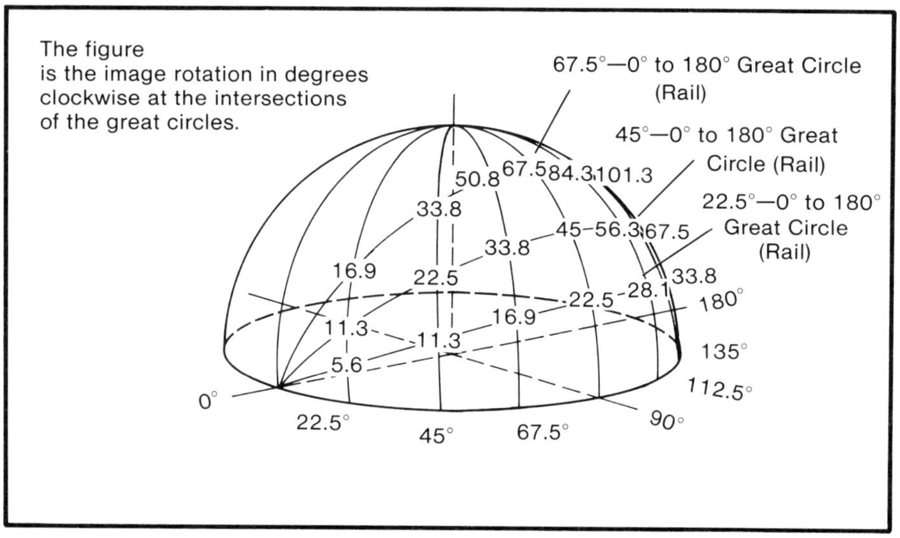

Figure 20-13. Hemisphere With Polar Great Circles And 0°—180° Great Circles.

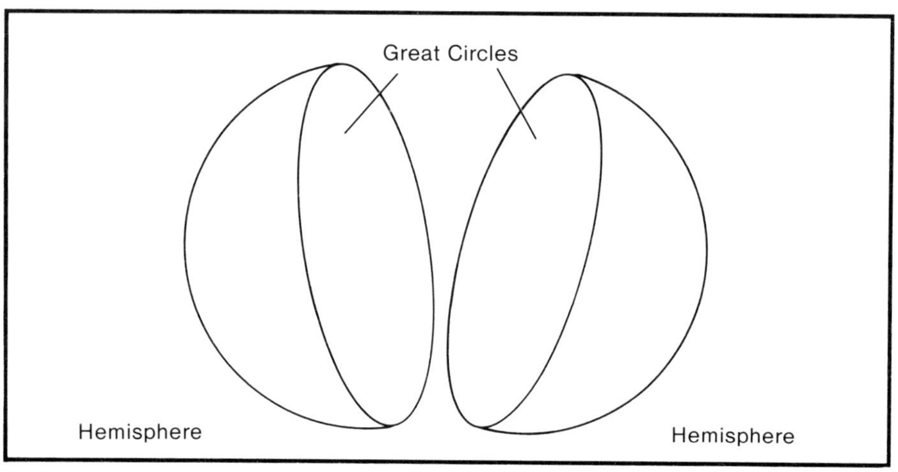

Figure 20-14. Sphere Cut In Halves.

(Example: The 262° polar great circle is 360°−262° = 98° polar great circle.)

The video lesson shows a reference hemisphere with a metal rail that pivots at 0° to 180° on the equator (same principle as an ice cream scoop). Using such a device, it is relatively simple to locate any point on the hemisphere. The rail forms a plane that intersects the equatorial plane along the 0° to 180° axis. The angle that this plane makes with the equatorial plane (the hemisphere's base) can be called the elevation.

Excursions That Are Not Great Circles

We have called the routes that the rays of light travel "paths." We can call the routes that the image travels as the mirror is moved in a particular way "excursions." All of the excursions that have been discussed so far have been great-circle excursions. That is, if the excursions were followed all around the sphere (of which the reference hemisphere is half), its diameter would be equal to the sphere's diameter. If the sphere were cut at a great circle, the two hemispheres would be equal in size. It's as though an orange were cut exactly through the middle (Figure 20-14).

Not all excursions are great-circle excursions. Figure 20-15 shows a sphere with several cuts made in it. The great circle cuts are shown as solid lines; cuts that are not great circles are shown as dashed lines. In solid geometry a cut is described as the intersection of a plane with a sphere. A great circle is the intersection of the surface of the sphere with a plane that goes through the center of the sphere.

If a mirror is set at 0° rotation on its horizontal axis and at 45° on its vertical axis, the image falls on the equator at 90° and is vertical. If the vertical axis remains fixed at 45°, but the mirror is rotated around the horizontal axis, the excursion which the image makes is a circle, but it is

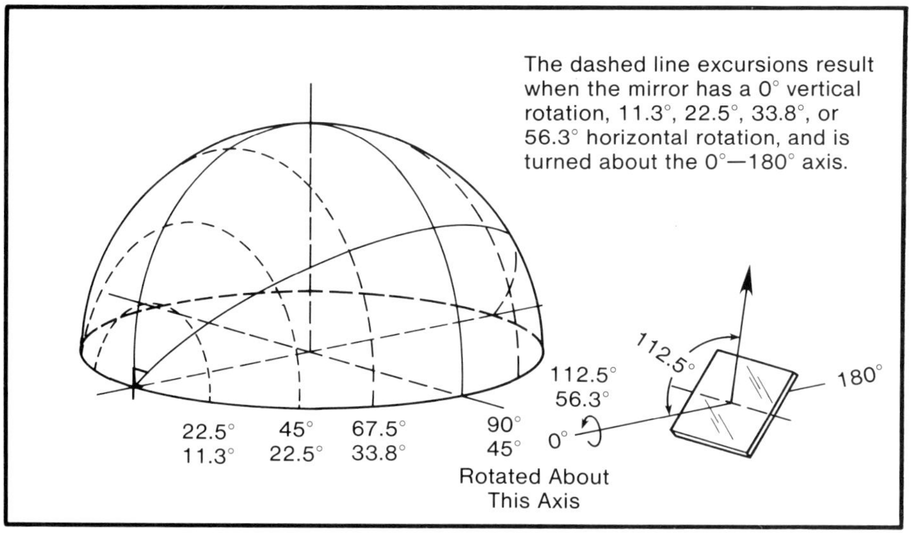

Figure 20-15. Excursions That Are Not Great Circles.

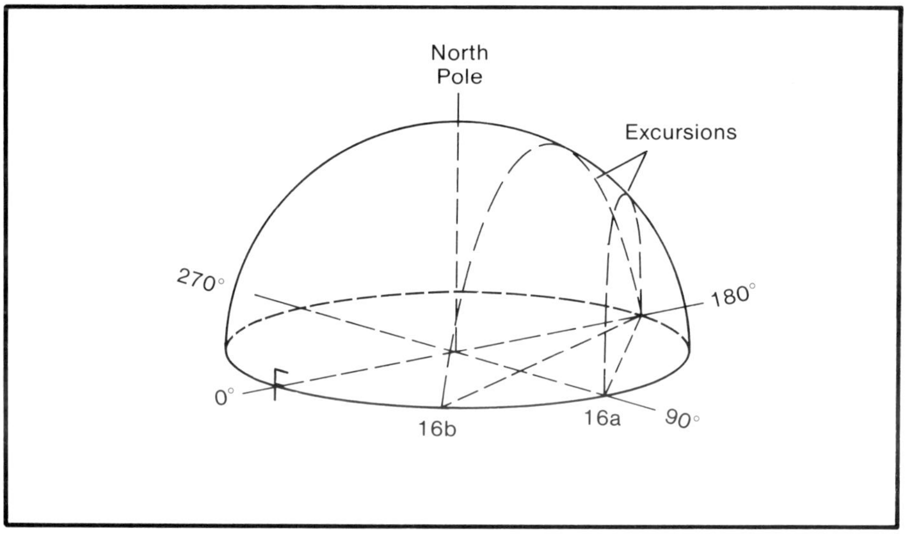

Figure 20-16. Excursions That Are Not Great Circles.

not a great circle. Another way of looking at it is that the mirror is set on the equatorial plane and gradually tilted back until it is horizontal without shifting the position of the mirror edge on the equatorial plane. The excursion moves up on the hemisphere, but it does not reach the North Pole. As the mirror approaches the horizontal position, it approaches the grazing angle. Therefore, when it is laid flat, the image must go through the 180° position on the equator. The plane formed by this excursion is vertical; it is 90° to the equator plane. The excursion is the intersection of the sphere surface with a plane that does not go through the center of the sphere. It is as though you had taken a cleaver and cut the hemisphere vertically without passing through the center of the sphere.

If the original angle of rotation around the vertical axis is 22.5°, the original equatorial position of the image is at 45° (Figure 20-16). If the mirror is rotated on its horizontal axis without changing the vertical axis position, the excursion path will be as shown. The three excursion circles will meet at 180°, but their planes will not be parallel. All the planes will be vertical, however.

Thc mathematics required to find the orientation of the image at positions on such excursions is complicated. In practice, it is easier to find the position on the hemisphere and rotate the rail (the great circle excursion that goes through the 0° and 180° positions on the equator plane) until it lies on the image on the hemisphere. Thus the number of degrees along the rail is found and the orientation of the image can be calculated by the equations given earlier.

A Caution About Optical Drawings

Many drawings that involve optical parts and layouts are straightforward to make and interpret. However, some caution should be used when dealing with certain problems. In an earlier section concerning anamorphic lenses made with cylindrical curves, the drawings were made as square rather than round lenses, as they actually are, because when drawn round they look like conventional spherical lenses.

Some problems also exist when drawing mirrors and the paths of light reflections.

A transparent cube is shown in Figure 20-17. An isometric view is given and most people can readily interpret the few lines as a cube. A line is drawn from one corner, A, diagonally across the cube to another corner, B. This could be the path of a light ray. Think of it as a wire.

It is common engineering practice

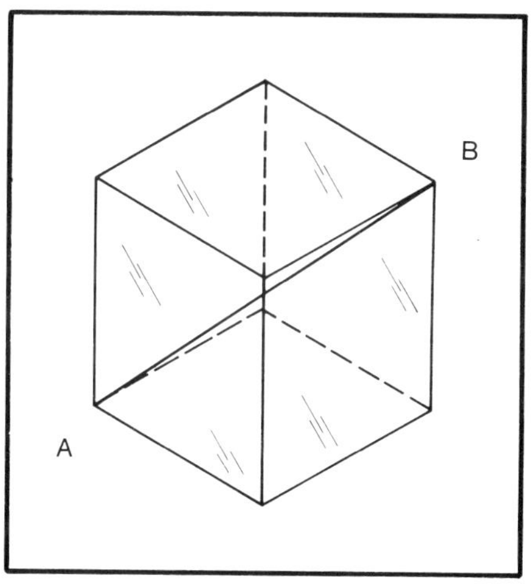

Figure 20-17. Isometric View Of Cube With Wire From Diagonally-Opposite Corners.

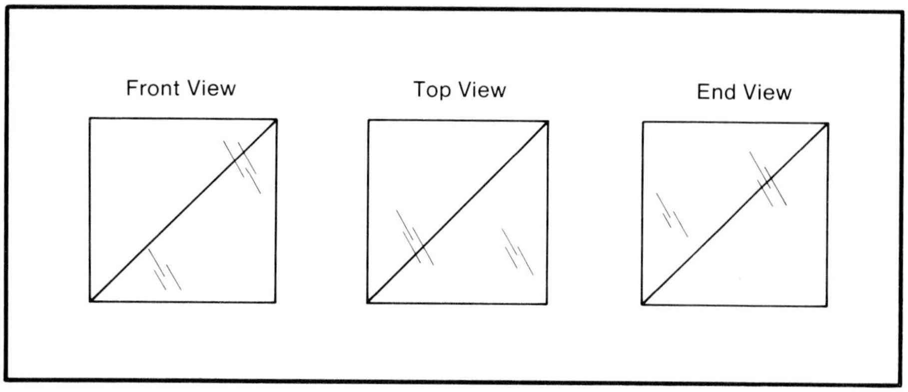

Figure 20-18. Plan View Of Cube With Wire.

to draw three plan views of parts—a front view, top view, and end view. The three plan views of the cube with the wire are shown (Figure 20-18). It is apparent that all three views are identical. Without the isometric view, these three plan views would be almost impossible to interpret.

In the tape session, a three-wire device is shown representing a reflected light ray and the normal to the mirror surface. Looked at straight on, it is apparent that the angle of incidence equals the angle of reflection (Figure 20-19a). However, when viewed from other angles, angle I no longer looks as though it equals angle R (Figure 20-19b). In plan drawings of three-dimensional light paths, care must be taken so the actual situation is apparent to those who look at the drawings.

Paper Path Folding

When two or more mirrors are used in a system, it is usually known where the object is and where the image is wanted. The question is: How many mirrors are needed to get the image in the specific location with the correct orientation?

Knowledge of the laws of reflection, the use of cascaded reference cubes, and the reference hemisphere can aid a designer considerably. However, there is a time-saving method that is very useful and can be referred to as the folded paper path.

Use of the paper method works with two-dimensional systems; that is, systems in which a plan view will show all that is needed to know about the optical layout (the vertical axes of all mirrors are parallel). Fortunately, this type of layout is much more common in practice than the three-dimensional layout in which the vertical axes of the mirrors are not parallel. In the two-dimensional layout, the plan view may be any of the three views discussed earlier; front, top or side.

The first step is to lay out a straight-line drawing of the system without mirrors. In a projection system, for example, there will be an object (perhaps a slide), a lens, and a screen. If it is a printer layout, there will be a negative plane, lens, and paper plane. In a folded camera system the object may be at infinity, so all that is needed is the lens and the film plane. The example used here is a projection system (Figure 20-20).

The object size and the lens focal length are known. The projected image size determines the screen

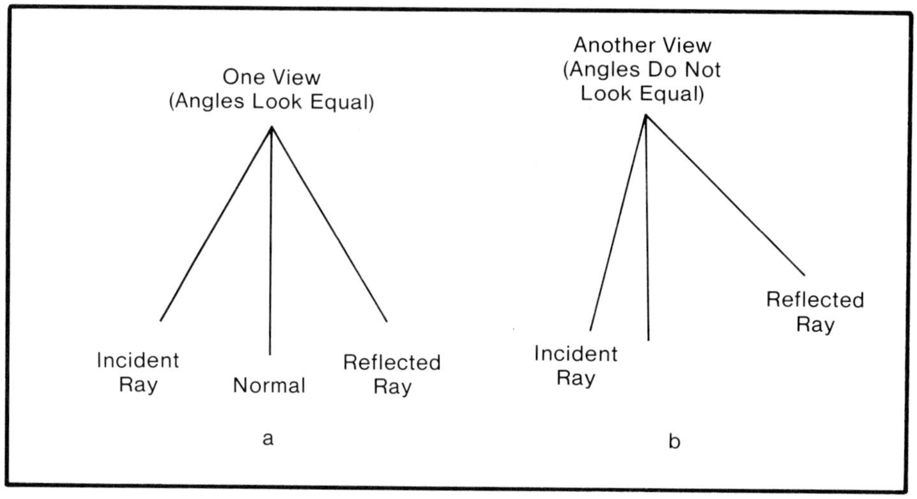

Figure 20-19. Wire Device That Illustrates The Law Of Reflection.

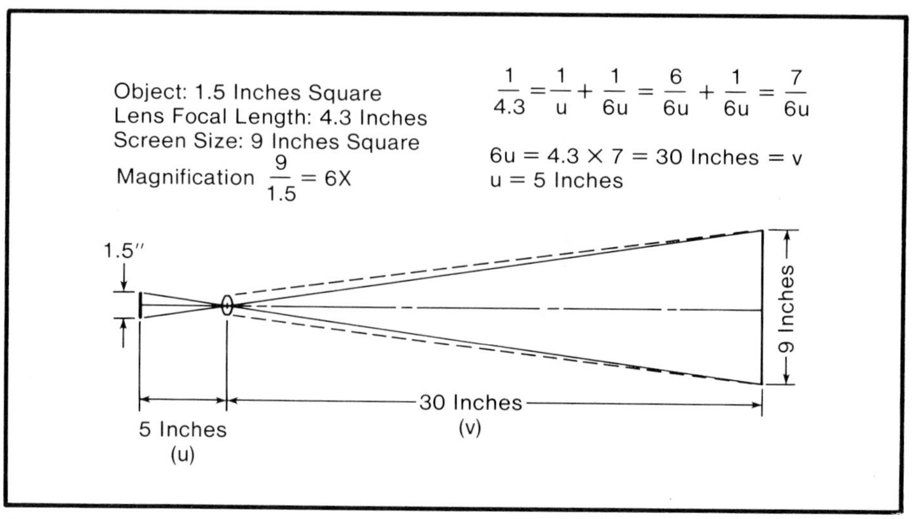

Figure 20-20. Optical Layout For Projection System.

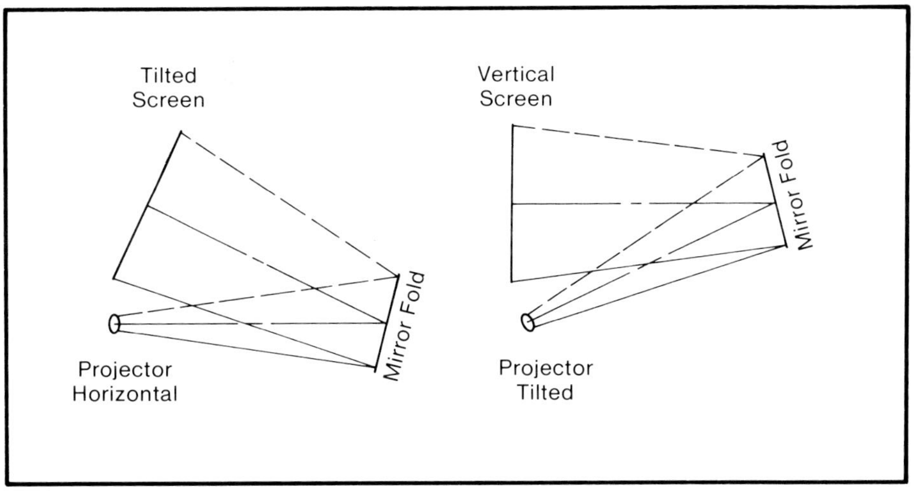

Figure 20-21. One Mirror Folds.

size. Using the equation:

$$\frac{1}{f} = \frac{1}{u} + \frac{1}{v}$$

and the magnification desired (screen size divided by slide size), the equation becomes:

$$\frac{1}{f} = \frac{1}{u} + \frac{1}{m \times u}$$

where m x u is the object distance (magnification times the image distance). Thus, the placement of the object and image can be drawn, as well as their sizes. Two rays are drawn through the optical center of the lens from the edges of the object to the edges of the image. Then two additional rays are drawn from the periphery of the lens to the edge of the image. These outer rays will help determine the size of the mirrors needed. One is drawn as a solid line, and the other as a dashed line. This helps to keep track of the image orientation. The image plane is, of course, perpendicular to the optical axis.

The paper is cut along the outside lines. In most cases, the paper can be cut across the principal plane of the lens. The object side is not needed. When the paper is folded, every fold represents a reflection. Two folds in the paper represent two mirrors. Each fold line is a mirror's position. The path of the axial ray is readily seen, especially if it is drawn on both sides of the paper.

In our example, an optical arrangement is needed in which a rear-projection screen will be vertical and above the projection lens (Figure 20-21). If one fold is tried the screen cannot be made vertical unless the projector is tipped upward. If two mirrors are tried, it will be found that the image is upside down (Figure 20-22). Therefore, at least three mirrors must be used to accomplish this. These conclusions can be reached in minutes with the paper light beam. The slides could be inverted and the two-mirror arrangement used. This becomes inconvenient if the slides already in the tray have to be inverted everytime the system is changed from a standard projector to a two mirror projection system.

The light beam must be fairly long and narrow in order to obtain the desired three reflections. If this cannot be achieved because there is insufficient room for the large mirrors required, the path will have to be lengthened. This can be done by using a longer focal length lens. The object and image can then stay the same size, but the u and v are lengthened. The magnification remains the same.

Finding the Mirror Size

One solution of the mirror placement is shown in Figure 20-23. The length of the fold is the easiest dimension to find. The other dimensions are found by unfolding the paper path. The mirrors are found to be trapezoidal in shape. If the light beam is square, the small dimension is the width of the light path at the ends of the mirror closest to the lens. The large dimension is the width of the light path nearer to the screen. By measuring these dimensions on the paper, the shape of each mirror can be closely approximated. If the beam is round, the mirrors are elliptical and fit within the square beam trapezoids.

If the screen is rectangular, the beam will also be rectangular. Cut a second paper light path using the other dimension of the screen for the

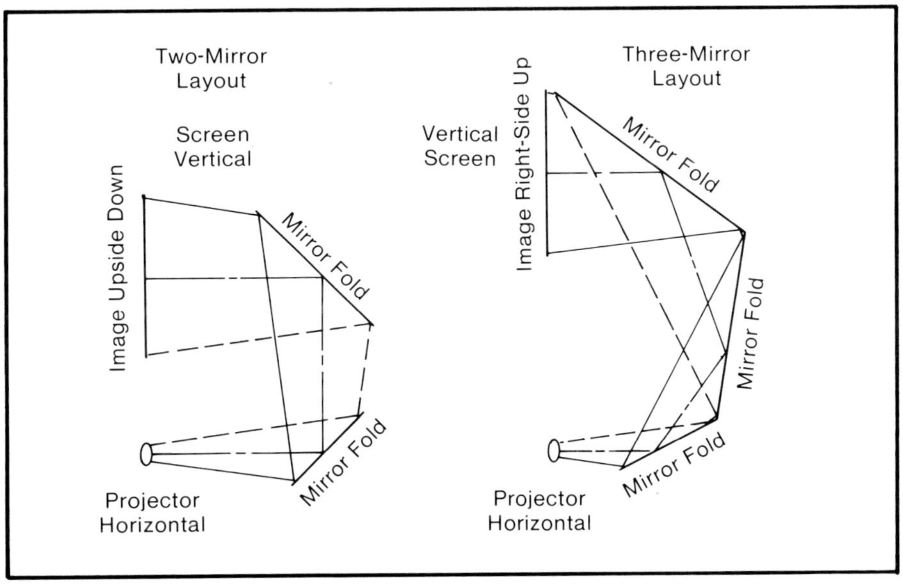

Figure 20-22. Two- and Three-Mirror Projection Layouts.

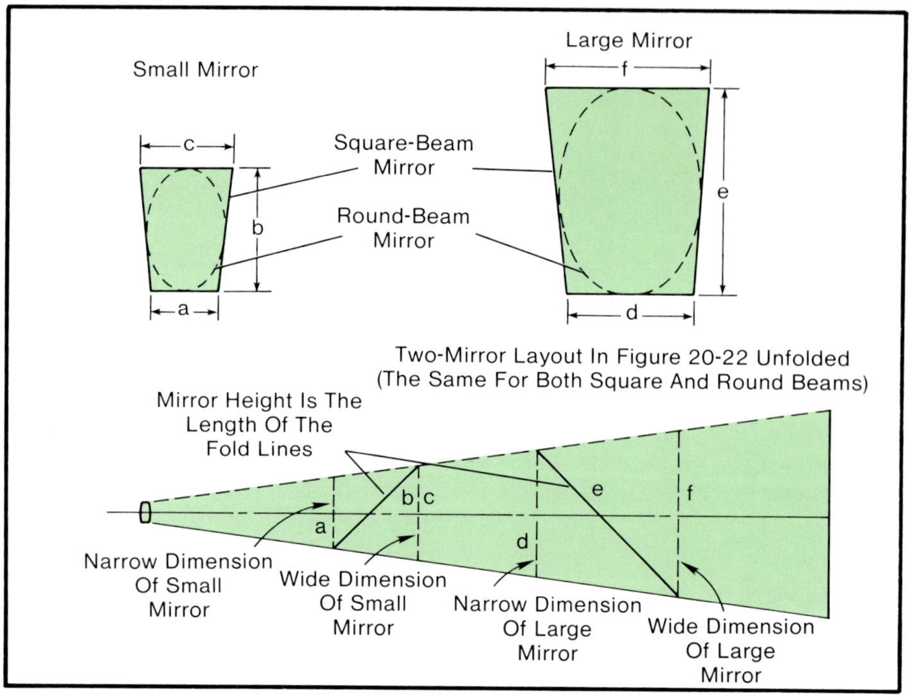

Figure 20-23. Finding Minimum Mirror Size Of Square-Beam Layout and Round-Beam Layout.

width of the light path. The lens end stays the same. The positions of the ends of each mirror are marked on this new light path, and their widths at each position are measured. This gives the sizes of the mirrors (Figure 20-24).

A Three-Dimensional Folded Light Path

A plastic foam core material with thin cardboard laminated to both surfaces is normally used as a lightweight picture mounting board. It can be used as a material for finding three-dimensional mirror light paths.

The material is laid out (Figure 20-25) and cut into the light-path shape in the same way the paper light path was done. At each reflection, the board is cut almost through with a razor blade or sharp knife leaving one cardboard surface intact. The uncut cardboard acts as a hinge for folding. When the paper path is folded flat, the foam core material can be partly folded to provide the three-dimensional aspect to the path. Several cuts can be made at different angles if the first angle is not quite right. When a desired path is found, the core-board can be fixed by taping it at each bend. The angle of the bend in one dimension is measured from the angle of the cut. In the other dimension, it is calculated from the angle the path makes with itself. The mirror surface is placed perpendicular to the bisector of the light path's angle.

Three-dimensional thinking combined with the trial and error process are needed to use this type of light-path finding device. However, it can be used to solve some tricky mirror placement problems.

The videotape lesson shows a metal jointed device for finding image rotation in mirror systems. Since the details of construction cannot be given, its existence shall only be mentioned here.

Signal Mirrors

Signaling mirrors are used to attract attention in emergency situations. Aircraft pilots and sailors in lifeboats at sea use them to attract search parties. If the sun is shining and the people in the boat have a mirror, they can reflect the sunlight toward the plane. The flash of the mirror is easily seen from great distances.

A plane in the sky is a small target for the beam of light flashed by a mirror, and on a clear day the beam of light itself cannot be seen by those using the mirror.

However, a student sitting in a shaft of sunlight in a schoolroom can easily flash the light of the sun all around with a mirror. The spot of light that the beam makes on the walls, floor, or other students can easily be seen. The beam can be

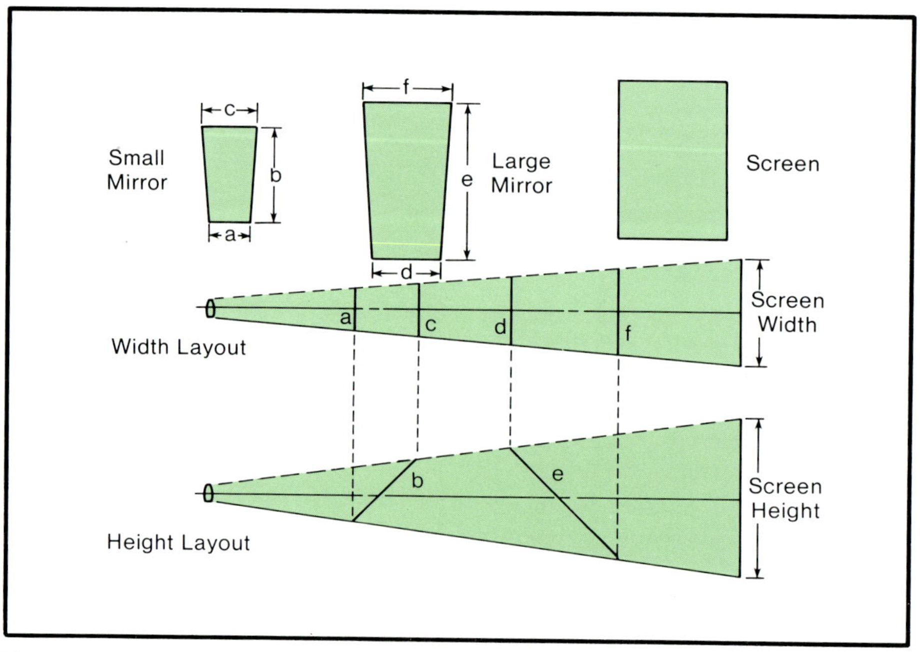

Figure 20-24. Finding Minimum Mirror Size Of Rectangular-Beam Layout.

made to move toward the target by watching where the spot of light goes.

There are no walls to tell the people in a lifeboat where the beam is. They are fortunate if they get the beam close enough to the search craft for the rescuers to see the flash of the mirror. That is, if it is an ordinary mirror.

However, special mirrors are made and are a part of the equipment stocked on life rafts, along with water, food, etc. These mirrors, made of polished metal, are unbreakable and reflect from both sides. A small hole is drilled in the center of the mirror (Figure 20-26). The hole allows a small beam of sunlight to pass through the mirror. If a person holds the mirror up between his eye and the plane, he sees the reflection of the small spot of light the sun makes on his cheek coming through the hole. As the mirror is angled, the reflected spot moves. If the mirror is angled so that the spot of light appears to be on the plane, the beam of sunlight reflected from the mirror will be aimed at the plane, and the rescuers will be able to see the flash.

An improved version of the signal mirror (Figure 20-27), allows the user to look through the screen of retroreflector beads located on the rear surface of an unsilvered portion of a mirror. A dimmed image of the sun is internally reflected from the clear front surface and appears to emanate from the direction of the reflected rays. If the mirror is held up on a sunny day, a small spot of light will be seen in the retroreflective material. If the mirror is angled so that the spot of light lines up with the plane, the mirror will be reflecting a beam of sunlight toward the plane.

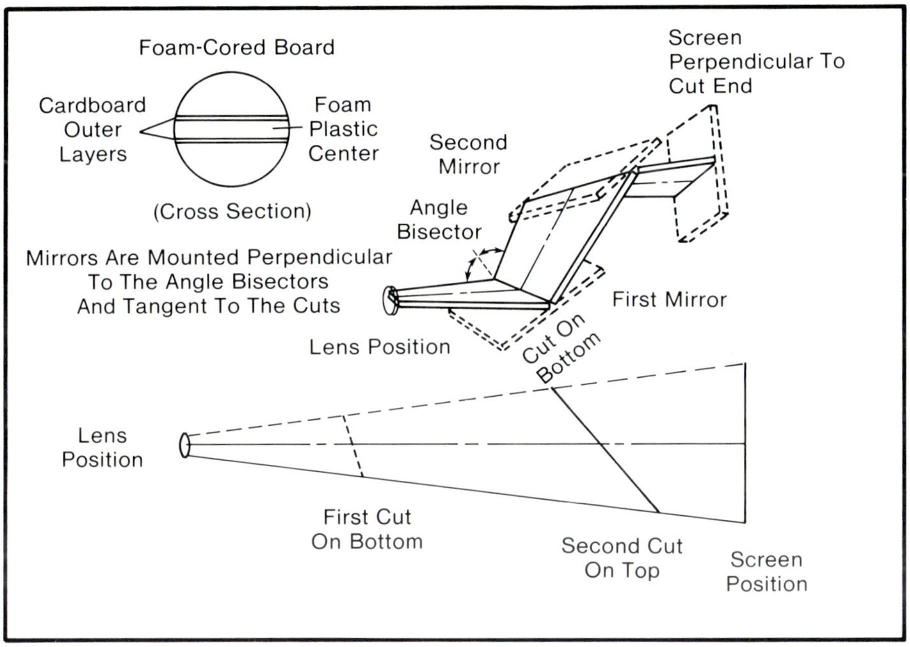

Figure 20-25. Mirror Placement With Folded Optical Beam.

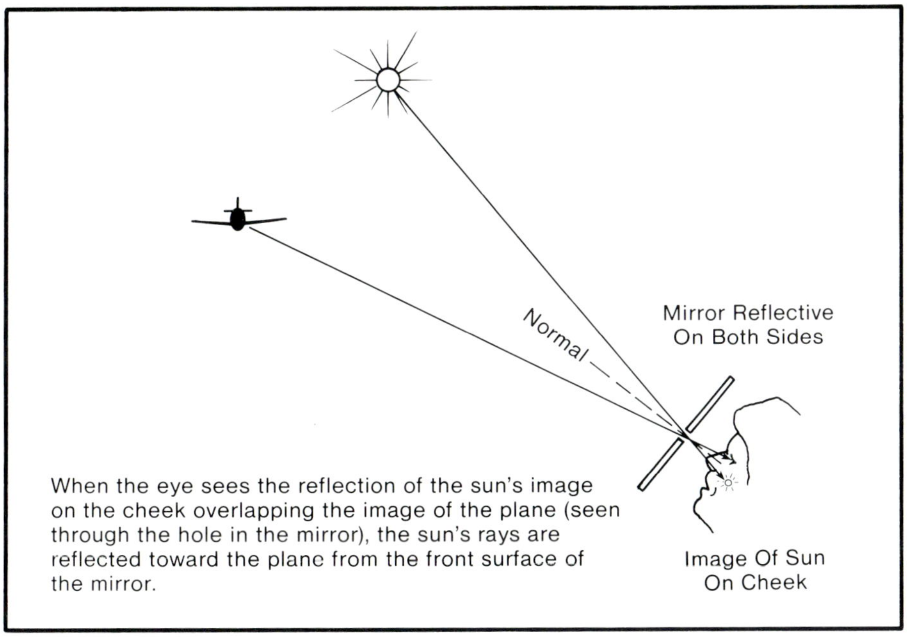

Figure 20-26. A Signal Mirror.

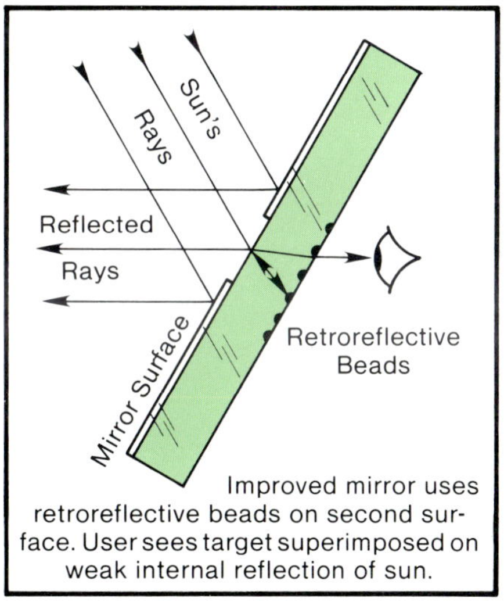

Figure 20-27. Improved Signal Mirror.

Problem

Draw a reference hemisphere. Place a mirror in the hemisphere so that it is rotated about its horizontal axis by 22½°. The image is on the 0° great circle, 45° up from the equator. Now, leaving the horizontal tilt fixed, rotate the mirror on its vertical axis so that the excursion is on the right half of the hemisphere. Rotate so that the excursion goes halfway around the hemisphere.

Locate the excursion path and describe it. Also give the orientation of the image at various locations.

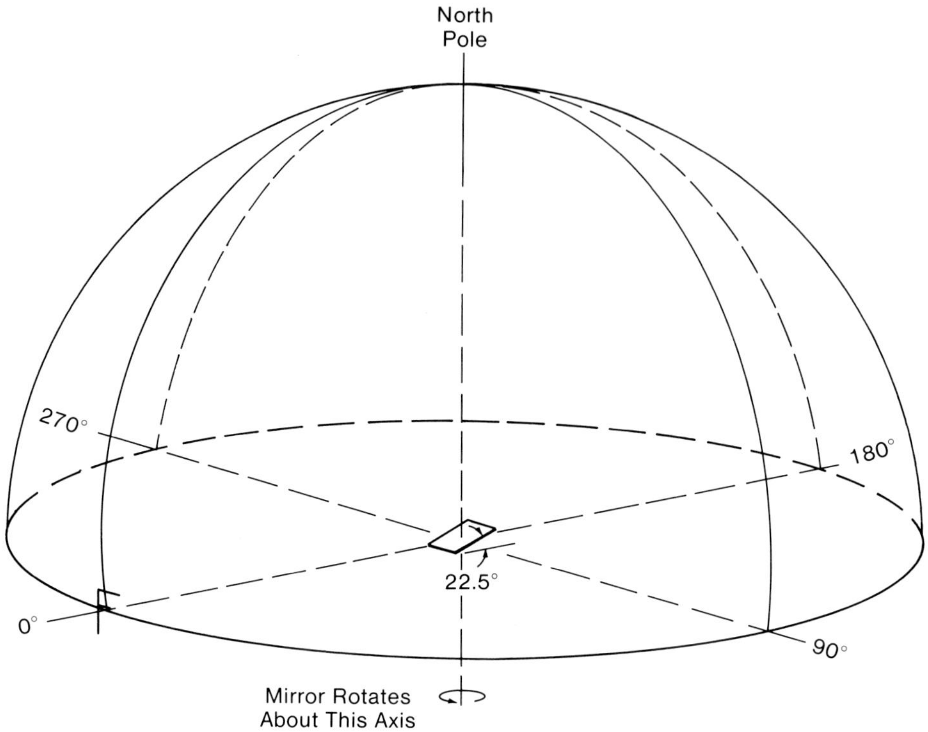

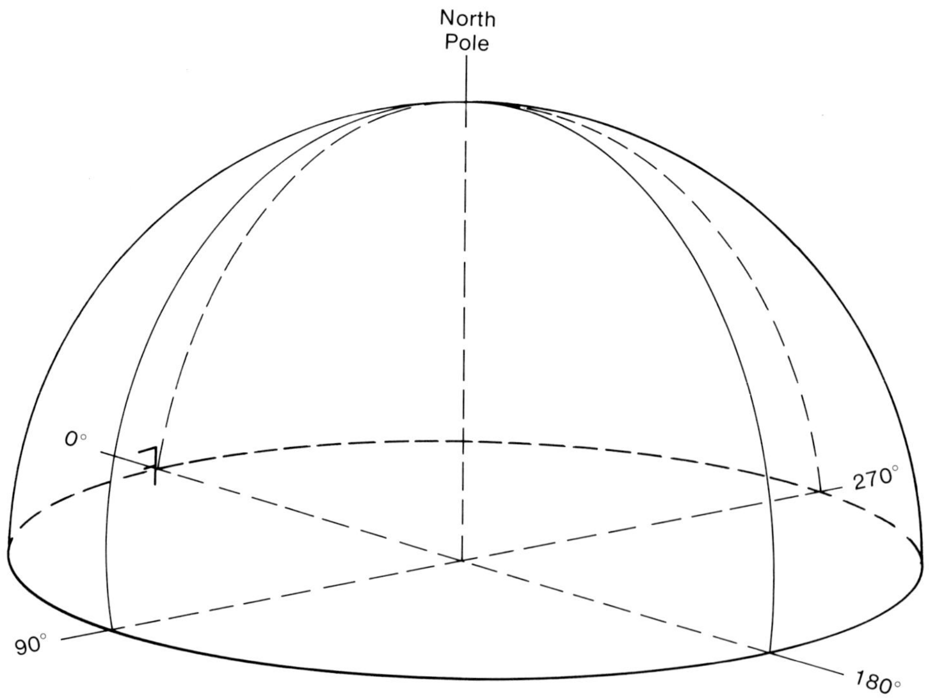

Problem
Chapter 20
Video Lesson 15

Chapter 21

VIDEO LESSON 16

Excursions, Mirrors at a Fixed Angle Retroreflection Devices and Projection Systems

Summary

When a mirror is placed in a reference hemisphere, rotated around its horizontal axis 22½°, and then rotated around its vertical axis, the excursion is known at two points (established in Chapter 20). The 0° position has an elevation of 45° and at 180° it has an elevation of 0°. The rotation at 0° is 0°, while at 180° it is 45°. The path is found empirically using the hemisphere, and goes at 45° elevation for quite a distance, lowers to 35° at the 90° position, and then slowly lowers to 0° when it reaches the 180° position. A fixed-position, two-mirror device reflects at twice the angle between the mirrors. Such a device becomes the basis of a sextant used for measuring the elevation of sun and stars for navigation. A similar device is used for calculating the distance to ships. A 90°-corner mirror device always returns light along the path on which it came. Lenses which utilize this principle are used for road signs and very bright, narrow reflection-angle screens for TV and studio photography. Using the principles learned in this course, the optical system of a slide projector is designed, step by step. A review of the current state of optics is given.

Excursion of a Rotating Mirror

In the problem assigned in the last lesson, the spatial reference hemisphere was to be used. A mirror was rotated around its horizontal axis by

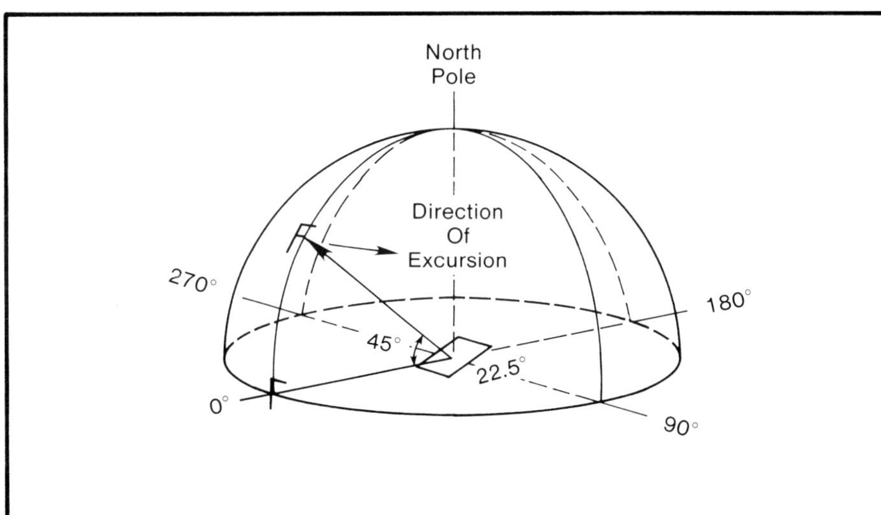

Figure 21-1. Mirror Rotation About Vertical Axis = 0°: Mirror Rotation About Horizontal Axis = 22.5°.

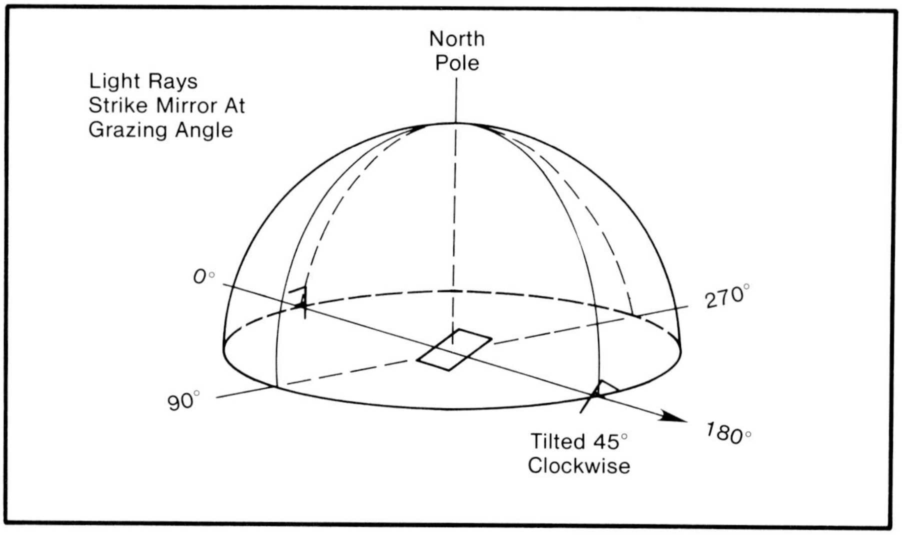

Figure 21-2. Mirror Rotation About Vertical Axis = 90°.

22½° causing the image to move along the great circle towards the North Pole. Next it was rotated around the vertical axis to the right or counterclockwise as viewed from the North Pole. The problem was to find its approximate excursion (path of the image as reflected onto the surface of the hemisphere). The second part of the problem was to find the approximate orientation of the image at several places along the excursion.

Only two positions of the excursion are known. When the rotation around the vertical axis is 0°, that is, when the mirror's optical axis is aligned along the 90°-270° line of the hemisphere base, the reflection is along the 0° to the North Pole great circle. Since the mirror is, in essence, tipped back by 22½°, the reflection will be at 45° or halfway up the great circle (Figure 21-1).

The direction of the excursion is also known. As the mirror is rotated on its vertical axis around the 0° position, the excursion path is horizontal. The excursion is close to a circular path parallel to the hemisphere base.

The image orientation up the 0° great circle has no rotation, so that the image lies along the great circle in its original position at 0° rotation. Using the letter F as the object, the vertical member of the image letter F is parallel to the 0° to north pole great circle.

The other location on the hemisphere where the position of the excursion is known is at the 180° position on the equator. When the mirror has been rotated 90° on its vertical axis, it reaches the grazing position. At grazing position, the image is always on the equator so that it has a 0° elevation (Figure 21-2).

In the last lesson it was found that the image at grazing position rotates twice as fast as the mirror when the mirror is rotated around an axis that lies on the 0°-180° line along the base of the hemisphere. For example, when the mirror is rotated 45°, the image has rotated 90°, so it lies on its side. When the mirror is rotated 90° so that it is vertical, the image has rotated 180° so that it is upside down. In the present problem at the grazing position, when the

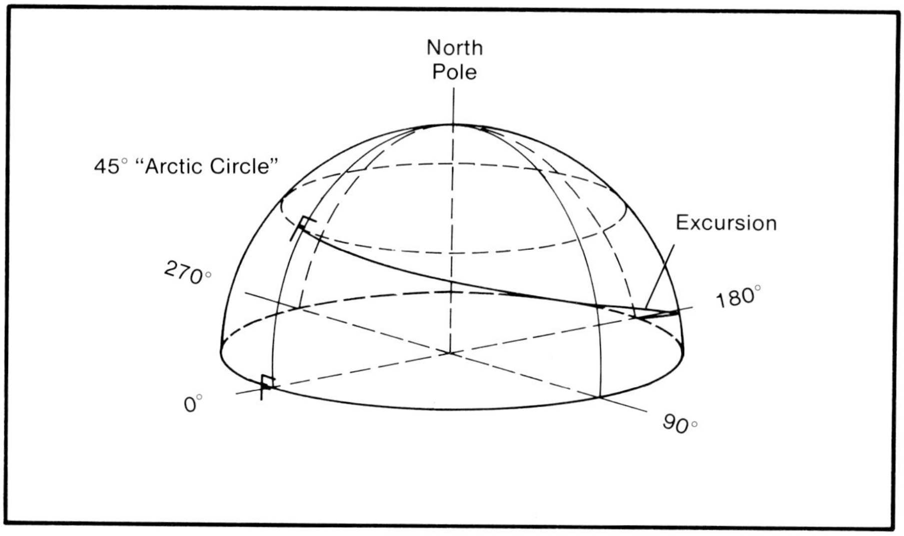

Figure 21-3. Excursion From 0° To 180° As 22½° Mirror Rotates 90°.

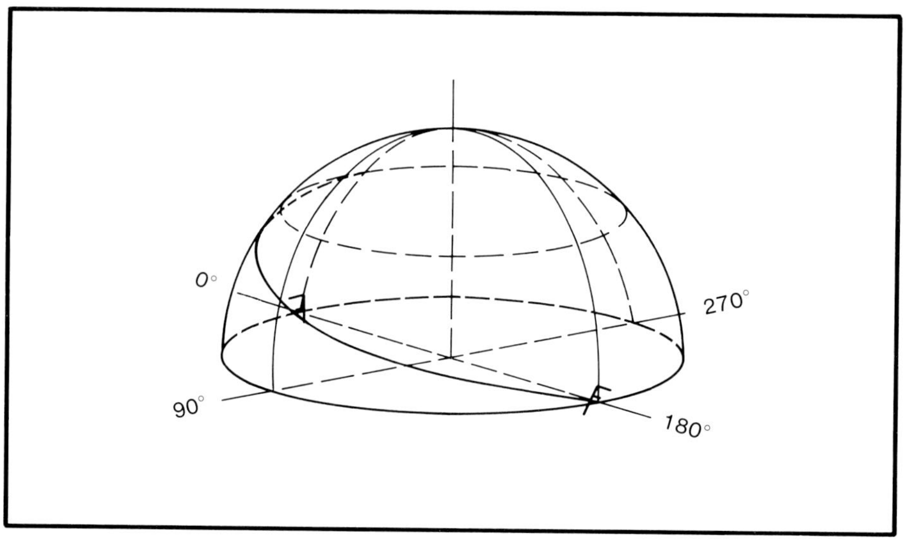

Figure 21-4. Back View Of Excursion From 0° to 180°.

mirror has been rotated 90° on the horizontal axis, the mirror will be rotated 22½° around the 0°-180° line axis. Therefore, the image at grazing position has rotated 45° or twice the 22½° mirror angle.

These two positions are the only ones known. However, knowing that the excursion is horizontal at the 0° position, it can be assumed that the excursion becomes horizontal at the elevation of 45° for a distance and gradually arcs downward until it reaches the 180° position, where it reaches the equator.

It can also be assumed that the image starts at 0° rotation and gradually rotates clockwise until it reaches 45° rotation angle at the 180° equator position.

This is essentially what happens. This problem was set up on a reference hemisphere, and the following approximate elevations and angles of image rotation were found empirically.

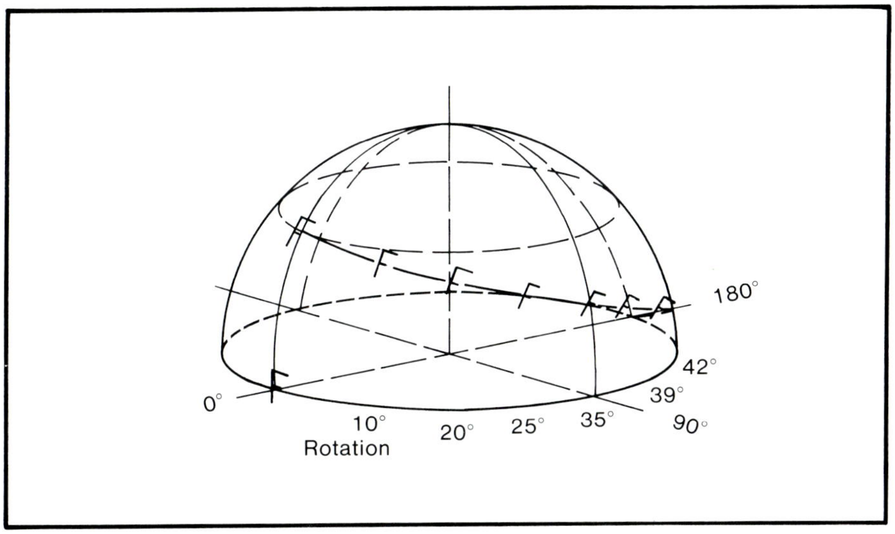

Figure 21-5. Image Orientation Along The Excursion 0° To About 135°.

Rotation of Mirror on Vertical Axis	Position of Image	Elevation of Image above Equator	Rotation of Image (Clockwise)
0°	0°	45°	0°
11¼°	22½°	45°	10°
22½°	45°	42°	20°
33¾°	67½°	40°	25°
45°	90°	35°	35°
66¼°	112½°	28°	39°
77½°	135°	21°	42°
88¾°	157½°	10°	44°
90°	180°	0°	45°

Figures 21-3 and 21-4 show this excursion on the hemisphere, while Figures 21-5 and 21-6 show the image orientation at several locations. If the mirror is rotated to the left on its vertical axis, it has the same excursion, but in mirror image. Its path, as it reaches the 180° mark, is not the same as the right excursion—it is the mirror image of it. In Figure 21-7, the rotation of the image is counterclockwise on the left side of the hemisphere.

In Figure 21-8 the lower hemisphere has been added to the upper hemisphere normally used. If the mirror had been rotated forward 22½° instead of backward, the excursion would be along the lower hemisphere. The excursions are shown—they are the same number of degrees from the equator at each angular position, but the path in the lower hemisphere is shown in minus degrees.

This problem reveals the limitations in finding mirror paths theoretically. In many instances empirical tests must be made to find the image's location and rotation after reflection from a mirror rotated around both the vertical and horizontal axes.

Mirrors at a Fixed Angle

If two mirrors are mounted with a permanent angle between them a light ray incident to one mirror reflects to the other. It then reflects from the second mirror on a path that is twice the angle between the mirrors to its new path.

Figure 21-9 shows two mirrors mounted at 45° to each other. The solid line trace shows a ray striking the first mirror at an angle of in-

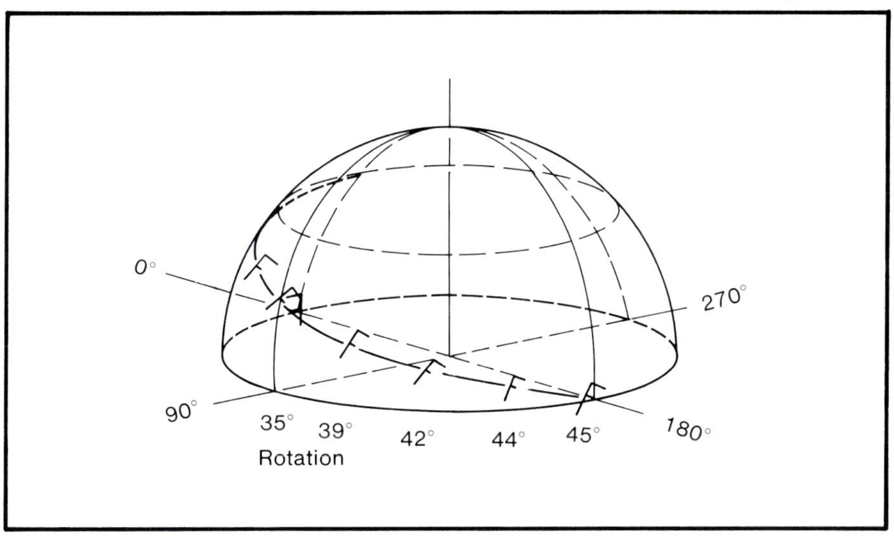

Figure 21-6. Back View Of Image Orientation Along The Excursion.

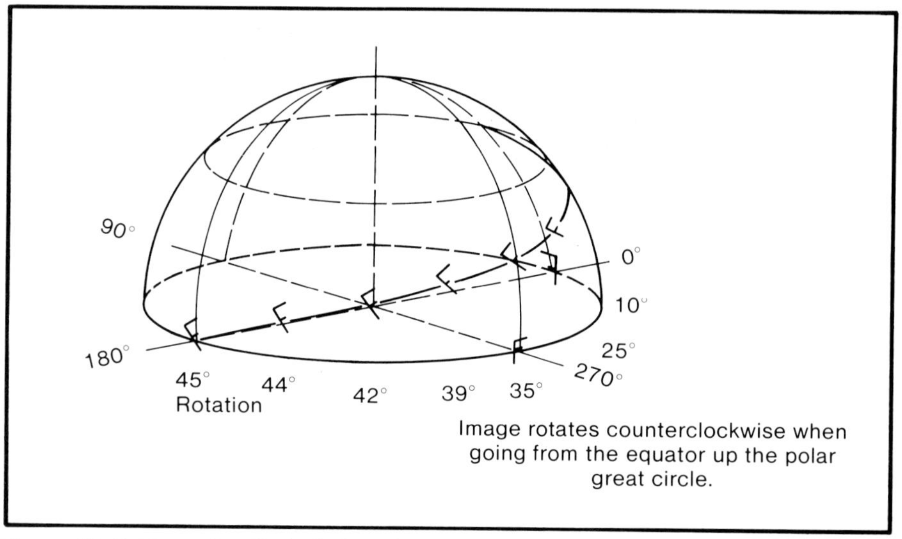

Figure 21-7. Excursion On Left Side Of Hemisphere (0° Through 270° To 180°) When Mirror Is Rotated Clockwise On Its Vertical Axis.

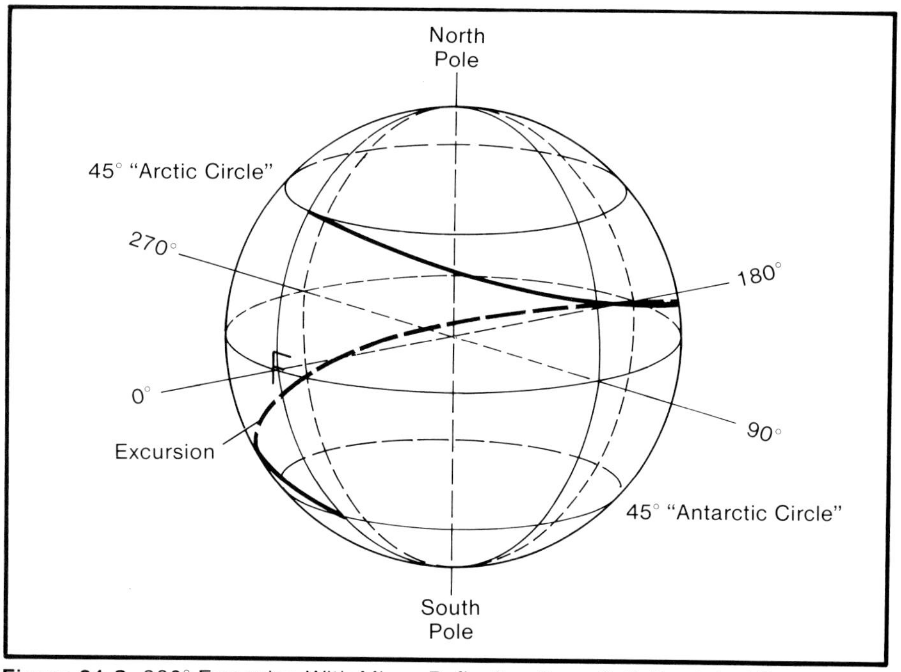

Figure 21-8. 360° Excursion With Mirror Reflecting From Both Front And Back Surfaces. Two Hemispheres Required. Mirror Rotated Counterclockwise.

cidence of 22½°. It reflects at 22½° and strikes the other mirror at an angle of 22½° and its angle of reflection is the same. The angle between the entering ray and the final reflecting ray is 90°.

The dashed line shows another ray at any odd angle. Its exit path from the two reflections is also at 90° to its entering path.

Figure 21-10 shows two mirrors fixed at 60° to each other. The ray path shows that the exiting ray is 120°, or twice the angle between the mirrors, to the entering ray. Therefore, when a fixed angle of reflection is needed, no matter what direction the incoming ray takes, the fixed angle two-mirror system is useful.

The Cross-staff Navigational Instrument

For a long time the world has been artificially zoned into degrees of longitude and latitude. By knowing the longitude and latitude of a place, its location on the globe can be readily determined (Figure 21-11).

Early navigators found that when they knew the elevation above the horizon of a star or the sun and the exact date and time, they could calculate the longitude and latitude of their ship. They could determine their exact location. The calculation involves spherical trigonometry, and details of navigation which are beyond the scope of this course.

The method of determining the elevation of a star or the sun only involves the use of simple optical instruments. The horizon and the celestial body are both visible from a ship. What was needed was an instrument to measure the angle of elevation of the stars above the horizon.

The original instrument used is called a cross-staff. As Figure 21-12 shows, it consists of a fixed length crosspiece mounted on a rod so that it can slide back and forth. The rod is calibrated in degrees of elevation. The navigator aligns the bottom of the crosspiece with the horizon and the top of the crosspiece with the sun (star). The crosspiece is moved back and forth on the rod until the angle between the eye and the bottom and top of the crosspiece are equal to the elevation of the sun (star). On a calm day, when the ship is motionless, this is fairly simple. However, when the ship is rolling it is extremely difficult to make this measurement.

The invention of the sextant made reading the elevation of the sun (star)

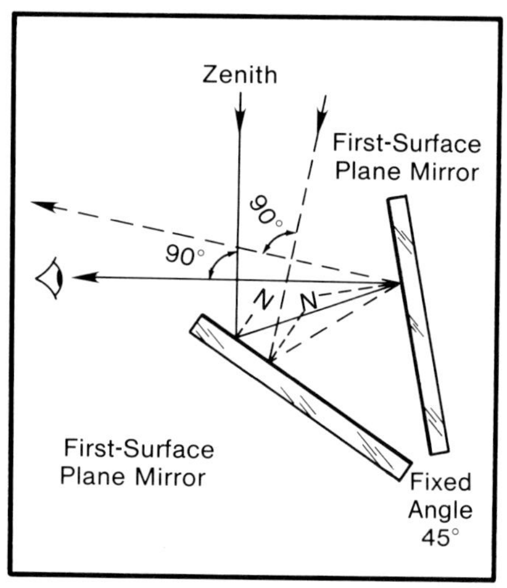

Figure 21-9. Fixed Angle Between Mirrors.

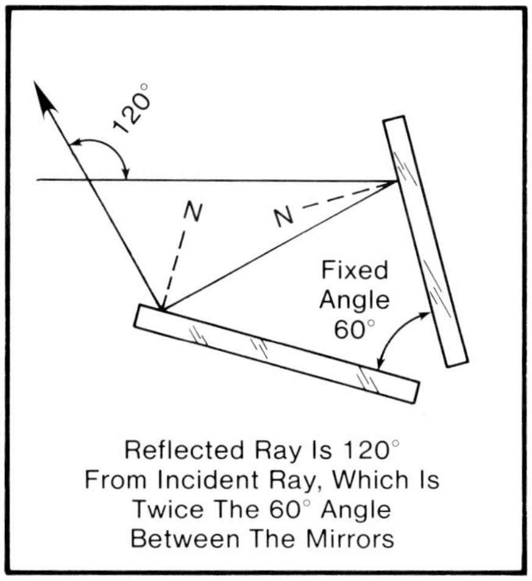

Figure 21-10. Fixed-Angle Mirrors At 60°.

much easier. The sextant uses two mirrors with a movable angle between them. Their total angle of reflection remains constant, although the actual angle of the incoming ray may vary.

The Sextant

The sextant is an optical instrument that uses the principle of a constant angle of reflection of two fixed mirrors to measure the elevation of the sun (star) above the horizon when the ship is rolling (Figure 21-13). The fact that the angle of reflection is constant when the aim angle is changing makes it easy to use

The two-mirror device where the two mirrors are fixed at 45° to each other serves as an example. In Figure 21-9, the sun is at the zenith; that is, straight up. Its angle of elevation is 90°. The eye looks straightforward at the horizon, at 0°. The eye sees the sun. The fixed mirrors can rotate without changing the position of the image of the horizon.

The sextant has a triangularly shaped metal frame. The bottom of the triangle is curved and calibrated in half degrees. This scale is called the limb. A metal arm containing a mirror at its upper end is pivoted to the frame near its top. It has an opening in it with a marker that runs along the half-degree scale. Attached to the frame is a low powered telescope—which is usually Galilean—and has a positive objective lens and negative eyelens.

The telescope looks at two images through a mirror/window. One half is silvered and is a mirror. The other half is clear and is a window. This is not a beam splitter.

Through the telescope, the viewer can see the horizon through the clear portion and the sun (star) in the mirrored portion. The arm is moved carefully until the lower edge of the sun is seen to align with the horizon. The image of a star, which is a point, can also be aligned with the horizon. The star's elevation can be read from the scale on the limb.

On the outline of the sextant in Figure 21-13, angle B is the angle between the two mirrors. Angle A then is the elevation, and it is twice angle B. The scale is calibrated in half degrees but marked in full degrees so that reading it will give the elevation directly.

Of course, the navigator must know which star is being looked at and the exact time of the measure-

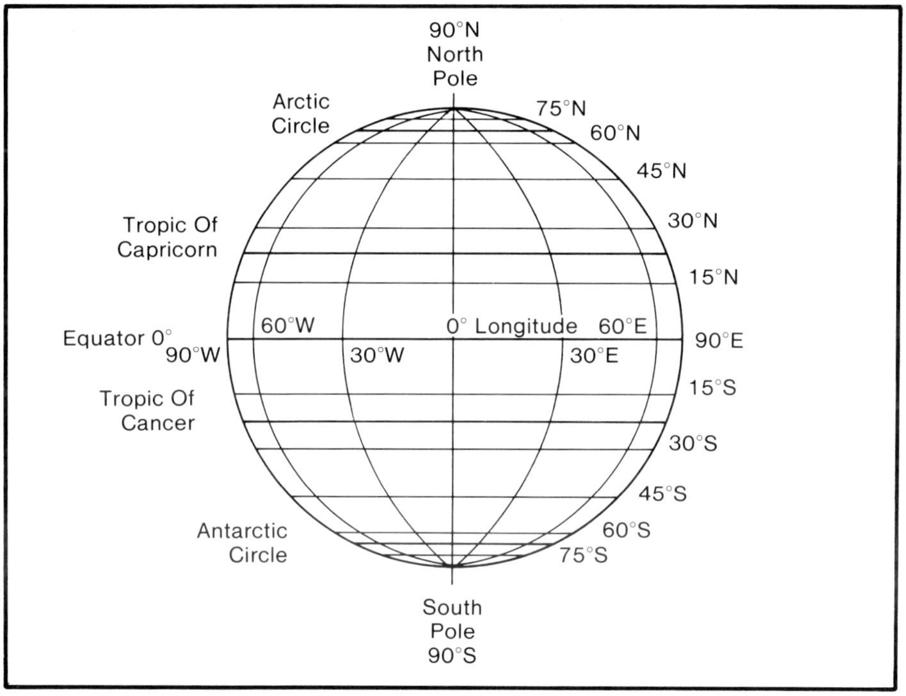

Figure 21-11. Global Lattitude And Longitude.

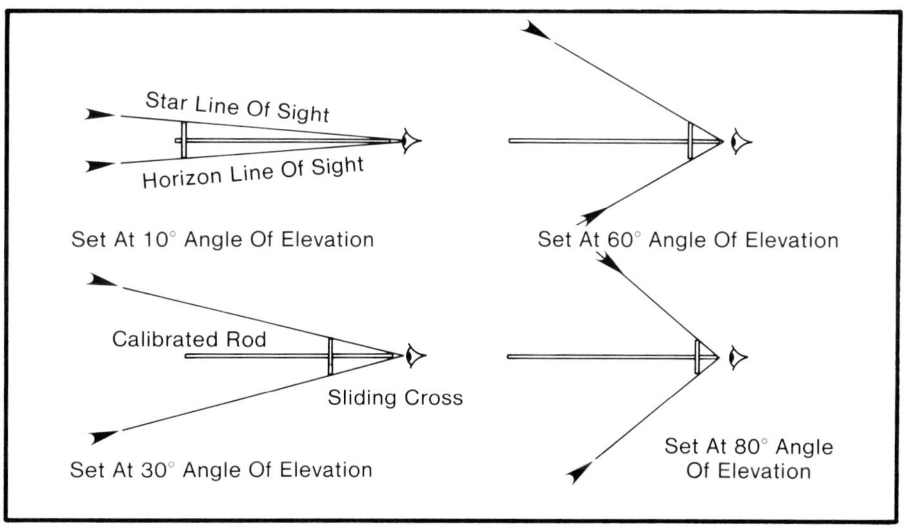

Figure 21-12. Cross-staff For Determining The Elevation Of A Celestial Body Above The Horizon.

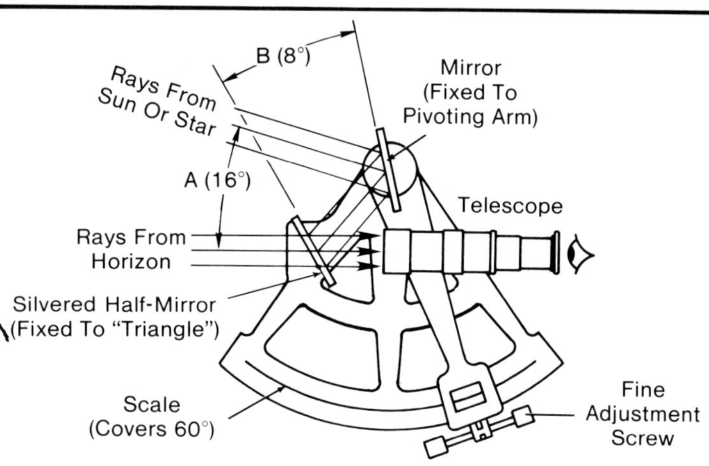

The silvered half-mirror allows the observer to see the horizon directly together with the reflection from the second mirror. When the superimposed images are observed through the telescope the subtended angle may be read directly from the scale. Since the reflected image is deviated by twice the angular displacement of the movable mirror, each degree on the scale is equivalent to two degrees of angular measurement.

Figure 21-13. Sextant.

ment so that the latitudinal-longitudinal position of the ship can be calculated.

Since the sun is seen as a reflection from two mirrors with an angle between them, the sextant can move without changing the position of the image of the sun. The rolling of the ship does not move the sun rapidly across the objective lens of the telescope, because of the two-mirror arrangement. The sextant is, therefore, much easier to use than the crossbar.

A Nautical Rangefinder

An instrument similar to the sextant has been used to locate the distance between a ship and its observer (Figure 21-14). This is useful in nautical military actions. Instead of measuring elevation, the instrument finds the angle between the bow and the stern of a ship some distance from the observer. The ship must be broadside to the line of sight. The bow of the ship is seen directly and the stern in the reflection, or vice versa. The arm is moved until the two align. The distance is calculated using the length of the ship and triangulation. If the ship is not broadside, the instrument can measure the height of the top of the ship's mast above the waterline, and the same calculations can be made. Obviously, the observer must be familiar with the dimensions of enemy ships for the method to be useful.

The drawing shows the triangle that is used. If the ship is 1,000 feet long, and the angle is 2°, then the half angle is 1°. The tangent of the half angle is:

$$\frac{500\text{ ft.}}{\text{distance}}$$

(500 feet is one half the ship's length.)

$$\text{Tangent } 1^\circ = \frac{500\text{ ft.}}{\text{distance}}$$

$$\text{Distance} = \frac{500\text{ ft.}}{\text{tangent } 1^\circ} = \frac{500\text{ ft.}}{.017455}$$

28,645 ft. or about 4.7 nautical miles (6076.1 ft./nmi.).

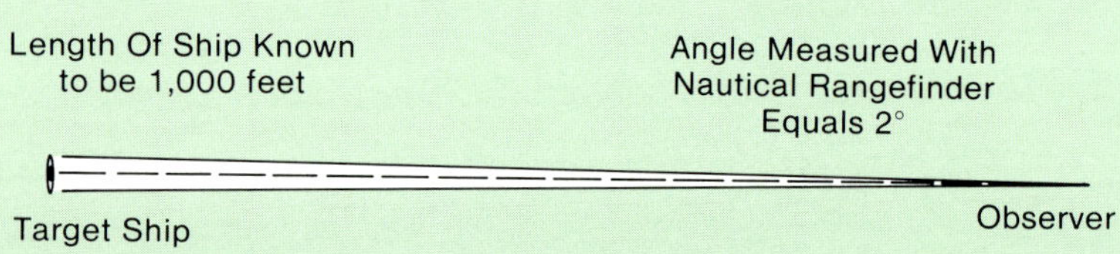

$$\text{Tangent of Half Angle} = \frac{\text{Half The Ship's Length}}{\text{Distance}}$$

$$\text{Distance} = \frac{\text{Half Length}}{\text{Tangent of Half Angle}}$$

$$\text{Distance} = \frac{500\text{ Ft.}}{.017455} = 28{,}645\text{ Feet} = 4.71\text{ Nautical Miles}$$

Figure 21-14. Nautical Rangefinder.

Figure 21-15. Right-Reflecting Mirror.

The Corner Mirror

The angle of reflection from the two mirrors is twice the angle between the mirrors. If two mirrors are at 90° to each other, the angle of reflection is 180°. The light entering returns toward the direction from which it came.

The right-reflecting mirror covered earlier was an example of this (Figure 21-15). By placing two mirrors at right angles you see yourself as others see you, not as you see yourself in a single mirror. However, this right-reflecting mirror produces a two dimensional effect. Light going in at any angle that is not 90° to the line joining the two mirrors does not come back on itself—it obeys the simple law of reflection, and the angle of reflection is equal to the angle of incidence at each mirror surface. This means that the light is reflected in such a way as to allow the subject to see his entire reflection.

If, however, a third mirror is added at right angles to the other two, the three-mirror combination returns light from any direction. Such a mirror is called a corner mirror because the mirrors are placed as they would be to form the corner of a cube (Figure 21-16).

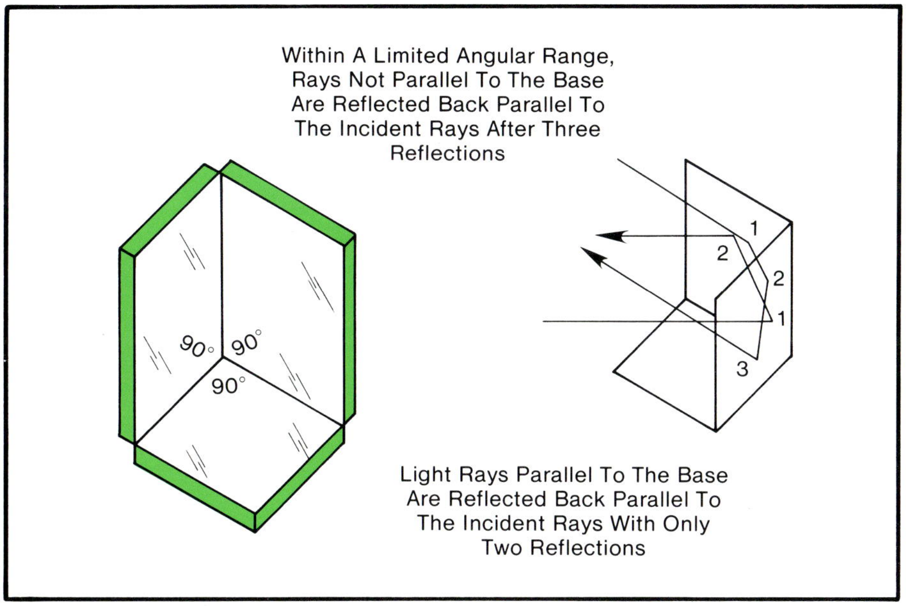

Figure 21-16. Corner Mirror.

The corner mirror arrangement is very efficient over a few degrees. Light aimed into the corner mirror near its axis is returned on the axis from which it came.

The reflecting arrangement of the "corner mirror" can be made by using mirrors, or more commonly, prisms. Perhaps you will recall that at 45°, the critical angle makes a mirror of the reflecting surface of a prism. Corner mirrors are most often made as multiple prisms (Figure 21-17).

If one surface of a sheet of plastic is molded into many tiny, prism-corner reflectors (retro-reflector prisms), the plastic sheet becomes very efficient as a surface that sends light back in the same direction from which it came.

Such material is used in road signs. As a car moves down the highway at night, the headlights send light ahead and to the sides. If a sign is placed near the roadway and aimed approximately perpendicular to the approaching headlights, the reflection from the corner mirror material will appear very bright to the occupants of the car. If the material is cut into letters advertising a product, the name of the product will appear bright to the car's occupants, as though illuminated from within. Because the reflection in each corner mirror prism is at about 45°, the reflections are greater than the critical angle, so they are complete. No silvering of the surfaces is needed. This type of surface is also molded in red plastic and used as reflective taillights on automobiles.

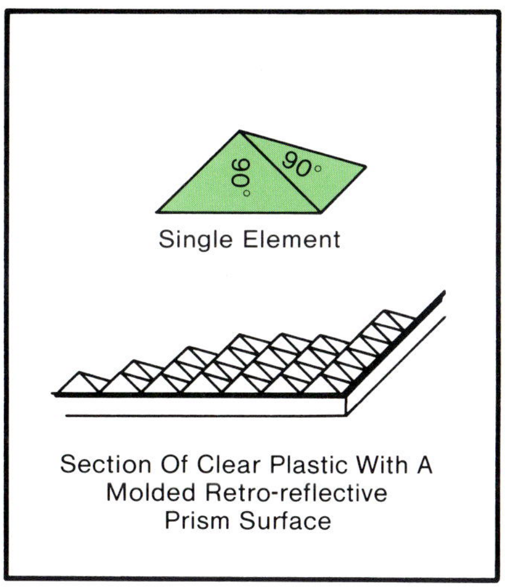

Figure 21-17. Retro-Reflector Prisms.

Additional Retro-Reflector Applications

Direct reversal reflectors can be made in the shape of strong lenses (retro-reflective lenses). Because the optical quality needed is not great, the surfaces are easily molded from either glass or plastic.

The front surface is a strong lens curve. Figure 21-18 shows how a ray of light enters, is reflected by the rear surface, and returns in a path 180° to its original direction. Such strong curves have images with field curvature, so the rear surface is also curved to compensate for this aberration which makes the return beam brighter. In larger size road sign buttons the rear surface is silvered because the angle of some of the rays is less than the critical angle and they can escape through the rear of the lens.

Such optical devices can be made very small. In the smallest size, about like grains of sand, they become tiny spheres. The spheres are cemented to a screen background with transparent cement. The whole screen then becomes highly reflec-

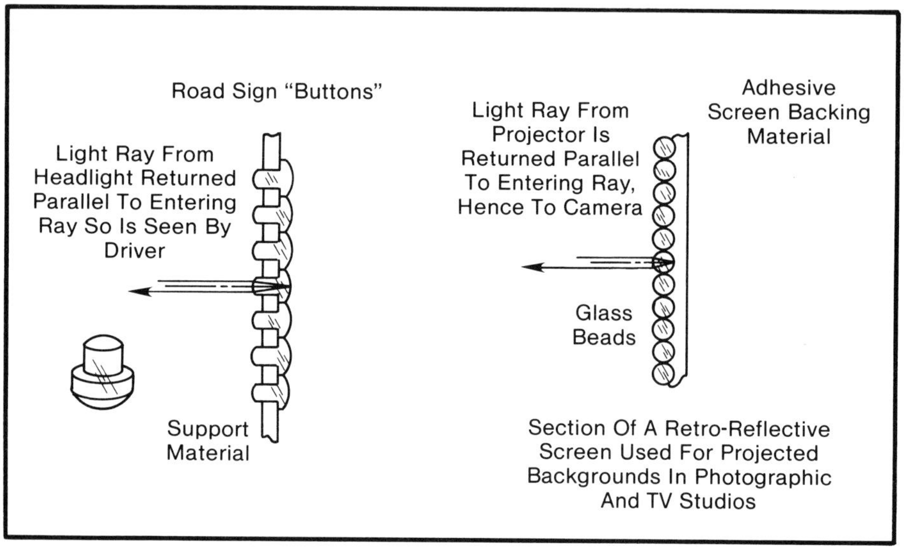

Figure 21-18. Retro-Reflective Lens Applications.

tive within a very narrow-reflection angle. A major use of this material is in both conventional and TV photography. The actor stands in front of a screen made of the tiny sphere material (Figure 21-19) and a projector is arranged so that it projects onto a 45° beamsplitter in front of the camera lens. The slide in the projector is focused on the screen. Because the projector beam axis coincides with the camera axis, the image is reflected back and is imaged by the camera. The actor is also imaged by the camera, and appears to be right in front of the projected image. The image projects on the actor, but because the reflection from the actor is diffuse, it is so much dimmer than the studio lights shining on the actor that it does not show. The shadow of the actor on the background image is not visible to the camera because the projector axis and the camera axis are the same. Care is taken in lighting the actor so that no light from the studio lights falls on the screen. The direction of studio light on the actor must be the same as the direction of light in the projected scene so that the illusion appears natural.

Designing the Optical System of a Projector

As a last lesson in this program, a number of the individual optical principles will be brought together as the factors involved in designing a projection optical system. The system may be considered as a transparency (slide) projector. The same technique can be used for a motion picture projector.

Figure 21-20 shows the transparency with the frame indicating the object size. This is then shown in a two-dimensional cross section. In order to project the transparency, a projection lens is added. As has been the practice in this program, the projection lens is represented as a single lens, although in reality it must be made of at least three elements to provide adequate corrections of aberrations. The projected image on the screen is determined by the object size, projection distance and the lens focal length.

The next thing needed is a projection lamp (Figure 21-21). The filament of the lamp is the part most involved in the optical system so it is added and shown in the drawing as a short, straight line. If the lamp is turned on and the image of the transparency will be focused on a screen, only a small central area of the transparency is illuminated. Something is needed to spread the light evenly over the entire transparency area. If the projection lens is removed and the eye is placed in the position of the projection lens, the coils of the lamp filament can be seen in the central area—and no light can be seen coming through the transparency's outer area. The image of the filament in the projection lens

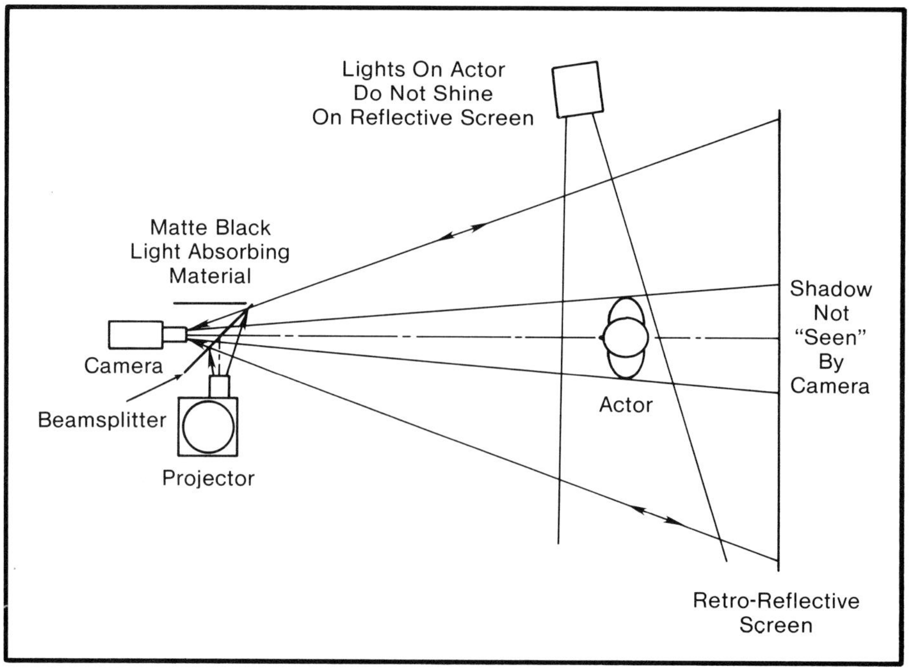

Figure 21-19. Front Projection System.

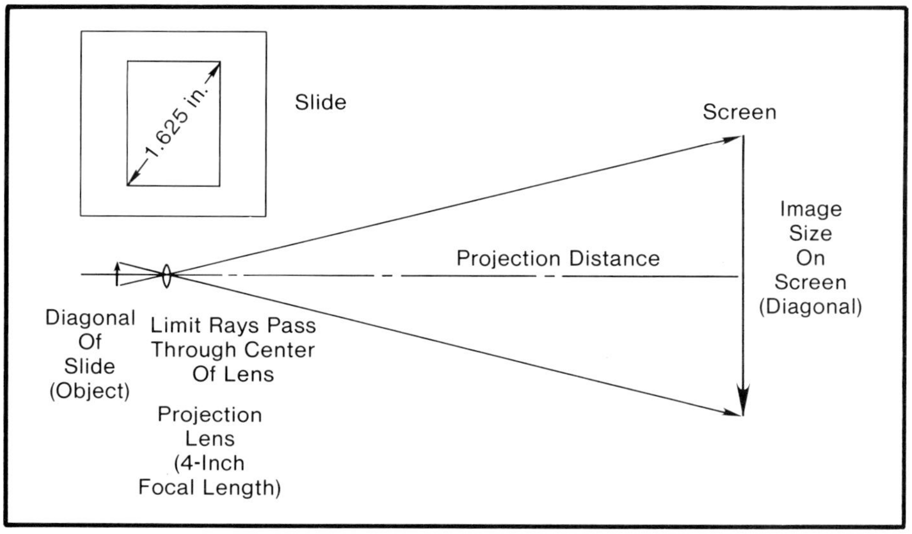

Figure 21-20. Projection Optical System (Slide, Lens, And Screen).

appears as an 0.125 x 0.25 inch rectangle. If it could be doubled to an 0.25 inch square, the light output of the projector could be nearly doubled.

In an earlier discussion of spherical mirrors, it was learned that light coming from the center of curvature would be reflected right back at the center of curvature. Thus, if a spherical mirror is placed behind the lamp so that the center of the filament is at the center of curvature, it will reflect the image of the filament right back on itself. Then, if it is tilted at a slight angle, the image of the filament will be slightly displaced. By watching the image in the projector lens, a mirror angle can be found that will place the image of the filament coil right next to the real filament. This reflected light goes on through the lamp and the optical system and adds to the illumination on the screen—nearly doubling it. An 85% increase is not unusual.

A condenser-lens system can be added to spread the light evenly over the entire transparency area. The condenser lens works on the same principle introduced earlier in the program. A lens was found to be filled with light from a star, if held between the eye and the star, and if the lens was placed at a distance equal to its focal length from the pupil of the eye.

If the eye is placed in the position of the projection lens, and a proper condenser lens system is placed between the slide and the filament, the condenser lens will appear filled with light from the filament, and the slide aperture will be fully illuminated.

Now we must decide what constitutes the important attributes of the condenser system. Figure 21-21 shows how the diameter of the condenser lens can be found by schematic ray tracing. The rear aperture of the projection lens should be approximately filled with the image of the filament. The diameter of a 4-inch *f*/2.8 projection lens is about 1.4 inches. For mechanical reasons (to allow room for the transparency mount) the front condenser lens is placed about 0.375″ behind the transparency. By tracing the outer rays, the minimum permissible diameter of the condenser lens is found to be about 2″. In practice, a 2.25″ diameter is commonly used to allow for slight mounting and alignment errors.

In order for the projection lamp filament to be focused on the rear lens of the projector lens, the focal length of the condenser lens system and the position of the filament must be determined. To keep the projector compact, the distance must be as small as practicable, yet there must be room for the heat absorbing glass between the lamp and the rear condenser lens.

Assume that the overall distance between the transparency and the lamp filament is about 3.8″. The overall distance between the filament and the projection lens is then about 8 inches. The distance between the transparency and the pro-

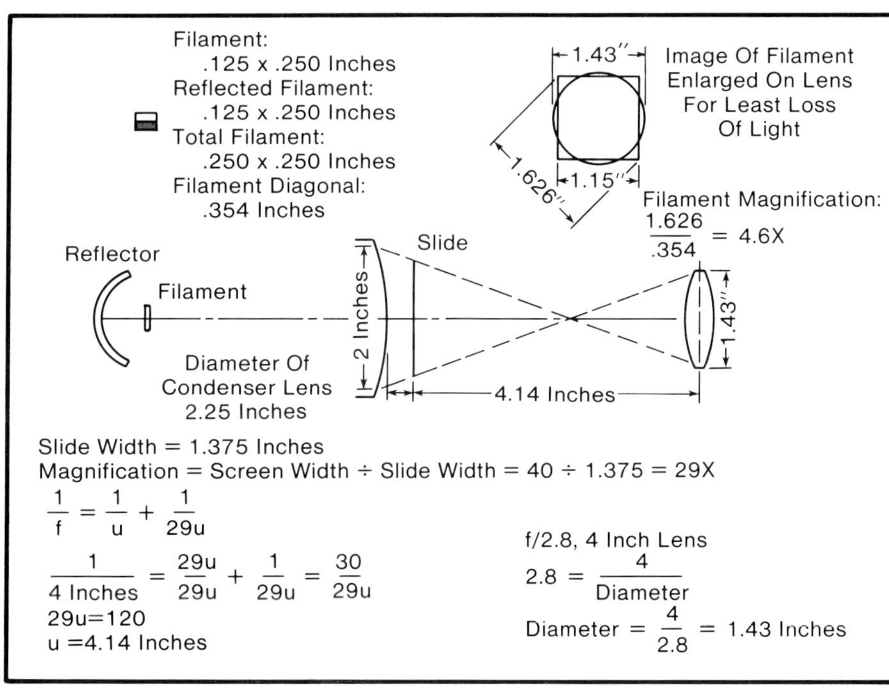

Figure 21-21. Projection System.

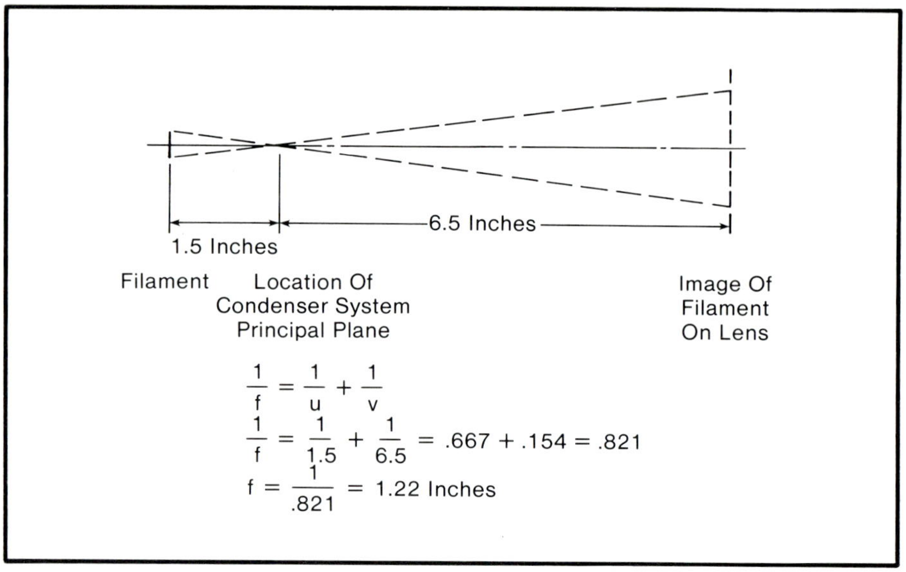

Figure 21-22. Finding Focal Length Of A Condenser System.

jection lens is about 4.1″. The focal length of the condenser system can be found by applying the formula:

$$\frac{1}{f} = \frac{1}{u} + \frac{1}{v}$$

The principal plane of the condenser lens is about 1.5″ from the filament and about 6.5″ from the rear element of the projection lens (Figure 21-22).

$$\frac{1}{f} = \frac{1}{1.5} + \frac{1}{6.5} =$$

$$.667 + .154 = .821$$

$$f = \frac{1}{.821} = 1.22 \text{ inches}$$

The drawing shows how the focal length can be found by ray tracing. The filament is .35″ across, and the image of the filament on the lens is 1.625″ across. Rays are drawn from the edges of the filament to the edges of the image. The principal plane is located where the rays cross.

A single condenser lens with this focal length and diameter is impractical to make because of the strong curves required. It is common practice to use two lenses to obtain a short focal length (Figure 21-23). A better light distribution is achieved by making one aspheric surface that can be easily molded. A strong curve is no more costly to mold than a weaker one so the other curve on this lens, and the two curves on the other lens, do not have to be as strong and will be less costly to grind and polish.

The optical system is almost complete. If the parts are assembled on an optical bench, and a low voltage applied to the filament, the image of the filament should focus on the rear surface of the projection lens, and should be visible by looking into the projection lens (the low voltage is used to protect the eye from brilliant light). If the projection lens is removed and the eye takes its place, the condenser lens appears filled with light.

Now, if the system is put together with a transparency in place, and the voltage turned up to normal, the transparency can be projected on the screen with bright, even illumination. However, the film soon "pops" from the heat, which throws the screen image out of focus.

A slide or motion-picture frame taken on a reversal color film is projected with the emulsion side toward the projection lens. When acclimated to room temperature and humidity, the film is fairly flat. When heated suddenly, the film soon "pops" to a concave shape because, as the moisture is driven rapidly out of the emulsion by the heat, the emulsion shrinks. Because the transparency is held in the projector by its edges, the center moves away from the projection lens, causing the out-of-focus condition.

This is the reason that room was provided in the condenser-lens system for a heat-absorbing glass. The heat-absorbing glass is a window with plane-parallel sides made of glass with a high iron content. It absorbs infrared radiation, and converts it into heat (Figure 21-24). The mount that holds the heat-absorbing glass is made of heat conductive metal to carry the heat away from the

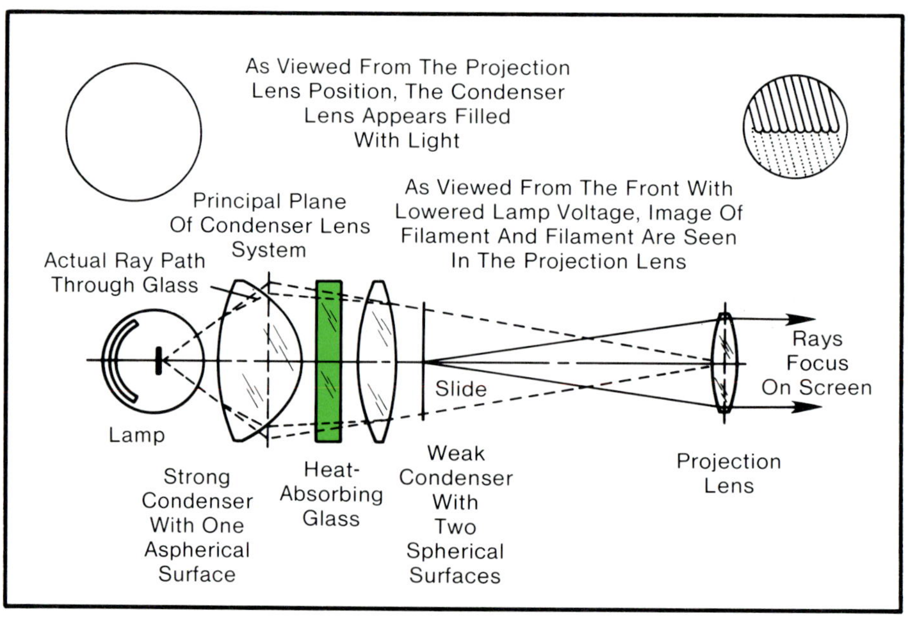

Figure 21-23. Completed Projection Optical System.

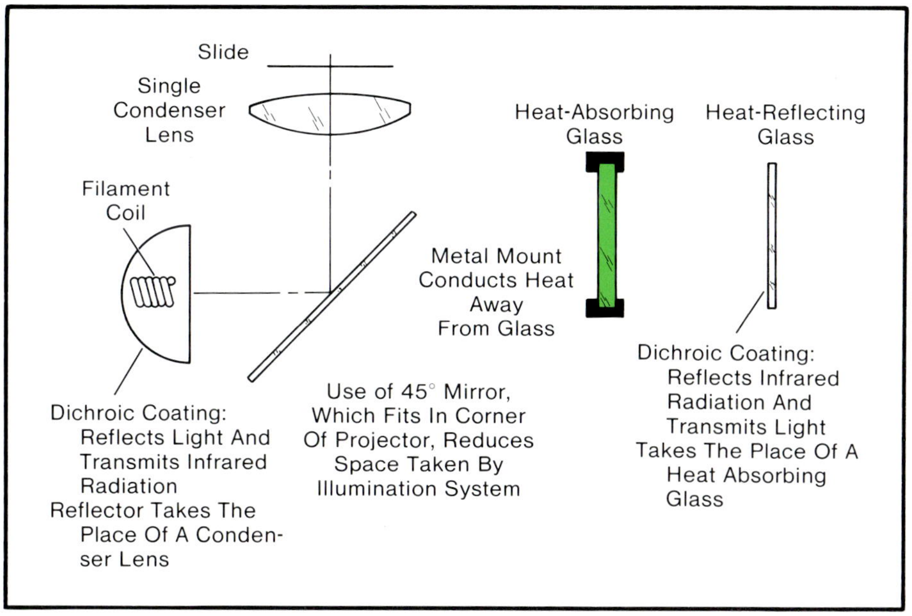

Figure 21-24. Methods Of Reducing Heat On The Slide.

glass.

Some projectors have a window with a dichroic multiple-layer coating that reflects infrared radiation and transmits visible radiation (light). The infrared rays are reflected toward the lamp and do not reach the slide. A blower conducts the hot air away from the chamber containing the bulb, the heat absorbing glass, and the condensers. A special duct may be used to blow air directly over the slide, keeping it cooler.

A third arrangement for heat control is common today. The reflector can be built as part of the lamp, and is coated with a dichroic layer that reflects light and transmits infrared radiation. Much of the heat goes out the back of the lamp and is not reflected forward toward the transparency. A heat-absorbing glass is still necessary, however, to reduce the amount of heat contained in the direct radiation from the filament.

In slide projectors, the reflector can take the place of one of the condenser lenses, and can be placed at quite a distance from the single condenser lens. A 45° mirror can be used to shorten the overall length of the optical system.

In Super-8 projectors, a lamp with a built-in reflector can eliminate the need for condensers altogether.

Some projection lamps have a coiled filament in a small quartz envelope. The envelope is filled with a halogen gas. Most lamps grow dim with time because tungsten is evaporated from the filament and gradually condenses on the inner surface of the glass envelope, acting as a neutral density filter. In the quartz-halogen lamp, the gas causes the vaporized tungsten to condense back on the filament, which prevents the dimming and lengthens the life of the filament. The reflector is built around the tiny lamp as shown in Figure 21-24.

If you look at the image of the filament in the lens of a projector that contains this type of bulb, you will see a bright circular light with a dark center, the image of the end of the filament's coil support.

Screens

Most slide projectors give enough illumination on the screen for satisfactory viewing by moderate sized groups. However, if the projection distance is long, and the screen large, the screen image looks too dim.

The reflectance of the screen material makes a difference in image brightness. White painted screens do not reflect as much light as silver (aluminum) painted screens. Beaded screens, like tiny road sign buttons, give a high reflectance, but only within a very narrow-viewing angle about the optical axis (Figure 21-25). Lenticular silver screen material is the best compromise. Tiny concave reflectors are molded into

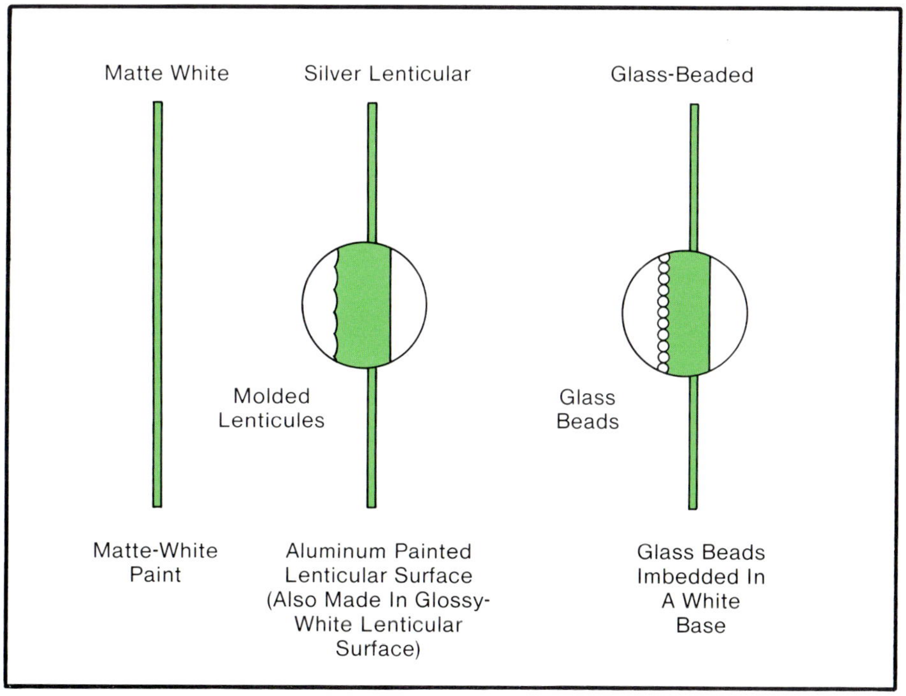

Figure 21-25. Types Of Screens.

the screen surface and painted with a semi-diffuse aluminum paint. The reflectivity is high and the molded reflectors are designed to cover a reasonable viewing angle. The lenticules can cause poor image sharpness when the screen size is small and the viewing distance is short.

Higher Voltage

The normal projection lamp is designed to be used with standard 120-volt electricity. Transformers are sometimes used to increase the voltage which increases the light output but shortens the life of the bulb. The output increases rapidly with small increases in voltage. The formula used to calculate the light increase is:

$$\text{Voltage}^{3.7} = \text{Light}^{13}$$

When the voltage is increased by a power of 3.7, the light increases by a power of 13.

The Value of This Course

Optics is the study of light and its behavior as it is reflected, absorbed and transmitted by various materials. Optical engineering makes use of the laws of optics to design instruments that use light to serve useful purposes. This course has shown, in a practical way, how schematic-ray tracing, graphic-ray tracing, and some mathematical equations can be used to quickly solve some of the more common problems encountered in optical engineering.

During the period between the First World War and the Second World War, optics reached a development plateau. New instruments were designed, but the same methods were used to design them that had been in use during the previous 25 years. Then, about 1950, a whole new series of techniques was developed.

The study of the ability of lenses to form images changed to include such terminology as spatial frequencies, sine waves, modulation transfer functions, and many others. Computers were developed that took much of the drudgery out of the mathematics of lens design. They could be programmed to create a design, calculate the quality of the image and re-design to improve the quality. Now they can even print a large-scale drawing of the star image at any field angle. As a result, the quality of lenses has improved and new concepts, such as zoom lenses, have become possible. A whole new series of optical glasses, called rare-earth glasses, were developed and gave the designer a much wider scope in his design. Plastics became available that were suitable for use as a lens material and techniques were developed to mold plastic lenses that produced quite good images. As a result, the cost of cameras and other optical devices has been kept reasonable.

In the field of physical optics, similar revolutions are taking place. An optical device which produces laser light has been developed. Many uses have been found for lasers—surgery, metal welding, drilling, cutting, hardening, and extremely accurate alignment and leveling of large structures. Lasers are being considered as prime defensive weapons. Their use as an energy source is just starting, and laser fusion is not too far off. Laser inferometers are used to measure curves on large lenses to a tiny fraction of a Newton's ring. There is even a new field called laser chemistry. Lasers are used for the extremely rapid recording and display of digital information. One form of video disc uses a laser, as do the scanners used at the supermarket which read and record the prices of purchases. A whole new field of photography and holography is possible because of the laser. And all these fascinating and useful devices became possible because imaginative optical engineers saw needs and put the principles of optics to use in solving those needs.

If you are faced with a need that the application of optics can solve, you have been given a number of tools in this program that can help you solve them. Put your imagination to work. One of these tools is to imagine yourself as part of the system. Look backward in a system and imagine what you would see if you were a lens, a prism, or a mirror. What would you have to do to accomplish a result? Would you have to image? Where would you have to reflect? Would you invert the image or leave it right side up? Analyze the situation in this manner, and you may surprise yourself with the ideas you can produce.

Decide whether the situation calls for schematic or graphic-ray tracing, or for solving the problem mathematically. Perhaps you need an aid, such as a reference hemisphere to help you.

And if, at times, you find yourself upsidedown, or reversed, left to right, stay with it. It can be done. And you'll find that using practical optics to solve problems can even be fun.

IMP®

GLOSSARY

Abbe Dispersion Number:
A commonly used number which indicates the dispersive power of a glass formulation.

$$V = \frac{n_D - 1}{n_F - n_C}$$

Aberration:
The inherent characteristics of a lens that result in imperfect imaging of object points (e.g., star images). Some manufacturing defects, such as decentered or tilted lenses, or those which form a curved image plane or cause a geometric distortion, are considered aberrations. Field curvature and distortion are also considered aberrations. (See Astigmatism, Diffraction, Chromatic Aberration, Coma, Spherical Aberration.)

Absorptance:
That part of radiant energy which is absorbed by a surface or transmitting material. It is usually expressed as a decimal value or percentage.

$$\text{Absorptance} = \frac{\text{Absorbed Light}}{\text{Incident Light}}$$

Achromatic:
Colorless, white or gray. An even mix of wavelengths, as an achromatic light source or as an achromatic reflecting surface.

Achromatic Lens:
A lens that has been partially corrected for chromatic aberration and is made of at least two elements. Typically the positive element is crown glass, with low index of refraction and low dispersion, and the negative element is flint glass, with high index and dispersion.

Actual Angle:
The maximum field half-angle of a telescope derived from the focal length of the objective lens and the diameter of the field lens. Used with the apparent angle to calculate magnification.

Acute Angle:
Any angle less than 90 degrees.

Acute Triangle:
A triangle which has all angles less than 90 degrees.

Aerial Lens:
A lens designed for flat-field use with objects at an infinite distance. Typically, a lens used to photograph the earth from high altitudes as an aid to preparing maps.

Afocal:
Without a focus. An optical device in which parallel light rays entering, leave as parallel light rays.

Albada Finder:
A viewfinder in which the negative surface of the finder lens reflects a white border to the eyelens to provide a parallax-free frame used as an aid when aiming the device.

Aluminizing:
The process of coating a polished glass or plastic surface with aluminum to make a highly-reflective first-surface mirror. The aluminum, evaporated in a vacuum, condenses on the polished surfaces. The aluminum is usually overcoated with silicon monoxide to protect the fragile metal layer.

Amici Prism:
A right-angle roof prism.

Anamorphic:
Image distortion by compression or expansion in only one direction at right angles to the optical axis.

Anastigmat:
A compound lens that has been corrected for astigmatism; an anastigmatic lens.

Angle:
A plane figure formed by two intersecting straight lines.

Angle Of Deviation:
The angle formed between the incident ray path and the exiting ray paths as a ray passes through a lens or wedge prism. A measure of the angular deviation from its original path.

Angle Of Incidence:
The angle formed by an incident ray and a reference line drawn perpendicular to the optical surface at the point of incidence.

Angle Of Minimum Deviation:
The smallest angle of deviation of a light ray which can be achieved by a given wedge prism. This angle occurs when the internal ray path is perpendicular to the wedge-angle bisector.

Angle Of Reflection:
The angle formed by a reflected ray and a reference line constructed perpendicular to the surface at the point of reflection.

Angle Of Refraction:
The angle formed by a refracted ray and a reference line constructed perpendicular to the surface at the point of penetration. The ray is "refracted" (bent) while passing from one optical medium to another.

Angle Of Rotation:
The angle through which an image has turned about the optical axis while passing through an optical element or system (e.g., dove prism).

Angstrom:
An angstrom is a unit of linear displacement sometimes used to describe the wavelength of light. An angstrom is one-tenth of a nanometre (10^{-10} metre).

Angular Size:
The angle that an object at a given distance subtends.

Annealing:
A manufacturing process used to relieve strain in glass. The temperature of glass is raised to slightly less than the melting point, and then slowly returned to room temperature. The cooling process may take days or weeks.

Antireflection Coating:
One or more thin layers of an optical material applied to an optical surface to reduce reflectance and increase transmittance.

Aperture:
An opening through which light can pass, most commonly the lens opening or the opening in the diaphragm.

Aperture, Clear:
The (central) part of a lens opening which is not covered by the lens mount.

Aplanatic Condition:
A condition in which light rays are bent equally by each surface of a positive lens element used as a magnifying lens. In this condition distortion is minimized. The aplanatic condition in a wedge prism minimizes chromatic aberration.

Apochromatic Lens:
A lens corrected for three colors (wavelengths) of light, commonly red, green and blue.

Apparent Angle:
Half of the maximum angle of view visible through the eyepiece of a telescope.

Arc:
A small, very bright light source in which light is generated by electricity passing from one electrode to another through air or other gas.

Aspect Ratio:
The ratio of length to width in rectangular camera formats and picture shapes (e.g., 5/4 or 1.25). In motion pictures, the frame width divided by the frame height.

Aspheric:
Non-spherical. A lens surface that is not spherical. Usually the curve is a surface of revolution about the lens axis, as a paraboloid or ellipsoid.

Aspheric Mirror:
A concave image-forming mirror that is non-spherical. Usually the shape is a paraboloid or ellipsoid.

Astigmatism:
A fundamental lens aberration in which rays from an object point are imaged as a tangential line before reaching the image plane, as a cross at the image plane, and as a radial line beyond the image plane. Astigmatism is zero on axis and increasingly evident as the distance from the axis is increased.

Astronomical Telescope:
An optical instrument used to observe objects at infinite distances. The image of an astronomical telescope is usually inverted (which would be unacceptable for a terrestrial telescope).

Atom:
The smallest particle that an element can be broken into without changing the nature of the element. Consists of a nucleus of protons and neutrons surrounded by "shells" of electrons. The number of protons equals the number of electrons. If one or more electrons is added or taken away, an ion is formed. The atom is electrically neutral while the ion is electrically charged.

Axis:
A reference line passing through all the centers of curvature of the lens surfaces in an optical system.

Back-Coated Mirror:
A mirror which has the reflective coating on the second surface.

Baffle:
A black shield used to absorb unwanted light rays. Used inside lens mounts to reduce flare.

Barrel:
A tubular lens housing or mount which usually includes a diaphragm or fixed aperture but not a shutter.

Barrel Distortion:
A lens aberration caused by differences of magnification at different field angles. Barrel distortion is negative distortion; the magnification becomes less as the field angle increases. A square object is imaged as barrel-shaped.

Beam:
A bundle of light rays that are diverging, converging or parallel.

Beamsplitter:
A reflecting surface which divides a beam of light into two or more beams.

Bevel:
An angled edge that is ground on an optical part in order to remove a sharp edge and to prevent chipping.

Biconcave Lens:
A negative, diverging lens which has two inward curving surfaces.

Biconvex Lens:
A positive, converging lens which has two outward curving surfaces.

Binocular Magnifier:
A magnifying device that permits two-eyed viewing.

Binocular Vision:
Two-eyed vision with stereopsis or three dimensional relief perception.

Binoculars:
An optical instrument composed of two terrestrial telescopes mounted so that a distant object can be viewed simultaneously by two eyes, one telescope for each eye. Each telescope portion has an objective lens, a pair of Porro prisms to orient the image correctly, a field lens, and an eyelens.

Blackbody Radiation:
Incandescent radiation emitted by a theoretical, perfectly-black object with a nonreflecting surface heated to incandescence.

Blank:
A roughly shaped piece of glass from which a lens or other optical element can be made. May be molded, cast or sawed from bulk glass.

Block, Lens:
A metal supporting tool used to hold lenses during grinding and polishing.

Block Prism:
A thick piece of glass having at least two plane-polished surfaces.

Brass Gauge:
A brass template used to check the radius-of-curvature of a ground lens.

Brightness:
That aspect of color vision that relates to the apparent lightness or darkness of an object.

Brilliant Reflex Finder:
An optical viewfinder with a positive finder lens and a rectangular positive lens used as a viewing screen.

Bundle Of Rays:
A collection of parallel, diverging, or converging light rays with a defined outer limit. A light beam.

Button:
A molded lens that returns light rays back to its source by refraction and internal reflection. Used to make reflective road signs.

Button, Pitch:
A circular piece of thermoplastic material used to bond glass lens blanks to a metal positioning and supporting block during grinding and polishing.

Camera:
An optical device used to form images on a sensitized material or viewing screen.

Canada Balsam:
A clear thermoplastic tree resin used as a cement for combining optical elements. It is being replaced by synthetic bonding agents such as methyl methacrylate.

Candela:
A unit of luminous intensity.

Cascade:
An arrangement of devices, each of which feeds into the next. Cascaded-cube mirror devices consist of a series of real or imagined cubes containing plane mirrors as an aid to finding the orientation of an image as it is reflected through a series of mirrors. Also a mathematical procedure by which the values of two interrelated functions are found.

Cassegrain Telescope:
A two-mirror telescope. The objective mirror has a hole in the center behind which the image is formed.

Cementing:
The process of bonding two or more optical elements to form a single optical device without a central air space.

Center Of Curvature:
A point on the optical axis which is equidistant from all points on a lens surface.

Centering:
A manufacturing process used to form the edge of a lens equidistant from its optical centerline (axis).

Cerium Oxide:
A polishing compound consisting of a powdered mineral which polishes faster than common red rouge (ferric oxide), and called white or pink rouge in the trade.

Chamfer:
A small bevel ground along the edge of a lens element to make it less sharp and more chip resistant.

Chip:
A physical defect resulting when a flake of glass is broken from the edge or corner of an optical element.

Chroma:
The Munsell color system term for chromatic saturation (color intensity).

Chromatic Aberration:
A fundamental lens aberration which causes a lens to have a shorter focal length for short wavelengths of light than for long wavelengths.

Chromatic Reflection:
A nonuniform reflection of light wavelengths which causes a color difference between the incident and reflected ray. A colored or selective reflection.

Chromatic Transmission:
Transmission by a material that absorbs more of some wavelengths than others. The transmitted light appears colored. Example: A color filter.

Circle Of Confusion:
The circle of light formed where the image plane (film plane) intersects the conic bundle of rays forming an out-of-focus image point.

Clear Aperture:
The diameter of the bundle of rays that can pass through a lens element. The clear aperture is less than the lens diameter because the lens mount blocks the outer rays.

Coating:
One or more thin layers of optical material applied to an optical surface to reduce reflection, create a mirror surface, absorb light or protect the surface. May be applied chemically or by evaporation/condensation in a high vacuum.

Cold Mirror:
A mirror with a dichroic coating that reflects light and transmits infrared radiation. (See Dichroic)

Collimator:
A reverse telescope which projects parallel light rays from a point light source. Used to test and align lenses and other optical devices.

Color:
The visual response to light (perception) of different wavelengths or combinations of wavelengths. A single wavelength or combination of wavelengths has three visual attributes: hue (color), brightness (value) and saturation (chroma).

Color Temperature:
The temperature in Kelvin (absolute) degrees of a blackbody that produces light having the same proportion of wavelengths as a given light source.

Coma:
A fundamental lens aberration that occurs in the field (off axis) and not on the axis. It is caused by the failure of concentric light rays, passing through different concentric zones of the lens, to form a common image point. A comatic point image may appear pear shaped, have two tails, or have a head and a tail like a comet.

Component:
One or more optical elements in a lens system which are treated as a single optical unit.

Compound Lens:
A lens composed of two or more elements which may be separate or cemented.

Compound Microscope:
A microscope having several lens components (i.e., an objective lens, field lens and eyelens).

Concave:
Curved inward. A lens surface shaped like the inner surface of a hollow sphere.

Concentric:
Having the same or a common center.

Condenser:
A lens element or lens system used to refract light rays from a source to evenly illuminate an area.

Conjugate Distances:
The object and image distances measured on the optical axis of a lens.

Conjugate Points:
The intersections of the image and object planes with the optical axis.

Constant Deviation:
The property of a prism or fixed-mirror device that maintains a constant angle between the entering and exiting rays as the angle of incidence changes.

Contrast:
The difference in brightness between the brightest and darkest areas in a visual field or image.

Converging Light Rays:
Non-parallel light rays approaching the point of their intersection.

Convex:
Curved outward. A lens surface shaped like the exterior surface of a sphere.

Corner Reflector:
Three mirrors mounted at right angles to each other to form a reflector that returns light rays toward their source parallel to the incident rays.

Corrected Lens:
A compound lens designed to reduce aberrations.

Cosine-Fourth Law:
The illuminance on a surface varies as the fourth power of the cosine of the incident angle of the illumination.

Critical Angle:
The largest possible incident angle for a light ray passing from a more-dense to a less-dense optical medium.

Cross-staff:
A primitive navigation instrument used to find the angles of elevation of stars, the moon or the sun above the horizon. It has been replaced by the sextant.

Crown Glass:
The basic optical glass made of sand and soda ash (sodium carbonate). It has a low index of refraction and low dispersion (high Abbe number).

Curvature:
A term used to describe the spherical surface of a lens. The spherical surface of a strong lens surface has a short radius of curvature, and the spherical surface of a weak lens surface has a long radius of curvature.

Curvature Of Field:
See Field Curvature.

Curve:
A line that is constantly changing direction. Example: a section of a circle, elipse, or parabola.

Cylindrical Lens:
A lens that has a positive or negative cylindrical surface.

Decentered Lens:
A mounted lens in which the physical (geometric) center of the mount and the optical center of the lens do not coincide.

Definition:
The degree of clarity of detail in an image. It is determined by acutance (sharpness of edges), resolving power (limit of the ability to distinguish closely spaced or small objects) and modulation transfer function (degree of contrast imaged at different image frequencies).

Densitometer:
A device for measuring density (the light absorbing characteristics of transmitting or reflecting materials).

Density:
A measure of the light absorbing characteristics of a material.

Depth Of Field:
The range of object distances from a camera lens within which the objects at a given image distance are imaged as "acceptably sharp."

Depth Of Focus:
The range of image distances from a camera lens within which objects at a given object distance are imaged as "acceptably sharp."

Depth Perception:
See Stereopsis.

Deviation:
See Angle Of Deviation.

Diamond:
The hardest known naturally occurring substance, formed of pure carbon, with the highest index of refraction of any known material.

Diamond Dust:
Small particles of diamond used as cutting or grinding material.

Diaphragm:
An aperture that limits the amount of light going through a lens.

Dichroic:
Exhibiting the property of dichroism, a phenomenon which produces two colors (usually one reflected, the other transmitted) by mutual interference of light waves as they pass through (and are reflected back-and-forth between) two closely-spaced optical surfaces. If a glass plate is coated with a partially-reflecting mirror surface, a clear spacer layer, and a second partially-reflecting mirror surface, the closely-spaced mirror pair will reflect one color and transmit its complement; thus a dichroic (two-color) filter is formed.

Dichroic Mirror:
A beam splitter mirror used to reflect some wavelengths while transmitting others. Made by coating very thin alternating layers of reflecting and clear materials on a transparent substrate.

Diffraction:
An optical effect in which the wave nature of light causes light rays to bend in direct proportion to their wavelengths as they pass an opaque edge. (Compare to Refraction.)

Diffraction Grating:
An optical device, finely grooved, used to form spectra by reflection or transmission from incident light by diffraction.

Diffuse:
Disorderly reflection or transmission of light rays.

Diffused Light:
Light rays going in many directions—disorderly light rays.

Diffuse Reflector:
A surface which is microscopically rough and reflects light rays in many directions.

Diffusion:
Causing light rays to travel in many directions in a disorderly arrangement.

Diopter:
A unit of lens power. The diopter power is the focal length in millimeters divided into 1,000 mm.

$$\Delta = \frac{1{,}000}{f}$$

A triangle is used as the symbol for diopter or diopter power. A 2 Δ lens is a lens with 2 diopter power.

Dispersion:
Variation of refraction by wavelength. All transparent materials have different indexes of refraction for light of different wavelengths.

Displacement:
The linear or angular measure of the distance of an image or ray from its original optical axis.

Distortion:
An aberration of magnification which does not affect image focus.

Diverging Light Rays:
Non-parallel light rays leaving the point of their intersection or leaving a point light source.

Doublet:
A lens made of two cemented lens elements. Both inner surfaces must have the same radius of curvature: one concave, the other convex; or both plano.

Dove Prism:
A segment of a right-angle prism that is a trapezoid in side view. It inverts images.

Drawn Glass:
Sheets of glass drawn up from molten glass to be used for window panes.

Effective Aperture:
The diameter of the diaphragm of a lens as measured on the front lens surface. The effective aperture determines the largest bundle of rays that the lens will use to form an image. Used to calculate the *f*-number.

$$\text{effective aperture} = \frac{\text{focal length}}{f\text{-number}}$$

It is larger or smaller than the diaphragm opening depending on whether the lenses in front of the diaphragm have positive or negative power.

Effective Focal Length:
The distance from the equivalent refracting plane, or the second principal plane, to the focal plane when the lens is focused on an infinitely distant object. The effective focal length is what is meant when the simple term "focal length" is used.

Electromagnetic Radiant Energy:
Energy that is given off by electrically charged particles when their speed or direction is changed. Such energy radiates from the source in all directions and travels in straight lines through air, space, and transparent objects.

Electromagnetic Spectrum:
An orderly arrangement of all the forms of electromagnetic radiant energy by wavelength. The visible spectrum can be formed from white light by diffraction or by dispersion.

Electron:
One of the very small parts of an atom. Electrons are negatively charged.

Element, Optical:
A lens, window, mirror, prism, wedge or other unit, usually made of glass, that is used in an optical instrument to control light.

Emery:
An abrasive material made of crushed minerals such as corundum, magnetite, or hematite. Used with water as a slurry for grinding glass.

Energy:
The physical ability to do work. Energy due to movement is called kinetic, and stored energy is called potential.

Entrance Pupil:
A circular area on or near the front surface of a lens, defined by the rays of light that are limited by an internal stop, usually the diaphragm.

Envelope:
The outer limit of a bundle of light rays. Also, the glass housing that surrounds the filament in an incandescent light bulb.

Equatorial Telescope Mount:
A mount with a polar axis and a declination axis. It usually is power driven around the polar axis at the rate of one revolution in 24 hours so that a celestial field will remain stationary in the field of view.

Equivalent Focal Length:
The focal length of a compound lens. The distance between the second principal plane and the focal plane as measured on the axis when the lens is focused at infinity. Also called effective focal length.

Equivalent Refracting Plane:
The plane in a lens where the bending of light rays seems to take place. It is formed by the intersection of construction lines that are parallel to the axis and the lines that converge at the focal point. It is usually a spherical surface rather than a plane.

Erect Image:
An image whose orientation is the same as that of the object.

Erecting Lens:
A lens within a telescope used to invert the image. Usually a symmetrical compound lens with a magnification of 1X that also acts as a relay lens.

Erecting Ocular:
An eyepiece having an erecting lens, field lens and eyelens. Used in terrestrial telescopes to provide a right-side-up image.

Evaporative Coating:
A high-vacuum process for applying very thin coatings to the surface of glass and plastic optical elements. The coating material is heated until it evaporates. In the absence of air the evaporated molecules travel at high speed and strike the glass surface with enough energy to adhere.

Excursion:
A path traced on the surface of a reference sphere by the image of an object reflected by a mirror located at the center of the sphere.

Exit Pupil:
A circular area of light at a given distance behind a lens. It is the limiting stop in air.

Exposure:
The amount of light striking a reference surface multiplied by the time of the illuminance.

$$E = I \times t$$

Exposure Meter:
A device having a light-sensitive cell which reads luminance (reflection meters) or illuminance (incident meters) and is an aid in calculating camera exposures.

Eye Adaption:
The ability of the eye's lens to change focus by changing its shape.

Eye Distance:
That distance between the farthest extension of the eyelens (usually the back surface of the lens) and the exit pupil in an instrument such as a telescope or microscope.

Eyelens:
The lens nearest the eye in an optical instrument. Usually a magnifying lens used to view a real image formed by an objective lens.

Eyepiece:
The mounted eyelens and field lens of an optical instrument. Also known as the ocular.

***f*-Number:**
A measure of a lens aperture. Its numerical value is found by:

$$f\text{-number} = \frac{\text{focal length of lens}}{\text{diameter of lens opening}}$$

The *f*-numbers usually used are in a series: 1, 1.4, 2, 2.8, 4, 5.6, 8, 11, 16, 22, 32. Each *f*-number in the series represents a lens opening that lets in ½ the amount of light of the preceding one.

Fiber Optics:
The transmission of light through strands of glass or other transparent materials. Coherent strands of fiber optics retain the same relative positions of each fiber at the image end as at the object end which permits the transmission of images. Incoherent strands are used to transmit illumination. Each fiber uses the critical-angle reflection principle to minimize light loss.

Field:
The area of the object or image covered by a lens or optical instrument. It is nearly always circular and is measured in degrees. Also called the field of view.

Field Coverage:
Usually refers to the image plane of a camera lens or magnifier, and means the angular or linear diameter measurement of two circles: the circle of illumination and the circle of acceptable definition.

Field Curvature:
Because a simple lens typically has the same focal distance at various field angles, the image field of an uncorrected positive lens is a spherical surface.

Field Flattener:
An optical element used in front of the film plane to change a curving image field to a plane image field coincident with the film.

Field Glasses:
A pair of Galilean telescopes used to view distant objects with low magnification (e.g., opera glasses). (Compare with Binoculars.)

Field Lens:
A converging lens used in front of the eyepiece of an optical instrument to increase the field of view by imaging the entire objective lens on the exit pupil.

Field Of View:
The maximum field angle that can be seen through an optical instrument. The angle imaged at the corner of a film format by a camera lens.

Field Stop:
A fixed aperture that limits the field of view to its corrected, illuminated area. It is seen as a sharp circular edge in the image.

Filament:
The tungsten wire in a light bulb that incandesces when electricity flows through it.

Film Plane:
The position in a camera where the film is located. It coincides with the focal plane when the lens is focused and is located at the best-compromise focus position for the image, usually favoring the central area.

Filter:
A transparent optical element that absorbs some wavelengths more than others.

Finder:
An aiming and framing device attached to a camera or telescope which is used to aim and/or determine the field coverage of the instrument.

Finite Distance:
Any distance less than infinity.

Fire Polish:
To smooth a glass surface by flame-heating the glass until it melts and then cooling it until it hardens with a microscopically-smooth surface.

First-Surface Mirror:
A mirror with a reflective coating placed so that the incident rays are reflected without going through the glass.

Fish-Eye Lens:
A lens with a field angle of about 180 degrees.

Fission (Nuclear):
The splitting of atomic nuclei with the release of enormous amounts of energy. The source of energy in the sun that causes the emission of tremendous amounts of electromagnetic radiant energy.

Fixed-Angle Mirrors:
Two mirrors placed with a fixed angle between them. The angle of reflection from the mirrors is twice the angle between the mirrors.

Flange Focus:
The distance between the seating surface of the flange that holds a mounted lens and the film plane.

Flare:
Non-image forming light which may produce ghost images and a reduction of image contrast. Usually the result of mount-surface and lens-surface reflections.

Flat Field:
An image that is sharply focused in a plane perpendicular to the lens axis. A flat-field lens is fully corrected for field curvature.

Flint Glass:
Crown glass made with lead salts added. Flint glass has a higher index of refraction and higher dispersive power than most crown glasses.

Focal Length:
The distance between the second principal plane or equivalent refracting plane of a lens and the lens focal point when the lens is imaging an object at infinity. In a positive lens, the focal length is measured on the side of the lens opposite the object. In a negative lens, the focal length is measured on the same side as the object.

Focal Plane:
A plane, perpendicular to the axis, which intersects the focal point.

Focal Point:
The point on the optical axis where light rays from a distant object point, also on axis, converge after being refracted by the lens.

Focus:
The place at which light rays from object points form a sharp image after being refracted by a lens. Also, the act of moving the lens or image receptor (film) to obtain a sharp image.

Foot Candle:
A unit of illuminance. The illuminance falling on a surface placed one foot from a source of one candela. One lumen per square foot.

Fovea:
The small central portion of the eye's retina where greatest sharpness and color are seen. Located on the axis of the eye's lens. The fovea, a small dimple in the retina, is packed with cones and has no rods.

Free Aperture:
Another term for clear aperture.

Frequency:
The number of waves that pass a given point in a unit of time. Usually stated as cycles per second or Hertz. Also the number of cycles of a wave in a given distance (e.g., cycles per millimetre).

Fresnel Lens:
A thin equivalent of a thick lens achieved by segmenting the spherical surface and moving each concentric segment toward the plano surface.

Front-Meniscus Lens:
A positive-meniscus lens used as a camera lens with the convex surface forward and the diaphragm behind the lens. Has relatively short back focus and considerable field curvature.

Front-Projection Screen:
A plane sheet of reflective material used for viewing projected images. The image is projected onto the side from which it is viewed.

Fusion:
The combining of atoms (or ions) of hydrogen to form helium atoms (or ions) with enormous amounts of energy given off.

Galilean Telescope:
A telescope of moderate power consisting of a positive objective lens and a negative eyelens.

Geometric Optics:
That part of optics dealing with light rays as straight lines traveling from a source which are bent by reflection or refraction.

Gauss Point:
The intersection of the equivalent refracting plane or principal plane with the lens axis. In thick lenses there are two Gauss points, one for each principal plane. Also called cardinal points and for a lens in air, nodal points.

Ghost Image:
A type of flare. An out-of-focus image in the image plane (usually of a light source) caused by reflection from a lens surface acting as an image-forming concave mirror.

Glass:
A transparent, non-crystalline, fused mixture of inorganic salts, primarily oxides. Common glass is a mixture of sand (primarily silicon dioxide), soda ash (sodium carbonate), and lime (calcium oxide).

Glass Type Number:
A six digit number that identifies the characteristics of an optical glass. The first three digits are the decimal value of the index of refraction to the sodium line (n_D) and the last three digits are the Abbe number with one decimal value. A glass with an index of refraction of 1.5127 and an Abbe number of 60.5 has a glass type number of 512605.

Gloss:
A highly reflective surface condition that results in the specular reflection of light.

Graphic Arts:
The craft, field of study, and methodology of printing by mechanical and photomechanical means.

Graphic Ray Tracing:
A technique for tracing rays through optical interfaces using a geometric method wherein circular arcs represent the indexes of refraction of the two optical media. If the angle of the incident ray is known the angle of the refracted ray can be found by accurate drawing.

Grating:
See Diffraction Grating.

Grazing Emergence:
When a ray leaving a high-index medium pierces the surface at the critical angle, the ray emerges at a 90 degree angle to the normal, i.e., parallel to the surface.

Great Circle:
A circle on the surface of a sphere whose center coincides with the center of the sphere.

Grinding:
A stage of optical manufacturing during which the surfaces of optical elements are shaped. Utilizes wet emery or diamonds as grinding abrasives.

Ground Glass:
Glass or plastic with one polished surface and one roughened surface which is used as a focusing aid.

Gunsight:
An optical device, commonly a unity telescope, that aids in aiming a gun.

Half-Silvered:
Optical surface that has been coated with a thin metallic layer that reflects about 50% of the incident light. The term originated when mirrors and pellicles were chemically coated with silver, but is used generally to describe any metal-coated beam-splitting surface.

Hard Coating:
A term applied to modern anti-reflection coatings because they are coated in such a way as to make them resistant to mechanical damage as from cleaning.

Heat-Absorbing Glass:
Glass which has a high iron content, transmits light and absorbs infrared radiation. Usually has a greenish color.

Heat Filter:
A dichroic filter that reflects infrared radiation and transmits visible light.

Hertz:
The unit used to define frequency. 1 Hertz equals one wave or cycle per second.

Highlight: The part of the object with the greatest luminance, imaged with the greatest illuminance.

Horizon:
A line formed where the sky and earth or sea appear to meet. Used as a reference in navigation when establishing the elevation of a star, the moon or the sun.

Hot Mirror:
A dichroic filter that reflects infrared radiation and transmits visible light.

Hue:
That aspect of color vision that can be related to regions of the spectrum. It is the aspect that is normally thought of as color, e.g., red, blue, green, etc.

Hyperfocal Distance:
The object distance at which a camera is focused so that the far limit of depth-of-field is at infinity. The near limit of depth-of-field is then located at half the hyperfocal distance.

Hyperopia:
Farsightedness. A condition of the eye in which the eye's lens-to-retina distance is shorter than normal. The lens focuses near objects behind the retina rather than on it. Far objects can be imaged sharpely but near objects cannot.

Hypotenuse:
The side of a right triangle located opposite the right angle.

Illuminance:
The measure of light falling on objects. Measured in either foot-candles or metre-candles (lux).

Illumination:
Light falling on objects or coming through translucent objects making them visible. The measure of light falling on objects is illuminance and the measure of light coming through translucent objects from behind is luminance.

Image:
A likeness of an object formed by an optical element or system.

Image Distance:
The distance between the equivalent refracting plane or second principal plane and the focal point measured on the optical axis.

Image Plane:
A more or less flat plane that is the focus of an optical system. Usually slightly concave (toward the lens) in the central zone and convex in the outer zone.

Image Point:
An infinitely small area within an image. The image is a mosaic of an infinite number of image points.

Image Size:
The linear or angular measure of the image of an object.

Incandescence:
The emission of light by a material that has been heated to a high temperature.

Incandescent Lamp:
A light source with a transparent envelope containing an inert gas or vacuum and a resistance-wire filament. An electrical current heats the wire so that it incandesces.

Incident Light:
The light falling on a surface. The energy of incident light is measured in foot candles or lux (metre candles). It is the illumination.

Index Of Refraction:
A measure of the light-bending property of an optical medium. It is a ratio of the speed of light in air divided by the speed of light in the medium.

Infinity:
The object distance at which light rays coming from each object point to the lens are essentially parallel; thus large finite distances can be considered "infinite" for specialized optical systems such as cameras. The object distance at which the image distance equals the focal length. Infinity-object-distance increases with increasing focal length and decreases with decreasing aperture diameter.

Infrared Radiation:
Radiation which has wavelengths longer than those of visible radiation.

Integrating Sphere:
A hollow sphere, or part of a sphere, internally coated with white, very diffuse paint and used to measure the diffuse reflection or transmittance of materials or as part of a very even, diffuse light source for enlargers.

Interference:
A phenomenon of the wave nature of light as it passes through closely-spaced partially-reflective optical surfaces. The waves are reflected back and forth upon themselves, enhancing and cancelling the wave energy depending on the wave lengths and the surface spacing. (See Dichroic).

Interocular Distance:
The average distance between the pupil centers of human eyes when the axes of the eyes are parallel. About 60 mm.

Inverse Square Law:
The illumination of a surface by light from a point source varies inversely as the square of the distance from the source.

$$\text{Brightness} \sim \frac{1}{\text{distance}^2}$$

When two surfaces are at different distances from the source, their comparative brightnesses are:

$$\frac{\text{brightness a}}{\text{distance b}^2} = \frac{\text{brightness b}}{\text{distance a}^2}$$

Iris Diaphragm:
An adjustable lens aperture that is made of thin opaque overlapping leaves, and actuated by a rotating ring.

Lambert:
A unit of luminance equivalent to $1/\pi$ candelas per square centimetre. A footlambert is equal to $1/\pi$ candelas per square foot.

Laser:
Light Amplification by Stimulated Emission of Radiation. An optical device that produces monochromatic, coherent, collimated light.

Lateral Chromatic Aberration:
One of the fundamental lens aberrations caused by dispersion. Images formed by light of short wavelengths are smaller than those of longer wavelengths and therefore are seen as a spectrum along tangential edges in the off-axis image.

Lens:
One or more optical elements having flat or curved surfaces. If used to converge light rays, it is a positive lens; if used to diverge light rays, it is a negative lens. Usually made of optical glass, but may be molded of transparent plastic or made of a natural or synthetic crystaline substance to transmit very short wavelengths (UV) or very long wavelengths (IR).

Lens Bench:
A precisely made laboratory device to evaluate the image forming qualities of lenses. Consists of a distant point-source or collimator, sliding lens holders for positioning, and a microscope for examining the image point formed by the lens. May also have a T-bar arrangement for examining the lens field.

Lens, Block:
A metal supporting tool used to hold lenses during grinding and polishing.

Lens Form:
The shape of a lens as shown by its cross-section. The basic lens forms are: bi-convex, plano-convex, concave-convex (meniscus) bi-concave, plano-concave.

Lens Mounts:
The metal tubes (barrels) and rings (retainers) that hold lens elements in alignment with proper spacing.

Lens Opening:
The aperture or hole in the diaphragm of a lens.

Lens Power:
A measure of the relative magnification of a lens. Power is defined as the focal length of the lens divided into 1 metre (1,000 mm). For example, a lens with a focal length of 100 mm has a power of $1000/100 = 10$ diopters. (See Diopter).

Lens Speed:
A number that indicates the relative amount of light a lens will transmit at its maximum opening. Its smallest *f*-number. The term comes from photographic applications where a large-diameter lens can achieve a suitable film exposure in a shorter time, hence, a "faster" lens.

Lenticular:
Divided into small lens-like areas that form an array. Said of reflecting surfaces such as projection screens.

Lenticular Screen:
A front-projection screen with spherical or cylindrical lenticules molded on the surface. Provides a bright image over a predetermined viewing angle.

Lenticule:
One unit area in a lenticular array. A lenticule resembles the dimple formed when a convex lens is pressed into a soft surface.

Light:
The form of electromagnetic radiation to which the human eye is sensitive. It generally travels in straight-line paths, exhibiting the characteristics of both wave and particle motion. Light has wavelengths from about 400 to 700 nanometres. Ultraviolet and infra-red radiation are imaged by some optical devices (their wavelengths are close to the visible wavelengths of the spectrum) but they are not visible radiation and therefore cannot be considered "light."

Light Funnel:
A four-or-more sided conic device with a mirror coated interior surface designed to concentrate light at its narrow end. There are narrow limits on the effectiveness of such devices.

Light Ray:
The path of a small-diameter beam of light. In graphical ray tracing, the straight line that represents the path along which light travels.

Limit Of Resolution:
The smallest width white and black line pairs a lens can image as separate lines. In the extreme, the Rayleigh limit is expressed by the equation:

$$\text{Rayleigh Limit} = \frac{1.22\lambda}{d}$$

Where λ is the wavelength of light, d is the diameter of the effective aperture and the limit is the angular separation of two just-resolved object points. See Resolving Power.

Loupe:
A relatively-high power magnifying glass. If made to fit the eye as a monocle, or onto a pair of glasses, may be called an eye loupe.

Lumen:
A unit of luminous flux. One lumen is the luminous flux emitted per unit solid angle by a source having an intensity of one candela (standard candle).

Luminance:
The measure of the brightness of an object. It is the luminous flux per unit solid angle per unit area measured at right angles to the direction of the flux lines. Common units are candelas per square foot and foot lamberts.

Lux:
One metre-candle. A unit of illuminance equal to one lumen per square metre.

Macro Lenses:
Lenses specifically designed for taking photographs of objects at close distances. Commonly, they are well corrected for aberrations and have a flat field.

Macrophotography:
Photography at close distances, with magnifications of about 1X.

Magnesium Fluoride:
A clear, crystalline mineral used as a lens coating to reduce surface reflections. Has an index of refraction of $n_D = 1.38$. Because such coating is most effective when the index of refraction is the square root of the glass index, this material is most effective with glass that has an index $n_D = 1.90$.

Magnification:
The ratio of image size to object size. In magnifying lenses and microscopes, the magnification refers to how large the image looks compared to how large it would look to the unaided eye at 10 inches. In enlargers and projectors, it is the projected-image-size to original-image-size ratio. In camera lenses, "magnification of a lens" is a comparison of the image size created with a given lens to the image created by a normal-focal-length (equal to the diagonal of the film format) lens.

Magnifying Lens/Magnifying Glass:
A positive lens used to form an enlarged image to aid viewing.

Magnifying Power:
A measure of the ability of a lens or optical instrument to make an object appear larger. The magnifying power of a magnifying lens is the focal length of the lens divided into 10 inches (250 mm). The magnifying power of a telescope is the apparent field angle divided by the actual field angle.

Marginal Rays:
The light rays that pass through an optical system close to the edge of the aperture.

Mask:
A usually plane device with opaque areas that cover parts of an image and transparent areas that allow other parts of the image to be seen.

Medium, Optical:
Any transparent material commonly used to pass light rays. Includes glasses, plastics, liquids and mineral crystals.

Meniscus Lens:
A lens with one concave and one convex curve. If the lens is thicker at the center than at the edges it is a positive meniscus lens. If the lens is thinner at the center than at the edges it is a negative meniscus lens.

Metallic Coating:
A thin metal layer applied to a polished surface to provide high reflectivity or to create a mirror.

Metrogon Lens:
A wide-angle lens with relatively low distortion and fairly flat field used for aerial photography.

Microphotography:
Photography in which small images are formed from large objects as in microfilming of documents. A very short focal length objective lens is used. (See Photomicrography.)

Microscope:
An optical instrument used to view small-object detail with great magnification.

Millimeter:
A unit of measure equal to 10^{-3} metre or 1/25.397 inch (.03937 inch). Frequently used in optical measurement.

Minimum Angle Of Deviation:
The smallest angle of deviation achievable by a wedge prism. When the angle of incidence equals the angle of emergence, the angle of deviation is minimized.

Mirror:
An optical element with a smooth, polished surface used to reflect light specularly.

Mirror Lenses:
Image-forming optical systems that use curved mirrors, often along with lenses, to form images. Used in telescopes and as telephoto lenses for cameras.

Molded Blank:
A piece of molded glass or other optical material from which a lens or other optical element can be made by grinding and polishing.

Molded Lens: A lens in which one or both surfaces have been made by molding rather than grinding and polishing. Road-sign buttons, condensers, finder lenses, and some camera lenses are examples.

Monochromatic Light:
Light consisting of a single wavelength (color) or a very narrow band of wavelengths. Literally: one color.

Monocular:
Pertaining to one eye. Monocular is the name often given to a low-power terrestrial telescope that uses a pair of Porro prisms to erect the image. Half of a binocular.

Multilayer Coating:
Alternating layers of relatively high- and low-index material applied to an optical surface to reduce reflection or to act as a dichroic filter.

Myopia:
Nearsightedness. A condition of the eye in which the eye's lens-to-retina distance is longer than normal. The lens focuses objects in front of the retina rather than on it. Near objects can be imaged sharply but far objects cannot.

N:
Index of refraction. Unless otherwise indicated, N refers to the index of refraction of the optical material to the sodium line at 589.6 nm. When specifying other wavelengths, the lower-case letter is usually used with a subscript letter of the spectral line at which the index was measured.

Nanometer:
The unit usually used in specifying the wavelength of light. One nanometre is equal to 10^{-9} metre.

Nearsightedness:
See Myopia.

Negative Curve:
A concave curve. A curve which, when used as a lens surface, diverges light rays.

Negative Lens:
A lens that is thinner in the center than at the edges. A lens that diverges light rays.

Newton's Rings:
Interference patterns formed by light passing through two closely spaced partially reflective surfaces. See Dichroic.

Newtonian Telescope:
A telescope with a parabolic mirror objective, a plane 45-degree reflector to divert the converging rays 90 degrees from the axis, and an eyepiece to magnify the image.

Nodal Points:
The intersection of the principal planes of a lens with its axis. Also called principal points if the lens is operated in air. A thin lens has one, a thick lens has two.

Nomograph:
A diagram that enables one, by using a straightedge, to locate the value of one variable when the values of two or more related variables are known.

Normal:
A reference line constructed perpendicular to a surface.

Object:
The source of light rays imaged by an optical system.

Object Distance:
The distance from the first principal plane of a lens or optical system to the object point.

Object Size:
The linear or angular dimension of an object measured perpendicular to the optical axis.

Objective:
A lens or mirror used to form a real image.

Objective Lens:
The lens that forms the first image in an optical system.

Ocular:
The eyelens, often combined with a field lens, of an optical instrument such as a microscope or telescope. It sometimes contains a reticule.

Opal Glass:
A diffusing glass formed by mixing many tiny colorless particles in molten glass.

Opal Lamp:
A light bulb with an envelope made with a highly diffuse coating of opal glass.

Opacity:
A measure of the light-absorbing quality of a transmitting material. The reciprocal of the transmittance.

$$\text{opacity} = \frac{1}{\text{transmittance}} = \frac{\text{incident light}}{\text{transmitted light}}$$

Opaque:
Describes a material that does not transmit light.

Opaque Projector:
An optical instrument used to project the image of an opaque object, usually a photograph or text, on a screen.

Optical Axis:
A straight line joining the centers of curvature of the lens surfaces.

Optical Center:
In a thin lens it is the intersection of the equivalent refracting plane and the optical axis. The point in the lens where rays are transmitted without bending. In a thick lens, the second principal point. The Gauss point.

Optical Density:
See density.

Optical Element:
Any single lens, prism, window, mirror, wedge or other transmitting or reflecting part of an optical system.

Optical Fiber: A relatively long, small-diameter optical element made of a transparent material that transmits light efficiently along its length.

Optical Flat:
An optical element with plane-parallel surfaces. Usually has surfaces flat to within fractions of a Newton's ring, and has very little wedging.

Optical Glass:
Glass made to have a specific index of refraction and dispersion. It is made as colorless, and as free of impurities as possible.

Optical Materials:
The materials from which optical elements are made.

Optical Plastic:
Plastics such as the acrylics that are suitable for making optical elements.

Optical Scale:
A measuring reticle used in magnifiers, loupes and microscopes to measure lengths, angles, and the radii of small arcs.

Optical Surface:
A reflecting or refracting surface of an optical element.

Optical Viewfinder:
A camera aiming and framing device that contains optical elements. Usually a reverse Galilean telescope with a negative finder lens and a positive eyelens.

Optically Flat:
A term which describes a surface that is ground and polished so that it is extremely flat.

Optics:
The science of light, how it travels and how it is transmitted, absorbed, reflected, and refracted, and how images are formed by lenses. Includes physical optics, biological optics, geometrical optics and physiological optics. Also used to refer to the glass parts of an optical instrument as in "telescope optics."

Overcoat:
A thin coating applied over a reflecting or transmitting surface to provide physical protection. It must be transparent and of such thickness that it will not adversely affect the main coating.

Paper-Path Folding:
A method for finding an appropriate optical path for an optical layout in which mirrors or prisms are used. Each fold in the paper pattern indicates the location and size of a plane reflecting surface.

Parabola:
A geometric curved figure that is a conic section created at the intersection of a plane parallel to a cone surface.

Paraboloid:
A three-dimensional figure formed by rotating a parabola about its axis. So-called parabolic mirror surfaces are actually paraboloids.

Parallax:
The apparent shift in position of objects at different viewing distances with changing viewpoint.

Parallel:
Equidistant at all points, as in parallel lines or parallel planes.

Pellicle:
A beam-splitting mirror consisting of a thin membrane which has been stretched tightly on an open frame with an optically flat surface and coated with a partially-reflecting and partially-transmitting metallic coating.

Penta Prism:
A five-sided prism. Usually internally reflects light at a 90 degree angle.

Penta Roof Prism:
A penta prism with a roof along one of the reflecting surfaces. Commonly used in the finder system of single-lens reflex cameras.

Periscope:
An instrument used to optically shift the viewpoint.

Periscopic Lens:
A simple camera lens consisting of two positive meniscus lenses placed symmetrically about a diaphragm.

Perpendicular:
Positioned at 90 degrees to a reference line or surface.

Perspective:
The angular placement of object points in a field of view. The perspective of a three-dimensional fixed object is always constant from one point of view, but is different from a viewpoint that is closer to or farther from the object.

Phase:
In wave motion, the spatial or temporal relationship of peaks and valleys of two waves in the same region. It is the phase relationship of a light wave and its reflection between closely spaced reflecting surfaces that causes interference phenomenon.

Photogrammetry:
Photographic measurement, as in aerial photography for the purpose of map making.

Photographic Exposure:
See exposure.

Photographic Field:
See field of view.

Photography:
The system or process of recording images on light sensitive materials. Literally: light drawing.

Photomacrography:
See macro photography.

Photometer:
A device that measures illuminance. Can also be used to measure the luminous intensity of a source.

Photomicrography:
The photography of very small objects as through a microscope.

Photon:
A quanta or particle that carries radiant energy.

Physical Optics:
The study of light and other forms of electromagnetic radiant energy having to do with their wave-like nature (e.g., interference, diffraction).

Physiological Optics:
The study of the biological aspects of vision. This includes primarily the study of the structure of the eye, optic nerves, vision center of the brain, and various aspects of perception such as color.

Pincushion Distortion:
A lens aberration caused by variable image magnification with changing field angle such that a square object on axis is imaged as a pincushion shape. Pincushion distortion is positive distortion because the magnification increases as the field angle increases.

Pinhole:
A small diameter hole in a thin material, such as aluminum foil, that can be used to form images.

Pinhole Camera:
A camera in which a small hole is the image-forming device. Because the pinhole has a very high *f*-number (e.g., *f*/256) it does not have a conventional focal length. It has great depth of focus and depth of field, requires long exposures and suffers from diffraction.

Pitch:
Any of a variety of viscous substances used to bond optical elements to supporting blocks during grinding and polishing.

Pitch Button:
A circular piece of thermoplastic material used to bond glass lens blanks to a metal positioning and supporting block during grinding and polishing.

Plane (Plano):
A flat surface.

Plane-Parallel Optic:
An optical element such as a window or filter that has two plane surfaces that are parallel.

Plano:
Refers to a flat surface of an optical element.

Plastic Lens:
A lens made of optical quality plastic.

Point:
A geometric entity that has neither length, width nor thickness. Identifies a location.

Point Light Source:
A very small light source. May have a small filament, as in a flashlight bulb or in an enlarging bulb used in a specular enlarger, or may be an arc light.

Polarization (Linear):
Conversion of light with waves that vibrate in all directions normal to the ray path to light with waves that vibrate in only one direction.

Polarized Light:
Light with waves that vibrate in only one direction perpendicular to the light ray path. Unpolarized light waves vibrate in all directions perpendicular to the light ray path.

Polarizer:
A polarizer is an optical device that changes ordinary light into polarized light, such as a polarizing screen or filter.

Polish:
To make a surface microscopically smooth so that it reflects light specularly.

Polished Plate Glass:
Glass that is hot rolled, then machine ground and polished to final thickness and surface finish in huge sheets.

Polisher:
A tool with an exact curve used to polish glass lens surfaces.

Portrait Lens:
A lens with residual spherical aberration used to produce slightly soft unsharp images. Used in portraiture to obscure unflattering skin texture, wrinkles and small blemishes.

Positive Curve:
A convex curve.

Positive Lens:
A lens that is thicker in the center than at the edges. A lens that converges light rays and can be used to form real images.

Pot Glass:
Optical glass allowed to harden in the crucible (ceramic pot) in which it was prepared. It is broken into chunks, sawed into slabs, then cut to form cubes from which lens blanks are molded.

Pressing:
A molded blank used to make an optical element.

Primary Light Source:
Any object that emits light. Common primary sources are: the sun, light bulbs, neon signs, flash lamps, fluorescent tubes, CRT displays, arcs, and chemoluminescent objects such as firefly tails.

Principal Planes:
The two planes at which light rays appear to bend as they pass through a lens. In a thin lens there is only one, sometimes called the equivalent refracting plane.

Principal Points:
The intersection of the principal planes of a lens with its optical axis.

Prism:
An optical element used to change the direction of a light beam, or to change image orientation. A prism has polished plane faces which are used to transmit or reflect light. Also may be used to form a spectrum from white light.

Prism Power:
The measure of wedge deviation, calibrated in prism diopters. One prism diopter is the deviation that displaces a ray one centimetre at a distance of one metre, an angular displacement of 0.573 degrees.

Process Lens:
A long focal length, flat-field lens used at magnifications near 1X, such as 0.2X to 2.0X. Used to make photo copies, graphic arts "line" and "halftone" negatives.

Projection:
The process of forming an enlarged screen image of a small illuminated object, usually a slide, transparency, or motion picture frame.

Projection Lens:
A lens designed to produce enlarged images of flat illuminated objects on a reflective or translucent screen.

Projection Lens Bench:
A laboratory device with an illuminated slide object, a screen, and an adjustable lens holder.

Projector:
An optical instrument used to project an image.

Protective Coating:
See Overcoat.

Pupils:
The opening in the human eye through which light passes during vision. (See also Entrance Pupil and Exit Pupil.)

Quantum:
The smallest unit of electromagnetic radiant energy. Equivalent to a photon.

Radar:
A distance or speed measuring system which makes microwave echoes visible on CRT (cathode ray tube) displays.

Radial Astigmatism:
A lens aberration that causes an off-axis object point to be imaged as a short line parallel to a field radius.

Radiant Energy:
Emitted emergy that travels radially from its source.

Radiation:
The process of emitting energy radially. That energy which is radiated.

Radius of Curvature:
A measure of the shape of a lens surface. If the surface is spherical, the radius of curvature is the radius of a sphere of which the lens surface is a portion.

Rangefinder:
An optical instrument for measuring the distance between an object and its observer. Used on cameras to aid focusing and on weapons to determine the distance to a target.

Rare-Earth Glass:
Made of boric acid with rare-earth salts added to increase the density and the index of refraction.

RAT Formula:
Explains what happens to light falling on a surface.

$$I = R + A + T$$

Which means that all the Illuminance is accounted for by Reflection, Absorption or Transmission.

Ray:
The path of a small diameter beam of light.

Ray Tracing: The analysis of an optical system by observing the paths of selected light rays passing through it.

Rayleigh Scattering:
Scattering of light waves by particles much smaller than the light's wavelength resulting in the separation of colors. Responsible for blueness of sky and redness of sunsets.

Reading Glass:
A low-power magnifying lens large enough to permit binocular viewing and usually mounted in a frame with a handle.

Real Image:
An image formed by converging light rays. The image may be formed on a ground glass, projected screen, photographic film, etc.

Rear-Meniscus Lens:
A positive-meniscus lens used as a camera lens. The diaphragm is in front of the lens.

Rear-Projection Screen:
A screen made of a translucent material. The image is projected on the rear surface of the screen and viewed from the front surface.

Reference Sphere:
A real or conceptual sphere or hemisphere used to find the locations and orientations of images reflected by a plane mirror located at the center of the sphere.

Reflectance:
A ratio describing the relationship between incident light and reflected light.

$$\text{Reflectance} = \frac{\text{Reflected Light}}{\text{Incident Light}}$$

Reflection:
The change in direction of light rays bouncing off a surface.

Reflection, Law Of:
The angle of reflection equals the angle of incidence. Both angles are measured from the ray to the normal. The reflected ray lies in the plane of incidence and the normal.

Reflection Telescope:
A telescope in which the objective is a parabolic or spherical mirror. A catoptric (mirror) telescope or catadioptric (mirror and lens) telescope.

Reflectivity:
See Reflectance.

Reflector:
An optical element which is coated on one surface with a metal or multiple-layer coating and is used to reflect light. A mirror. Used to divert light from a source. It may be formed of metal or plastic.

Reflex Finder:
A viewfinder with a positive finder lens, a plane mirror at 45 degrees to the lens axis, and a translucent viewing screen at 90 degrees to the lens axis.

Refracting Telescope:
A telescope that has a lens objective rather than a mirror objective. Forms images entirely by refraction. A dioptric telescope.

Refraction:
The bending of light rays in inverse proportion to their wavelengths as they pass from one transparent medium to another of different refractive index. (Compare to Diffraction.)

Refractive Index:
See Index Of Refraction.

Relative Aperture:
The focal length of a lens or concave mirror divided by the effective diameter. The *f*-number.

Relay Lens:
A lens used to re-image a real image formed by another lens. Often used in telescopes to invert and relocate the image.

Relief:
The distance between the eyelens and the viewing position.

Resolution:
The ability of a lens to image closely spaced objects without merging them. Commonly measured using test objects having alternating equal-width light and dark lines.

Resolving Power:
See Resolution.

Resolving Power Targets:
Reflection or transmission test objects having a series of progressively smaller sets of alternating black lines and clear or white spaces of equal width. Sometimes made with several contrasts and with other element shapes (e.g., letters, circles).

Reticle:
A pattern located in the focal plane of an optical instrument so that it coincides with the image formed by the objective lens. Used to aim the instrument, or to make measurements on the object.

Retina:
The light-sensing layer at the rear of the eye.

Retrofocus Lens:
A camera lens that has a negative lens component in front and a positive lens component in the rear. The second principal plane is behind the lens, giving increased back focus. Used in wide-angle lenses for single-lens reflex cameras to provide clearance for the moving mirror.

Retroreflective Optical Devices:
Optical reflecting devices which return rays in the direction from which they came, or very nearly so. Example: roadside reflectors.

Reversed Telephoto Lens:
See Retrofocus Lens.

Reversibility Principle:
The ray path through an optical system is independent of the direction of the ray.

Right-Angle Prism:
A prism whose cross-section is a right triangle with one 90-degree and two 45-degree angles. The prism faces that are at right angles are transmitting surfaces, while the hypotenuse face is a reflecting surface.

Right-Reflecting Mirror:
Two plane mirrors placed at 90 degrees to correct the right-to-left image reversal seen in a single plane mirror.

Rolled Glass:
Optical glass that is poured from a crucible while still molten and rolled under steel rollers to yield a given width and thickness. The rolled glass is cut into cubes which are molded into lens blanks.

Roof Prism:
A prism in which one reflecting face has two surfaces at 90 degrees. (See Penta Roof Prism.)

Rotating Wedges:
A pair of circular wedges which rotate in opposite directions about the optical axis. In one position the wedges form a plane parallel window and do not deviate light. When rotated so that the wedge bases are aligned, they form a wedge prism with twice the wedge angle of a single prism. Used in some types of rangefinders for triangulation.

Rouge:
A compound used to polish the glass surfaces of optical elements. Red rouge is iron oxide, and white rouge, used almost exclusively today, is cerric oxide.

Saturation:
That aspect of a color that relates to the purity of a color. If a color is pastel or grayish, it is low in saturation, if the color is intense, it is high in saturation.

Scattering:
A phenomenon that occurs when the light energy is absorbed by air molecules and other small particles causing them to radiate the absorbed energy in wavelengths shorter than those absorbed. Scattering accounts for blue sky and reddish sunsets.

Schematic Ray Tracing:
A method of finding where light rays go using drawings. It uses the location of the principal plane of a lens and the focal point.

Schmidt Camera:
A reflection telescopic camera with relatively wide field and large aperture. The objective is a spherical mirror.

Screen:
A plane surface used for viewing real images.

Secondary Light Source:
An object that reflects light from a primary light source.

Second-Surface Mirror:
A mirror in which the light rays pass through the glass support before striking the reflective surface.

Selective Reflection:
See Chromatic Reflection.

Selective Transmission:
See Chromatic Transmission.

Sextant:
A navigation instrument which incorporates a mirror and beamsplitter to measure the angle of elevation of a star, or the limb (edge) of the moon or the sun.

Sharp Image:
An image having good definition. An image that is properly focused and in which fine detail can be seen.

Sharpness:
The definition or clarity of detail in an image, primarily judged by the edges of the imaged objects.

Shutter:
A light valve. An opaque device used to control the time that light is allowed to go through an aperture. Used in a camera to determine exposure time.

Signaling Mirror:
A mirror with a hole in it. Used by seamen in lifeboats to signal their location to rescue aircraft.

Silicon Dioxide:
A mineral used to make glass. Occurs naturally as quartz, and is a major constituent of sand. Often called silica.

Silicon Monoxide:
A protective overcoat for aluminum mirror surfaces. It is oxidized to a thin layer of quartz which provides excellent protection for the soft metal surface.

Sine:
A trigonometric function of an angle. The ratio of the side opposite the angle divided by the hypotenuse in a right triangle. Abbreviation: sin.

Sine Wave:
A wave whose energy, if plotted on cartesian coordinates, approximates a sine curve.

Single-Lens Reflex Camera:
A camera in which the objective lens also serves as the viewfinder lens. The image on the ground glass usually has the same size and focal position as the photographic film. When a roof penta prism is used, the viewfinder image has the same orientation as the object.

Slide:
A transparent positive photographic image on film that is projected for viewing.

Slide Projector:
An optical device used to project the images of slides onto a screen.

Slope:
A measure of the angle that a line makes with a horizontal base. Defined as the rise divided by the run.

Snell's Law:
Describes the path that a light ray takes as it goes from one optical medium to another–as from air to glass.

$$n \text{ Sin } I = n' \text{ Sin } I'$$

Where n is the index of refraction of the first medium, and I is the angle of incidence measured from the normal to the interface, while n′ is the index of refraction of the second medium and I′ is the refracted angle.

Sodium Line:
A spectral line with a wavelength of about 589.6 nm emitted by electrically stimulated sodium vapor. The standard D line used to establish the index of refraction of glass, noted n_D.

Source:
An object which originates electromagnetic radiation, such as a light source.

Spatial Frequency:
The number of repeated patterns in a given distance. Used in measuring resolving power, e.g., line pairs per millimeter.

Spectrum:
A display, ordered by wavelength, of electromagnetic radiant energy.

Specular:
A reflection in which parallel incident rays emerge as parallel reflected rays. Also emission from an illumination system in which the rays are orderly.

Speed (of a Lens):
Refers to the maximum aperture of a photographic objective.

Speed of Light:
See Velocity of Light.

Sphere:
A solid geometric figure in which all surface points are equidistant from a common center point. The distance from the center point to each surface point is the radius. A cross-section of a sphere is a circle.

Spherical Aberration:
A fundamental aberration that causes a loss in image sharpness. It is caused by lenses having more power (shorter focal length) at their periphery than on axis. An axial object point is focused closely to the lens by the peripheral rays and farther from the lens by the axial rays. Spherical aberration of a given lens is reduced by stopping down the lens.

Split Field:
An instrument in which the viewer sees two images divided across the center of the image. In a rangefinder when the lower image is aligned with the upper image the instrument has been set for the range of the object. Also used in optical instruments for comparing the color of two objects or the brightness of two sources.

Spotted Block:
A metal tool to hold lenses during the grinding and polishing operations. The spots are seats cut into the block to hold the lenses.

Star Image:
The image formed by a lens or objective mirror of a point of light located at a great distance. Used to study lens aberrations and decentering on a lens bench.

Stereo Effect:
Stereopsis or three-dimensional visual effect. One of the components of depth perception.

Stereopsis:
The illusion of three-dimensional depth that results from binocular vision. Also called stereoscopic vision.

Stop:
A lens diaphragm or aperture or its setting.

Stopping Down:

Reducing the aperture of a camera lens to reduce the brightness of the image. Also can improve image quality by reducing some aberrations.

Strain:
An uneven distribution of molecular forces within glass. It can be seen as an irregular display of spectral colors if the glass is placed between crossed polarizers. Can be reduced by annealing.

Subtend:
To be opposite to and establish limits. The side of a triangle subtends the angle opposite it.

Symmetrical Lens:
A compound lens in which the front elements are the same as the rear elements (or nearly so). In a cross section of a symmetrical lens, the left half appears to be a mirror image of the right half.

Tangent:
A trigonometric function. In a right triangle, the tangent of an angle is the ratio of the opposite side divided by the adjacent side. Tangent also means "touching." A straight line is tangent to a curved line if it touches at only one point and does not pass through the curved line.

Tele-Extender:
A negative lens that lengthens the focal length and thereby increases the image size. Sometimes called a Barlow lens.

Telephoto Lens:
A lens whose focal length is greater than the diagonal of the film format, and whose second principal plane is in front of the lens so that the lens has a short back focal length compared to its effective focal length. The front lens components have positive power, while the rear lens components have negative power.

Telescope:
An optical instrument used to view distant objects with magnification.

Telescope Magnification:
The ratio of the image size as seen through the telescope to the apparent size as seen by the unaided eye. Can be found three ways:

$$M = \frac{\text{objective focal length}}{\text{eyelens focal length}}$$

$$M = \frac{\text{diameter of objective lens}}{\text{diameter of exit pupil}}$$

$$M = \frac{\text{apparent angular field of view}}{\text{real angular field of view}}$$

Television:
A system of broadcasting or CCTV visual images. A television camera converts visual images into electronic signals that are broadcast using radio waves. The television receiver amplifies the wave signals and converts them into electron streams that stimulate light-emitting phosphors on the face of the television tube. Television signals can be recorded on Video Tape or Video Disc systems.

Terrestrial Telescope:
A refracting telescope that permits viewing distant objects right-side-up. The image formed by the objective lens is inverted by an erecting lens or prism system.

Tessar-Type Lens:
A four-element camera lens derived from a triplet design in which the rear positive element is a cemented doublet. The Tessar lens was originated by Carl Zeiss.

Test Glass:
A glass lens-like tool used to measure the spherical surface of lenses during manufacture. The test glass has an extremely accurately made spherical surface just opposite of that desired on the lenses. The other surface of the test glass is a polished plano surface to see through. The test glass surface is placed on the lens surface in monochromatic light and examination is made of Newton's rings formed.

Thermoplastic Cement:
An optical cement that is a solid at room temperature, but that changes into a viscous liquid at elevated temperatures. Canada balsam, which was used for years as optical cement, is a thermoplastic cement as are some methacrylate cements.

Thermosetting Cement:
An optical cement that initially is liquid at room temperature and permanently solidifies at a higher temperature. Often a methacrylate plastic material.

Thick Lens:
Any real lens in which the two principal points do not coincide. Includes lenses with an appreciable thickness and all compound lenses.

Thin Film:
Any coating applied to the surface of an optical element to perform an optical function. Such functions include raising or lowering reflectivity (e.g., anti reflection coatings), providing a combination of transmission and reflection (e.g., beam splitters), and separating light from infrared radiation (e.g., hot mirrors).

Thin Lens:
A hypothetical lens with no thickness, equivalent to the refracting plane used in schematic ray tracing. The thin lens is usually assigned a diameter and focal length, and may be either positive or negative. Many thin, single-element lenses are, for practical purposes, thin lenses.

T-Number:
A number which can be used in place of the *f*-number and takes into consideration the percent transmission of the lens. Generally the T-number is:

$$\text{T-number} = \frac{f\text{-number}}{\text{square root of transmittance}}$$

Total Internal Reflection:
When light rays in a high-index medium exceed the critical angle (to the normal to a surface), they are totally reflected within the material. This principle is used in prisms to change the angle of a light beam without having to metallize the reflecting surface.

Translucent:
A property of transmitting optical materials in which parallel entering rays exit in disorderly paths.

Transmission:
Light passing through a transparent or translucent material.

Transmittance:
The ratio of transmitted to incident light.

$$\text{transmittance} = \frac{\text{transmitted light}}{\text{incident light}}$$

Transparency:
A photographic image on a transparent base viewed by transmitted light. Small transparencies mounted for projection are called slides.

Transparent:
The characteristic of a material that allows light rays to pass through without diffusion.

Triplet:
A lens with three elements. Usually consists of a front crown positive lens, flint negative lens, and back crown positive lens.

Tungsten Lamp:
A light source with a filament made of tungsten wire. The most common type of light bulb.

Twin-Lens Reflex Camera:
A camera with matched objective lenses whose focusing movements are coupled. The lower lens is the camera objective lens, the upper lens is the finder lens.

UHF:
Ultra High Frequency. A band of electromagnetic radiant energy used for television transmission.

Ultra-Violet Radiation:
Electromagnetic radiant energy with wavelengths shorter than those of violet light. Wavelengths range from about 10 nm to 400 nm.

Unit Magnification:
A magnification of 1X. The image is the same size as the object.

Unit Power:
Having a magnification of 1X.

Unpolarized:
Ordinary light. Light traveling with its wave vibration in all directions at right angles to the light path.

Vacuum:
Absence of gas molecules. Space outside Earth's atmosphere.

Variable-Focus Lens:
A lens system made so that one or more components can be moved to change the focal length. Similar to a zoom lens, but in a variable-focus lens the back focus varies with change in effective focal length, while in a zoom lens, the back focus remains constant.

Variable-Power Wedge Prism:
See Rotating Wedges.

Velocity Of Light:
In a vacuum, approximately 186,283 miles per second. In air, approximately 186,000 miles per second.

View Camera:
A camera, usually of large size, that focuses on a ground glass viewing screen. View cameras have interchangeable lenses, bellows, use sheet film, and have lens movements for correcting image keystoning and optimally using the depth of field.

Viewfinder:
The device by which a camera is aimed and sometimes focused. The field of view of the viewfinder may be slightly smaller than that of the camera to compensate for aiming errors.

Viewing Screen:
A translucent screen placed in the focal plane so that the image can be seen. Example: rear projection screen.

Vignetting:
The shadow of an obstruction in the light path, usually at the edges of the image area. It may be partial, so that the image is dimmed, or complete so that the image is eliminated.

Virtual Image:
An image formed by diverging light rays formed by a negative lens or convex mirror. Virtual images can be seen but not formed on a viewing screen.

Visible:
Can be detected by the eye.

Visible Spectrum:
A spectrum of the wavelengths from 400 to 700 nm contained in white light formed by diffraction or dispersion. The visible spectrum is real and can be seen by the eye. The colors normally seen in the visible spectrum are violet, blue, blue-green, green, yellow green, yellow, orange and red.

Wave:
One form of movement which light exhibits as it travels. One measure of wave movement is frequency.

Wavelength:
The distance from the peak of one wave to the peak of the next wave. The wave period.

Wedge Prism:
A prism used to deviate a light beam by refraction. Usually an isosceles triangular prism with a fairly small wedge angle—30 degrees or less.

White Light:
Colorless illumination that has a fairly even distribution of wavelengths from 400 nm to 700 nm.

Wide-Angle Lens:
A lens with a focal length shorter than the diameter of the circle of good image definition.

Window:
An optical element, usually made of transparent glass, used to protect other optics. Most windows have plane parallel surfaces and are called optical flats. A few windows have concentric spherical surfaces, one convex and one concave, and are powerless lenses. If a window is deeply curved it is called an optical dome.

X-Axis:
The horizontal axis in Cartesian coordinates.

X-Rays:
Electromagnetic radiant energy with wavelengths from about 0.1 nm to 10 nm.

Y-Axis:
The vertical axis in Cartesian coordinates.

Zoom Lens:
A variable focal length lens with a constant back focus.

INDEX

D

E

F

G

N

O

P

Q

R

S